Contested Russian Tourism

Cosmopolitanism, Nation, and Empire in the Nineteenth Century

Contested Russian Tourism

Cosmopolitanism, Nation, and Empire in the Nineteenth Century

Susan
Layton

BOSTON
2021

Library of Congress Cataloging-in-Publication Data

Names: Layton, Susan, author.
Title: Contested Russian tourism : cosmopolitanism, nation, and empire in the nineteenth century / Susan Layton.
Other titles: Imperial encounters in Russian history.
Description: Boston : Academic Studies Press, 2021. | Series: Imperial encounters in Russian history | Includes bibliographical references and index.
Identifiers: LCCN 2021004459 (print) | LCCN 2021004460 (ebook) | ISBN 9781644694206 (hardback) | ISBN 9781644694213 (adobe pdf) | ISBN 9781644694220 (epub)
Subjects: LCSH: Russians--Travel--Europe, Western--History--19th century. | Heritage tourism--Europe, Western--History--19th century. | Cosmopolitanism--Russia--History--19th century. | Tourism in literature. | Cosmopolitanism in literature. | Russian literature--19th century--History and criticism. | Travelers' writings, Russian--19th century--History and criticism.
Classification: LCC G155.R8 L39 2021 (print) | LCC G155.R8 (ebook) | DDC 914.04/280899171--dc23
LC record available at https://lccn.loc.gov/2021004459
LC ebook record available at https://lccn.loc.gov/2021004460

ISBN 9781644694206 (hardback)
ISBN 9781644694213 (adobe pdf)
ISBN 9781644694220 (epub)

Book design by Lapiz Digital Services
Cover design by Ivan Grave

Published by Academic Studies Press
1577 Beacon Street
Brookline, MA 02446, USA
press@academicstudiespress.com
www.academicstudiespress.com

Contents

Acknowledgments

This book is the product of research that began many years ago and benefited from stimulating comments by the organizers and participants at several conferences. Four were outstanding. Long before COVID-19 made Zooming ubiquitous, Diane Koenker and Anne Gorsuch organized a "virtual conference" for brainstorming among all the contributors to their pioneering collection of essays *Turizm: The Russian and East European Tourist under Capitalism and Socialism* (2006). Ron Suny and Olga Maiorova helped refine my thinking about tourism in relation to imperialism at the conference "Imperial Nation: Tsarist Russia and the Peoples of Empire" that they and their colleagues organized at the University of Michigan in 2011. Olga, again, and Deborah Martinsen spurred improvements in my treatment of Dostoevsky as travel writer at the conference held in 2013 at Columbia University in preparation for their co-edited volume of articles *Dostoevsky in Context* (2015). Finally, I thank Zaal Andronikashvili, Susanne Frank, and Giorgi Maisuradze for inviting me to speak on tourism at the meetings they organized on "Imperial Romanticism" in Tbilisi and Kazbegi in 2015, an unmatched opportunity to mix business with pleasure.

On that latter occasion, too, Emzar Jgerenaia, deputy director of the Georgian National Library, kindly supplied me with a copy of a rare mid-nineteenth-century Russian account of travel to Georgia. With respect to my sources, I also wish to express my gratitude to Timothy S. J. Phillips for having made available to me his excellent doctoral dissertation on the sociocultural history of nineteenth-century Russian resorts, a study prepared at Oxford University under the supervision of Catriona Kelly.

My most immediate debt is to Academic Studies Press and the referees of the manuscript. The series editor Michael Khodarkovsky began shepherding this book into print a couple of years ago. I thank him for his invaluable suggestions in conversations and in the course of reviewing successive drafts. I also benefited tremendously from Derek Offord's meticulous review of the manuscript. Thanks to his labors, big ideas as well as fine points came into better focus. Working with the press has been delightful, from initial contact with the director Igor Nemirovsky to the weekly exchanges with Kate Yanduganova, the acquisitions editor for Slavic, East European, and Central Asian Studies. With patience and efficiency, Kate solved many problems that arose along the way. Thanks, too, to Stuart Allen and the other copyeditors for their painstaking work.

At a late stage in the preparation of this book, the University of Illinois at Urbana-Champaign provided me generous support as a participant in its Open Research Laboratory. My fellow Slavist John Randolph and the administrator James Fleener were much appreciated hosts. The vast knowledge and tireless assistance of the chief Slavic librarian Joe Lenkart allowed me to profit to the maximum from the rich collection. I am equally grateful to the research librarian Jan Adamczyk who has for many years facilitated my work by email. Most importantly, he set up the digital search engine that served me for locating all references to cosmopolitanism in the writings of the nineteenth-century critic Vissarion Belinsky. Among Joe's other colleagues, I give special thanks to Maeve Presnick who prepared the illustrations from the Russian magazine *Niva* and from books in the University of Illinois collection. I am grateful to the university for permission to reproduce those images here. Bella Irvine and Kevin Adams, other members of the library team, provided online assistance in the past and were ever ready to solve practical problems I encountered using materials in Urbana-Champaign.

Many thanks to the History and Literature Museum in Yalta for preparing and granting permission to publish the image of the postcard reproduction of Ilya Repin's painting of a Russian woman tourist with a Tatar guide in the Crimea. The Russian National Library courteously supplied the 1857 picture of Prince Mikhail Vorontsov's Alupka palace in the Crimea.

Bits and pieces of this book have appeared in the following articles: "Colonial Mimicry and Disenchantment in Alexander Druzhinin's 'A Russian Circassian' and Other Stories," *Russian Review* 60, no. 1 (January 2001): 56–71; "The Divisive Modern Russian Tourist Abroad: Representations of Self and Other in the Early Reform Era," *Slavic Review* 68, no. 4 (Winter 2009): 848–71; and "Our Travelers and the English: A Russian Topos from Karamzin to 1848," *Slavic and East European Journal* 56, no. 1 (2012): 1–20. An earlier version of my discussion of Lidia Veselitskaya's *Mimi at the Spa*, prepared for a 2010 anniversary conference on Tolstoy, is due to appear in *Tolstoi: Art and Influence*, edited by Joe Andrew and Robert Reid (Leiden: Brill, 2021). My thanks to one and all for permission to use the selected material in the present new framework.

Illustrations

Note on Transliteration, Translation, and Abbreviations

I have followed a dual system of transliteration. Reference matter, parenthetical notations, and titles of Russian journals follow the Library of Congress guidelines, while the main body of the text, explanatory footnotes, and the index use anglicized forms of Russian proper names and place names. Translations are mine unless otherwise indicated. The following abbreviations are used in the notes:

PSS—*Polnoe sobranie sochinenii*
PSSiP—*Polnoe sobranie sochinenii i pisem*
SS—*Sobranie sochinenii*

Introduction

Nikolai Karamzin (1766–26) was the first Russian to gain national celebrity for recounting a tour of Western Europe. After visiting the German lands, Switzerland, France, and England during 1789–90, he published *Letters of a Russian Traveler* (1791–1801), a literary triumph that made him the inescapable ancestor of Russian travel writers ever after. Should we call this gentleman traveler a "tourist?" Certainly not, said the Russian historian Dmitry Likhachev in a speech of 1992. Likhachev asserted that the *Letters* represents the author "not as a 'tourist' but rather as the cultural equal" of Immanuel Kant and the other great minds Karamzin sought out abroad.[1] In other words, Karamzin was too cultivated, too erudite to count as a tourist. Likhachev's observation sprang from a milieu in which the state had long sought to control the meanings and practice of tourism. In the 1920s, Soviet tourist officials launched a campaign to root out the "bourgeois" (or, synonymously, the "petty bourgeois") vacation mentality. The new ideal was the "socialist" tourist venture formulated as a strenuous "circuit (tour) using one's own physical locomotion." The Soviet tourist base resembled a "scout camp," tents trumped hotels, and there were official merit badges for hiking, mountain climbing, cycling, and so forth. Even if inferior in athletics, the socialist tourist was supposed to be working at physical, moral, and intellectual self-improvement and contributing to the building of communism. That regime tightly restricted leisure and rest, both associated with

1 D. S. Likhachev, "Vstupitel'noe slovo na torzhestvennom vechere posviashchennom 225-letnemu iubileiu so dnia rozhdeniia N. M. Karamzina," in *N. M. Karamzin. Pro i Contra. Lichnost' i tvorchestvo N. M. Karamzina v otsenke russkikh pisatelei, kritikov, issledovatelei: antologiia*, compiled, introduced, and commentary by L. A. Sapchenko (St. Petersburg: RKhGI, 2006), 747.

medical needs.[2] With de-Stalinization came more lenient policies toward leisure and consumption, a prolonged development that occurred partly in response to consumer pressure.[3] All the same, tourist officials never ceased promoting purpose over pleasure. That tradition resounded in Likhachev's view of Karamzin.

In broader perspective, Likhachev's judgment of Karamzin stands on the negative side of disagreements about tourism in our time. The anthropologist Ulf Hannerz characterized tourism as "largely a spectator sport" attracting a passive hoard of incompetents who travel to enjoy "home plus" wherever they go: "Spain is home plus sunshine, India is home plus servants, and Africa is home plus lions." Hannerz's antithesis of the "tourist" is the "real cosmopolitan," eager and able to "engage alien systems of meaning" and experience immersion in foreign cultures.[4] Challenging such dichotomies, Dean MacCannell's seminal sociological study *The Tourist* salvaged tourism as a quintessentially modern pursuit that engages many intelligent, resourceful people who actively make meaning of their experiences.[5] The sociologist John Urry took a similarly positive view of tourists as "aesthetic cosmopolitans," the term he chose to designate self-confident, middle-class citizens participating in the "massive global flows of tourists" as of the 1970s. Urry's tourists believe they have the right to travel anywhere, and they like to get outside the tourist "bubble." They display openness to "other peoples and cultures" and possess at least "a rudimentary ability" to locate the foreign "historically, geographically, and anthropologically" in relation to

---

2 Diane P. Koenker, *Club Red: Vacation Travel and the Soviet Dream* (Ithaca and London: Cornell University Press, 2013), 3–6, 53; Diane P. Koenker, "Travel to Work, Travel to Play: On Russian Tourism, Travel, and Leisure," *Slavic Review* 62, no. 4 (2003): 657–59; Anne E. Gorsuch, *All This Is Your World: Soviet Tourism at Home and Abroad after Stalin* (Oxford: Oxford University Press, 2011), 6–10; Anne E. Gorsuch, "'There's No Place Like Home': Soviet Tourism in Late Stalinism," *Slavic Review* 62, no. 4 (2003): 760–85; and Anne E. Gorsuch and Diane P. Koenker, "Introduction," in *Turizm: The Russian and East European Tourist under Capitalism and Socialism*, ed. Anne E. Gorsuch and Diane P. Koenker (Ithaca: Cornell University Press, 2006), 6–8.

3 Gorsuch, *All This Is Your World*, 1–5, 130–67.

4 Ulf Hannerz, "Cosmopolitans and Locals in World Culture," in *Cosmopolitanism: Critical Concepts in the Social Sciences*, 4 vols., ed. Gerard Delanty and David Inglis (London: Routledge, 2011), 3:28–30. The cited source of the "home plus" idea is the travel writer Paul Theroux.

5 Dean MacCannell, *The Tourist: A New Theory of the Leisure Class*, 3rd ed. (Berkeley, Los Angeles and London: University of California Press, 2013 [1976]), esp. 1–16. See also Donald Horne, *The Intelligent Tourist* (Smithfield, NSW: Margaret Gee Publishing, 1989) and *The Great Museum: The Re-presentation of History* (London: Pluto Press, 1984).

home.[6] To a great extent, Urry's model overlapped the "real cosmopolitan" that Hannerz posited as the *opposite* of the "tourist." A convergence toward Urry occurred in the anthropologist James Clifford's essay "Mixed Feelings" (1998). In an earlier article, Clifford noted tourism's reputation of being a practice "incapable of producing serious knowledge."[7] "Mixed Feelings," however, located "cultural tourists" in international "contact zones" where "cosmopolitan competences" are exercised. On this view, cultural tourism counts as one of today's "discrepant cosmopolitanisms."[8]

These incompatible assessments of the competences of tourists have deep roots in the past. Jean-Didier Urbain, James Buzard, and others have described "tourist phobia" or "anti-tourism" as the denigration of tourists on the part of observers laying claim to superior learning, sophistication, resourcefulness, and sensitivity.[9] In his influential study *The Beaten Track: European Tourism, Literature, and the Ways to 'Culture', 1800–1918*, Buzard showed that tourist phobia had arisen among the English by the eighteenth century and evolved as a constitutive component of tourism. Citing a published example from 1800, the first edition of the *Oxford English Dictionary* glossed "tourist" as a plain synonym for traveler. By the late 1790s, however, William Wordsworth and others were already making the word a pejorative. For Wordsworth, "the tourists" were detestable day-trippers overrunning the Lake District he cherished for his own rambles.[10] That case exemplifies

---

6 John Urry, *Consuming Places* (London and New York: Routledge, 1995), 165–67.

7 James Clifford, *Routes: Travel and Translation in the Late Twentieth Century* (Cambridge, MA and London: Harvard University Press, 1997), 65.

8 James Clifford, "Mixed Feelings," in *Cosmopolitics: Thinking and Feeling Beyond the Nation*, ed. Pheng Cheah and Bruce Robbins (Minneapolis and London: University of Minnesota Press, 1998), 368–69. See a similar use of "cosmopolitanism" as a description of "individual behaviours, values or dispositions manifesting a capacity to engage cultural multiplicity," in Steven Vertovec and Robin Cohen, "Introduction: Conceiving Cosmopolitanism," in *Conceiving Cosmopolitanism: Theory, Context and Practice*, ed. Steven Vertovec and Robin Cohen (Oxford and New York: Oxford University Press, 2008), 1.

9 On *touristophobie* (tourist phobia) and *le mépris antitouristique* (anti-tourist contempt), see Jean-Didier Urbain, *L'idiot du voyage. Histoires de touristes*, 2nd ed. (Paris: Plon, 2002), 19–20, 37–45. "Anti-tourism" is the preferred term in James Buzard, *The Beaten Track. European Tourism, Literature, and the Ways to 'Culture', 1800–1918* (Oxford: Oxford University Press, 2001). See also Marc Boyer, *Histoire générale du tourisme du XVIe au XXIe siècle* (Paris: Harmattan, 2005), 5, 45, 53–54; Jost Krippendorf, *The Holiday Makers: Understanding the Impact of Leisure and Travel* (London: Heinemann, 1987), trans. Vera Andrassy, 41–43; and Jonathon Culler, "The Semiotics of Tourism," in *Framing the Sign: Criticism and Its Institutions* (Oxford: Basil Blackwell, 1988), 153–57.

10 Buzard, *The Beaten Track*, 1, 19–28.

the false distinction that self-styled "travelers" draw between themselves and "tourists." As a negative value judgment, "the tourist" has always signified somebody else, the "other guy," or simply the "traveling idiot" (Urbain's *l'idiot du voyage*). Wherever they crop up, those degraded others comprise the sheep-like crowd of mediocrities, lacking imagination, initiative, clear priorities, good breeding, and knowledge of the languages, cultures, and history of foreign countries.[11] The falseness of the tourist-traveler dichotomy has become a commonplace in the scholarly literature. In contributing to that discussion, two leading Russian historians in the field—Diane Koenker and Louise McReynolds—have illustrated anti-tourism through secondary references to Henry James's perceptions of "vulgar" American holidaymakers in Switzerland and France in the 1870s.[12]

The question that initiated my research for this book is whether tourist phobia has a more extensive imperial Russian history than previously indicated. Although some work suggests that the official Soviet disapproval of hedonism had antecedents in imperial Russia, that question has not received the special attention it deserves.[13] Russian contributions in particular have tended to ignore the issue, as illustrated by G. P. Dolzhenko's pioneering monograph *The History of Tourism in Prerevolutionary Russia and the USSR*. Concentrating on the rise of organized, collective undertakings, the study argues that the pursuit of "knowledge and the improvement of health" (*poznanie i ozdorovlenie*) motivated imperial Russian as well as Soviet tourism. On this view, a thirst for knowledge sent Russians to sites close to their homes, to "distant corners of our country," and to "centers of world culture abroad." The physical fitness regime found expression in organized group hiking in the Crimea as of the 1870s and participation in mountaineering clubs and bicycle touring societies established in the 1880s and 1890s.[14] In sum, Dolzhenko's history gives the impression that no matter what destination or activity imperial Russian tourists chose, they always had a positive, wholesome public image.

11 Ibid., 1–3.

12 Koenker, *Club Red*, 6; and Louise McReynolds, *Russia at Play: Leisure Activities at the End of the Tsarist Era* (Ithaca and London: Cornell University Press, 2003), 156n7.

13 Louise McReynolds, "The Prerevolutionary Tourist: Commercialization in the Nineteenth Century," in Gorsuch and Koenker, *Turizm*, 18.

14 G. P. Dolzhenko, *Istoriia turizma v dorevoliutsionnoi Rossii i SSSR* (Rostov-on-Don: Izdatel'stvo Rostovskogo universiteta, 1988), 8, 20–45.

From an entirely different standpoint, the chapter on tourism in McReynolds's delightful social history *Russia at Play* also gave the impression of Russian tourists breezing along from one pleasure to the next without incurring much in the way of bad publicity. In parallel to Stephen Lovell's history of the dacha, published the same year, *Russia at Play* made entertainment and leisure the solutions to the old problem of searching for a Russian bourgeoisie that appeared to be "missing" if judged by political structures that existed in Western Europe.[15] *Russia at Play* located the "bourgeoisie" (or, synonymously, the "middle classes") not in their politics but rather in "their cultural activities."[16] Lovell's history of the dacha similarly identified the Russian middle classes by how they spent their leisure time. *Summerfolk*, however, delved into the "increasingly serious image problem" that plagued the imperial Russian dacha as of the 1870s, the beginning of the boom in summering in the suburbs: "Run-of-the-mill comfortable middle-class pleasure seekers were anathema both to wealthy conservatives and to the Russian intelligentsia, being seen by the former as vulgar upstarts and by the latter as unbearably complacent and materialistic."[17] In treating tourists as "protagonists of modernity" who emerged and proliferated with the development of a commercialized tourist industry as of the 1870s, McReynolds implied that they, too, offended the traditionally anti-bourgeois intelligentsia and the equally anti-bourgeois elitist guardians

---

15 For classics on the "problem of the middle" and the "search for the 'Russian bourgeoisie,'" consult Edith Clowes et al., *Between Tsar and People: Educated Society and the Quest for Public Identity in Late Imperial Russia* (Princeton: Princeton University Press, 1991), 3–89, 183–211.

16 McReynolds, *Russia at Play*, 3–4. For precursors of the cultural optic, see discussion of the "rich cultural life" that helped unify the middling ranks "despite the limited consolidation of social and political associations," in Elise Kimerling Wirtschafter, *Social Identity in Imperial Russia* (Ithaca and London: Cornell University Press, 1997), 97; and essays on Russian painting, merchant art patronage, the art market, theater, and architecture in Clowes et al., *Between Tsar and People*, 93–128, 271–87, 308–39.

17 Stephen Lovell, *Summerfolk: A History of the Dacha, 1710–2000* (Ithaca and London: Cornell University Press, 2003), 104. At last count, there were over sixty definitions of the Russian intelligentsia. Unless otherwise indicated, I will use the term to mean politically engagé intellectuals "possessed of social conscience," "a secular code of ethics, a commitment to social justice," "a critical attitude toward conditions in society and government," and "a desire to change those conditions": see discussion with voluminous bibliography in Elise Kimerling Wirtschafter, "The Groups Between: *Raznochintsy*, Intelligentsia, Professionals," in *The Cambridge History of Russia*, vol. 2, ed. Dominic Lieven (Cambridge: Cambridge University Press, 2006), 251–52.

of highbrow culture.[18] *Russia at Play*, though, chose not to explore the theme that will preoccupy me: the disapproval that hounded Russian tourists both in the West and in their favorite domestic playgrounds, the Caucasus and the Crimea.

A close look at the anti-tourist discourses shows that Russian tourism was a divisive pursuit setting cosmopolitans against nationalists and homebodies, metropolitans against "provincials," and men against women. In the latter third of the nineteenth century, anti-bourgeois sentiment also assumed a considerable role and eventually yielded an image of the Russian tourist *burzhui*, a derogatory term for the bourgeois as the personification of greed and vulgarity.[19] Social identities are inevitably on the line when tourists congregate and vie for cultural capital.[20] The present history of Russian tourism from Karamzin into the fin de siècle has a different focus, however. The central thematic axis explores how the Russian pursuit of tourism as an extension of the "civilizing mission" in the Caucasus and the Crimea interacted with the exaltations, envy, and resentment that cultural tourism aroused in Russian visitors of Western Europe. What were the specifics of that West/East dialectic?

---

18 McReynolds, *Russia at Play*, 5–9, 155.

19 On Alexander Herzen's influential conception of the bourgeoisie as "not so much a social class as an aesthetic and ethical type," see Martin Malia, *Alexander Herzen and the Birth of Russian Socialism* (New York: The Universal Library, Grosset and Dunlap, 1961), 359, and more generally 351–55. Other representative discussions of anti-bourgeois sentiment are James H. Billington, *The Icon and the Axe: An Interpretative History of Russian Culture* (New York: Vintage Books, 1970), 233, 327, 447–48; Derek Offord, *Journeys to a Graveyard. Perceptions of Europe in Classical Russian Travel Writing* (Dordrecht: Springer, 2005), 161–65; and Sidney Monas, "The Twilit Middle Class of Nineteenth-Century Russia," in Clowes et al., *Between Tsar and People*, 28–37. *Meshchanstvo* (lesser townspeople) was an official social category introduced under Catherine II. In popular parlance, however, it came to signify bourgeois philistinism. Poems by Pushkin were revelatory. Near the end of his life, when he was struggling financially and Russian aristocrats shunned him as a scribbler, Pushkin, ever proud of his noble lineage, defiantly sculpted a poetic persona as an honorable *meshchanin* in opposition to Russian parvenus who acquired aristocratic titles through ingratiating service to tsars. The episode prompted Dmitry Mirsky's 1934 description of Pushkin as "the poet of the youth of Russian bourgeois culture": see Catriona Kelly, "Pushkin's Vicarious Grand Tour: A Neo-Sociological Interpretation of 'K vel'mozhe' (1830)," *Slavonic and East European Review* 77, no. 1 (January 1999): 1–5, 21–28.

20 Cf. Buzard's disputing MacCannell's interpretation of tourism as a builder of "social cohesion in an otherwise fragmented modernity," in *The Beaten Track*, 9. See pertinent formulations in MacCannell, *The Tourist*, 1–9, 176–77, 203. For additional critique of MacCannell's model of modernity, see Jean-Paul Dumont, review of *The Tourist*, by D. MacCannell, *L'homme—Revue française d'anthropologie* 17, nos. 2–3 (1977): 190–91.

And what new nuances might they add to the established picture of Russian imperial-national self-definition?

The investigation's methodological inspiration comes mainly from *The Beaten Track*. Buzard treats the development of steam transport, the emergence of the guidebook enterprises of John Murray III and Karl Baedeker, and Thomas Cook's bringing tourism to the British masses. For the most part, however, Buzard's book is a literary study of the "cultural construction" of English (and some American) tourists on the Continent between 1800 and 1914. Buzard argues that tourism "fundamentally engages and tests cultural *representations*" and therefore invites approaching "mainly from the direction of literary analysis."[21] This reflects the fact that personal documents, guidebooks, and imaginative literature are our main sources of knowledge about leisure travel in the "pre-statistics age," the whole time prior to the founding of data gathering national tourist boards shortly after World War I.[22] Setting no restrictions on genre, the literary line of analysis ranges over the entire realm of print culture: travel writing, prose fiction, poetry, drama, journalism, guidebooks, and memoirs. The main quarry is the public image of the tourist, pro and contra, although private letters and diaries serve, too, as corroborative evidence. Authors of some studies of literary representations of tourism call their projects "literary tourism."[23] My book, however, follows the more common practice of construing "literary tourism" as travel to places associated with the life and works of a writer: his or her homes, "haunts," and "shrines," such as the birthplace and the grave.[24]

---

21 Buzard, *The Beaten Track*, 13.

22 John Towner, "Approaches to Tourism History," *Annals of Tourism Research* 15 (1988): 49.

23 Sanna Turoma, *Brodsky Abroad. Empire, Tourism, Nostalgia* (Madison: University of Wisconsin Press, 2010). See also George G. Dekker, *The Fictions of Romantic Tourism: Radcliffe, Scott, and Mary Shelley* (Stanford: Stanford University Press, 2005).

24 Alison Booth, "Author Country: Longfellow, the Bröntes, and Anglophone Homes and Haunts," *Romanticism and Victorianism on the Net* 48 (2007): 1–29, https://doi.org/10.7202/017438ar; and Ian Ousby, *The Englishman's England: Taste, Travel and the Rise of Tourism* (London: Pimlico, 2002), 17–43. See also Nicola J. Watson, *The Literary Tourist: Readers and Places in Romantic and Victorian Britain* (Basingstoke and New York: Palgrave Macmillan, 2006); Watson, "Shakespeare on the Tourist Trail," in *The Cambridge Companion to Shakespeare and Popular Culture*, ed. Robert Shaughnessy (Cambridge: Cambridge University Press, 2007), 199–226; and Stephanie Sandler, *Commemorating Pushkin: Russia's Myth of a National Poet* (Stanford: Stanford University Press, 2004), 47–84.

Defining the Tourist and *Turizm*

The first order of business is to define the Russian "tourist," a task embroiled in the pesky question of origins.[25] Tourism is understood today as a pleasure trip, a holiday, a vacation affording "recuperation and liberation from the stresses of daily life."[26] But when did this modern phenomenon begin in Russia?[27] While disagreeing about origins, scholars have at least reached a consensus on the attributes and circumstances of modern Russian leisure travel. As elsewhere, modern tourism in Russia arose with industrialization and developed along with steamboats and railroads.[28] Equally important were the growth of cities, the spread of commerce, the creation of new wealth, and the sharing of that wealth by the expanding Russian middle class.[29] Vital as well was the creation and circulation of a "body of images and descriptions" of places to entice readers and "induce them to travel."[30] Imaginative literature, travel writing, guidebooks, and visual art

---

25 On the general issue, see Judith Adler, "Origins of Sightseeing," *Annals of Tourism Research* 16, no. 1 (1989): 7–29.

26 Shelley Baranowski and Ellen Furlough, "Introduction," in *Being Elsewhere: Tourism, Consumer Culture, and Identity in Modern Europe and North America*, ed. Shelley Baranowski and Ellen Furlough (Ann Arbor: University of Michigan Press, 2001), 4–7, 19 (quote 5). On tourism as "a matter of being elsewhere," see also A. J. Burkart and S. Medlik, *Tourism: Past, Present, and Future*, 2nd ed. (London: Heinemann, 1981), 3. Note also the definition of *turist* as "a traveler, especially a travel lover" (*liubitel'-puteshestvennik*), in Vladimir Dal', *Tolkovyi slovar' zhivogo velikorusskogo iazyka*, vol. 4, ed. Baudouin de Courtenay, 3rd ed. (St. Petersburg-Moscow: Tovarishchestvo M. O. Vol'f, 1903–1909; repr. Moscow: Progress/Univers, 1994), 871. *Turist* does not appear in the earlier editions.

27 On precursors of modern tourism in ancient Greece, ancient Rome, and medieval religious pilgrimages, see Harmut Berghoff, "From Privilege to Commodity? Modern Tourism and the Rise of the Consumer Society," in *The Making of Modern Tourism: The Cultural History of the British Experience, 1600–2000*, ed. Harmut Berghoff et al. (New York: Palgrave, 2002), 161–62.

28 On steamboat services, see McReynolds, *Russia at Play*, 165–66. For pertinent railroad history, consult Richard Mowbray Haywood, *Russia Enters the Railway Age, 1842–1855* (Boulder, CO and New York: East European Monographs; distributed by Columbia University Press, 1998).

29 Andreas Schönle, *Authenticity and Fiction in the Russian Literary Journey, 1790–1840* (Cambridge, MA: Harvard University Press, 2000), 203–4, 207–8; Christopher Ely, "Origins of Russian Scenery: Volga River Tourism and Russian Landscape Aesthetics," *Slavic Review* 62, no. 4 (Winter 2003): 675; McReynolds, "The Prerevolutionary Russian Tourist," 24–25 (foreign travel) and 27–42 (developments mainly in the empire); and Dolzhenko, *Istoriia turizma*, 30–63.

30 On the general point, see John F. Sears, *Sacred Places: American Tourist Attractions in the Nineteenth Century* (New York: Oxford University Press, 1989), 3.

all contributed to that process.[31] Reading and pictures served to map itineraries, "mark" sights and sites to see, and prepare a "tourist gaze," be it "romantic" (solitary visual enjoyment) or "collective" (convivial sharing of visual consumption).[32] Related cultural factors were the development of art education, the opening of museums to the public, and the proliferation of spectacles in the city, all of which contributed to the development of an aestheticizing tourist way of looking, a "pure," morally unencumbered gaze of disinterested contemplation.[33] These various forces all came together in Russia only in the mid-1880s, to establish "a consumer market" for tourism.[34] A landmark year was 1885 when a certain Leopold Lipson established in Petersburg Russia's first company modeled on the business of Cook.[35]

In addition to the consensus on the general contours of modern Russian tourism, the scholarly literature has treated the reign of Peter the Great as a pre-modern watershed. Peter sent state servitors to Western Europe, he made his own first trip abroad in 1697–98 (the Grand Embassy), and Russian aristocrats of the time started emulating the English practice of the Grand Tour. This typically occurred in conjunction with official missions, diplomatic postings, or study excursions. Whatever the motivation for the travel, Russian participants in Grand Tourist activities were labeled "tourists" engaging in "tourism" in Sara Dickinson's *Breaking Ground*, a study of Russian travel literature from Peter I into Alexander Pushkin's time.[36] Dickinson's terminology chimes with Gorsuch's and Koenker's assertion that when considered from a wide historical perspective, "the Russian word *turizm* possesses both a broad

31 For special attention to guidebooks, see William George Lywood, "From Russia's Orient to Russia's Rivera: Reimagining the Black Sea Coast/Caucasus from Romantic Literature to Early Tourist Guidebooks" (MA thesis, Ohio State University, 2009. Available at rave.ohiolink.edu/etdc/view?acc_num=osu1236622370.), esp. 33–50.

32 On "marking" as the constitution of a sight as such, see the classic MacCannel, *The Tourist*, 41–42, 116–17; Culler, "Semiotics of Tourism," 160–61; and elaborations of both in Schönle, *Authenticity and Fiction*, 57–60. On varieties of the "tourist gaze," see Urry, *Consuming Places*, 137–38 and Carol Crawshaw and John Urry, "Tourism and the Photographic Eye," in *Touring Cultures: Transformations of Travel and Theory*, ed. Chris Rojek and Urry (London: Routledge, 1997), 176–77.

33 Schönle, *Authenticity and Fiction*, 208.

34 Ely, "Origins of Russian Scenery," 675.

35 Leopol'd Lipson, *Pervoe v Rossii predpriiatie dlia obshchestvennykh puteshestvii, 1885–86* (St. Petersburg: V. Kene, 1885).

36 Sara Dickinson, *Breaking Ground: Travel and National Culture in Russia from Peter I to the Era of Pushkin* (Amsterdam: Rodopi, 2006), 15, 22, 24, 29. *Puteshestvenniki* (travelers) is mistranslated as "tourists" (143) in discussion of a letter Konstantin Batyushkov wrote in Paris in 1814.

and a narrow definition." In the narrow sense, the tourist was the purposeful, athletic type promoted in official Soviet discourse. But in the broad sense, "a tourist was anyone who followed a leisure travel program of visual, cultural, and material consumption," whether in imperial or Soviet Russia.[37]

Other commentators, however, have sought to pin down more precise origins of the modern. Dolzhenko's history posited a solid sightseeing link between modern tourism and the travels of Russian traders and merchants of the sixteenth century or "even earlier."[38] For their part, the organizers of a conference in Petersburg in 1995 proclaimed the "commencement of tourism in Russia" in 1895, the founding year of the bicycle touring society that evolved into the Russian Society of Tourists (*Rossiiskoe obshchestvo turistov*).[39] Calling that suggestion "utterly incomprehensible," the Dolzhenko-Putrik handbook dated the "prehistory of *organized* Russian tourism" (italics mine) from 1696 when Peter I sent more than sixty Russians abroad to study navigation, to 1878 when the Society of Nature Lovers/Caucasus Alpine Club (*Obshchestvo liubitelei prirody i al'piiskii kavkazskii klub*) was established in Tiflis in affiliation with the local chapter of the Russian Geographical Society.[40] In designating the Russian "tourist" a bourgeois identity contingent on the emergence of a tourist industry, McReynolds considered the period from Peter I to 1875 "the prehistory of tourism" (the age of "the noble 'traveler'") and suggested dating "the arrival of the tourist" in 1875, the year that a Petersburg newssheet, "financed completely by advertisements," began publication with the aim of providing information about hotels in the Russian capitals.[41]

---

37 Gorsuch and Koneker, "Introduction," 2.

38 Dolzhenko, *Istoriia turizma*, 9.

39 Cited in Christopher Ely, "The Picturesque and the Holy: Visions of Touristic Space in Russia, 1820–1850," in *Architectures of Russian Identity: 1500 to the Present*, ed. James Cracraft and Daniel Rowland (Ithaca: Cornell University Press, 2003), 223n3. Ely observed that any number of other steps along the way might qualify as the starting point of modern Russian tourism.

40 G. P. Dolzhenko and Iu. S. Putrik, *Istoriia turizma v Rossiiskoi Imperii, Sovetskom Soiuze i Rossiiskoi Federatsii: 1696 g.–sovremennost'* (Moscow and Rostov-on-Don: Izdatel'skii tsentr "MarT," Feniks, 2010), 3, 12–13, 26–28. Never boasting more than forty members, the association was largely devoted to nature study rather than tackling the Caucasus mountain peaks that Englishmen had begun scaling in the 1860s. On the latter, see Charles King, *The Ghost of Freedom: A History of the Caucasus* (Oxford: Oxford University Press, 2008), 124–33.

41 McReynold, *Russia at Play*, 164. While remarking the discredited status of the traveler-tourist dichotomy, McReynolds argues that her opposition between the "noble 'traveler'" and the "bourgeois 'tourist'" nonetheless "makes a useful social distinction" in "The Prerevolutionary Tourist," 18.

While it is logical to seek the origins of modern tourism in commercialization and the emergence of associations such as the cycling club, the literary lens shows that in parallel to what Buzard found for the British, Russians also started worrying about tourism "long before any coherent tourist industry was in place."[42] Seeds of Russian tourist phobia were already visible in Karamzin's *Letters of a Russian Traveler*, a narrative produced well before the words *turist* (feminine *turistka*, plural *turisty*) and *turizm* entered the Russian language. Karamzin's gentleman traveler persona regards traveling others as recreational defilers of Enlightenment ideals of the Grand Tour, and his cautionary examples are the English. That self/other dynamic took lexicological form in the 1830s when the word *turist* began gaining currency in Russian. Apparently first past the post in printing the foreign borrowing, *Literaturnaia gazeta* (*Literary Gazette*) in 1831 explained *turist* as what "the English call their compatriots who roam around Europe out of idleness or to dispel their spleen." A Russian glossary of 1837 similarly defined the tourist as "an Englishman traveling about the world." E. N. Kologrivova's obscure novel *Golos za rodinu* (*A Vote for the Homeland*, 1842) continued to identify tourists as travelers "bred solely in the British isles, principally in the English counties."[43] Based on the author's trip of 1842, Countess Evdokiya Rostopchina's lyric poem "More on Naples" (written in 1846) remained in the rut. The work pitted the sensitive Russian self against a vulgar *turist* other—a rich Englishman with his family, swaggering around Naples as though they own the place and inducing local people to abandon Neapolitan clothing in favor of English attire.[44]

By the time Rostopchina went abroad, however, Russian writers were in fact already using *turist* to designate themselves and their compatriots visiting the West. In that decade, too, the idea of a holiday—"vacation days"—also became current in Russia with reference to a break in the academic year.[45] An outstanding early expression of the vacation mentality was Pavel

42 Quote from Buzard on the British tourist rush to the Continent immediately after the Napoleonic Wars, in *The Beaten Track*, 32.

43 M. P. Alekseev, "Pushkin i angliiskie puteshestvenniki v Rossii," *Literaturnoe nasledstvo* 91 (1982): 579, 638n2. On Kologrivova as a translator of Dante, see commentary in F. M. Dostoevskii, *Polnoe sobranie sochinenii*, 30 vols. (Leningrad: Nauka, 1972–1990), 19:324.

44 E. P. Rostopchina, "Eshche o Neapole," *Stikhotvoreniia, Proza, Pis'ma* (Moscow: Sovetskaia Rossiia, 1986), 212–16.

45 See Stepan Shevyryov's account of his 1847 trip to a monastery on Lake Silverskoye, around 260 miles east of Petersburg: *Poezdka v Kirillo-Belozerskii monastyr': Vakatsionnye dni professora S. Shevyreva v 1847 godu* (Moscow: Universitetskaia

Annenkov's *Letters from Abroad* (1840–43), a travel account that locates the author among "tourists" in Venice and amicably remarks other Russian sightseers in Rome.[46] Compatriot tourists are enjoying Italy alongside the author but not in sufficient numbers to trigger a phobic reaction. Annenkov's nationalization of tourism on the printed page reflected the fact that Russians were becoming more numerous on the Continental leisure travel circuit. Prior to the revolutionary upheavals in the West in 1848, the 1840s were a time when cultivated Russians of means took to journeying abroad as "an indispensable part" of their education.[47] A time, too, when Russian tourists caught the satiric eye of an English author. Bearing names such as Princess Fuskymuskoff, the Russian women tourists in a story set in Baden-Baden stand out at for their "magnificent diamonds" and clothes of the sort that Thorstein Veblen's *The Theory of the Leisure Class* would term "conspicuously wasteful apparel."[48] The Russian *turist* cropped up on the domestic scene in the 1850s in stories set at the North Caucasus spas Pyatigorsk, Kislovodsk, Zheleznovodsk, and Essentuki. Although the so-called "tourist" might or might not be a *kurist* (a person taking a water cure), none of them was literally touring but rather clustering at the resorts. With clustering came crowd phobia, so that in Russia's case, MacCannell's truism that "tourists dislike tourists" was first borne out at home rather than abroad.[49]

The initial shocks of modernization in the early reform era (1855–70) during the reign of Alexander II (1855–81) gave Russians the impression that leisure travel, along with so much else, had suddenly shaken loose from the national culture's and society's traditional moorings and therefore needed a new name. That name was *turizm*, a word that made what seems to have been its sole appearance in print in the imperial age in 1859. The term gained currency only in the Soviet era, but its fleeting appearance in the nineteenth century significantly announced the social recognition of a

tipografiia, 1850)—cited in Christopher Ely, *This Meager Nature: Landscape and National Identity in Imperial Russia* (Dekalb: Northern Illinois University Press, 2002), 119–21. Shevyryov was a poet, critic, and Moscow University professor of literature.

46 P. V. Annenkov, *Pis'ma iz-za granitsy*, in his *Parizhskie pis'ma*, ed. I. N. Konobeevskaia (Moscow: Nauka, 1983), 19.

47 Derek Offord, *Portraits of Early Russian Liberals: A Study of the Thought of T. N. Granovsky, V. P. Botkin, P. V. Annenkov, A. V. Druzhinin and K. D. Kavelin* (Cambridge: Cambridge Universisty Press, 1985), 109.

48 Frances Trollope, *The Robertses on Their Travels*, 3 vols. (London: Henry Colburn, 1846), 2:190; and Thorstein Veblen, *The Theory of the Leisure Class*, ed. and intro. Martha Banta (1899; repr., Oxford: Oxford University Press, 2007), 112.

49 MacCannell, *The Tourist*, 10.

new phenomenon, a "phenomenon of modern times."[50] As we will see, circumstantial and social criteria characterized *turizm*: industrialized transport, a huge wave of travelers, their increased social diversity, and more women among them. The upheavals of the early reform era entangled tourism in controversies about culture and aesthetics, political ideologies, gender, morality, social identity and *embourgeoisement*, the Russian Empire's relation to the capitalist West, and ethnic Russians' relation to the native peoples of the Caucasus and the Crimea. Modern tourism's divisive impact on the educated classes was vividly dramatized in the writings of Ivan Turgenev, Leo Tolstoy, Fyodor Dostoevsky, and lesser authors, as well as the commentaries of the radical journalists Nikolai Chernyshevsky (1828–89), Nikolai Dobrolyubov (1836–61), and Dmitry Pisarev (1840–68). My study centers on controversies about tourism that arose in the turbulent early reform era and then observes how they played out into the fin de siècle. Throughout the book, I will stress the context of domestic rather than international politics, to show that divisions manifested among Russian tourists in foreign lands reflected the home society to which they had to return.[51]

Two factors motivated closure of the study at the turn of the century. The first was the wealth of illuminating neglected writings for the reign of Alexander II. The second consideration was the embarrassment of riches in the period of Russian Modernism, loosely dated from the early 1900s to the late 1920s. During that time, cosmopolitan tendencies revitalized Russian literature, art, and music. The Ballets Russes took Paris by storm, and virtually every literary genius of the era, including Alexander Blok (1880–1921), Anna Akhmatova (1889–1966), and Osip Mandelshtam (1891–1938), traveled abroad and wrote about it. Imperial borderlands, especially the Crimea and Georgia, also attracted members of the Modernist avant-garde, including Mandelshtam, who visited Georgia and wrote an account of his trip to Armenia.[52] The poet Maximilian Voloshin's (1877–1923) home in Koktebel in the Crimea became an artistic colony of sorts, a retreat for his fellow Russian writers. The very richness of the Modernist flowering and the complexities of its political contexts—the revolutions of 1905 and 1917, the

50 The text was N. A. Mel'gunov [N. Livenskii, pseud.], "Turisty voobshche i osobenno russkie. (Pis'mo v redaktsiiu iz-za granitsy)," *Otechestvennye zapiski* 123 (March 1859), section 6: 1–11, treated below, in ch. 5.

51 Cf. Buzard on the British, *The Beaten Track*, 154.

52 For an overview of this "Russo-Georgian encounter," see Harsha Ram, "Modernism on the Periphery: Literary Life in Postrevolutionary Tbilisi," *Kritika: Explorations in Russian and Eurasian History* 5, no. 2 (Spring 2004): 367–82.

Russo-Japanese War, and World War I— make it unwieldly in my framework. Modernist Russian tourism demands a special study and shall receive only cursory treatment in my book.[53]

The Tourist as Cosmopolitan

Cosmopolitanism opens up important issues surrounding tourism. Kwame Anthony Appiah has written that "Cosmopolitanism and patriotism, unlike nationalism, are sentiments more than ideologies."[54] To quote him again, that status has made "cosmopolitanism" a "shape shifter" signifying many things to many people. "Cosmopolitanism" circulates today among other plastic terms such as "globalization" and "multiculturalism."[55] Another shape-shifting alternative is "transnationalism." "Globality" has become yet another contender.[56] But unlike "transnationalism" and other terms of our time, "cosmopolitan" and "cosmopolitanism" were the words imperial Russians used, and they attributed positive as well as negative meanings to them.[57] My study's theme of provinciality particularly brings out the positive connotations that cosmopolitanism acquired in imperial Russia. This in

---

53 Among the stimulating work relevant to such a study are N. P. Komolova, *Italiia v russkoi kul'ture Serebrianogo veka. Vremena i sud'by* (Moscow: Nauka, 2005); Georges Nivat, "L'Italie de Blok et celle de Gumilev," *Revue des études slaves* 4, nos. 3–4 (1982): 697–709; and Irina Shevelenko, "Imperiia i natsiia v voobrazhenii russkogo modernizma," *Ab Imperio* 3 (2009): 171–206.

54 Kwame Anthony Appiah, "Cosmopolitan Patriots," in *For Love of Country: Debating the Limits of Patriotism*, ed. Martha C. Nussbaum and Joshua Cohen (Boston: Beacon Press, 1996), 23. On literature as a "moral and sentimental domain" that "is a constituent and precondition" for developing "an internationalist ethic or culture," see Bruce Robbins, *Feeling Global: Internationalism in Distress* (New York: New York University Press, 1999), 17.

55 Kwame Anthony Appiah, *Cosmopolitanism: Ethics in a World of Strangers* (London: Penguin Books, 2006), xi. For illuminating discussion of "cosmopolitanism" and related terms, see Lauren M. E. Goodlad and Julia M. Wright, "Introduction and Keywords," *Romanticism and Victorianism on the Net* 48 (November 2008): 1–16. https://doi.org/10.7202/017435ar. On "cosmopolitan" as an "internationalist perspective," see also James Buzard, "Race, Imperialism, Colonialism, Postcolonialism, and Cosmopolitanism," in *The Oxford Handbook of Charles Dickens*, ed. John Jordan et al. (Oxford: Oxford University Press, 2018), 531.

56 Gayatri Chakravorty Spivak, "Cultural Talks in the Hot Peace: Revisiting the 'Global Village,'" in Cheah and Robbins, *Cosmopolitics*, 329.

57 See similar argument concerning the Victorian period in Tanya Agathocleous and Jason R. Rudy, "Victorian Cosmopolitanisms: Introduction," *Victorian Literature and Culture* 38 (2010): 390.

itself corrects the contention that "cosmopolitan" was essentially and already an anti-Semitic slur in mid-nineteenth-century Russia, as it so notoriously became at the time of the "doctors' plot" under Stalin.[58] Anti-Semitism and anti-cosmopolitanism have crisscrossing histories in imperial Russia but never achieved a definitive merger. Anti-Western discourse rather than anti-Semitism was in fact the steadier strand of anti-cosmopolitanism in imperial Russia as of the eighteenth century.

If we substitute "their lives" for "life abroad," Jeffrey Brooks's formulation about Russian popular culture vis-à-vis the West provides a good point of departure for setting the positive pole of my field of inquiry. In his chosen context, Brooks defined Russian cosmopolitanism as "a sympathetic orientation toward foreigners and life abroad," "increased interchange with other nations and receptiveness to the surrounding world."[59] Studies of Russian travel writing and tourism have assumed an equivalent concept of cosmopolitanism. In this context, "cosmopolitan" signifies worldliness and sophistication, especially the linguistic and cultural competences that allow a person to feel "at home" in a foreign land. The most frequently referenced

58 See Frank Grüner, "'Russia's Battle against the Foreign': The Anti-Cosmopolitanism Paradigm in Russian and Soviet Ideology," *European Review of History* 17, no. 3 (June 2010): 445–72. Grüner cites only one example from that period: the identification of a traveling Jew as a "cosmopolitan" in Ivan Goncharov's *The Frigate Pallas* (1856–58). Grüner jumps to the conclusion that "obviously at that time already the concepts 'cosmopolitan' and 'kike' (Russian '*zhid*,' pejorative appellation for 'Jew') had a largely similar meaning in Russia" (446). Let us note some immediate counter-evidence. In reviewing an installment of *The Frigate Pallas* serialized in 1856, the critic Stepan Dudyshkin declared that Goncharov himself made "cosmopolitan" choices in visiting and writing about Japan. Dudyshkin thought Japan sounded interesting but remote from Russia's urgent national need to rebuild after the Crimean War: see his review of "Russkie v Iaponii," *Otechestvennye zapiski* 104, nos. 1–2 (January 1856), section 3: 49–50. See also the lyric poem "Summertime" ("Leto," 1856) by Alexander Herzen's friend and collaborator Nikolai Ogarev. The "cosmopolitans" there are larks and sparrows that the exiled poet observes behaving in the fields of England just as they do in Russia. He envies their capacity to make homes in two places, while he pines for Russia, symbolized by "primeval forest" and birch trees: N. P. Ogarev, *Izbrannye stikhotvoreniia i poemy* (Moscow: Gosudarstvennoe izdatel'stvo khudozhestvennoi literatury, 1938), 145–46.

59 Jeffrey Brooks, "Official Xenophobia and Popular Cosmopolitanism in Early Soviet Russia," *American Historical Review* 97 (December 1992): 1431. See also treatment of Sergei Eisenstein and other culture producers in Stalinist Russia as "cosmopolitan patriots," loyal to the regime but actively engaging with Western arts, in Katerina Clark, *Moscow, the Fourth Rome: Stalinism, Cosmopolitanism and the Evolution of Soviet Culture, 1931–1941*(Cambridge, MA: Harvard University Press, 2011), 30–34. At the time of writing, Clark was unaware of Appiah's "Cosmopolitan Patriots."

individual is Karamzin, who fashioned himself in his *Letters* as "a citizen of the world" (*grazhdanin vselennoi*).[60] Travel accounts about Western Europe in Pushkin's era provide additional examples of confident, "cosmopolitan" navigation of the foreign.[61] The same criteria have prompted description of the well-traveled, multilingual Turgenev as "a supremely cosmopolitan writer," a writer who was regarded in his time as a "Russian European" and who served in the Soviet era as an inspiration for diasporas of Russian writers and intellectuals seeking to forge new links between European culture and their national heritage.[62] McReynolds's survey of prerevolutionary Russian tourism conjured cosmopolitanism on a massive scale: "In the last years of the imperial era, Russia's tourists represented progress, modernity, and a multicultural nationalism driven by a capitalist engine that would not be stalled by traditional prejudices. Annually by the tens of thousands, they sought cosmopolitan and adventurous identities through the multiple opportunities for contrast offered by travel."[63] This formulation evokes sophistication and also implies the practice of cosmopolitanism "in everyday life," "ways of living abroad at home" through reading, fashion, shopping, cuisine, home décor, and other forms.[64] My examination of tourism shall corroborate all those positive meanings of cosmopolitanism but shall expand and weigh them against anti-cosmopolitan, anti-Western discourses.

---

60 N. M. Karamzin, *Pis'ma russkogo puteshestvennika*, ed. Iu. M. Lotman et al. (Leningrad: Nauka, 1984), 321, 331. For representative remarks on Karamzin, see Schönle, *Authenticity and Fiction*, 45–46; Offord, *Journeys to a Graveyard*, 90; Dickinson, *Breaking Ground*, 110, 112, 149; and Ingrid Kleespies, "East West Home Is Best: The Grand Tour in D. I. Fonvizin's *Pis'ma iz Francii* and N. M. Karamzin's *Pis'ma russkogo putešestvennika*," *Russian Literature* 52, nos. 1–3 (2002): 262.

61 Dickinson, *Breaking Ground*, 175.

62 Quote from Anne Lounsbery, *Life Is Elsewhere. Symbolic Geography in the Russian Provinces, 1800–1917* (Ithaca and London: Northern Illinois University Press, an imprint of Cornell University Press, 2019), 119; see also Greta Slobin, "Turgenev Finds a Home in Russia Abroad," in *Turgenev: Art, Ideology, and Legacy*, ed. Robert Reid and Joe Andrew (Amsterdam and New York: Rodopi, 2010), 190, 193, 216.

63 McReynolds, *Russia at Play*, 192.

64 Sheldon Pollock et al., "Cosmopolitanisms," in *Cosmopolitanism*, ed. Carol A. Breckenridge et al. (Durham and London: Duke University Press, 2002), 11. See also Delanty and Inglis, "Introduction: An Overview of the Field of Cosmopolitan Studies," in Delanty and Inglis, *Cosmopolitanism*, 1:7–8; Appiah, *Cosmopolitanism*, 113; Mica Nava, "Cosmopolitan Modernity: Everyday Imaginaries and the Register of Difference," *Theory, Culture & Society* 19, nos. 1–2 (2002): 81–99; Gorsuch and Koenker, "Introduction," 5; and Rebecca L. Walkowitz, *Cosmopolitan Style: Modernism Beyond the Nation* (New York: Columbia University Press, 2007), 2.

Considerable complications already begin with Karamzin. He produced his *Letters* during the reign of Catherine the Great, the time when the word *cosmopolite* first appeared in print in Russia. As James Billington explained, the responsible author was Theodore Henri Chudi, a Swiss actor who wormed his way into the entourage of Catherine's court and made a "successful career as a gigolo and glorified gossip columnist" after reinventing himself as the Chevalier de Lussy. As editor of the first French-language journal in Russia *Le Caméleon littéraire*, Chudi published an article designating his "frivolity" a characteristic of the French nation and by extension a qualification for his personal claim to "the title of Cosmopolitan." With a fellow such as Chudi waving the banner, it is no wonder Russians began associating "cosmopolitanism" with hedonism, sensuality, superficiality, and, in general, "the virus of Voltaire."[65] In comic representations of the everyday life of the westernized nobility, Denis Fonvizin (1745–92) and other alarmed dramatists of the period made the mockery of Russian Gallomania a central literary tendency. As brilliantly portrayed by those writers, the francophone nobles with their "craze for French fashion" appeared not "as admirably cultivated cosmopolitans but as superficial and frivolous men and women prone to personal infidelity and national disloyalty."[66]

The patriotic, proto-nationalist currents of Karamzin's *Letters* protected his account from falling under suspicions of disloyalty. All the same, the era's arbiters of genteel conduct did find his gadding about Europe with the apparent purpose of *writing* about it shockingly hedonistic and vainglorious.[67] Furthermore, Karamzin's Europeanization of Russian and his more general openness to foreign literatures and culture provoked resistance from Admiral Alexander Shishkov (1754–1841), the early nineteenth century's leading champion of the hopeless project of guarding Russian, Russian literature, and Russian life against foreign influences. And from Shishkov's pen, too, in the mid-1820s came a patriotic denunciation of self-described citizens of the world as people lacking proper love of their

65 Billington, *The Icon and the Axe*, 234 (Chudi's French text is not supplied).

66 Derek Offord, "Linguistic Gallophobia in Russian Comedy," in *French and Russian in Imperial Russia*, vol. 2: *Language Attitudes and Identity*, ed. Derek Offord et al. (Edinburgh: Edinburgh University Press, 2015), 94–95. See also discussion of Russian literary images of the French tutor or governess in Russia as an incarnation of "some combination of arrogance, indolence, ignorance, cynicism and contempt for Russia and its vernacular," in Offord et al., "Introduction," in ibid,. 10.

67 Schönle, *Authenticity and Fiction*, 230–31n57.

families and their homeland, the national community that unites its members through language and faith.[68]

Shishkov's opposition between cosmopolitanism and patriotism brings us into Pushkin's time when Russian pleasure trips to the West grew more problematic than ever in the public eye. On the one hand, an upsurge of Russian nationalism followed the Napoleonic Wars. But on the other hand, the period brought to Russia a new, "romantic cosmopolitanism," a sensibility most radically represented by the "heroic individualism" of Lord Byron performing wanderlust and disaffection to home, hearth, and nation.[69] Within this matrix, imperial Russian tourism abroad began its history of becoming deeply politicized and subject to public moralizing that entailed labeling tourists dangerous "cosmopolitans." This discourse stood in continuity with Shishkov's conservative nationalist worries about disloyalty and a perceived lack of respect for Russian and Russian culture, but it also anticipated the Soviet linkage of "cosmopolitanism" to "bourgeois decadence."[70]

In providing patriotic alternatives to foreign travel, tourism within the Russian Empire offered safe havens from anti-cosmopolitan attacks. Spas, mountains, and sea made the non-Slavic territories of the Caucasus and the Crimea a prized "pleasure periphery" for Russian leisure.[71] In those domestic venues, however, metropolitan tourists snubbed their "provincial" compatriots, meaning residents of the towns and culturally impoverished country estates in the empire's vast central sphere, the "non-exotic, non-borderland Russian space" outside Petersburg and Moscow.[72] There was thus some tension between the hegemonic imperial consciousness that

68 G. M. Hamburg, "Language and Conservative Politics in Alexandrine Russia," in Offord et al., *French and Russian in Imperial Russia*, 125–28.

69 Delanty and Inglis, "Introduction," in Delanty and Inglis, *Cosmopolitanism*, 1:4–5. See also Esther Wohlgemut, *Romantic Cosmopolitanism* (Basingstoke: Palgrave Macmillan, 2009), 95–118; Kirsten Daly, "Worlds beyond England: *Don Juan* and the Legacy of Enlightenment Cosmopolitanism," *Romanticism* 4, no. 2 (1998): 189–201; and Goodlad and Wright, "Introduction and Keywords," 9.

70 On the Soviet perceptions, see Gorsuch, *All This Is Your World*, 69, 140, 151–52. The *Oxford English Dictionary*'s inclusion of the official Soviet characterization of "cosmopolitanism" is cited in Agathocleous and Rudy, "Victorian Cosmopolitanisms," 389.

71 Term borrowed from Louis Turner and John Ash, *The Golden Hordes: International Tourism and the Pleasure Periphery* (New York: St. Martin's Press, 1976).

72 Lounsbery, *Life is Elsewhere*, 13. On nineteenth-century Russian and Soviet time-space dichotomies between Moscow and the provinces, see also Katerina Clark, "Political History and Literary Chronotope: Some Soviet Case Studies," in *Literature and History:*

united Russians and the social antagonisms that divided them. In renaming the Russian *tsarstvo* (tsardom) the *Rossiiskaia imperiia* and assuming the title of *imperator* (emperor), Peter the Great established "the imperial idea" that would percolate down "irregularly and unevenly" from the dominant elites into the mass of the population via highbrow belles-lettres, the mass press, and popular literature.[73] Functioning as "a form of imperialism," Russian domestic tourism, its guidebooks and souvenirs greatly contributed to that percolation.[74] Russia's status as a multinational continental empire gave domestic tourism an extraordinary, indeed "uniquely imperialist role" that shored up "relations of power and privilege" between ethnic Russians and the empire's minority peoples.[75] No matter what their social standing at home, Russian men and women tourists could experience themselves as colonizer kings and queens with regard to the "Asian" natives of the Caucasus and the Crimea. The bonding force of the imperial idea was unbreakable but weakened by socioeconomic divisions and snobbery in the domestic tourist arena. Social underdogs, however, seem to have found a way of boosting their self-esteem. Though limited, my material on this issue would suggest that Russian tourists low in the pecking order at home got a special thrill out of playing hegemonic lords and ladies of empire on vacation in "Asia."

Whether abroad or on the southern periphery, cosmopolitan tourists in this book fit Urry's profile of the "aesthetic cosmopolitan," minus the accretions of modern-day mass tourism and with a stronger emphasis on the consumption of art as the aim of going to Western Europe. Appiah's presentation of Sir Richard Burton clarifies the type. Commanding many languages, including Persian, Arabic, and Hindi, Burton roamed India, the Middle East, Africa, "large swaths of Asia and Latin America," and even the United States. His "voracious assimilation" of religions, literature, art,

---

Theoretical Problems and Russian Case Studies, ed. Gary Saul Morson (Stanford: Stanford University Press, 1986), 234–46.

73 Alfred J. Rieber, "The Sedimentary Society," in Clowes et al., *Between Tsar and Society*, 348–49. On relevant popular entertainments such as circus acts, see Thomas M. Barrett, "Southern Living (in Captivity): The Caucasus in Russian Popular Culture," *Journal of Popular Culture* 31 (Spring 1998): 83–88.

74 Quoted concept from Dennison Nash, "Tourism as a Form of Imperialism," in *Hosts and Guests: The Anthropology of Tourism*, ed. Valene E. Smith, 2nd ed. (Philadelphia: University of Pittsburgh Press, 1989), 37–52.

75 McReynolds, "The Prerevolutionary Tourist," 18; and Gorsuch and Koenker, "Introduction," 8. See also Nash, "Tourism as a Form of Imperialism," 39.

and customs from around the world marked him as "someone who was fascinated by the range of human invention, the variety of our ways of life and thought." And yet, his writings expressed racial prejudice and no sign of compunction about buying native slaves to carry his luggage in Africa. The ethical insensitivity, typical of his time, placed Burton outside what Appiah considered "the first strand of cosmopolitanism: the recognition of our responsibility for every human being." On this view, Burton participated only in "the second strand of cosmopolitanism: the recognition that human beings are different and that we can learn from each other's differences."[76] The tourists I treat belong overwhelmingly to the second strand. However, a few of their stories demonstrate that aestheticism and ethical, moral engagement are tendencies that may exist in the same person.[77]

Authenticity and Culture

While there can be no disputing the general proposition that a desire to be "elsewhere" motivates tourism, there is a huge amount of theoretical discussion about how much *difference*, how much "authentic" foreignness tourists actually encounter (especially in our time), how they measure it, and how much they really want. Salman Rushdie has popularized the notion that one of the "great struggles" of humankind is the "mighty conflict between the fantasy of Home and the fantasy of Away, the dream of roots and the mirage of the journey."[78] The remark is catchy but sounds glib alongside Yury Lotman's monumental work devoted precisely to analyzing the semiotics of space that Rushdie evoked: the mutually illuminating visions of the native realm and the foreign. Lotman observed that every journey is about crossing boundaries, and the first one is the threshold of the traveler's own home. From that starting point, symbolic fields of meaning develop along an axis setting self against other, home against the world beyond.[79] Edward Said's analysis of the European construction of the Orient termed such processes the creation of "imaginative

76 Appiah, *Cosmopolitanism*, 1–8 (quotes 4, 7–8).

77 In imperial Russian writing and thinking, distinctions between ethics and morality were much blurrier than posited in Kwame Anthony Appiah, *The Ethics of Identity* (Princeton: Princeton University Press, 2005), 230–37.

78 Cited in Craig Calhoun, "The Class Consciousness of Frequent Travellers: Toward a Critique of Actually Existing Cosmopolitanism," in Vertovec and Cohen, *Conceiving Cosmopolitanism*, 86.

79 Jurij Lotman, *The Structure of the Artistic Text*, trans. Ronald Vroon (Ann Arbor: University of Michigan Press, 1971), 217–31, 237–38; and Yuri M. Lotman, *Universe*

geography" (although much of what he posited might be more accurately described as imaginative ethnography).[80] Scholars in the Russian field have used alternative terms such as "imaginary," "imagined," and "symbolic geography."[81] Whatever we call them, such schemes extend Lotman's fundamental opposition between Russian *svoi* and *chuzhoi*: "one's own," "ours," versus the polysemantic *chuzhoi*—"somebody else's," "foreign," "alien." The *svoi/chuzhoi* division and an accent on "alien" were fundamental to Slavophilism and conservative Russian nationalism. Cosmopolitans, on the other hand, resist the binary and especially the slippage from "foreign" into "alien" in responding to cultures and natural environments *different* from home.

In encountering foreign peoples both abroad and on the southern pleasure periphery, nineteenth-century Russian tourists took an essentialist view, attributing to each people a sui generis, authentic cultural core. The assumed authenticity was always a delusion. "Cultural purity is an oxymoron," wrote Appiah: to seek "some primordially authentic culture" is comparable to "peeling an onion."[82] In developing his argument about cultural "contamination" as the counter-ideal to the flawed "ideal of cultural purity," he referred to Rushdie's oft-quoted celebration of "mongrelization." Appiah went on, however, to stress that "cosmopolitan contamination" is not modern but ancient:

> Alexander's empire molded both the states and the sculpture of Egypt and North India; first the Mongols then the Mughals shaped great swaths of Asia. [. . .] Islamic states stretched from Morocco to Indonesia. [. . .] The traders of the Silk Route changed the style of elite dress in Italy.[83]

---

of the Mind: A Semiotic Theory of Culture, trans. Ann Shukman, intro. Umberto Eco (London and New York: I. B. Tauris, 2001), 171–77.

80 Edward W. Said, *Orientalism* (New York: Vintage Books, 1979), 49–73; see also Susan Layton, *Russian Literature and Empire: Conquest of the Caucasus from Pushkin to Tolstoy* (Cambridge: Cambridge University Press, 1994), 36–53.

81 Dickinson, *Breaking Ground*, 20–21; Edith W. Clowes, *Russia on the Edge: Imagined Geographies and Post-Soviet Identity* (Ithaca: Cornell University Press, 2011); and Lounsbery, *Life Is Elsewhere*.

82 Appiah, *Cosmopolitanism*, 107.

83 Ibid., 112. See also the investigation of "cosmopolitan literary production" in Sanskrit and other languages beginning before the first millennium, in Sheldon Pollock, "Cosmopolitan and Vernacular in History," in Breckenridge et al., *Cosmopolitanism*, 15–53 (quote 19). Pollock contends that popularization has made mongrelization "a banal concept and a dangerous one, implying an amalgamation of unalloyed, pure forms, whether vernacular or cosmopolitan, that have never existed," 47.

Among the current objections to essentialist concepts of culture, Clifford's formulation strikes a particularly fitting note for tourism: culture is not a fixed entity bounded in time and space but rather a work in progress, an itinerary, a route crisscrossing the "global world of intercultural import-export." Cultural production and consumption are thus about the interplay between roots and routes.[84] M. Crick's like-minded view, quoted in Urry's *The Tourist Gaze*, states that culture is constantly "invented, remade and the elements reorganized."[85] In the Russian field, outstanding examples of such approaches to culture are studies of the development of Russian music and the visual arts in the imperial age.[86] As a broader phenomenon of everyday life, cosmopolitan cross-fertilization seems to have been "ubiquitous across [many Russian] imperial borderlands and often proved the source of real consternation among agents of empire attempting to consolidate administrative control of regions distance from the metropole."[87]

The notion of cultural "authenticity" has become so indefensible that tourism studies opt to speak of the "authenticity effect" or "perceived authenticity."[88] While tourists have always relished "being elsewhere" to escape the daily grind, their subjective perceptions of satisfying foreignness

84 Clifford, "Traveling Cultures," *Routes: Travel and Translation in the Late Twentieth Century*, 17–46, quote 25.

85 John Urry, *The Tourist Gaze: Leisure and Travel in Contemporary Society* (London: Sage Publications, 1996 [1990]) 9. See also Edward W. Said, *Culture and Imperialism* (London: Chatto and Widus, 1993), 271–79.

86 Richard Taruskin, *Defining Russia Musically: Historical and Hermeneutical Essays* (Princeton: Princeton University Press, 1997), 43; Rutger Helmers, *Not Russian Enough? Nationalism and Cosmopolitanism in Nineteenth-Century Russian Opera* (Rochester: Rochester University Press, 2014), 1–3; and Katia Dianina, *When Art Makes News: Writing Culture and Identity in Imperial Russia* (DeKalb: Northern Illinois University Press, 2013), 23–28.

87 See James F. Brooks, "Bondage and Emancipation across Cultural Borderlands: Some Reflections and Extensions," in *Orientalism and Empire in Russia*, ed. Michael David-Fox et al. (Bloomington, IN: Slavica Publishers, 2006), 292. For representative cases, see Daniel Brower and Susan Layton, "Liberation through Captivity: Nikolai Shipov's Adventures in the Imperial Borderlands," in David-Fox et al., *Orientalism and Empire in Russia*, 270–90; Michael Khodarkovsky, *Bitter Choices: Loyalty and Betrayal in the Russian Conquest of the North Caucasus* (Ithaca and London: Cornell University Press, 2011); Willard Sunderland, "Russians into Yakuts? 'Going Native' and Problems of Russian National Identity in the Siberian North," *Slavic Review* 55, no. 4 (Winter 1996): 824–53; and Yuri Slezkine, *Arctic Mirrors: Russia and the Small Peoples of the North* (Ithaca and London: Cornell University Press, 1994).

88 See respectively Buzard, *The Beaten Track*, 172; and Deepak Chhabra, "Staged Authenticity and Heritage Tourism," *Annals of Tourism Research* 30, no. 3 (2003): 703.

apparently differed as much in the past as they do today.[89] Less complicated than the issue of culture, the realm of nature provides a good starting point for clarification. Tourists seek nature for aesthetic pleasure and active leisure, two desiderata easily combined.[90] The home-plus tourists, however, like the comfort and convenience of having hotels and infrastructure near the mountains, lake, waterfall or whatever other natural attraction that they have come to consume. The topographical difference from home suffices to constitute "authenticity." On the other hand, the different-from-home kind of vacationer is quick to perceive any tourist development as overdevelopment, attracting unwelcome crowds and causing environmental pollution. From this standpoint, "authenticity" is to be found only in untrammeled nature, preferably enjoyed in solitude or with just a few other visitors.[91]

There are numerous definitions of "cultural tourism," but what works best for the Russian material is Buzard's two-part model that sets a narrowly aesthetic program against the broadly ethnographic urge to encounter a whole "way of life" different from home. In the narrow sense, cultural tourism concerns "aesthetic objects and their role in personal acculturation," primarily art, architecture, and monuments. In the ethnographic sense, "cultural tourism" is an extremely flexible concept, covering cuisine, customs, festivals, "sex tourism attractions," highbrow and popular entertainments, concerts, and theater.[92] For the surface-skimming tourist, just being "elsewhere" and passively soaking up the "local color" suffices to produce the ethnographic authenticity effect. The same sort of visitor dispatches the aesthetic program by ticking off the objects that guidebooks and other markers have designated the proper things to see. On the other hand, the

---

89 Buzard, *The Beaten Track*, 175–77, in dialogue with Urry, *The Tourist Gaze*, 11; and Erik Cohen, "Authenticity and Commoditization in Tourism," *Annals of Tourism Research* 15, no.3 (1988): 376–78.

90 MacCannell, *The Tourist*, 80.

91 Cf. Russian Sentimental literature's idea of "authenticity" as immersion in the "pristine innocence" of nature, enabling deeper self-knowledge than society affords, in Schönle, *Authenticity and Fiction*, 9–16.

92 For both definitions, see Buzard, *The Beaten Track*, 7, 174; the quote comes from M. Bauer, "Cultural Tourism in France," in *Cultural Tourism in Europe*, ed. Greg Richards (Wallingford, UK: CAB International, 1996), 147–48. The latter collection gives much attention to the difficulty of defining "cultural tourism." For further illustration of the problem, see Valene M. Smith, introduction to Smith, *Hosts and Guests*, 4–6. "Cultural tourism" is restricted there to trips to see "vestiges of a vanishing life-style," such as "peasant culture in Bali." "Ethnic tourism" specifies performances of folk dancing and the like, while "the Museum-Cathedral circuit" falls under "historical tourism," highlighting the "glories of the past."

aesthetic cosmopolitan experiences the ethnographic authenticity effect by getting off the beaten path, to ferret out a restaurant with menus exclusively in the local language, for instance. On the more demanding, art-centered track of cultural tourism, authenticity resides in exercising the competences that dictate a discriminating list of acculturating objects to see and afford a deep, rather than "merely" touristic, spurious response to them.[93]

Those self/other interplays in the quest for cultural capital began taking shape in Russian writings in the early reform era, when the social composition of tourists going abroad significantly expanded. In dealing with the British tourist rush to the Continent right after the Napoleonic Wars, Buzard showed that print culture represented "an aristocracy of inner feeling" supplanting money as the marker of the traveling elite.[94] Education and refinement were marshaled to establish a distinction from the new wave of middle-class people, demeaned as ill-bred, unsophisticated "cockney" types, muscling in on the Grand Tourist circuit. A similar contrast took shape in early reform Russian writings but with national peculiarities stemming from the Table of Ranks. Introduced by Peter the Great in 1722, the system allowed men to acquire nobility by advancing to the requisite rank in the parallel, fourteen-tiered hierarchies of military, civil, and court service.[95] The creation of the service nobility immediately produced social conflicts that still resonated in the nineteenth-century literary history of tourism. Ambitious commoners could rise to high rank through service to the state. So, too, might the tsar confer aristocratic titles for services. Both conditions aroused insecurity among the old nobility, the men of "good" families who sought to "either stop the access of commoners to service or eliminate the provisions for automatic promotion to nobility through the Table of Ranks." Stressing "the psychological and moral aspects of service," they argued that commoners lacked the "the refinements of heart and mind that come with European and literary education."[96]

Service rank alone could overawe social underdogs, keenly aware that the top four ranks were the generals (officially known as the *generalitet*):

93 Buzard, *The Beaten Track*, 80–81.

94 Ibid., 121.

95 For an excellent overview showing the mid-nineteenth-century Table, see Irina Reyfman, "Service Ranks," in *Dostoevsky in Context*, ed. Deborah A. Martinsen and Olga Maiorova (Cambridge: Cambridge University Press, 2015), 98–99.

96 Marc Raeff, *Origins of the Russian Intelligentsia: The Eighteenth-Century Nobility* (New York and London: Harcourt Brace Jovanovich, 1966), 53–70 (quotes 54, 66).

"Sensitive to the elite's scale of values, the common people spoke reverently of the 'general,' viewing this rank as the *summum bonum* in social status."[97] As of the early reform era, tourism narratives reflected Russian society's pervasive obsession with rank; and the self-important but uncouth general tellingly became an object of satire on the part of writers who saw themselves as members of a nobility of character and mind. Illustrating the heterogeneity of their social category, those authors included the wealthy landed noble Turgenev and Dostoevsky, a noble by virtue of his father's having attained the minimal, eighth rank conferring hereditary nobility through his service as a doctor in a public hospital in Moscow.

Russia's version of the traveling aristocracy of inner feeling also included *raznochintsy*, "(literally 'people of various ranks' or 'people of diverse origins')," meaning those outside the five officially designated social categories or "estates" (*sosloviia*): nobles, clergy, merchants, townspeople, and peasants.[98] The most famous *raznochintsy* were the politically engagé literary critics Vissarion Belinsky (1811–48), Chernyshevsky, and Dobrolyubov. As exemplified by Bazarov, the son of a rural doctor in Turgenev's *Fathers and Children* (1862), radical *raznochintsy* were often hostile to aristocratic culture and aggressively ill-mannered. Some of them, however, including Dobrolyubov, seized chances to pursue traditional high-society regimes of cultural tourism abroad. This makes them fascinating, uniquely Russian case studies of middle-class aesthetic cosmopolitans taking to leisure travel as an "emancipatory and democratizing" opportunity.[99]

Toxic Cosmopolitanism

In Russian imagination, the two tracks of cultural tourism tended to run along a geographical divide: imposing Culture was located in the West, while exotic cultures galore awaited vacationers in "Asia." From Karamzin's time into the late nineteenth century, Western Europe in general and Italy in particular were objects of Russian cultural envy. A borrowing from German, the very word "culture" (*kul'tura*) was not widely used in Russia

97 Ibid., 119.

98 Wirtschafter, "The Groups Between," 246.

99 Quoted concept borrowed from discussion of modern-day tourism in Gorsuch and Koenker, "Introduction," 1.

prior to the 1880s.[100] That peculiarity bespoke the national perception of culture deprivation. Isaiah Berlin observed that since the early nineteenth century, "Russia has in no sense been a cultural backwater."[101] The well-known point, though, is that Russians felt "an existential envy" of the West viewed as a superior civilization, richer by far in the arts and sciences.[102] As Berlin's essay indicated, literature was the first art to give Russians faith in their capacity to overcome "cultural marginality" in comparison with the West.[103] Gogol's brilliance added momentum, as did Belinsky's interpretive essays, most notably those on Pushkin, Gogol, and the young Dostoevsky. Thanks in part to Belinsky, Russian literature by Gogol's era had attained the public stature of "high art" but in a field of "thin culture"—a national culture with little to show in the other arts, especially painting and sculpture.[104] This juncture in Russian cultural self-assessment came in the late 1830s and early forties, just at the time when a new wave of Russian travelers such as Annenkov revived emulation of the Grand Tour. He and other art lovers began making Italy the Russian tourist's El Dorado, a development that reflected the big cultural problem that Billington encapsulated as the "missing Madonna."[105] Would Russia ever produce a Raphael? Would any Russian painting ever rise above the strictly national to attain the universal significance of the *Sistine Madonna*?[106] Or was national significance,

100 Dianina, *When Art Makes News*, 29.

101 Isaiah Berlin, "Introduction," in *Russian Intellectual History: An Anthology*, ed. Marc Raeff (New York, Chicago and Burlingame: Harcourt, Brace and World, 1966), 6.

102 Liah Greenfeld, "The Formation of Russian National Identity: The Role of Status and *Ressentiment*," *Comparative Studies in Society and History* 32, no. 3 (July 1990): 549–91. On Turgenev's and Henry James's shared theme of "provincial culture-seekers" abroad, see Dale E. Peterson, *The Clement Vision: Poetic Realism in Turgenev and James* (Port Washington, New York and London: Kennikat Press, 1975), 68. For more general investigation of Russian literature's trope of provinciality, consult Lounsbery, *Life Is Elsewhere.*

103 Michael David-Fox, *Showcasing the Great Experimen: Cultural Diplomacy and Western Visitors to the Soviet Union, 1921–1941* (New York: Oxford University Press, 2012), 13. For an overview of the issue of marginality in relation to travel writing, see Offord, *Journeys to a Graveyard*, 7–23.

104 Anne Lounsbery, *Thin Culture, High Art: Gogol, Hawthorne and Authorship in Nineteenth-Century Russia and America* (Cambridge, MA: Harvard University Press, 2007).

105 Billington, *The Icon and the Axe*, 346–51.

106 After the *Sistine Madonna* was transferred from a Piacenza cloister to the Dresden museum in 1753, it acquired an international "cult status" as an emblem of "the miracle of artistic creation," rather than an object of religious veneration: Laura Engelstein,

in fact, more important than the universal *because* it was national—"ours" not "theirs," *svoi* rather than *chuzhoi*?

Russians were not alone in experiencing culture envy of Western Europe—and Italy, in particular—but they singularly assigned a compensatory role to their empire. An English traveler in 1732 wrote that "though Italy be the country for the sights, England is the country to live in, for my money." Besides, "Florentine beef is not half so good as our English."[107] English tourists more seriously believed that although "the artistic treasures" of the Continent could liberate them from "cultural insularity," their own sociopolitical system could not be bettered. As the travel writer Frances Trollope stated in her book *A Visit to Italy* (1842), her "rambles through the world" had taught her that the Constitution of England "is the only one which appears to be formed in reasonable, honest, and holy conformity to the freedom of man as a human being."[108] Like the Englishman hungering for "our" beef in Florence, Russian tourists, too, missed home foods. In 1838, for example, Pushkin's friend the poet and critic Prince Pyotr Vyazemsky (1792–1878) wrote a travel lyric declaring the nonexistence of decent tea in Germany outside the Frankfurt home of the welcoming Russian ambassador.[109] As for politics, the Slavist Simon Karlinsky observed that many Russian writers, from Dostoevsky to Ilya Erenburg (1891–1967), turned into "superpatriots" as soon as they hit the West.[110] "Democracy?" they would sneer. "Just show me the slums." The point is well taken, but tourist writings, especially prior to the emancipation of the serfs, will also show us Russians learning just how far their homeland deviated from the democratic ideals of freedom and human rights.

However daunting the culture, or enviable the social democracy observed in Western Europe, Russians could always fall back on the imperial idea, the gratification of possessing their vast continental empire,

---

Slavophile Empire: Imperial Russia's Illiberal Path (Ithaca and London: Cornell University Press, 2009), 63–67.

107 Chloe Chard, *Pleasure and Guilt on the Grand Tour: Travel, Writing and Imaginative Geography, 1600–1830* (Manchester and New York: Manchester University Press, 1999), 45.

108 Buzard, *The Beaten Track*, 100.

109 P. A. Viazemskii, "Samovar," *Stikhotvoreniia* (Moscow: Sovetskaia Rossia, 1978) 167–71. The poem states that "the mind is cosmopolitan, but the heart a homebody" (*V nas um—kosmopolit, no serdtse—domosed*).

110 Simon Karlinsky, *Anton Chekhov's Life and Thought: Selected Letters and Commentary*, trans. Michael Henry Heim in collaboration with Simon Karlinsky (Berkeley: University of California Press, 1973), 184.

unmatched in modern times for its geographical and ethnocultural diversity. As "Russians sought to rationalize and master the overwhelming sense of inadequacy" vis-à-vis the West, the empire assumed significance as a "mechanism of compensation for backwardness."[111] In gaining a multinational empire before achieving a sense of nationhood, Russians developed the double identity of *russkii* and *rossiiskii*: ethnocultural and imperial Russianness. The self-definitions tended to coalesce, and during the latter half of the nineteenth century, Russians imagined the empire in increasingly nationalistic terms.[112] They took national pride in it as an achievement that placed them in "the club of European colonial powers."[113] Dostoevsky notoriously epitomized the imperialist-national sentiment in crowing about the Russian army's 1881 siege and destruction of the Turkmen fortress of Geok-Tepe, a major event in the conquest of Central Asia. The news that the operation had killed "14,000 Muslims in a single day" outraged many members of the Russian reading public. Dostoevsky, however, took the occasion to publish his "pithy justification of Russia's empire": "In Europe we were hangers-on and slaves, while in Asia we shall be the masters. In Europe we were Tatars, while in Asia we are the Europeans."[114] A parallel logic pertained in Russian tourism: in the West we are pilgrims seeking acculturation, but in the Caucasus and the Crimea we are the European civilizers.

---

111 The quoted concept is Adam Ulam's, cited in Marc Bassin, *Imperial Visions: Nationalist Imagination and Geographical Expansion in the Russian Far East, 1840–1865* (Cambridge: Cambridge University Press, 1999), 13. Classics on Russian "compensatory nationalism" are Hans Rogger, *National Consciousness in Eighteenth-Century Russia* (Cambridge, MA: Harvard University Press, 1960); and Malia, *Alexander Herzen and the Birth of Russian Socialism*, 292.

112 On "rethinking the empire in nationalistic terms" in the early reform era, see Olga Maiorova, *In the Shadow of Empire: Defining the Russian Nation through Cultural Mythology, 1855–1870* (Madison: University of Wisconsin Press, 2010), 6. See more generally Geoffrey Hosking, *Russia: People and Empire, 1552–1917* (Cambridge, MA: Harvard University Press, 1997) and commentary in Seymour Becker, "Russia and the Concept of Empire," *Ab Imperio* 3–4 (2000): 329–42; Bassin, *Imperial Visions*, 9–15, 45–68; Ronald Grigor Suny, "The Empire Strikes Out: Imperial Russia, 'National' Identity, and Theories of Empire," in *A State of Nations: Empire and Nation-Making in the Age of Lenin and Stalin*, ed. Ronald Grigor Suny and Terry Dean Martin (Oxford: Oxford University Press, 2001), 23–66; Robert P. Geraci, *Window on the East: National and Imperial Identities in Late Imperial Russia* (Ithaca and London: Cornell University Press, 2001); and Edyt M. Bojanowska, *A World of Empires: The Russian Voyage of the Frigate Pallada* (Cambridge, MA: Harvard University Press, 2018).

113 Olga Maiorova, "Empire," in Martinsen and Maiorova, *Dostoevsky in Context*, 96.

114 Robert P. Geraci, "Islam," in ibid., 218.

Russians took to vacationing in the South of the empire in the holiday spirit of innocent adventure akin to visiting a cultural heritage festival or theme park. In marginalizing imperialist violence and giving maximum credence to the official ideology of the civilizing mission, that typical attitude coincides with Peter Van der Veer's concept of "colonial cosmopolitanism." Van der Veer wrote that "cosmopolitanism is the Western engagement with the rest of the world, and that engagement is a colonial one." "Colonial officers," no less than missionaries, have displayed "a willingness to engage with the other." Furthermore, in the colonial context, "cosmopolitan openness to other civilizations" entailed the "desire to bring progress and improvement." This amounts to "a cosmopolitanism with a moral mission," the French ideal of *une mission civilisatrice*.[115] One should add, though, that when venturing forth to spread "enlightenment," the self-appointed civilizer took his gun as well as his Bible (the lesson of *Robinson Crusoe*).[116] A terse rejoinder to Van der Veer's narrative would be that European expansion all too often meant "traveling to distant places, meeting interesting people, and killing them."[117] In Van der Veer's telling, "colonial cosmopolitanism" presumes to bestow "civilization" as a "gift."[118] Many people on the

115 Peter Van der Veer, "Colonial Cosmopolitanism," in Vertovec and Cohen, *Conceiving Cosmopolitanism*, 166–67. Without mentioning alternative views such as Appiah's "ruthless cosmopolitanism," the editors of the "Victorian Cosmopolitanisms" special issue of *Victorian Literary Studies* find Van der Veer's essay stimulating: Agathocleous and Rudy, "Victorian Cosmopolitanisms," 392. The worrying plasticity of Van der Veer's oxymoron is elsewhere illustrated by the formulation of a "cosmopolitanism" that incorporates racism in Julia Kuehn, "Colonial Cosmopolitanism: Albert Smith and Rudyard Kipling in Victorian Hong Kong," *Studies in Travel Writing* 19, no. 3 (2015): 227–28, 236–39. Kipling loathed the Chinese he saw in Hong Kong in 1889: "I hate Chinamen," he wrote, these "unhuman" [sic] people with their "yellow faces," these "people that ought to be killed off because they are unlike any people I ever met before." On the other hand, remarks Kuehn, Kipling loved seeing the Hong Kong evidence of Britain's "colonial and economic power." Kuehn calls Kipling's outlook "economic cosmopolitanism," courtesy of Marx and Engels and John Stuart Mill, who observed in 1848 that capital was becoming "more and more cosmopolitan." What excited Kipling was British supremacy in the *globalization* of trade, and he expressed that while spewing racism. Is that not a stretch too far for the elastic idea of "cosmopolitanism?" And how helpful is Kuehn's conclusion that a "dialectic" is at work in "colonial cosmopolitanism"—a Hegelian dialectic in which the "players" are the two big "isms" shifting their relationship to one another under the pressure of changing times?

116 See Peter Hulme, *Colonial Encounters: Europe and the Native Caribbean, 1491–1797* (London and New York: Routledge, 1986), 175–222.

117 Pollock et al., "Cosmopolitanisms," 10.

118 See Bruce Grant, *The Captive and the Gift: Cultural Histories of Sovereignty in Russia and the Caucasus* (Ithaca and London: Cornell University Press, 2009).

receiving end, however, have experienced the civilizing mission as "ruthless cosmopolitanism," "toxic" cosmopolitanism bent upon establishing a coercive international order.[119]

Russian tourists in the Caucasus and the Crimea were beneficiaries of ruthless cosmopolitanism. During the Romantic era, they enjoyed their resorts in the North Caucasus while the Russian army was busy killing the people of nearby Chechnya and Daghestan, those very natives that poets such as Pushkin and Mikhail Lermontov made so "interesting" to the reading public. The Russian annexation of Georgia that began in 1801 was not an all-out military conquest, but it did entail violence—including the murder of a Georgian archbishop who had become a prominent articulator of the resentment and animosity that Georgians of all classes felt toward the Russian presumption to "civilize" them. The boom in Russian tourism in the Crimea did not occur until the 1870s, nearly a century after Russian campaigns cleared the way for Catherine the Great to annex the territory. Russian vacationers in the seventies were not directly following the troops, but they had come to enjoy a land whose Tatar population emigrated to Turkey en masse after the Crimean War, during which they fell under suspicion of disloyalty to Russia and suffered severe persecution.[120] Tourists at the time found a drastically diminished number of Tatars in the Crimea and rarely bothered to wonder about the reason for that demographic change.

Unsurprisingly, my research turned up little in the way of critical Russian perceptions of domestic tourism's indebtedness to colonial violence. My sources display Russians' intense aesthetic receptiveness to exotica. Tropes of spectacle and theatricality abound. Cropping up, too, are familiar stereotypes of the "lazy native," the native trickster, swindler, and thief.[121] There are even frank expressions of Russian complacency about waging war to clear the way for tourist development in places inhabited by "savages." Nonetheless, we will also observe an interesting collision between the Russian state's drive to modernize and Europeanize the southern borderlands and the different-from-home kind of tourist's craving for the "authenticity" of pristine nature and "uncontaminated" pre-capitalist

119 Appiah, *The Ethics of Identity*, 220–23.

120 For the remark that domestic Russian tourists generally followed the troops, see McReynolds, *Russia at Play*, 172.

121 On laziness, see the classic Syed Hussein Alatas, *The Myth of the Lazy Native: A Study of the Image of the Malays, Filipinos and Javanese from the 16th to the 20th Centuries and Its Functions in the Ideology of Colonial Capitalism* (London: Frank Cass, 1977).

foreign cultures. The literary lens also reveals resistance to the assumption that ethnic Russians have a natural right to dominate all the other peoples of the empire. An outstanding, if qualified, critique of ruthless imperialism occurs in the moderate liberal Evgeny Markov's (1835–1903) *Sketches of the Crimea* (1872; 2nd ed. 1884, 3rd ed. 1902), a literary travelogue classic reprinted in Simferopol in 2015.[122] On a smaller scale, some late nineteenth-century stories satirically dramatize outbursts of Russian tourist bigotry toward natives of the southern borderlands.

Outline

The book falls into three sections arranged chronologically. The first section narrates the process of Russians coming to recognize and fret about tourism as part of their national culture. Chapter 1 observes the nationalistic and aesthetic limits of Karamzin's enactment of Enlightenment world citizenry. The author of the *Letters of a Russian Traveler* situates himself in an international community of learned gentlemen, and he articulates sentimental humanism. But he cares little for the English tourists overrunning the Continent. He casts them as representatives of an empire that is more culturally advanced but morally inferior to Russia's. Through his imperial-nationalist optic, Karamzin implicates the English tourist passion for Italy in the disinterested, universalistic aestheticism that he rejects in favor of the moralistic, heartfelt judgment of art.

Chapter 2 maps the shift from Enlightenment to Byronic cosmopolitanism in tourism narratives of the 1840s. The overarching theme is the displacement of Karamzin's travel values. Italy and art consumption become Russian tourist priorities, the official patriotism of the regime of Nicholas I is parodied or ignored, and English tourists emerge as the experts to emulate. The social composition of Russian tourism expands, and steam transport starts supplanting the coach, most dramatically in the form of train travel, encountered abroad. The next chapter constitutes the nationalist counter-coup to the romantic vacation mentality in the context of the Slavophile-Westernizer controversy. The Westernizer Belinsky configured "cosmopolitanism" differently than did the conservative nationalists Mikhail Pogodin (1800–75) and Mikhail Zagoskin (1789–1852). All three,

122 Evgenii Markov, *Ocherki Kryma: Kartiny krymskoi zhizni, prirody i istorii,* 2nd ed. (St. Petersburg and Moscow: Tovarishchestvo M. O. Vol'f, 1884); 3rd ed., repr. (Simferopol': Terra-AiTi, 2015 [1902]).

however, considered tourism abroad a cosmopolitan practice threatening Russia's self-realization as a nation. Chapter 4 then examines the mini-boom in tourism in the North Caucasus that occurred when revolutionary unrest in Europe in 1848–49 prompted the Russian state to curtail leisure travel abroad. The main concerns are tourism in tandem with brutal empire building, Petersburg snobbery toward "provincials," and divergent perceptions of Asian cultural authenticity.

The book's second section investigates tourism's entanglement with social change, the rise of numerous political ideologies, developments in aesthetics, and debates about morality in the early reform era. The top-down cosmopolitanism of the post-Crimean War Russian state brought Russians the freedom to travel again, but with new freedom came public soul-searching about the proper aims of trips abroad. Chapter 5 identifies a paradigmatic, proto-Soviet feud of the 1850s. On the practical, knowledge building side stood the anti-tourist, anti-aesthetic writings of Chernyshevsky; and, on the side of pleasure, the art-oriented, elitist travel narrative *From Abroad* by the lyric poet Afanasy Fet (1820–92). Tensions between pleasure and purpose, nation and empire, sophistication and provinciality are further observed in the sixth chapter's analysis of the representation of a futile quest for cross-cultural brotherhood at a luxury vacation site in Tolstoy's *Lucerne*.

Chapter 7 examines the antagonisms that erupted among Russian tourists when an unprecedented number of them "inundated" the West after the Crimea War. Seeking to police Russian tourism from within, Turgenev and other aesthetic cosmopolitans fashioned a crowd of culturally incompetent, materialistic, ill-bred, and reactionary compatriots, often traveling with serfs. But to the minds of the radicals Chernyshevsky and Pisarev, the elitist cultural tourists were shirkers no less than the crowd. Representing a political anti-tourism peculiar to Russia, those socialist revolutionaries denigrated the cosmopolitan acculturation agenda, maligned pleasure trips to Italy in particular, and articulated a killjoy work ethic in furtherance of an increasingly overt program to overthrow the state. In tracking the mounting anti-cosmopolitanism of Dostoevsky's animus toward tourism, chapter 8 establishes a parallel between him and the Chernyshevskian cohort. Dostoevsky's conservative nationalism put him at political loggerheads with utilitarian socialist ideology. And yet, in casting Russian cultural tourism as quasi-colonial, "servile worship" of the West, Dostoevsky's travel account *Winter Notes on Summer Impressions* (1863) sacrificed the enjoyment of art, an outcome that converged on Chernyshevsky's anti-elite, anti-aesthetic outlook.

Part three seeks to shed new light on the *embourgeoisement* of tourism in the latter third of the nineteenth century. Chapter 9 investigates the arrival of the Russian bourgeois on the Continental tourist circuit in the late sixties and early seventies. Politically diverse writers, including Herzen, the Slavophile Ivan Aksakov (1823–86), Dostoevsky, and the populist revolutionary Pyotr Lavrov (1823–1900), represented Russian tourists in the West as arrivistes who put their homeland to shame. A big challenge, however, came from Markov, whose travel writings and journalism propagated "bourgeois ideals": commerce, the bourgeois work ethic, domestic comfort, cleanliness, and concern with décor; social democracy, and the democratizing value of vacationing in the West.[123] In addition to his 1869 account of visiting Switzerland and lesser travel narratives of the time, the mass-circulation magazine *Niva* (*Grainfield*, 1870–1918) provides an abundance of material inviting middle-class Russians to adopt the Cook attitude to tourism as a means of "leveling upward" to their social superiors.[124] Chapter 10 shows *Anna Karenina* disregarding the bourgeois influx into tourism, to reconsider instead the aristocratic tourist style that Tolstoy knew from his visits of Europe in the fifties and sixties. Related to controversies that pitted nativism against cultural preference for the West, Anna's and Vronsky's sojourn in Italy receives special attention and acquires new nuances against the backdrop of the little-studied, unabated Russian tourist passion for Italy at the time.

The last two chapters take us back to the empire's pleasure periphery from the 1880s to the turn of the century. Chapter 11 explores tensions between the Russification agenda of imperialist nationalism under Alexander III and the tourist zest for exotica. The central text is the second edition of Markov's *Sketches of the Crimea*. The liberal author casts doubt on the civilizing mission and seeks to learn from Tatars, but he also invests heavily in spectacle and conjures an idea of the Crimea as a tourist theme park. Tourist development arouses similar ambivalence. While Markov sees capitalist tourism spoiling the Crimea's natural environment

123 E. L. Markov, "Burzhuaznye idealy," *Russkaia rech'* (July 1879): 200–38.

124 On middle-class British tourists seeking "something of the gloss of their social superiors" through travel, see Edmund Swinglehurst, *Cook's Tours: The Story of Popular Travel* (Poole: Sterling Publishing Co., 1982), 34 (cited in McReynolds, *Russia at Play*, 164–65).

and corrupting Tatars, he dots his narrative with architectural symbols of the triumphant march of tourism as an arm of empire-building.

The final chapter examines the depiction of tourist decadence at the fin de siècle, a development that occurred in the spiritual orbit of Tolstoy. Dominating the field are Gleb Uspensky's (1843–1902) "The *Burzhui*" (1888), Chekhov's "The Lady with the Little Dog" (1899) and "Ariadne" (1895), and Lidia Veselitskaya's tourist adultery novella *Mimi at the Spa* (1891). Internationally acclaimed in its time, *Mimi at the Spa* was part of Veselitskaya's *Mimi* trilogy republished in Russia in 2003, a testimony to the work's anticipation of the consumerism of the luxury-loving *nouveaux riches* of post-Soviet society. Veselitskaya trained a Tolstoyan eye on the vacation fling, set against a proto-Soviet regime of self-improvement that includes hiking. But in its preoccupation with sex, *Mimi at the Spa* normalized hegemonic Russian stereotypes of Georgians and Armenians. On the other hand, Chekhov's short tale "A Loose Tongue" (1899 version) comically laid bare the racism of eroticized power relations between Russian women tourists and Tatar guides in the Crimea. The author and journalist Viktor Burenin (1841–1926) left an equally arresting representation of Russian vacationers insulting local people working in the tourist service sector in the North Caucasus.

The book closes with a recapitulation and suggestions for further research.

Part One
Becoming Tourists

1

Russia's Enlightenment Travel Model: Karamzin, the English, and Italy

After the disruption of the Seven Years' War (1756–63), the English resumed the practice of touring the Continent with a gusto indignant observers perceived as a degeneration of the Enlightenment travel ideal. As instituted in England in the seventeenth century, the Grand Tour was an educational project meant to prepare young men of property for assuming positions of leadership in government, diplomacy, and society. Robert Lassels's *An Italian Voyage* (1670) pioneered the Enlightenment program of erudite travel. Many similar books followed, including Thomas Nugent's four-volume classic *The Grand Tour* (1749). In accordance with the centrality of Latin and Greek in British education, the Enlightenment model enthroned Italy and France as the two obligatory destinations. One went to Italy (especially Rome, Florence, and Venice) for the art and the sites of classical antiquity in order to cultivate a historical consciousness of empire, develop connoisseurship, and purchase works of art. The French part of the tour centered on Paris as a finishing school for social graces. But the Louvre and other showcases of art furthered the goal of connoisseurship, while the Roman antiquities in Lyon, Arles, and other cities served, as they did in Italy, to refine aesthetic taste and stimulate historical imagination.[1]

1 James Buzard, "The Grand Tour and after (1660–1840)," in *The Cambridge Companion to Travel Writing*, ed. Peter Hulme and Tim Youngs (Cambridge: Cambridge University Press, 2002), 38–41.

By the late 1750s many travelers were purportedly violating the high-brow program. In Lady Mary Wortley Montagu's often quoted phrase, the "folly of British boys" abroad was giving "us the glorious title of Golden Asses all over Italy." Tobias Smollett similarly feared that Britain's "raw boys" on tour were creating false impressions of "British national character."[2] Targeting Smollett in the personage of Dr. Smelfungus, Lawrence Sterne's *A Sentimental Journey through France and Italy* (1768) mocked such worries. This narrative whose protagonist never reaches Italy contains a tongue-in-cheek typology of English travelers. Foremost are the "idle" and the "spleen-afflicted." Other categories include the "vain" and the "proud."[3] Had the number of feckless English tourists actually increased? One can only speculate. Perhaps Lassels and like-minded arbiters had set standards too high, perhaps the tourists' tutors had grown increasingly lax, or perhaps nothing much had changed. Maybe "drinking, gaming, and whoring always held more attraction for the young traveler than did the other, more exalted objectives of the tour."[4]

By the 1770s Russians, too, were questioning the value of foreign travel. During the reign of Peter I official missions to the West gave the Russian elite opportunities to pursue tourism. The prime example was the travel diary of the tsar's servitor Pyotr Tolstoy sent to Venice in 1697 to study navigation. Based in Venice for over sixteen months, Tolstoy visited many other Italian cities including Rome, Florence, Bologna, and Ferrara.[5] Diplomatic appointments or state-sponsored studies remained typical entrées into leisure travel for Russian nobles in the reign of Elizabeth (1741–61) when the Grand Tour gradually became an established practice among the crème de la crème of the Russian aristocracy. Peter III's brief reign (December 1761–June 1762) then proved a godsend to hedonists. In February 1762, five months before his assassination, he freed nobles from obligatory state and military service and granted them the liberty to travel abroad. This royal

2 Buzard, *The Beaten Track*, 97–98; Buzard, "The Grand Tour," 42; and Boyer, *Histoire générale du tourisme*, 45.

3 Lawence Sterne, *A Sentimental Journey and Other Writings*, ed. Ian Jack and Tim Parnell (Oxford: Oxford University Press, 2003), 8–10.

4 Buzard, "The Grand Tour," 42.

5 Max J. Okenfuss, "The Cultural Transformation of Petr Tolstoi," in *Russia and the West in the Eighteenth Century*, ed. A. G. Cross (Newtonville, MA: Oriental Research Partners, 1983), 231–33; James Cracraft, *The Petrine Revolution in Russian Architecture* (Chicago and London: University of Chicago Press, 1988), 132–36; and Offord, *Journeys to a Graveyard*, 25–47.

manifesto gave leisure a new legitimacy.[6] Pleasure-seeking travelers took the cue and became targets of satirists. Those writers included Fonvizin who first traveled abroad in 1777–78. Initially circulated among family and friends, Fonvizin's Gallophobic travel letters cast Paris as a magnet for Russians interested in nothing but *les spectacles et les filles* ("shows and the girls").[7] In 1779 the pioneering journalist Nikolai Novikov (1744–1818) published a satire further attesting to the Grand Tour's tarnished reputation. The author announced that readers could now see "free of charge" on the streets of Petersburg "a young Russian piglet who has been traveling in foreign parts for the enlightenment of his mind and having duly profited, has returned a perfect swine."[8]

Against the backdrop of this satiric tradition, Karamzin's *Letters of a Russian Traveler* sought to set the proper norms of a journey to Western Europe. This was a bold ambition for a fledgling journalist, poet, and translator. Karamzin came from a family of nobles near the mid-Volga town of Simbirsk and received his secondary education in Moscow. After brief service in the Preobrazhensky Guard in Petersburg, he returned to Moscow in 1785, began writing for Novikov's journals, and moved among Freemasons, some of whom played a role in organizing and perhaps financing his trip abroad.[9] The *Letters* put Karamzin on the path to becoming Russia's most eminent prose writer of his time. Beginning with its serial publication in 1791–92, the travel account struck readers as a "literary encyclopaedia" both *dulce* and *utile*.[10] In an innovative, elegant style patterned after French, Karamzin introduced Russians to the conventions of the hybrid genre of the West European travelogue. The *Letters* contains practical remarks and a wealth of knowledge but centers on the

6 Wladimir Berelowitch, "La France dans le 'grand tour' des nobles russes au cours de la seconde moitié du XVIIIe siècle," *Cahiers du monde russe et soviétique* 34, nos. 1–2 (1993): 194.

7 Ibid., 203. On vice charges that the Paris police leveled against aristocratic Russian visitors between 1761 and 1774, see also 208–9n66.

8 A. G. Cross, *"By the Banks of the Thames": Russians in Eighteenth-Century Britain* (Newtonville, MA: Oriental Research Partners, 1980), 232.

9 A. G. Cross, *N. M. Karamzin: A Study of His Literary Career, 1783–1803* (Carbondale and Edwardsville: Southern Illinois University Press, 1971), 1–34. Karamzin's trip cost 1800 rubles, a tidy sum at the time: see Konstantine Klioutchkine, "Sentimental'naia kommertsiia: 'Pis'ma russkogo puteshestvennika' N. M. Karamzina," *Novoe literaturnoe obozrenie* 25 (1997): 84.

10 T. Roboli, "Literatura puteshestvii," in *Russian Prose*, ed. B. Eikhenbaum and Iu. Tynianov (The Hague: Mouton, 1963 [1926]), 51.

personality of the author, a cultivated gentleman with a tender heart, propounding the new vogue for feeling. The "vaguely religious, sentimental, and cosmopolitan ideas" of Freemasonry impressed Karamzin and paved the way for his assimilation of Jean-Jacques Rousseau and Johann Gottfried Herder.[11] Additional preparation came from Sterne's *Sentimental Journey* and *Tristam Shandy* (1759–67), which tempered sentimentalism with irony and humor. Under these influences, Karamzin imported into Russia the West's "epidemic of tenderness" and a sentimentalized Enlightenment humanitarianism dramatized through enactments of compassion.[12] His fiction, too, proselytized the cult of feeling, as in "Poor Liza" (1792), the story of a Russian peasant who commits suicide after a Petersburg aristocrat seduces and abandons her.

In conformity to the Enlightenment conception of travel as "pleasurable instruction," a formidable intellect jostles against sentimentality throughout the *Letters*.[13] Karamzin's narrator pursues what he calls the pleasures of "the eyes and the ear."[14] A standard eighteenth-century regime of scenic and cultural sightseeing comprises the visual component that dominates the book. Throughout the *Letters*, however, the "seeing" motif intertwines with discourse. That category covers the many instances of learned conversation with the likes of Kant. But other displays of discursive concern also counterbalance the indulgence of visual pleasure: the author's acts of reading and writing, his abundant use of literary citations, and the prominence of the written word among his travel souvenirs. All things considered, the *Letters* produces an "unusually intellectual version of the Grand Tour."[15] As Likhachev insisted, the erudite, sensitive author of the *Letters* was not a raw youth requiring acculturation but rather a

11 D. S. Mirsky, *A History of Russian Literature from Its Beginnings to 1900* (New York: Vintage Books, 1958), 62–63.

12 Tanya Page, "Sentimentalism," in *Handbook of Russian Literature*, ed. Victor Terras (New Haven and London: Yale University Press, 1985), 396–97.

13 On the general concept see Charles L. Batten, Jr., *Pleasurable Instruction: Form and Convention in Eighteenth-Century Travel Literature* (Berkeley: University of California Press, 1978).

14 Karamzin, *Pis'ma russkogo puteshestvennika*, 321 (subsequent citations appear parenthetically). For an outstanding English translation, see Nikolai Karamzin, *Letters of a Russian Traveller*, trans. Andrew Kahn (Oxford: Voltaire Foundation, 2003).

15 Andrew Kahn, "Introduction: Karamzin and the Creation of a Readership," in Karamzin, *Letters of a Russian Traveller*, 1.

sterling representative of the Russian sociocultural elite of his time: he functions in the West as "an ambassador of Russian culture."[16]

Karamzin's account, however, configures Russian "culture" almost exclusively in the ethnographic sense of a "way of life"—the life of the educated classes. On the other hand, "culture" understood as the arts and sciences appears meager by comparison to what the traveler encounters in Western Europe, especially in England, the imperial state that the *Letters* pits against the Russian Empire. My investigation of this culture-centered clash focuses on the role that the English assume as foils to the authorial performance of edifying travel. On the whole, Karamzin's persona is tolerant, courteous, and sociable, and he expresses affinity to select Englishmen, especially William Coxe, the author of *Travels in Switzerland in a Series of Letters* (1789).[17] For the most part, however, the *Letters* typecasts English tourists as spleen-afflicted, recreational, profligate, and discourteous. Although presented as the fruit of direct observation, this portrayal is indebted to English satire. Sterne's *Sentimental Journey* was a major source. Given Karamzin's admiration for Alexander Pope, he probably also knew *The Dunciad* (1742), a poem in which drinking and sex loom large in the young Englishman's Grand Tour.[18] The *Letters* extends the satiric English tradition but makes it signify intellectual and moral inferiority to the Russian author's pursuit of pleasurable learning on the Enlightenment high road.

Italy operates as a major divider between Karamzin's persona and the English. After visiting country estates outside London, he remarks that the English have "long had a passion for going to Italy" to buy paintings and sculptures for their private collections (377). By the 1780s, Russian aristocrats, too, had developed the practice of cramming their homes with luxury goods purchased on their Grand Tours, mainly to France and Italy: paintings, furniture, artefacts, and *petits riens* (*bezdelushki*) such as bejewelled snuff-boxes.[19] In suppressing this aspect of contemporary elite Russian culture, the *Letters* places the English "passion" for Italy and Italian art on the same trivial level as the English "passion for horses" mentioned a couple of paragraphs later (378). Karamzin was certainly right to

16 Likhachev, "Vstupitel'noe slovo," 747.

17 Karamzin, *Pis'ma russkogo puteshestvennika*, 222–23.

18 See Alexander Pope, *The Dunciad*, vol. 5 of *The Poems of Alexander Pope*, 11 vols., ed. James Sutherland, 3rd ed. (London and New York: Routledge, 1993), 374.

19 Kelly, "Pushkin's Vicarious Grand Tour," 9.

attribute to British Grand Tourists a "passion" for Italy.[20] He, on the other hand, skipped Italy. What meaning does that deviation from the Grand Tourist norm acquire in the *Letters*? How in particular do the author's morally encumbered aesthetics look in relation to the English ardor for Italy and Italian art?

If Italian and Italianate art test the limits of Karamzin's aesthetic cosmopolitanism, the English, abroad and at home, most sorely test the limits of his moral cosmopolitanism. In connection with Peter the Great's modernization of Russia, Karamzin's persona asserts: "Everything *national* is trifling compared to the *human*. What counts is being *human beings* not Slavs" (254). The statement reflects Karamzin's affinity to the German philosopher Christoph Martin Wieland (1733–1813), featured in the *Letters* in conversation with the traveling author.[21] In essays he wrote in the 1780s, Wieland characterized cosmopolitans as *Weltbürger* who consider all peoples "branches of a single family." Karamzin speaks that idiom but does not explicitly subscribe to Wieland's moral conviction that a citizen of the world should seek "to reduce the sum of evils that weigh upon humanity [. . .] and to increase the sum of good in the world."[22] Cosmopolitanism in Karamzin's *Letters* is basically about world literature and membership in an international community of learned gentlemen, in line with the outlook of David Hume.[23] Even the idea of a global "family" is sorely tested by the nationalist dimension of Karamzin's account, especially in relation to the English. Claiming that the English disappointed him, the author recalls being an ardent Anglophile before his trip (380). Karamzin had in fact translated Shakespeare's *Julius Caesar* (1787) and James Thomson's poetry and had praised England's system of government. The *Letters* still admires much in England but depicts Englishmen as the West European people most alien to Russian national character.[24] In the process, the *Letters* extends its concep-

20 On British tourism, see John Pemble, *The Mediterranean Passion: Victorians and Edwardians in the South* (New York: Oxford University Press, 1987).

21 On Napoleon's view of Wieland as "the German Voltaire," see Pauline Kleingeld, "Christoph Martin Wieland," in *The Bloomsbury Dictionary of Eighteenth-Century German Philosophers*, ed. Heiner F. Klemme and Manfred Kuehn (London: Continuum Publishers, 2010), 851.

22 Pauline Kleingeld, "Six Varieties of Cosmopolitanism," in Delanty and Inglis, *Cosmopolitanism*, 1:174.

23 On Hume, consult Thomas J. Schlereth, *The Cosmopolitan Ideal in Enlightenment Thought* (South Bend, IN: University of Notre Dame Press, 1977), 1–24, 129.

24 Cf. Kleespies, "East West Home Is Best," 261–62. On Karamzin's positive as well as negative responses to England, see Offord, *Journeys to a Graveyard*, 90–101.

tion of the English tourist to the imperial level. Just as the Russian traveler towers in mind and spirit over English visitors of the Continent, so too does Russia emerge as the culturally backward but morally superior empire with regard to Britain.

English Literature and Peter the Great

The opening pages of the *Letters* announce the Russian author's appropriation of English literature. A prefatory paragraph first published in the book's 1801 edition defends his decision to produce a narrative that intermingles factual information with personal impressions and "idle details" he gathered as an "inexperienced" youth "in travel dress with a walking stick in hand and a knapsack on my shoulder." The faux-naïf novelistic image jars with the ensuing account of a learned gentleman traveling by coach with luggage, assisted by porters, and occasionally hiring guides. But emulating a novel is precisely the point. In justifying the subjective nature of his account and the inclusion of travel trivia, the author sets the novels of Samuel Richardson and Henry Fielding against Friedrich Büsching's *Geography*, a dull-sounding compilation of statistics and geographical data (393).[25] Independently of this preface, the opening installment of the *Letters* privileges English literature as the author's vade mecum. Addressed to "friends" left behind, his letter recalls "contagious" tears welling up in his eyes as he left his weeping household serfs in Moscow. As soon as his coach stirred, he felt "instantly orphaned in the world and in my soul!" His melancholy grew acute: "I felt like *crying out my heart*, as Shakespeare says" (6).[26] Shakespeare will remain a spiritual companion and cultural hero for the traveling author. In Paris he disparages Racine and Corneille by comparison to Shakespeare, providing in English (with Russian translation) a passage from *King Lear* to make his point (233–34). "Bow down, here is Shakespeare!" he commands his readers apropos of the poet's statue in Westminster Abbey (375). Karamzin's persona carries the deified poet in his head but complements him with two books in English purchased in Frankfurt-en-Main roughly eight weeks into his trip. To fill idle moments on the road, he selects Ossian's (James Macpherson's) *Fingal* and Oliver Goldsmith's *The Vicar of Wakefield* (68, 83). That cosmopolitan embrace

25 The academic edition prints the preface as an appendix.

26 Cassius speaks these words in Karamzin's translation of *Julius Caesar* (act 4, scene 3). For other Shakespeare references, see 90, 233, 345–46, 368.

of English writers resembles Goethe's advocacy of "world literature," transcending national boundaries.[27]

At the same time, however, the absence of a homegrown vade mecum points to Russia's cultural underdevelopment, all the more glaring in light of its geographical dimensions. The *Letters* conveys a "territorial consciousness" well established among the late eighteenth-century Russian elite. On this view, the Russian Empire was "a European country with an Asian extension," a vast "fatherland" where the "territorial habitat of the Russian nation overlapped with the territory of the state," making the two seem interchangeable.[28] After charting his passage from Tver to Petersburg, Karamzin's narrator moves through the western borderlands of the empire, specifying Narva, Riga, and Courland, and noting how the German language and German money gradually supplant Russian and rubles (7–12). In parallel to this departure, the traveler will stage his joyful homecoming at Kronstadt, an imperial site outside the Great Russian core (388). The western borders have special status as the points of exit and entry in Karamzin's round-trip scheme.[29] His territorial awareness nonetheless extends far east and south. Kazan, Simbirsk, the Caucasus, the Don River, Cossacks, Siberia, and Kalmyks cross his mind or figure in his conversation in the course of his journey (9, 30, 232, 298, 309, 380).

The empire is enormous but has few achievements in the arts and sciences, and none of universal significance. In a playful provocation to Western perceptions of Russians as Asian barbarians, Karamzin's narrator likens himself to the fourth-century Scythian traveler-hero of Jean-Jacques Barthélémy's *Voyage du jeune Anacharsis en Grèce* (1788).[30] And yet for all his wit, sophistication, and erudition, the *Letters* author cannot hide Russia's cultural marginality vis-à-vis Western Europe. He takes pride in the speed with which his homeland has assimilated imported enlightenment but is hard pressed to name a Russian culture producer besides himself. The exception proving the rule is the poet Mikhail Kheraskov (1733–1807), some of whose patriotic lines Karamzin's persona recites for

27 On Goethe, see Delanty and Inglis, "Introduction," in Delanty and Inglis, *Cosmopolitanism*, 1:4–5.

28 Williard Sunderland, "Imperial Space: Territorial Thought and Practice in the Eighteenth Century," in *Russian Empire: Space, Power, People, 1700–1930*, ed. Jane Burbank et al. (Bloomington: Indiana University Press, 2007), 33–66, esp. 46.

29 Ingrid Kleespies, *A Nation Astray: Nomadism and National Identity in Russian Literature* (DeKalb: Northern Illinois University Press, 2012), 25–45.

30 Karamzin, *Pis'ma russkogo puteshestvennika*, 61.

German hosts (66). After enumerating British luminaries in literature, history, science, and philosophy, the Russian traveler in England asserts that "geniuses are born in all countries, their fatherland is the world; and could one fairly say that Locke, for example, was more profound than Descartes or Leibniz?" (381) However, even in this global context, no Russians are cited. The deficiency of Russian culture makes Karamzin's narrator seem extraordinarily brilliant. His self-aggrandizement would become even plainer in his "Un mot sur la littérature russe" (1797), published in Hamburg and devoted mainly to summarizing and praising his own travel *Letters*.[31]

Against the backdrop of Karamzin's largely barren Russian cultural landscape, Peter the Great plays the compensatory role of a *political* genius of the Russian Enlightenment. The image of Peter in the *Letters* reflects the eighteenth-century cosmopolitan conviction that all nations move along a "common road" to the universal goal of enlightenment.[32] As the "glorious Monarch" who pursued Europeanization for the good of his country, Peter seamlessly blends cosmopolitanism and patriotism. The tsar has the stature of a giant who moved all Russia forward "with his mighty hand." He is Lycurgus bringing law to the empire and "young Mars in the armor of the Preobrazhensky regiment," a description of Gottfried Knellor's portrait (1697) of him in Windsor castle (253–54, 353).[33] In the eyes of Karamzin's persona, the tsar's transformation of Russia surpasses the achievements of the enlightened monarchs of Western Europe and even serves as a "presentiment of Russia's superiority" over the West.[34]

But for all Peter's superhuman stature, his legendary Grand Embassy to the West and his second trip abroad (1716–17) put him on Karamzin's existential plane and make him an accessible travel mentor. Eighteenth-century Russian panegyrics represented Peter suffering the tribulations of travel and working hard abroad to acquire knowledge to teach his subjects

31 Cross, *Karamzin*, 166.

32 Richard Pipes, "The Background and Growth of Karamzin's Political Ideas down to 1810," in *Karamzin's Memoir on Ancient and Modern Russia*, trans. and analysis by Richard Pipes (Cambridge, MA: Harvard University Press, 1959), 56.

33 On the prominence of Mars in Peter's public persona, consult Richard S. Wortman, *Scenarios of Power: Myth and Ceremony in Russian Monarchy*, 2 vols. (Princeton: Princeton University Press, 1995, 2000), 1:42–43, 45, 48, 76. Voltaire called Peter a new Lycurgus: see Nicholas V. Riasanovsky, *The Image of Peter the Great in Russian History and Thought* (Oxford: Oxford University Press, 1992) 19.

34 Offord, *Journeys to a Graveyard*, 97.

and put Russia on a par with the West.[35] Karamzin's *Letters* personalizes this tradition by making the author and the emperor doubles of one another. Karamzin's narrator in Paris sees a drama about Peter going to Western Europe incognito to live in a seaside village and learn the craft of shipbuilding. Echoing the authorial self-presentation in the prefatory paragraph of the *Letters*, Peter features as a "wanderer" with "staff" in hand. Karamzin's narrator takes in sites the tsar visited: l'hôtel des Invalides, the Church of the Sorbonne, and the veterans' hospital in Greenwich (England). In each case, he evokes the Russian emperor as a thoughtful sightseer (272, 281–82, 356). Peter has journeyed to the West in search of "flowers of Enlightenment" (239–40). Yet, he behaves there not as a passive pupil but rather as a confident guest courteously speaking his mind to hosts when he disagrees with them. Karamzin's persona likewise pursues a knowledge-building agenda but in a politely self-assertive manner, to show West Europeans the best qualities of mind, manners, and education that the Russian nobility has to offer.

Elect Self and the Crowd

The emulation of Peter as the Russian Enlightenment traveler par excellence belongs to Karamzin's larger strategy of placing his persona among the "elect" in opposition to the "insensitive crowd," the degraded visitors of the Continent who are mainly English.[36] Prior to encountering any English travelers, Karamzin's persona establishes his membership in an international community of learned gentlemen. Among Germans at the outset of his trip he emerges as a cultivated, multilingual traveler at ease with eminent philosophers, writers, and scientists. He also stresses the importance of extended stays in a place in order to maximize intellectual opportunities. These meetings are presented as occasions of discourse. The first example is the author's conversation with Kant in Königsberg (20–21). Elsewhere in German lands Karamzin's persona stages similar meetings with Professors Christian Beck and Ernst Platner, Wieland, and Herder (61–66, 71–77). In Switzerland he extends the pattern of pleasures of the mind in interaction

35 Riasanovsky, *The Image of Peter the Great*, 12–40.

36 For the argument that Karamzin's *Letters* and prose fiction introduced into Russian literature the division between the "elect" and the "insensitive crowd," see V. G. Belinskii, "Russkaia literatura v 1845 g." (1846), *Polnoe sobranie sochinenii*, 13 vols. (Moscow: Akademiia nauk, 1953–59), 9:384.

with the clergyman and physiognomy theorist Johann Kaspar Lavater and the scientist-philosopher Charles Bonnet (167–69, 184–85). The Russian traveler invariably receives a warm welcome. His erudition and his fluency in German and French amaze the natives; and he repeatedly orchestrates intellectual triumphs over ordinary West Europeans, to emerge as the most sophisticated person present.[37]

After fixing his image as a learned cosmopolitan, Karamzin's narrator pits himself against the tourist crowd. En route from Zurich to Baden, he begins to encounter coaches with "English, German, and French faces" peering out the windows. From June to October, Switzerland is "full of travelers" who have come to enjoy the scenery (125). The multitudes make finding accommodations in Bern difficult. Having settled in at last, the Russian traveler encounters carousing Englishmen in a tavern. The English drinkers beg a poor Italian street musician to sing about his life and then shower him with silver coins (130). This flashy display of wealth is alien to Karamzin's persona, who tends to mention his own expenditures only to stress his frugality or the emotional sweetness of a transaction, such as his modest recompense to a Leipzig flower girl who offers him a bouquet.[38] The theme of the crowd evolves in Lausanne, a place the Russian traveler claims is "always full of young Englishmen" meant to be studying French but in fact perpetrating "various stupidities and pranks." The author admits that "our obliging compatriots" there can be just as bad (148–49). Elsewhere Karamzin observes in passing that some Russian travelers spend money lavishly abroad, a practice from which he distances himself and associates primarily with the English (377). However, those evocations of rambunctious, spendthrift Russians are too incidental to disrupt the narrative's structure of national opposition between the traveling author and the primarily English tourist crowd.

English tourists also figure as shocking horseback riders. While strolling in refined local company in Geneva, Karamzin's narrator observes a visiting Englishman on horseback charging into the midst of the pedestrians and sticking out his tongue at their protests. The rider remains feisty toward the guardsmen who arrest him as an English-speaking woman weeps by his side (183). This meek femininity sits well with Karamzin's later images of English women as paragons of domesticity. But he learns that some women

37 Dickinson, *Breaking Ground*, 109–11.

38 Klioutchkine, "Sentimental'naia kommertsiia," 91.

do not conform to his ideal. Walking in a forest near Lyon, he comes upon four fashionably outfitted blonde English women on horseback. The sight so stuns him that he fails to remove his hat, a lapse for which they saucily rebuke him (193). Those assertive recreational tourists violate the conviction that the author subsequently ventriloquizes as the view of the wife of François de Vaillant (1753–1824), an explorer of Africa: the good Parisian housewife asserts that women are "created for immobility" (298). Nature meant them to stay home. Another episode dramatizes the author's active command of British literature in opposition to idle Englishmen. On a boat along the Saône from Lyon to Macon, one English passenger plays with a dog, while another dangles a walking stick in the water. In the meantime, the *Letters* narrator meditates upon the scenery, speculates about what the area must have been like in prehistoric times, and ruminates about Ossian (211–12). Here again the Russian author exhibits a lively, text-oriented imagination, to suggest English tourist insensitivity to their own magnificent literature.

The section on Paris accentuates pleasures of the eye, and yet an engagement with the written word still continues to distinguish the author from the English. In conformity to Nugent's classic Grand Tour agenda "to see and be seen," the Russian traveler experiences the solipsistic pleasure of contemplating himself in a new place when he reaches the French capital: "*I am in Paris!*" he exults twice, as he enters the city.[39] Still in the coach, he begins enjoying Paris as spectacle: "What I knew only from descriptions I am now seeing with my own eyes," the "living picture of the most magnificent, glorious city in the world" (217). During his stay in Paris from late March into June, he visits many standard sites, feasts his eyes on the socially varied street scene, conducts his readers on a walking tour, frequents the theatre, and looks at art. Nonetheless, he begins each day reading in his rented rooms and ends each evening there writing his travel journal. That daily engagement with discourse modulates the theme of visual consumption.

Along with the quiet evenings devoted to writing his journal, the author's frugality and moral inhibitions set him apart from dissolute English tourists in Paris. Karamzin's narrator observes that even if he had the requisite wealth of the English, he would not want to imitate their practice of transgressing the boundary of "permissible" pleasures, to frequent

39 Thomas Nugent, *The Grand Tour, or A Journey through the Netherlands, Germany, Italy, and France*, 4 vols., 3rd ed. (London: J. Rivington and Sons, 1778), 4:109.

gambling halls and pursue "singers and actresses" (244). In the accidental company of an English sightseer, the Russian author visits the house-museum of an actress where his eyes linger on a corset exhibited in the boudoir (255–56). This commercialized tourist gazing is the furthest he strays from "permissible" enjoyment. His final brush with Englishmen dramatizes their prodigality and rowdiness. In a Calais tavern the night before sailing to Dover, the author encounters English youths who have just crossed the Channel and plan to "wander all over Europe" for no particular reason. They order vast quantities of expensive wine. The author samples the pink champagne and then retires to his room, only to be awakened late at night by the Englishmen "screaming, stomping, and banging" next door (324–25).

Moralistic Aesthetics

While restricting the English to a lowbrow program with a hint of sex tourism, the *Letters* extends the author's moral concerns to issues of aesthetic judgment. Karamzin's persona often finds scenery "picturesque," a fuzzy aesthetic category "capable of running the gamut from relatively mild [rustic] landscapes to the breath-taking cataracts and chasms of the Alps in stormy weather."[40] But whatever the scenic object may be, his Rousseauesque belief in the goodness of nature makes this sphere of aesthetic pleasure morally unproblematic (other mediators were Salomon Gessner and Albrecht von Haller, two German poets Karamzin translated prior to his journey). Exemplifying a standard anti-tourist practice, the *Letters* narrator customarily contemplates nature in solitude, a posture that heightens the observer's claim to exceptional sensitivity.[41] Among sites receiving this treatment are the Rhine Falls, the Bois de Boulogne, London and its environs, and the Alps.

In the Swiss Alps the traveler's soul soars above earthly cares, to make him "a citizen of the universe," in touch with eternity. The experience of spiritual elevation allows him to restore the preconceived belief in Swiss innocence that was shaken when local lads pestered him for money along the

40 Buzard, "The Grand Tour," 45. On the literary aestheticizing of rural Russia as a site of the picturesque understood as "a view that resembles a picture," consult Ely, *This Meager Nature*, 61–86.

41 Buzard, *The Beaten Track*, 176–82; on Rhine Falls, see also Dickinson, *Breaking Ground*, 169.

road from Zurich to Baden. In an apostrophe to Swiss parents, Karamzin's persona admonishes them to nip the mercenary behavior in the bud, lest it spread and contaminate the pure "balsalmic air of your mountains and valleys." Hiking then affords his encounter with the shepherds he portrays as the authentic Swiss *Naturmensch*—the enviable remnants of a golden age of innocence, a time when "all people were shepherds and brothers!" (125–26, 133–34, 137). Paying an appropriate literary tribute, the author performs a pilgrimage to places that "immortal Rousseau peopled with his romantic lovers" Julie and St. Prieux (149). With Rousseau as his principal guide, the Russian traveler finds moral tranquility and experiences the goodness in nature and the natives as a harmonious whole.

While pristine nature invariably strikes receptive chords in Karamzin's narrator, art poses moral challenges. Early in his trip he writes of adding travel impressions to the "picture gallery of [his] imagination." Already stored there, however, is a catalogue of art works and artists. Some serve to aestheticize his experience. A beautiful traveler in Saxony resembles a "Michelangelo statue" and a "Raphael painting," while her aged, humpbacked husband appears as a "Hogarth caricature" (50). Similarly, the countryside around Frankfurt abounds in "landscapes worthy of the brush of Salvator Rosa or Poussin" (86). More often, the art references concern the author's visits to galleries, museums, and churches. He goes searching for art that guidebooks and other texts or images have signposted. His commentaries, however, demonstrate the "fledging state of aesthetic education in Russia" at the time.[42] In Dresden, for instance, he responds most enthusiastically not to the paintings in the Gemäldigalerie but rather to the Grünes Gewölbe, a collection of porcelain and other *objets d'art* (55). In writing of painting, the *Letters* author tends to do little more than name works and repeat bromides about artists. The *Sistine Madonna*, he writes, epitomizes the "beauty, innocence, and holiness" of Raphael's contributions to the genre; "Titian is considered the world's preeminent colorist;" "Rubens's pupil Van Dyck is, of course, the world's preeminent portrait painter" (52–54). Beginning in the Mannheim museum, containing a copy

42 Schönle, *Authenticity and Fiction*, 208; and Kahn, "Karamzin's Discourses of Enlightenment," in Karamzin, *Letters of a Russian Traveller*, trans. Andrew Kahn, 531. On Karamzin's generally "archaic" tastes in painting, see also Rodolphe Baudin, "Karamzine critique d'art: la peinture dans les *Lettres d'un voyageur russe*," *Revue des études slaves* 83, nos. 2–3 (2012): 734–37, 756.

of Phidias's *Laocoön*, Karamzin's persona proves more receptive to statuary.[43] But no matter what art work he beholds, he assesses it in moral, emotional terms. Referencing Professor Platner's lectures on aesthetics that he attends in Leipzig, the Russian traveler defines artistic genius as the ability to transmit lofty sentiment and tug the heart strings (63–64). His main examples of such achievements are Christian funerary monuments to virtuous wives and mothers. He lends a sentimental gloss to *Laocoön*, too, however. After quoting Virgil's description of the priest's death, the traveler casts aside classical restraint, to interpret the statue as a heartrending family drama for all times: a monument to the "sorrow of an unhappy parent who sees the destruction of his children but cannot save them!" (92–93)

The moral and humanitarian inflections of Karamzin's aesthetics grow strongest in relation to Italian or Italianate art. Upon arriving in France from Switzerland, the author first stops in Lyon, where a resident German poet takes him to the studio of a French sculptor trained in Italy. They admire a work in progress: a "beautiful little boy sleeping the quiet slumber of innocence beneath Minerva's protective shield." Immediately after viewing this "very artfully" executed Italianate sculpture, the German guide takes the author to a hospital on the banks of the Rhône. The "spectacle" of disease and pain lacerates the Russian traveler's heart. Using the vocabulary of painting, he perceives the wards as preparatory studies for "subtly hued pictures of suffering humanity" (201). The juxtaposition of the sculptor's studio and the hospital viewed through an aesthetic lens conveys authorial resistance to art's invitation to forget human misery.

Statues of naked women raise another moral issue. In Weimar during a conversation with Herder concerning ancient art that the latter recently saw in Italy, Karamzin's persona toys with the idea of going there: "A sudden thought crossed my mind. What if I were to make my way to Italy from Switzerland to look at the *Venus de' Medici*, the *Apollo Belvedere*, the *Farnese Hercules*, and the *Jupiter Olympus*, to look at the majestic ruins of Rome and sigh over the transience of everything earthly?" (75). A famous Venus tops this list of art works, but when the Russian traveler actually sees statues of naked women, he strikes a prudish stance consistent with sentimentalism's cult of feminine purity and innocence. In visiting the Church of the Celestins in Paris, he finds Germain Pilon's "three naked Graces" beautiful but inappropriate in the "temple of the true God" (283). Are the

43 Kahn, "Karamzin's Discourses of Enlightenment," 531–32.

Graces out of place because they are pagan, unclothed, or both? This is not clear, but a subsequent incident suggests that nudity is the more perturbing factor. Seeing a statue of Venus in the Versailles gardens, the Russian visitor imagines that the trees' protective shade makes her feel less "ashamed of her nudity" (296) than she would otherwise.[44] Even in the standard setting of an aristocratic garden, Venus radiates a troubling disunity between physical beauty and Christian morality.

Karamzin's performance of prudery before sculptures of naked women was archaic in relation to Peter the Great's modernization of Russian visual culture. The Russian state's pioneering collector of art and antiquities, Peter purchased paintings on his first journey abroad. His agents made many further acquisitions, including a collection of Italian marble statues put on display in the Summer Garden laid out in Petersburg in the early 1700s. Brought from Italy to Petersburg in 1720, Peter's most prized statue was a life-sized Hellenistic Venus modeled on the Knidian Venus of Praxiteles.[45] Now known as the *Venus of Taurida*, the work was initially exhibited in a guarded gallery inside the Summer Garden.[46] In accord with old Muscovy's prohibition of graven images, the conservative clergy considered the foreign art dangerously "seductive."[47] However, the social elite of Peter's time followed his lead, to adorn their residences with Western art (Pyotr Tolstoy even had a picture of a naked woman in the same room with an image of a Russian saint).[48] During Catherine II's reign, when Karamzin produced his *Letters*, "countless Russians" saw "Peter's collections of paintings, sculptures, medals, antiquities, and other art works at his Summer Palace and

44 Karamzin quotes here Jacques Delille's poem on a 1775 storm that destroyed many trees in the Versailles gardens.

45 On Peter as collector and the prized Venus in particular, see James Cracraft, *The Petrine Revolution in Russian Imagery* (Chicago and London: University of Chicago Press, 1997), 193–94, 221–24; and Margaret Samu, "The Nude in Eighteenth- and Nineteenth-Century Russian Sculpture," *Experiment* 18 (2012): 37–39.

46 Around 1790 the statue was moved to Grigory Potemkin's Taurida palace in Petersburg. An ancient name of the Crimea, "Taurida" was an epithet Potemkin earned for his role in the annexation of that territory.

47 Cracraft, *The Petrine Revolution in Russian Imagery*, 75, 91, 293; and Emmanuel Ducamp, "Private Galleries as a Public Service," in *Great Private Collections of Imperial Russia*, ed. Oleg Neverov and Emmanuel Ducamp, with introductions by Mikhail Borisovich Piotrovsky and Nicholas Romanov, Prince of Russia (London: Thames and Hudson, 2004), 15–16. On the lack of conclusive evidence concerning the significance of nudity in the clergy's reactions, see Samu, "The Nude in Eighteenth- and Nineteenth-Century Russian Sculpture," 36–37, 39–40.

48 Offord, *Journeys to a Graveyard*, 47.

Garden, in the palaces and grounds of Peterhof, and in his Kuntskamera." Moreover, those visitors comprised "a motley, fluid, relatively numerous elite" rather than "the tiny, self-perpetuating elite inherited from Muscovite times."[49] With respect to representations of naked women, Karamzin stands outside the contemporary Russian mainstream of westernized aesthetics. More sentimental than religious, his outlook expresses nostalgia for old Muscovy, a simpler, pre-modern time and place glorified in his popular historical tale "Natalya, the Boyar's Daughter" (1792).

In taking the moral pulse of Karamzin's persona as an art consumer, let us finally note the lurid light he casts on Italian painting in France. True, he remarks paintings of religious subjects by Veronese, Raphael, and other Italian artists at Versailles and other sites. But during his visit of Fontainebleau, he trots out racy anecdotes from Jacques-Antoine Dulaure's *Nouvelle description des curiosités de Paris* (1786). François I, we read, had the palace decorated with Italian paintings of "god and goddesses, men and women" engaged in debauchery so shocking that Henry IV's wife had them effaced. Purportedly destroyed as well was a Michelangelo painting of "naked Leda" represented so "vividly and in such a sensual position" that it kept a palace governor in a constant, crazed state of "temptation" (214). This material chimes with Karamzin's earlier commentaries on Italian artists viewed in Dresden. Drawing there on Antoine Joseph Dezallier d'Argenville's *Abregé de la vie des plus fameux peintres* (1745), the *Letters* states that although Raphael created ethereal art, he died prematurely of dissipation due to his "excessive predilection for the fair sex." The same set of annotations sees fit to repeat the "wholly implausible" rumor that Michelangelo had a model for Christ murdered in order to achieve a realistic painting of the Crucifixion (52, 53).

Discovering England

If some art in France strikes Karamzin's persona as immoral, the "permissible" enjoyments of seeing and discourse lead him to proclaim Paris his favorite foreign city. Karamzin disapproved the French Revolution, but Russian censorship limited the *Letters* to a euphemistic reference to public "turbulence." En route to Haut-Buisson, the author bids Paris a fond farewell: "I have left you, beloved Paris, left with regret and gratitude! Midst

49 Cracraft, *The Petrine Revolution in Russian Imagery*, 282.

your noisy occurrences I lived calmly and merrily as a carefree citizen of the world. I gazed upon your turbulence with a tranquil soul, just as a peaceful shepherd gazes at the stormy sea from a mountain" (321). He recalls France as a hospitable land where visitors are made to feel at home. In fact, the preceding letters express many negative opinions of the French. The author even states his preference for eating alone in restaurants instead of going to Parisian dinner parties where he must answer idiotic questions about Russia, such as whether Russians ride reindeer in winter (274). Glossing over the inconsistencies, the glowing memory of Paris prepares the ground for representing England as an inhospitable land. Shortly before crossing the Channel, the author reintroduces motifs of his departure from Russia, to stage a wrenching break with France. Travel, he laments, has become "wandering orphanhood," a phrase recalling his self-presentation as an "orphan" as he leaves Moscow. Viewed from a cliff near Calais, the sea distresses him. He feels he has come to the "end of the world" and begins crying just as he did in leaving home (322–23).

The author's little discussed inability to speak English appears to be a contributing factor to his complaints about the English. Upon arrival in London, he seeks to shore up his cosmopolitan image. He describes himself as a "trouble-free citizen of the world," grown accustomed to facing the unknown so "frightening to the stay-at-home" (331). He puts on a "Parisian frock coat" to stroll about London, radiating a sophistication that contrasts with the provinciality of a courier from Petersburg who arrives in a quaint Russian uniform that makes him the laughingstock of local urchins (332, 359). On the other hand, the author's linguistic handicap compromises his worldliness. His spoken English is so poor that on his first night in London he wanders around long after dark because no passer-by can understand his pronunciation of the name of the street where his inn is located (334). "I now deeply regret I know English so badly" (338), he confides after an English lady refuses to converse with him in French (an expression of pride in her mother tongue that he urges his Russian readers to emulate). The struggle to speak English puts the Russian traveler at a disadvantage for the first time in his trip. He is no longer the erudite, sparkling conversationalist winning admirers wherever he goes. He has some local contacts but socializes primarily with other Russians, especially the Russian ambassador to England, the unnamed Semyon Vorontsov.

Furthermore, some of the natives are now linguistically superior to Karamzin's persona. At a cafe in the stock exchange English merchants astonish him with their Russian gained in the course of doing business in Petersburg: "Imagine my amazement," he writes, "all these people began talking to me in Russian! I thought I had been transported to my homeland by the wave of a magic wand" (344). The language problem, the traveler ultimately suggests, lies in English phonetics and English speech mannerisms. "English pronunciation is so difficult," he laments, "it is a miracle to recognize by ear a word you know by sight! I understand everything written for me, but in conversation I am left guessing. It seems English mouths are bound or that a ministry has imposed a big tariff on opening them" (369).[50] As the learned conversation motif recedes, sightseeing comes to the fore. Karamzin's persona visits St. Paul's, the Tower of London, art galleries, museums, and an array of civic sites conforming to educational, Enlightenment norms: Newgate prison, the Bedlam asylum, the Houses of Parliament, and the law court Justice Hall (340–44, 346–48). Astounded to encounter so many locals in the galleries and museums, he presumes to monopolize a self-improvement agenda. In his view, the natives flock to tourist attractions not to learn anything but simply because they have time on their hands, especially the women, free until six o'clock when they have to be home to cook dinner (350).

Unable to shine in conversation with the English, Karamzin's narrator accentuates his command of the written word. This burnishes his image as an interpreter of texts, but in the process makes Britain appear the home of more geniuses than any other country he visits. In addition to Shakespeare, Thomson, Richardson, and Fielding, Karamzin's literary pantheon includes Sterne, Pope, Milton, Dryden, Addison, Thomas Grey, Ben Jonson, Edmund Spenser, Oliver Goldsmith, and Jonathan Swift. He dismisses Chaucer, however, as the author of "indecent" stories (368–69, 374–78). Alongside the creative writers stand the world-renown British historians, scientists, and philosophers: Hume, Gibbon, the Scottish Enlightenment figure William Robertson, Bacon, Newton, Locke, and Hobbes (252, 381). Although stressing his control of written English, the Russian traveler also enacts rapturous appreciation of Handel's *Messiah* (334), an auditory pleasure that parallels discourse in overshadowing vision. This account of the musical performance underscores his sensitivity, while adding another stroke to the panorama of bedazzling culture in Britain.

50 On English as a bastard tongue versus Russian as a mighty, uncontaminated language destined to produce a great literature, see *Letters*, 369–70.

Along with its monumental contributions to world culture, England impresses the Russian traveler as a wealthy imperial state. Perhaps not by chance, his ride from Dover to London ironically reverses the steady exposure of the misery of the serfs in Alexander Radishchev's (1749–1802) *Journey from St. Petersburg to Moscow* (1790), a book that led Catherine II to incarcerate the author. On his journey to London, Karamzin's narrator discovers a "land of beauty" (327), free of "poverty" (*bednost' chelovecheskaia*, 329). He sees redbrick villages, manors of "rich lords," tidy inns, efficient posting stations, and busy throngs of well-dressed local travelers of every class (327–30). In "beautiful" London, prosperity combines with manifestations of imperial power: the multitude of ships in the port ("number one in the world"), the headquarters of the East India Company, and the circulation of commodities from the empire's Asian and North American colonies (330–34, 336, 364, 367). In the Russian visitor's eyes, the British are the "proud tsars of the seas" (381) generating fabulous colonial wealth that produces public good at home. The tradesmen of the capital are dressed as well as the lords. The city's bright street lighting bespeaks lavish public spending, and the national level of education is high. "Even the tradesmen read Hume," asserts the Russian traveler, and everybody reads newspapers, even in the tiniest hamlet (381). The clean, broad streets and sidewalks make London the best city for pedestrians, civic institutions such as Greenwich Hospital are exemplary, and the bliss of English domesticity appears unsurpassable (336, 365). The superlatives keep coming, with only a few qualifications. The traveler recoils, for instance, from squalid London streets where prostitution thrives. He also contends that money allows the rich to pervert England's admirable legal system (371, 382).

Us Versus Them

But precisely because England enjoys such power, prosperity, and cultural abundance, the *Letters* seems compelled to assert Russian moral superiority. Empire is a major arena of competition. British history, asserts the author, abounds in "evil deeds," and present-day British empire-building fits the pattern. Drawing on a French source, Karamzin's narrator claims that English gentlemen, so respectable at home, behave like "beasts" in their colonies (348, 372). The British past is bad, contemporary British imperialism is bad, and there is apparently no reason to expect moral improvement in the future. On the other hand, the Russian Empire of the *Letters* is morally unassailable and having "nearly caught up" with Germany, France, and England appears on course to "surpass" them (254). As usual, though, the narrative glosses over the cultural lag: where is the evidence of "catching up," let alone "surpassing" the West in the arts, sciences, social thought, and philosophy? Significant, too, is the absence of culturally rich but politically weak Italy in this scheme of international competition. The authorial eye is trained on the big powers of the immediate time of Catherine II, when Russian military triumphs massively expanded the empire, a process stirring Karamzin's imperial patriotism.[51]

A parallel competition between "us" and "them" is staged on the level of "national character" (*narodnoe svoistvo*, 384), a concept the *Letters* first articulates in the section on England. Scattered throughout the narrative, the traits of Russianness underwrite Enlightenment notions of the formative influence of climate but exclusively concern Karamzin's own social class rather than the common people.[52] In providing written answers to a French lady's questions about the Russian way of life, the traveling author reports that Russians are blessed with an enlivening climate all year round. The sudden arrival of spring makes that season in Russia the best in the world, but equally special is the winter, when crisp air and bright sun bring a rosy bloom to the cheek of a fur-clad beauty dashing in a sleigh through the sparkling snow. Invigorated by the climate, Russians are warm-hearted, animated, and sociable. They are free of acquisitiveness: they are generous and extend hospitality to friends and foreigners alike. Russians also have a distinct "moral physiognomy," an apparently age-old, collective consciousness of right and wrong (254). On the other hand, the English have a wretched climate worsened by industrialization. Perpetual "fog, gloom, and coal dust" block the sun. This environment produces morose, taciturn, and phlegmatic personalities, made worse by the national habit of eating

51 The *Letters* nonetheless expresses ambivalence toward Catherine. In Voltaire's house in Ferney the author looks "with pleasure" at a silk-work portrait of "our empress" but remains silent when a librarian at the King's Library in Paris shows him a copy of Catherine's *Instruction* (*Nakaz*) that proclaimed Russia a European country (159, 270). After the overthrow and murder of Paul I in 1801, Karamzin composed a laudatory ode to Catherine but expressed moral disapproval of her in his *Memoir on Ancient and New Russia* (1812): see Riasanovsky, *The Image of Peter the Great*, 69n150.

52 A rare reference to the Russian folk (*narod*) evokes them as louts. Watching the Parisian working class drinking and dancing at a suburban *guinguette*, the author remarks: "So, the Russian folk is not alone in worshipping Bacchus. The difference is that the drunken Frenchman is noisy but does not get into fights," 269.

rare roast beef. Karamzin's English are honorable and enlightened but cold, arrogant, mercenary, and xenophobic, convinced their country is the best in the world and overly proud of their constitution (329, 380–83). Upon occasion an Englishman may display lavish hospitality but only in order to exhibit his wealth. The lack of moral fiber is collective: individualistic to a fault, the English sprout "like wild oaks," indulging eccentricities to a degree unmatched by any other people (384).

Materialism and ostentatious self-display are the English vices most pertinent to leisure travel. Both stem from boredom and are associated with wicked empire-building. England's idle rich live as "Croesuses" vainly seeking happiness in the accumulation of money and material goods. That way of living infects the soul with the "moral disease" of "spleen" (*splin*). Spleen was soon construed in England as the fashionable ennui of dandyism.[53] Karamzin, however, translates the word as "melancholy" (329) or else plain "boredom" (*skuka*), and characterizes it as a force that drives Englishmen to travel, squander money, hunt, marry, and commit suicide (383). Moreover, today's idle rich are profiting from fortunes their grandfathers made in India, a theater of the colonial "beast" scenario. The evil source of the money implicitly taints everything it buys, including the "treasure-troves" of art works and antiquities purchased in Italy (377).

The tainted art collections from Italy resonate in the Russian author's concluding assertion of the moral and intellectual superiority of his Enlightenment travel program. Taking leave of his readers after reaching Kronstadt, Karamzin's persona sums up his trip as a journey of inner enrichment. His most valued souvenirs are the very letters that have comprised his narrative (the "mirror of my soul"). He also has his expense records, books purchased abroad, pebbles, dried grasses, and twigs, including one from the willow in Pope's garden at Twickenham, the tree beneath which he "wrote his best poetry." Compared to me, declares Karamzin's narrator, "all the Croesuses of the world are poor!" (388). Now linking himself to Pope just as he did to Shakespeare in his first letter, the author cherishes souvenirs of literary and sentimental value. On the Russian side stand memories and symbols of discourse, on the English side—costly art and antiquities.

---

53 Ellen Moers, *The Dandy: Brummell to Boerbohm* (London: Secker and Warburg, 1960), 11–14, 70–72, 94.

As the first effort to formulate proper norms for Russian trips abroad, the *Letters* imbued competing travel styles with imperialist national significance. The English crowd has betrayed its Enlightenment forefathers' educational ideal of the Grand Tour, but Karamzin's persona has embraced it with vengeance. He stereotypes the English as idlers and wastrels, traveling out of boredom and favoring coarse pleasures. Among them are rowdy boys fond of the bottle, horsey women violating sentimental gender ideals, gamblers, and lechers. At worse, the English tourist is a morally diseased materialist implicated in brutish imperial rule. In effecting his regime of highbrow sightseeing and discourse, Karamzin's persona disparages the English to the credit of himself, his social peers, and the modernizing Russian Empire dating from Peter the Great, a political formation whose vastness and alleged moral purity assume the compensatory function of offsetting cultural marginality vis-à-vis the West.

In Karamzin's *Letters*, being a "citizen of the world" turns out to have nationalist and aesthetic limitations. Contrary to the author's claim that what counts most is being "*human beings* not Slavs," he envisions the West falling behind Russia, and he establishes a hierarchy of national character, placing Russians at the top and the English at the bottom. In between come the Swiss cast in a pastoral mold, the French painted as hospitable but shallow, and the Italians regarded as "crafty" (384), an assertion that the author never seeks to substantiate. The self-proclaimed "citizen of the world" also expresses archaic moral reservations about Western art. The artistic subject and its setting should not offend Christian standards of modesty. Furthermore, aesthetic pleasure should not compromise the trip's humanitarian dimensions. By dint of the suffering it contains, the Lyon hospital counts as a weightier site than the studio of the Italian-trained sculptor visited immediately before. As the *Letters* would have it, the disinterested contemplation of art and the acquisition of expensive paintings, sculptures, and antiquities are morally questionable, typically English travel priorities.

In the end, then, the English "passion" for touring Italy seems something Russians should not emulate. Karamzin's narrator skips Italy but suggests that he found adequate substitutes: Lyon satisfies the taste for Roman ruins, Richmond is the "English Frascati," Italian comic opera singers in Paris are deemed superior to the troupes in Italy, and ruins in the Ermenonville garden resemble the "Sybiline temple in Tivoli" (203–4, 241, 308, 378.). Italy haunts the *Letters*, but the author never expresses regrets about not going there. He returns home satisfied, feeling emotionally and intellectually enriched.

Karamzin naturally worried about the durability of his Enlightenment travel program but failed to anticipate the threat that the tourist allure of Italy would pose. Consider "My Confession" (1802), a self-parodic story that imagines a trip abroad accelerating moral degeneracy. Told in the first person, the tale carries to extremes the pleasure principles of Karamzin's *Letters*.[54] After visiting Leipzig, Paris, and England with a tutor, the dissipated young Russian count of "My Confession" returns home, marries, sells his wife to settle a debt but remains her lover and eventually runs away with her. His adulteries drive her to an early grave, and he becomes a pawnbroker contributing to the financial and moral ruin of others. What would actually undermine the Enlightenment travel regime was not depravity on the order of "My Confession" but rather the romantic cosmopolitanism that made Italy a Russian tourist priority and in the process cast off Karamzin's sentimental imperial patriotism, his moralistic aesthetics, and his wholly negative image of English tourists.

54 Gitta Hammarberg, *From the Idyll to the Novel: Karamzin's Sentimentalist Prose* (Cambridge: Cambridge University Press, 1991), 240–50; and Schönle, *Authenticity and Fiction*, 79–80.

2

The Romantic Vacation Mentality

Whereas cosmopolitanism and patriotism coexisted in Karamzin's *Letters*, they parted ways in his subsequent writings. Shaken by the French Revolution, he made his journal *Vestnik Evropy* (*European Herald*, 1802–3) an organ of monarchist politics. He upheld the supremacy of the Russian nobility, defended serfdom, and underwrote autocracy as the only proper form of government for Russia.[1] After Alexander I appointed him the state historiographer, Karamzin enhanced his reputation with his twelve-volume *History of the Russian State* (1816–29). Through its semi-novelistic representation of Russia's rulers from earliest times, the *History* contributed to the post-Napoleonic wave of Russian national pride. In contradistinction to the *Letters*, the preface to the *History*'s first eight volumes, initially published in 1818, declared love of the fatherland (*otechestvo*) incompatible with being a citizen of the world. The preface asserted that "the genuine cosmopolitan is a metaphysical being or such an unusual figure that there is no need to speak about him, either to praise or to condemn him."[2] True, Karamzin continued, "the family of mankind" contains many glorious members, such as the ancient Greeks and Romans. Yet, "everyone's individuality is closely bound to his fatherland." We love home best "because we love ourselves." Karamzin added that a historian should avoid "crude partiality" and yet play upon "the feeling of 'we,' 'ours.' [. . .] Where there is no love, there is no soul." As in the

1 Cross, *Karamzin*, 206–10.

2 On the evolving usage and connotations of Russian words for "fatherland" as opposed to "homeland," see Sara Dickinson, "*Otechestvo, otchizna, rodina*: Russian 'Translations' of *Patrie* in the Napoleonic Period," in Offord et al., *French and Russian in Imperial Russia*, 179–96.

travel *Letters*, "our" land is the empire imagined as a benevolent formation: Russians had won their vast, multinational state through "courage and fortitude," "without recourse to the violence and villainy to which other devotees of Christianity resorted in Europe and in America."[3] In characterizing cosmopolitanism as a lack of national attachment too rare to merit discussion, Karamzin presumed to lay the matter to rest.

In fact, however, a new Russian controversy over cosmopolitanism lay ahead, and tourism in the West helped stir it. After the Napoleonic Wars, Russian journeys to the West resumed in the 1820s and grew more common in the 1830s.[4] Russians went abroad to pursue tourism and to study, in Germany in particular, owing to Nicholas I's displeasure with France after the 1830 revolution in Paris. Russian passports could not even mention France at the time, but as Annenkov put it, everybody went there anyway, "secretly, like a thief."[5] Aside from the problematic status of France, Russian regulations on foreign travel were surprisingly lenient prior to the revolutionary upheavals in Europe in 1848–49. Nicholas I's reign began in 1825 with the suppression of the Decembrist insurrection organized by elite Russian officers who aspired to supplant the autocracy with a constitutional monarchy or republican form of government. After executing the five principal conspirators and exiling over one hundred others to Siberia or regiments in the Caucasus, Nicholas intensified censorship and established the state police, the Third Section. In the 1830s the authorities also formulated a patriotic ideology denigrating the West. Nevertheless, during the long decade ending in 1848, many prominent Russian writers and intellectuals exercised their privilege to go abroad.

In this period the Russian practice of tourism caught up with the Byronic travel revolution that swept England after the Napoleonic Wars. The writings of Rousseau and Goethe treated exploration of the Continent as a revitalization of the heart and soul, a tendency that arguably began undermining the prestige of the Enlightenment model of travel as discourse.[6]

---

3 N. M. Karamzin, "Predislovie k *Istorii gosudarstva Rossiiskogo*," in his *Sochineniia*, 2 vols. (Leningrad: Khudozhestvennaia literatura, 1984), 2:233–34, 238 (translation in Marc Raeff, *Russian Intellectual History: An Anthology* (New York, Chicago, and Burlingame: Harcourt, Brace, and World, 1966) 118, 122.

4 On travel writing of the period, see Dickinson, *Breaking Ground*, 164–65, 170–75, 180–92.

5 Malia, *Alexander Herzen and the Birth of Russian Socialism*, 343.

6 Judith Adler, "Travel as Performed Art," *American Journal of Sociology* 94 (May 1989): 1372.

Karamzin's eclectic *Letters* extended Rousseau's idiom, particularly in depicting the Swiss Alps. By Pushkin's time, the preeminent promoter of travel as emotional and spiritual renewal was Madame de Staël's *Corinne, ou l'Italie* (1807), widely read in Russia as well as Western Europe.[7] Byron then made the decisive break with Enlightenment travel norms and the literature of sentiment. Beginning with the pan-European sensation *Childe Harold's Pilgrimage* (1812–18), Byron's poetry and public persona fashioned a new version of the cosmopolitan as a combination of the "willing exile," the "pilgrim," and the "vagrant," voraciously consuming the culture and scenery of foreign lands, after fleeing his nation's insular prejudices, stodgy patriotism, and conventional morality.[8] As received in England in the immediate aftermath of the Napoleonic Wars, *Childe Harold's Pilgrimage* acquired the electrifying allure of escape from the cares and strictures of home. The young were impatient to take their tours and yearned to do so in the new Byronic fashion. By the early 1820s, a conception of tourism as "an ameliorative *vacation*" had supplanted the Enlightenment's educational program. Travel was now conceived as a "tonic" to counteract the stress and constraints of everyday life. The journey beckoned as a time and space "apart" where one might freely yield to the promptings of imagination, experiment with potentials of the self thwarted at home, and multiply events in a spirit of innocent adventure.[9]

The most superficial, readily assimilated dimensions of Byronic cosmopolitanism—the escapism, aestheticism, disaffection to patriotism, and naughty flouting of conventional morality—entered into what were apparently the first representations of Russians as "tourists." One was Myatlev's feuilleton in verse *Sensations and Observations of Madame Kurdyukova Abroad, dans l'étranger* (1840, 1843, 1844) and the other—Annenkov's journalistic *Letters from Abroad*, addressed to Belinsky and serialized in *Otechestvennye zapiski (Annals of the Fatherland*, also translated as *Notes of the Fatherland)*. A wealthy Petersburg aristocrat with connections at court, Myatlev was an art connoisseur with a private collection in his apartment

7 Buzard, *The Beaten Track*, 110–13; Dekker, *The Fictions of Romantic Tourism*, 223–24; and Vladimir Nabokov, in Aleksandr Pushkin, *Eugene Onegin: A Novel in Verse*, trans. with a commentary by Vladimir Nabokov, 2 vols. (Princeton: Princeton University Press, Bollinger paperback 1981), 2.2:96.

8 Wohlgemut, *Romantic Cosmopolitanism*, 99–112; "willing exile" figures in Goodlad and Wright, "Introduction and Keywords," 9.

9 Buzard, *The Beaten Track*, 102–3; and Baranowski and Furlough, "Introduction," 5–6, 19.

on St. Isaac's Square, an elegant quarter of the city. Prior to visiting Western Europe in 1836–39, he published some poetry but went largely unnoticed until *Sensations and Observations*. The poem's narrator is a sophisticate, while the heroine is a yokel who introduces herself and her serf maid as "tourists" in conversation with an Englishman visitor in Switzerland.[10] Kurdyukova hails from Tambov, a town Pushkin and Lermontov had made synonymous with provincial tedium and banality.[11] A macaronic *tour de force*, *Sensations and Observations* saturates the yokel's speech and thought with French, plus smatterings of English, German, and Italian, all phonetically transliterated into Cyrillic. The linguistic mix creates a funny-looking text with clever multilinguistic rhymes. Among the charmed listeners of Myatlev's recitations of *Sensations and Observations* in literary salons was Lermontov, who wrote a lyric honoring Kurdyukova as an endearing alternative to Petersburg's bluestockings.[12] The Tambov tourist's fame grew not only through publication of the poem but also stage adaptations performed in Petersburg.[13] Enjoyed by visiting country squires as well as residents of the capital, Kurdyukova appears to have served as an object of cathartic laughter for Russians aware that in the eyes of the West, all Russia could seem provincial.[14]

While lacking literary pizzazz, Annenkov's *Letters from Abroad* complements Myatlev in constituting a Byronic vacation mentality. From a middle-ranking noble family in the Simbirsk region, Annenkov received his education in Petersburg, gave the civil service a try, and then settled into the life of a literary dilettante. A prolific but undistinguished critic, his lasting reputation rests on biographies and memoirs. His classic essay *A Remarkable Decade* (1880) described the development of the intelligentsia between 1838 and 1848, when Annenkov—a gregarious, portly *bon vivant*—formed friendships with eminent writers and social thinkers of the time,

10 I. P. Miatlev, *Stikhotvoreniia. Sensatsii i zamechaniia gospozhi Kurdiukovoi*, 2nd ed. (Leningrad: Sovetskii pisatel', 1969), 311 (subsequent citations appear parenthetically).

11 A. S. Pushkin, *Evgenii Onegin*, *Sobranie sochinenii*, 10 vols. (Moscow: Gostudarstvennoe izdatel'stvo khudozhestvennoi literatury), 4:5.27; and M. Iu. Lermontov, *Tambovskaia kaznacheisha* (*The Tambov Treasurer's Wife*, 1838), *Sobranie sochinenii*, 4 vols. (Moscow: Khudozhestvennaia literatura, 1983–84), 2:20–41.

12 Lermontov, "I. P. Miatlevu," *SS*, 1:303.

13 Belinskii, "Peterburgskii sbornik" (1846), *PSS*, 9:550.

14 Anon.,"Predislovie k izdaniiu 1904 g.," in I. P. Miatlev, *Sensatsii i zamechaniia gospozhi Kurdiukovoi za granitseiu, dan l'etranzhe*, 2nd ed. (St. Petersburg: Izdanie A. S. Suvorina, 1907), iii–iv, xvii. See also N. A. Kovarskii, "Poeziia I. P. Miatleva," in Miatlev, *Stikhotvoreniia*, 10.

including Belinsky, Gogol, Herzen, and Turgenev. During his first sojourn in the West from October 1840 to November 1843, Annenkov visited Italy, Sicily, France, Switzerland, Germany, Austria, Holland, Belgium, Denmark, England, Scotland, and Ireland. His *Letters from Abroad* meanders and shies away from "serious questions," a general tendency of his travel writing for which Gogol chided him.[15] Isaiah Berlin was more charitable: if not a profound thinker, Annenkov "was an agreeable, intelligent, and exceedingly civilised man, [. . .] an eager and observant intellectual tourist."[16]

Several shared themes of Myatlev's poem and Annenkov's *Letters* combine to dramatize Russian tourism of the 1840s as a rupture with Karamzin's *Letters of a Russian Traveler*. Prominent in both of these works of the Romantic era is Byronic receptiveness to Italy, the Grand Tour Mecca that Karamzin skipped. In addition, Karamzin's imperial patriotism, his morally encumbered judgment of art, his antipathy to English tourists, and his frugality are relegated to the past. In contesting Karamzin's patriotism, the writers of the forties simultaneously resist the official nationalism of their time. Finally, altered circumstances of travel complete the break with the past. Aristocratic privilege yields to a greater social range of Russian tourists, Kurdyukova defies Karamzin's gender bias, and steamships compete with the coach.[17] Trains, too, become part of the tourist experience, and they would divide Russians into enemies and partisans of the railroad.

The Byronic Tourist Regime

The opening stanzas of *Childe Harold's Pilgrimage* introduce the hero as a dissolute rebel. Having run "through Sin's labyrinth," he yearns for a revitalizing "change of scene." As he sails away, he bids good riddance to England and "bloated" homebodies glued to their comfy chairs.[18] He visits Portugal, Spain, Albania, Greece, Belgium, Switzerland, and Italy. Harold reflects on politics, including Spain's struggle for independence, the Ottoman

15 Consult the survey of Annenkov's life and writings in Offord, *Portraits of Early Russian Liberals*, 106–43, quote 119.

16 Isaiah Berlin, *Russian Thinkers*, ed. Henry Hardy and Aileen Kelly, with introduction by Aileen Kelly (Harmondsworth: Penguin Books, 1979), 114.

17 Regular steamship service to and from Petersburg began in the 1830s: Billington, *The Icon and the Axe*, 365.

18 Lord George Gordon Byron, *Childe Harold's Pilgrimage*, in *Poetical Works*, ed. Frederick Page, 3rd ed. (Oxford: Oxford University Press, 1979), 1.5, 6, 30 (subsequent citations appear parenthetically).

oppression of Albania, the monstrosity of war at Waterloo. His program nonetheless excludes educational and civic sites typical of Enlightenment travel (university lectures, law courts, hospitals). His priorities are aesthetic: the charm of Spanish women, the "majestic Rhine," the Alps, Lake Leman, and finally the art and scenery of Italy, the subject of the poem's concluding canto. Here the wandering hero's exalted aestheticism reaches its peak. Formulating tropes imitated ever after, the arriving traveler first glimpses Venice as "a sea Cybele," a "queen" of the seas, "rising with her tiara of proud towers." Her palaces are crumbling, "the Austrian reigns," the glory days are gone, "but Beauty still is here" (4.1–3, 11–13). Continuing his journey, Harold represents the banks of the Brenta canal at sunset as an "island of the blest" and declares that even the "weeds" of Italy are more magnificent than the "fertility" of other countries (4.26, 27). Florence makes him "drunk with beauty," and he greets Rome as "my country! city of the soul!" (4.50, 78). Having proclaimed Italy the "Mother of Arts," he marvels at *Laocoön* and the *Apollo Belvedere* in the Vatican museum (160–61).[19]

The Italian canto of *Childe Harold's Pilgrimage* acquired a racy comic companion in *Beppo: A Venetian Story* (1818). Describing himself as "a broken Dandy lately on my travels" (52), the English poet tells the story of Lara, a glamorous middle-aged Venetian adulteress whose sea-faring husband Beppo goes missing for years but eventually returns after escaping from pirates. The poet presents Venice as the seat of delightfully lax sexual morality rampant in Italy (41), and he recommends cuddling in a gondola: a boat "just like a coffin clapt in a canoe,/Where none can make out what you say or do" (19). The dandy traveler also eroticizes the museum-goer's gaze. The Venetian women he ogles merge with gorgeous paintings, especially Titian's *Urbino Venus* at the Uffizi:

> And like so many Venuses of Titian's
> (The best's at Florence—see it, if you will),
> They look when leaning over the balcony,
> Or stepp'd from out a picture by Giorgione. (11)

Collectively, the female population of Venice incarnates Eve in the Garden of Eden:

19 On the poem as stimulus to art-oriented tourism, see Rosemary Bechler, "Lord Byron's Grander Tour," in *Transports, Travel, Pleasure, and Imaginative Geography*, ed. Chloe Chard and Helen Langdon (New Haven: Yale University Press, 1996), 55.

> Eve of the land which still is Paradise!
> Italian beauty! didst thou not inspire
> Raphael, who died in thy embrace, and vies
> With all we know of Heaven, or can desire,
> In what he hath bequeath'd us? (46)

Speaking as a connoisseur of both art and feminine beauty, the poet intermingles the sacred and profane, art and tarts, aestheticism and sensuality.

The eroticized aestheticism renders England inferior to Italy. England is the home of "chilly women" and boring, bashful maidens who "smell of bread and butter" (39, 49). The poet loves Italian as a language "which melts like kisses from a female mouth," unlike the "harsh northern whistling, grunting guttural" of English (44). Italy's scenery has more appeal than England's, and in Italy the traveler sees the real sun, so unlike the dim orb—"weak as a drunken man's dead eye"—that rises over "reeking London's smoky caldron" (43). In mock apology, the poet exclaims: "England! with all thy faults I love thee still." But ensuing stanzas poke fun at parliamentary government, freedom of the press, habeas corpus, "Regent, Church, and King!" and "our little riots just to show we are free men" (47–49). The English dandy in Italy enjoys cutting loose from home and making a joke of domestic attachment. Beginning with the assault on patriotism in *Childe Harold's Pilgrimage*, these attitudes made Byron anathema to conservative English nationalists. To the mind of one such reviewer, *Beppo* was a propagator of pernicious "cosmopolitan liberality," "a tale of pollution, dipped in the deepest die of Italian debauchery."[20]

The pronounced intellectual dimension of Karamzin's prestigious *Letters* retarded the Byronic foreign travel regime's transfer into Russia. Known in Russia primarily through the French prose translations of Amédée Pichot, Byron represented the avant-garde trend in Russian literature as of the early 1820s.[21] Karamzin nevertheless retained authority as the founding father of edifying tourism, Russian style. Striking evidence comes from Vyazemsky, who endorsed Byronic Romanticism in essays on Pushkin's narrative poems of Russian wanderers in the Caucasus (*The Prisoner of the Caucasus*, 1822), the Crimea (*The Fountain of Bakhchisarai*, 1825), and Bessarabia (*The Gypsies*, 1826). But on the other hand, Vyazemsky upheld

20 Wohlgemut, *Romantic Cosmopolitanism*, 114–15.
21 Nabokov, commentary in Pushkin, *Eugene Onegin*, 2.1:159–61.

the Enlightenment's educational ideal for foreign travel. In an article of 1823 he argued that "our travelers" in the West should be looking "for the sources of commonweal of peoples and governments, to learn the secrets of political science in those countries, where it has long been taught to everyone."[22] Vasily Zhukovsky (1783–1852) and other Russian authors of the 1820s and 1830s expressed a similar conviction that a journey abroad was a momentous undertaking that required careful preparation through reading, in part to master the travelogue genre as practiced in the West and then adapt it by extending Karamzin's strategies of self-fashioning as a Russian sophisticate abroad.[23]

The entrenchment of Byronism in Russia nevertheless came to erode the authority of Karamzin's anti-English educational model. Pushkin's *Eugene Onegin* (1825–32) played a pivotal role in altering the register of "spleen," stigmatized as an English "moral disease" in Karamzin's *Letters*. In an effervescent style patterned after *Beppo*, Pushkin's novel equates *splin* (spleen) to *khandra* (depression, melancholy), cast as the glamorous predicament of the Russian "dandy," the English word used in the text.[24] Onegin enters the poem as a Petersburg man of fashion who has mastered the Byronic pose of ennui, has an English-style wardrobe and toiletry articles, a taste for rare roast beef, and wants to travel to Italy with the poet. "A Muscovite in Harold's coat" (7.24), Onegin has his volumes of Byron's poetry with notes in the margins, and he imitates Harold's wanderlust after killing Lensky in the duel. Confined to the empire's southern periphery, Onegin's itinerary reflected Pushkin's failure to receive permission to travel beyond Russia's frontiers. The *Onegin* poet can only fantasize about Western Europe, as illustrated by his Byronic description of the Brenta canal and a steamy gondola ride with a local woman in Venice (1.49).

The Allure of the West

As Pushkin's Venetian flight of imagination suggests, Byron helped engineer an Italian turn in Russian travel, a development that further sidelined

---

22 Schönle, *Authenticity and Fiction*, 270n4.

23 Dickinson, *Breaking Ground*, 174–75.

24 Pushkin, *Evgenii Onegin*, 1.4, in *SS*, 4:12 (subsequent references to the cantos and stanzas appear parenthetically). On Pushkin's awareness of *Beppo*, see Nabokov, commentary on *Eugene Onegin*, 2.1, 10–11. Other English words in *Onegin* are "beefsteaks" and "vulgar," 1.37, 8.15.

Karamzin. Images of Italy as an aesthetic paradise began emerging in Russian painting in the 1810s. Russian poetry took up the theme, to develop a cult of Italy as the "happy country" (*paese felice*), a land of the blest.[25] Even writers who had never been to Italy professed nostalgia for it, a fad in which Gogol participated, well before living mainly in Rome between 1836 and 1848.[26] A central figure in the Italian turn was Princess Zinaida Volkonskaya, an accomplished society hostess known in Russia as the "Corinne du Nord" for her patronage of the arts. Throughout the Romantic era, Russia's most prominent painters, including Karl Bryullov (1799–1852), Alexander Ivanov (1806–58), and Fyodor Bruni (1799–1875) went to live and work in Italy, and by the late 1830s Volkonskaya's villa in Rome had become a gathering place for the artistic community and Russian travelers.[27] Among those who visited Italy at this time was Vyazemsky during his first journey abroad in 1834. Another participant in the trend was Vasily Botkin (1811–69), the wealthy eldest son of a tea merchant who along with his brother Dmitry was an art connoisseur. Following an 1835 trip, Botkin published his "Fragments of Travel Notes about Italy" (1839) and "A Letter

---

25 Ely, *This Meager Nature*, 63–64, 242n11. For other representative studies of the Russian aesthetic passion for Italy in the first half of the nineteenth century, see Ettore Lo Gatto, *Russi in Italia* (Rome: Editori riuniti, 1971), 61–216; Patrizia Deotto, "Materialy dlia izucheniia ital'ianskogo teksta v russkoi kul'ture," *Slavica tergestina* 6 (1998): 197–226; Deotto, *In viaggio per realizzare un sogno. L'Italia e il testo italiano nella cultura russa* (Trieste: Università degli Studi di Trieste, 2002); Renato Risaliti, *Russi in Italia tra Settecento e Novecento* (Turin: Centro interuniversitario di ricerche sul viaggio in Italia, 2010), 16–47; and V. N. Toporov, "Italiia v Peterburge," in *Italiia i slavianskii mir. Sovetsko-ital'ianskii simpozium in honorem Professor Ettore Lo Gatto*, ed. N. M. Kurennaia et al. (Moscow: Akademiia nauk, 1990), 49–59, 73–74, 80.

26 On Gogol and Italy, see analysis of the "eye of the tourist," in Robert A. Maguire, *Exploring Gogol* (Stanford: Stanford University Press, 1994) 115–34; Lo Gatto, *Russi in Italia*, 125n6; *Gogol' i Italiia. Materialy Mezhdunarodnoi konferentsii "Nikolai Vasil'evich Gogol': mezhdu Italiei i Rossiei,"* ed. Mikhail Vaiskopf and Rita Giuliani (Moscow: Rossiiskii gosudarstvennyi gumanitarnyi universitet, 2004); and Stiliana Milkova, "From Rome to Paris to Rome: Reversing the Grand Tour in Gogol's 'Rome,'" *Slavic and East European Journal* 59, no. 4 (2015): 493–516.

27 John E. Bowlt, "Russian Painting in the Nineteenth Century," in *Art and Culture in Nineteenth-Century Russia*, ed. Theofanis George Stavrou (Bloomington: Indiana University Press, 1983), 122. On Volkonskaya, see also Rosalind P. Blakesley, "Art, Nationhood, and Display: Zinaida Volkonskaia and Russia's Quest for a National Museum of Art," *Slavic Review* 67, no. 4 (Winter 2008): 916–19; and Maria Fairweather, *The Pilgrim Princess: A Life of Princess Zinaida Volkonsky* (London: Robinson, 2000), 219–78.

from Italy" (1842).[28] Other Russians who traveled in Italy in this period included Turgenev, Herzen, and the poet Evgeny Baratinsky, who died suddenly in Naples in 1844 at the age of forty-four. In testimony to the destination's prestige, Nicholas I made his first journey to Italy in 1845.[29]

Several factors attracted Russian travel lovers to Western Europe. First, there was the dismal nature of travel at home. Although some nationally affirmative accounts of journeys in the provinces appeared in this period, "the general consensus held Russian travel to be a special national form of torture."[30] One man summed up the situation in recounting his 1845 trip from Moscow to Georgia. In central Russia, he wrote, "there is virtually nothing pleasurable to offset the inconveniences and deprivations. Nature and art are meager everywhere. There is no food for the eyes, the imagination, or the stomach."[31] Travelers had to contend with bad roads, bridges flooded or in disrepair, inefficient, grubby posting stations, a scarcity of provisions, and primitive inns, few and far between. Leaving behind the "sombre landscapes, provincial philistinism, and physical discomforts" of their homeland, Russians proved receptive to the scenery, art, architecture, and tourist amenities of the West: "the delights of foreign cuisines and the simple luxury of a good dinner and a comfortable bed with clean sheets after a long day's journey in the Mediterranean summer."[32]

The public display of artistry and the accessibility of art museums in the West greatly heightened the allure of foreign travel as a form of "serious leisure," contributing to acculturation and refinement.[33] Karamzin's son Andrei (1815–54) left an interesting testimony. Writing to his mother from Genoa in 1837, he confessed that along with the bedazzling scenery, flowers, and glorious weather, the "artistic wealth" and "refined luxury" of the city—the churches, palaces, marble terraces, and mosaics—filled him

28 V. P. Botkin, "Otryvki iz dorozhnykh zametok po Italii" and "Pis'mo iz Italii," in his *Pis'ma ob Ispanii*, ed. B. F. Egorov and A. Zvigil'skii (Leningrad: Nauka, 1976), 202–20.

29 Before Nicholas became tsar, he visited France, Switzerland, the Netherlands, parts of Germany, and England: see Nicholas V. Riasanovsky, *Nicholas I and Official Nationality in Russia, 1825–1855* (Berkeley: University of California Press, 1959), 27, 29; and Wortman, *Scenarios of Power*, 1:257–59.

30 Ely, *This Meager Nature*, 140. On positive accounts of traveling in central Russia, see Dickinson, *Breaking Ground*, 211–18, 220–26.

31 Ia. A__v, "Zametki na puti iz Moskvy v Zakavkazskii krai," *Otechestvennye zapiski*, no. 7 (1845), section 8: 11.

32 Offord, *Portraits of Early Russian Liberals*, 109, 110.

33 On the general concept, see Robert A. Stebbins, "Cultural Tourism as Serious Leisure," *Annals of Tourism Research* 23, no. 4 (October 1996): 948–50.

with "a profound envy, not personal but national."[34] He felt bad for Russia. But instead of resenting Italy its riches, he yielded to the pleasure of aesthetic self-expansion in Genoa and continued to do so in the churches and museums of Florence and Pisa. With respect to art museums, Russia's lag behind the West was tremendous. Into the mid-nineteenth century, art in Russia remained largely concentrated in private collections inaccessible to the public. This included the royal collection of the Hermitage which required special permission to visit until 1852 and even then placed restrictions on entry. Free access was instituted only in 1865. On the other hand, Western Europe's museum culture dated back to the eighteenth century. The royal collection of Dresden opened to the public in the 1760s, and similar developments followed in Munich (1779), Florence (1789), and Vienna (1792). The French revolutionary assembly decreed the Louvre a national collection in 1793; and the Rijksmuseum, the Prado, and London's national gallery opened to the public, respectively, in 1808, 1809, and 1838.[35] The stark contrast with Russia was underscored in the well-received *Sketches of Southern France and Nice* (1844), a travel account by the prominent writer Maria Zhukova (1805–55). Germany's art museums impressed Zhukova not only for their holdings but for their democratic admissions policies. She asserted that those collections welcome the whole German nation, whereas gaining access to art in Russia was so difficult that the treasures might as well be encased in the volcanic dust that had hidden Pompeii for centuries.[36] While not specifying the Hermitage, Zhukova clearly had it in mind as a royal preserve apparently bent upon remaining exclusionary.

The allure of leisure travel abroad stood in tension with the official view of Russia as a land more powerful, safer, and morally superior to the West. Encapsulated in the slogan "Orthodoxy, autocracy, and nationality," the patriotic creed was formulated in 1833 by the minister of education, Sergei Uvarov, a man who "had never read a Russian book and continuously used French and German in his own writings."[37] The overthrow of Charles X in France in 1830 and the subsequent revolutionary turmoil in north-western Europe, Italy, and Poland convinced Uvarov that Russia needed to shore

---

34 Risaliti, *Russi in Italia*, 35–40.

35 Dianina, *When Art Makes News*, 74, 136–40.

36 M. S. Zhukova [M. Zh-k-va], *Ocherki iuzhnoi Frantsii i Nitstsy. Iz dorozhnykh zapisok 1840 i 1842 godov*, 2 vols. (St. Peterburg: Izdatel'stvo A. Ivanova, 1844), 1:24.

37 The remark is attributed to the historian Sergei Solovyov in Michael T. Florinsky, *Russia. A History and an Interpretation*, 2 vols. (New York: Macmillan, 1965), 2:799.

up the status quo. As a professor of history at Moscow University, Pogodin played a catalytic role. In a lecture delivered in Uvarov's presence, he argued that the Russian people summoned the Varangians (Normans) to rule over them in ancient times. Pogodin's interpretation posited a popular foundation of autocracy, an idea that Uvarov seized upon and made central in the state ideology.[38] The official outlook sanctioned autocracy as the will of God, "the Russian God" that Nicholas I evoked as the supreme power guiding his rule. Unlike the plain meanings of Orthodoxy and autocracy, the ideology's third principle—"nationality" (*narodnost'*)—was notoriously woolly. Russia was a multinational empire where non-Russian elites participated in governance. What made autocracy and Orthodoxy "national" in this context was their long-standing existence in the Russian domain. In imagining "nationality" in terms of the Russian people's devotion to the God-appointed autocrat, the state interpreted the concept as a dynastic justification. On the other hand, as elaborated by Pogodin and other journalists and writers, "nationality" assumed the character of a romantic myth, attributing to the Russian people a unique spiritual profundity and a messianic calling to revitalize Christendom.[39]

The public persona of Nicholas I incarnated the official ideology. The tsar performed Orthodox piety at his coronation ceremony and other events staged in cathedrals and covered in the press. He cultivated the image of a chivalrous military emperor and made a hallmark of his reign the staging of spectacles that showed off the army's parade ground skills and splendid uniforms. The cult of the monarch simultaneously cast Nicholas as a devoted husband and father, and the happy family became a metaphor of the state under his rule.[40] That scenario associated the West with political violence, social conflict, and declining spirituality, while representing Russia as a harmonious, tranquil land with the world's most fortunate and virtuous population. Russia had in fact suffered no revolution, the epitome of which remained in most minds the French Terror. In playing upon those themes of national security and stability, official nationalism gained considerable support. But on the other hand, during that "remarkable decade" that Annenkov commemorated, the Russian intelligentsia began formulating

38 Wortman, *Scenarios of Power*, 1:380–81.

39 Riasanovsky, *Nicholas I and Official Nationality*, 117–29, 145; W. Bruce Lincoln, *Nicholas I: Emperor and Autocrat of All the Russias* (Bloomington: Indiana University Press, 1980), 239–52; and Maiorova, *From the Shadow of Empire*, 195–96n25.

40 Wortman, *Scenarios of Power*, 1:260–95.

what in their view was a more "genuine" nationalism, a populist nationalism oriented toward the culture and needs of the common folk.[41] For those liberal-minded Russians, traveling abroad was a precious respite from the drumbeat of official patriotism, at once militaristic and sentimental. Herzen's *Letters from France and Italy* summed up the relief in recalling his 1847 exit from Russia: "Königsberg was the first city where I found repose after nineteen years of persecution, the place where I felt free at last."[42]

Converging on the English

As the most numerous national group on the Continental tourist circuit, the English figured in Russian journalism of the late 1830s as the standard against which to measure contemporary Russian leisure travel practices, for better or worse. English travelers seemed more ubiquitous than ever. In the second part of Goethe's *Faust* (1832) Mephistopheles is surprised to find no holidaying Britons at the festivities of Walpurgis-Nacht. Heinrich Heine similarly wrote that there seemed to be an Englishwoman sniffing every lemon tree in Italy.[43] Yet, according to the mean-spirited satire "The Traveler of Our Time" (1838), Russians, too, were flocking abroad and in a vulgar manner (the author of the piece was Ivan Golovin, a repulsive Russian emigré that Herzen once encountered whipping an adolescent chimney sweep in front of a hotel in Turin and tossing a coin with each lash).[44] The specimen "traveler of our time" sounds English in many respects but lacks English sophistication. Golovin maintains that ennui and idleness motivate the journeys of most of his compatriots. Culturally unprepared and lacking clear priorities, they are hungry eyes in frantic motion. In their effort to see *everything*, they "may stay in a different town every night" and find they have only "ten hours" for Vienna. They crave sightseeing so much because "in Russia there is nothing to see." But whereas the English travel mainly to *see* things, Russians travel primarily to be *seen*. In an effort to radiate

---

41 Bassin, *Imperial Visions*, 40–41.

42 A. I. Herzen [Gertsen], *Pis'ma iz Frantsii i Italii*, *Sobranie sochinenii*, 30 vols. (Moscow: Izdatel'stvo Akademii nauk, 1954–65), 5:79. For an English translation, see Alexander Herzen, *Letters from France and Italy, 1847–1851*, ed. and trans. Judith E. Zimmerman (Pittsburgh and London: University of Pittsburgh Press, 1995).

43 Harmut Berghoff and Barbara Korte, "Britain and the Making of Modern Tourism: An Interdisciplinary Approach," in Berghoff et al., *The Making of Modern Tourism*, 2.

44 Herzen, *SS*, 11:404, 406. Written between 1847 and 1851, the *Letters* first appeared in a Russian book edition in London in 1854, misdated 1855.

"oriental luxury," they overdress and grossly overspend. Behind the showiness, however, is a lack of English self-assurance.[45] Particularly in Germany, Russian visitors are regularly fleeced, failing to understand there is a dual price structure: one for Germans, the other for foreigners. English travelers, too, are sometimes cheated in Germany but are never at a loss for a withering remark if handed an outrageous bill.

The theme of enviable English savoir faire also figured in "English Tourists" (1839), a satire supposedly translated from a 1755 article in a magazine called the *Adventurer.* Whether real or fabricated, this viewpoint presented from inside British culture posits an essentialist, two-sided English tourist character fixed since the mid-eighteenth century. The author typifies English tourists as "vagabondizing Sybarites" who regard the Continent as a "public playground" (*gul'bishche*) and go roaming in hopes of curing a "disease"—the "tedium that comes from idleness." At home the English are purportedly hospitable, good-natured, and kind. But on tour they tend to be arrogant, materialistic, insensitive, and boring. They think of "traveling" as "paying" and seek to assert their superiority on financial grounds. They grab the best seats in a coach, secure the best hotel rooms, monopolize the service personnel, and hog choice food at tables d'hôte, the common tables in hotels. In preparation for trips, the English spend "mountains of gold" on clothing and accessories. Dressed in their finery, they arouse "envy and hatred" throughout Italy and other relatively poor countries. In thrall to guidebooks, the English see all the sights, yet display "wooden insensitivity" to art and scenery. On excursions to the top of Etna, for example, only the English will ignore the sunrise because they are too busy making their tea. But on the other hand, brags the author, "nobody knows how to travel as well as the English," nobody has made travel such a "science" and an "art."[46] Their savoir faire extends to elegant, well designed briefcases, suitcases, and the "holy of holies in the tourist's arsenal"—the case for toiletries and medicines, all impeccably organized and deemed vital for grooming and health abroad (102).

---

45 Ivan Golovin, "Puteshestvennik nashego vremeni," *Syn otechestva* 5 (1838): 69–70. Golovin (1816–83) emigrated to the West in 1845. His other writings include the Swedish travelogue *Poezdka v Shvetsiiu v 1839* (1840) and *La Russie sous Nicolas I* (1845).

46 Anon., "Angliiskie turisty," *Otechestvennye zapiski* 5, no. 9 (September 1839), section 8: 100–7.

Kurdyukova, Karamzin, and Monsieur Byron

In conjunction with the power of Byron, the enviable aspects of those journalistic images of English tourists infiltrate *Sensations and Observations* and Annenkov's *Letters*. The Kurdyukova saga stages a complicated collision between the author and his *turistka* protagonist. Her voice usually befits a yokel but sometimes blends into the poet's, especially in artful descriptions of scenery in the Alps and Italy, where Kurdyukova affirms the truth of the adage "Vedi Napoli e poi muori!" (548). The unsteady relationship between the consciousness of the author and the heroine facilitates her function as a "laughing double" of Karamzin, a tool of affirmative laughter underwriting aesthetic cosmopolitanism.[47] Kurdyukova's mobility alone defies the multilingual eighteenth-century gentleman traveler's conviction that women should stay home. But Myatlev goes beyond gender to make his heroine a semi-educated *provincial* (adding a hoax to artistry, Myatlev had the Petersburg press print "Tambov" as the book editions' place of publication). The poem strikes the vacation note in the opening lines when a doctor advises the widow Kurdyukova to go abroad for a rest. She leaves home with two serfs (the house maid Annette and the tin smith Khariton) and tons of baggage, including a "little travel samovar" (*dorozhnyi samovarchik*, 205). From Petersburg the travelers head for Germany on a steamboat, and Kurdyukova thinks of Karamzin when they pass the Isle of Bornholm (188), the setting of a Gothic tale he wrote. But the competing mediator of her journey, particularly in Italy, proves to be "Monsieur Byron, *le poète*" (465). Kurdyukova cannot play Byron but remarks many Russian men tourists attempting to do so: they seek "to resemble a foreigner, / A Frenchman, an Italian, / Or an Englishman, *surtout*" (239–40).

Kurdyukova's political dimension simultaneously travesties Karamzin's imperial patriotism and the official nationalism of Nicholas I. The Tambov *turistka* believes that Western Europe envies Russia for its vastness, military might, the grandeur of "our God," our "wonderful *narod*" devoted to the tsar, and the national harmony that Russians enjoy, living together "as one big family" (293). That patriotic outlook adheres to the state ideology, plus

47 On the "laughing double for every serious form" in the context of "the parodic-travestying forms" of medieval Europe, see M. M. Bakhtin, "From the Prehistory of Novelistic Discourse," in *The Dialogic Imagination: Four Essays*, ed. Michael Holquist, trans. Caryl Emerson and Michael Holquist (Austin: University of Texas Press, 1981), 79. The concept of doubling is missing in the loose application of Bakhtinian "carnivalization" to Myatlev's poem in Kovarskii, "Poeziia Miatleva," 45.

the family theme of Nicholas's public persona: big empire, mighty army, love of the monarch, and religious faith uniting the tsar, the folk, and the educated classes. Myatlev's political parody also addresses the ethnocultural and imperialist mishmash of what the authorities called "nationality." The ethnocultural strand of Kurdyukova's Russianness is symbolized by Tambov, the travel samovar, and the national foods she recalls abroad: kvas, kasha, *kisel'* (a sort of blancmange), aspic, cabbage soup, suckling pig with horseradish, and *rassol'nik* (meat or fish soup with pickled cucumbers). Kurdyukova loves Russian cuisine but scoffs at food patriots. "*Le* patriotism," she insists, lies in the heart not in eating (294–95). Myatlev takes a dig here at Zagoskin's nationalistic novel *Homesickness* (1839). In Spain, Zagoskin's Russian hero misses his native vodka and pickled cucumbers so much that he bursts into tears.[48] As a riposte to such a standard of patriotism, plump Kurdyukova is a comic cosmopolitan eater, relishing foie gras, *dinde truffé*, *bœuf salé*, pasta, Italian ice cream, champagne, Château Lafitte, and much else (295, 459, 468).

Along with the ethnocultural markers, Kurdyukova glows with imperial consciousness. During a boat ride along the Rhine, hills in Neuwied intoxicate her fellow passengers. Why the fuss? she wonders. They should come to "our country" (*u nas*) to see "*les* Urals and *le* Caucasus" (230). Kurdyukova also claims to have family links to Kirghiz and Cossacks (403). Myatlev thus parodies Russians' enjoyment of their Eurasian empire as an arena of cultural and genetic intermingling that gave them an exotic edge over West Europeans. Finally, the poem's imperial theme lampoons Karamzin's sentimental evocation of Kronstadt as the returning patriot's first contact with home. Kurdyukova extols "our Kronstadt" as the "granite" edge of the empire, symbolizing Russia's supremacy over the West: just as the sea's crashing waves fail to erode the mighty fortress, so too are Western critics of Russia doomed to impotent fuming (294).

Perhaps echoing the tearful paternalism that Karamzin's narrator displays toward his household serfs, Kurdyukova is blind to the shame of serfdom, an attitude that combines with social pretentiousness to intensify the satiric thrust of her patriotism. Proud to belong to the serf-owing strata, she claims her great-grandfather was a duke (403), and she basks in the gross flattery of service people abroad who address her as "countess" (203, 247). She likewise takes a snooty attitude toward foreigners she considers

48 See commentary on Zagoskin's *Toska po rodine* in Miatlev, *Stikhotvoreniia*, 622.

her social inferiors. She recoils, for instance, from "slovenly" common folk selling vegetables in Basel (283) and prefers fancy urban hotels to village inns where the chambermaids are "peasants" (*baby*, 268–9). Kurdyukova likewise deplores the traveling English "*prolétaires*," the "hoi polloi" in Baden-Baden (263–64). Those English tourists, she believes, are cobblers and tailors passing as English lords. They have forgotten their place, thus threatening to destabilize the domestic class structure, an event Kurdyukova would not want to see happen in Russia.[49]

English tourist hegemony bruises the yokel's inflated self-image as a Russian landowner. Her interaction with the English is mixed, entailing a comic motif of tourism as international competition, another reminiscence of Karamzin. Myatlev's stereotypical English travelers first crop up on the Rhine: "an extremely affected English miss" and an old English lord "roaming around from boredom" (227). True to form, the English people Kurdyukova meets at her table d'hôte in Mannheim all have guidebooks they know by heart. They are nonetheless a convivial lot. In dining with them, she grows so tipsy she leaps to her feet to toast them in fractured English: "I veel keess yoo foreffer!" (258). Kurdyukova also seeks "English comfort" in hotels (308).[50] Even in Germany, however, English tourists begin to get on her nerves. Baden-Baden is so crowded with English visitors that the Tambov *turistka* fails to find accommodations (264). By the time she reaches Lausanne, she is "sick" of the English, the women in particular (326). At Ferney she proclaims *les Anglais* a "cheeky people" and accuses them of cutting off pieces of Voltaire's bed curtains as souvenirs (362).

Competition with the English continues in Italy, but cooperation occurs as well. English visitors are "always in the forefront," setting the standard of tourist savvy (460). In Florence, Kurdyukova finds no seats left in the taverns: "The English have taken them all." In flight from their foggy homeland, "the damned English" are omnipresent, to the "woe of *voyageurs* from other countries" (474–75). Yet however much Kurdyukova may bristle at English pre-eminence, she cannot escape its sway. Byron and Shakespeare organize her perceptions of Venice (451, 465), where she and Annette take a

49 On British anxieties concerning middle-class tourism as a threat to social stability at home, consult Buzard, *The Beaten Track*, 95–96.

50 *Komfort*, *feshenebel'nyi* (fashionable), and other English words associated with the lifestyle of the English aristocracy entered Russian in the 1820s and 1830s: see M. P. Alekseev, "Angliiskii iazyk v Rossii i russkii iazyk v Anglii," *Uchenye zapiski Leningradskogo universiteta, Seriia filologicheskikh nauk* 9 (1944): 100–101.

1. The serf maid Annette keeps looking at Venice, as her mistress Kurdyukova waddles onto the steps of a palazzo. This was one of the illustrations that the eminent artist Vasily Timm did for the republication of the Kurdyukova poem in 1857, when the post-Crimean War Russian tourist rush to the West had begun.

gondola ride (fig. 1). In Rome, Kurdyukova visits the Colosseum in moonlight, a customary English practice featured in *Childe Harold's Pilgrimage* (521). In Naples she joins a group of English people to ride to Pompeii on mules, and she takes a typically English, nighttime excursion to the summit of Vesuvius to watch the sunrise (570–73). In a familiar letter from 1817, the genius that Kurdyukova called "Monsieur Byron, *le poète*" described English tourists overrunning Italy as a pestilential "parcel of staring boobies."[51] However, for the Tambov yokel, the authoritativeness of Byron and the pleasures of the touring style that the English invented offset the annoying aspects of their ubiquity. Kurdyukova curses the English but desires to master their art and science of travel.

Art, Comfort Food, and Shopping

In striving to meet the norms of savvy tourism, Kurdyukova uses an unidentified guidebook, keeps a travel journal, hires local *cicerone*, and hunts

51 Buzard, *The Beaten Track*, 84.

down art in museums, palaces, and churches, especially in Italy. Her visit of the Uffizi alone takes up fifteen pages, naming and describing numerous paintings and sculptures. Many other collections receive similar, if briefer treatment: the Breira gallery in Milan, the Naples national museum, and in Rome—the Vatican, the Borghese, the Capitoline museums, the Quirinale, the Barberini, and other palaces. The yokel assiduously seeks cultural capital, but one of the poem's running jokes is her lack of aesthetic training and consequent shock at representations of nudity. In comic dialogue with Karamzin's reactions to statues of naked women, this theme encompasses Kurdyukova's confrontation with the *Venus de' Medici* in the Uffizi:

Голая стоит Венера,
И при ней два гладьятера,
Также каждый без штанов,
Без рубашки, и таков,
Ну . . . как мать на свет родила!
Я бы дочку не пустила,
Не пустила б и жену,
Муж была бы; *антр ну*,
Только так из опасенья,
Всё боялась бы сравненья!

Standing here is Venus naked,
Beside her are two gladiators,
Both like her are missing pants,
No shirts either . . . well, in fact,
They are outfitted in their birthday
Suits! I would never let a daughter
Visit this, and *entre nous*,
Fear of competition, too,
Would make me prohibit entry
To my wife, were I a husband! (507–8)

In Kurdyukova's eyes, Titian's Venuses are "hussies," "excessively naked" (508–9). On the other hand, the Tambov tourist in Cologne approves Rubens's paintings of corpulent women, a body type she identifies as her own (224). In the national museum in Naples, blushing Kurdyukova sees the *Farnese Hercules* as a Volga barge hauler just emerged from the river and content to stand there naked until "God dries him off" (591). Defying the sign barring entry to women, Kurdyukova and Annette barge into the same museum's *gabinetto segreto*, the "secret cabinet" established in the 1820s to house erotic art unearthed in Pompeii and Herculanum. There the Tambov visitors are "horribly embarrassed!" by the statue of Pan copulating with a nanny goat, one of the collection's most famous items (586–87). In showing the provincial at sea in the museum world of the West, Myatlev makes fun of Karamzin's moralistic aesthetics and the persistent underdevelopment of art education in Russia.

Along with the travesty of the tourist art pilgrimage, the other running jokes of the poem are Kurdyukova's fixations on food and her *toilette*. Meals often serve to soothe sightseeing fatigue. Tired out by her visit of the ducal palace in Venice, for example, she yearns to get back to her hotel to eat and then itemizes her lunch:

> Вермишели две тарелки,
> Да ещё две-три безделки,
> *Эн* каплун, *авек дю ри*,
> Штучек шесть *де пуасон фри*,
> И полблюда макаронов,
> И десятка два *маронов*.
>
> Two platefuls of vermicelli,
> Two or three other nibbles,
> *Un* capon *avec du riz*,
> Followed by six *poissons frits*,
> Half a portion of macaroni,
> And chestnuts by the dozen. (468)

To no surprise, Kurdyukova confesses, "I am stout," too broad in fact to climb a staircase in the Borromeo villa on Isola Bella (410). The Tambov *turistka* also has a wart on her face, big enough to warrant official mention on her passport (421). All the more reason, perhaps, for her desire to look fashionable. She takes much care with her outfits and hairdos. Authorial irony, however, reveals the comical fussiness of her wardrobe, coiffures, malachite combs, and other hair ornaments. Being *seen* abroad is a big part of her trip. "Here I am in Milan!" she squeals, a mocking echo of the authorial entry into Paris in Karamzin's *Letters* (419).

Kurdyukova is an avid shopper, a theme inverting Karamzin's frugality. The shopping sprees start in Hamburg, the first major city on the yokel's itinerary. There she buys two riding habits for herself, a mantilla for Annette, seven pairs of trousers for Khariton, and presents for folks back home. The poet characterizes the splurge as typically Russian. Without considering their budget for the rest of the trip, Russian travelers, claims the author, squander money in Hamburg where everything strikes them as "so cheap, so pretty," all presented by such courteous sales clerks that "it is impossible not to buy." Kurdyukova conforms to the rule, to leave Hamburg with "an awful *trou* [hole] in her purse" (202). Be they luxurious or modest, shops continue to captivate her in Frankfurt, Switzerland, Milan, Venice, and Naples. Among other items, she buys a small terracotta bust of Voltaire (a parodic debasement of the Enlightenment quest for contact with great men), painted plates showing Swiss scenery (a classic imagistic souvenir), bracelets, watches, wine, and a moustache comb (241, 319, 330, 359, 364, 436–37, 559). Although Myatlev had envisioned a chapter on France, *Sensations and Observations* ends with the yokel in Italy, competing and cooperating with the English, puzzling over art, enjoying the cuisine, and shopping.

Annenkov's Cosmopolitan Liberation

Whereas Kurdyukova's tour is a fun-house mirror reflection of Karamzin's journey, Annenkov's *Letters* performs the Byronic revolution undistorted. In sailing from Petersburg, Annenkov's narrator recalls Karamzin as the boat passes Bornholm. By the time he reaches Berlin, however, he is disavowing his Enlightenment ancestor. He calls himself a "profane" traveler, skipping preparatory reading to plunge directly into the "town squares, the wine cellars, picture galleries, palaces, museums, theaters, etcetera."[52] Beginning with German wine, the theme of gustatory pleasure evolves, to culminate in Paris, where Annenkov's narrator pens a hymn to the food, cites several restaurants, and asserts that "whoever has never dined at the Palais Royal has never dined." The recollection of his first twelve days in Paris even makes eating the metaphor of his spectatorial style of tourism: "theaters, squares, dinners, newspapers, books, shops—I gobbled it all down in one go" (40). While not a big shopper, he nonetheless checks out the men's clothing in stores in Hamburg and remarks the favorable quality-price ratio by comparison to home.

52 Annenkov, *Pis'ma iz-za granitsy*, 10 (subsequent citations appear parenthetically).

In contrast to Myatlev's parody of patriotism, Annenkov's persona makes this not-Karamzin point indirectly: he expresses indifference to his national identity and cultivates instead "a trans-European class consciousness" that tourism fosters.[53] He writes that summertime brings a "flood of foreigners" to German spas to enjoy the natural beauty, the castles, hiking, music, and dinners with champagne. "These towns—Ems, Wiesbaden and so forth—are the capitals of cosmopolitanism [. . .] they belong to everybody." As such the spas seem to herald the "rapprochement of all peoples and the rapid elimination of their innate distinctions in the future." Continuing in this vein, the author declares: "It is not the festivals, the balls, or fireworks that constitute the principal charm of these spas but rather the facility with which they bring a person into contact with European society" (72). When asked where he is from, Annenkov's narrator states, "I am Russian," but he attaches no special importance to this fact. Treating national identity as a confinement, he enjoys plunging into the holiday "flood," to become part of a supranational elite. Even at the urban level no stable concept of "home" fetters him. He calls either Petersburg or Moscow his hometown, depending on who is asking. If the question comes from a young bachelor, the answer is Petersburg; but to a man traveling with his wife and children—the answer is Moscow (77). This split reflects the cultural mythology of the two capitals already codified in *Eugene Onegin*: Petersburg as the westward-oriented realm of the state bureaucracy, the *beau monde*, and dandyism versus Moscow as the pre-Petrine "heart" of Russia, abounding in Orthodox churches and associated with the patriarchal family. But the point is that Annenkov's persona expresses no attachment to either city. For him, the query "Where are you from?" becomes an invitation to suspend reality and exult in the vacation freedom of shifting his self-presentation to strangers as he chooses.

This ludic attitude conforms to the Byronic agenda of roaming to escape the confinements of home. Reflecting Byron's influence, William Hazlitt declared in 1822:

> We go [sic] a journey chiefly to be free of all impediments and of all inconveniences; to leave ourselves behind, much more to get rid of others [. . .]. We are not the same, but another, and perhaps

53 Buzard, "The Grand Tour," 41.

> more enviable individual, all the time we are out of our own country. We are lost to ourselves, as well as our friends.[54]

Annenkov's persona similarly relishes release from his home-bound self.[55] The intoxicating feeling of liberation from home infects the travelogue's style. Annenkov chronologically maps his departure from Russia into Germany but then often abandons linear movement. Situated for long stays in Paris and Rome, for example, he retrospectively treats other places he has visited. In a further deviation from Karamzin's round-trip structure Annenkov's final letter leaves the author in Paris, stimulated by the Louvre and with no plans to go home.

Byronic Aestheticism Versus Politics

The prominent museum theme of Annenkov's *Letters* promotes aestheticism liberated from moral and humanitarian concerns. Wherever the author goes, he makes a beeline for art collections, beginning in the Dresden picture gallery where he seeks out the *Sistine Madonna*, Titian's *The Tribute Money*, Correggio's *Holy Night*, and other religious paintings. Those "marvelous countenances," he asserts, remain long in the viewer's mind. Profane art, too, rivets his eye. With an anti-Karamzin, *Beppo* inflection, Annenkov's persona strikes an exaggerated pose of erotic exhilaration at the sight of copies of the *Venus de' Medici*, the *Callipygian Venus*, the *Sleeping Hermaphrodite*, and additional Venuses in the Dresden collection: "Hot blood rushed to my head and heart [. . .]. What beauty! What splendor! What pleasure! [. . .] An accidental brush up against this naked beauty nearly drove me mad." After contemplating the perfection of classical sculptures, the author recoils from the "repugnant mugs" of ordinary women and men in the streets outside the museum (14).

However, to the extent that the urban masses of Western Europe display a taste for art, Annenkov's persona welcomes them into his supranational aesthetic community, specifically in Paris. Karamzin's *Letters* features compassionate images of poor people in the St. Antoine quarter of Paris and remarks other seamy sides of the city. Annenkov's narrator excludes

---

54 William Hazlitt, "On Going a Journey" (1822), in his *Table Talk; or Original Essays* (New York: Chelsea House, 1983), 261; Buzard, *The Beaten Track*, 103–4; and Dekker, *The Fictions of Romantic Tourism*, 26, 46, 101.

55 Annenkov might have known Hazlitt through Pushkin's discussion of his *Table Talk*.

such material and even portrays members of the Parisian working class as fellow art consumers. Referring to recent journalism in Paris, he asserts that local tradesmen and factory workers frequent the Louvre where they can be overheard voicing insightful responses to the art (83). The aestheticizing tourist eye thus paints a rosy picture of the Paris streets and presents the Louvre as an exemplary nation-building institution. In Annenkov's telling, the museum breaks down class barriers to popularize high culture and thereby help foster social cohesion.

The Louvre functions in the *Letters* as a refuge from the political clamor of newspapers, brochures, and broadsheets that assail the vacationing author in Paris. As a city of politics, Paris stands as the polar opposite of Rome, represented as an exalting aesthetic enclave: "a monument to art and to the eternal value of artistic beauty."[56] In Annenkov's Italian capital, every conversation concerns "a statue, a painting, some new find in this land already full of the masterpieces of the ancients" (27). Surrounded by ruins of antiquity, this "wonderful city" resembles "something solemn and lofty like an out-of-town house where some great man has worked," "a villa" to which "Europe sends its sons to calm down after turmoil, anxieties, factions, and any disquiet" (29). Published in 1841, Annenkov's vision of Rome as a paradise for art-lovers begs comparison with Gogol's unfinished story "Rome" (1842). The two writers spent much time together in Rome and undoubtedly shared their impressions. Gogol, too, cast Rome as an eternal monument to art, but his Paris is not so much about politics but rather capitalist materialism and consumer culture with its glittery shop windows.[57]

Annenkov's aesthetic passion for Italy reflects Byron's central role as his travel mentor. Sailing into Venice at sunrise, the *Letters* author identifies the Lido as the place where "Byron kept his horses and strolled along the shore." Venice comes into view "floating on the water like a marble boat, in Pushkin's phrase," and the Ponte dei Sospiri resembles a bridge crossing a canal to "our Hermitage" (19). Those details establish Russian affiliations to Venice, alongside Byron's presence, but Byron prevails. Annenkov proclaims nearly "unbearable for the northern eye" the resplendence of "fantastic structures" along the sun-drenched Grand Canal. Then, as the tourist sees boarded-up windows and palaces transformed into post offices, tribunals, and police stations, his "soul begins to tune itself in the Byronic key of

56 See translation and commentary in Offord, *Portraits of Early Russian Liberals*, 111. The two passages I discuss are translated there, 111–12.

57 Lounsbery, *Thin Culture, High Art*, 226–28. See also Maguire, *Gogol*, 132–33.

2. Another Timm illustration done for the 1857 edition of the Kurdyukova poem, this scene of Russian tourists captures considerable architectural detail of buildings on the Piazza San Marco.

lamentation" over faded glory (20). An Italian guide in Ferrara points out Byron's name "incised with a nail by his own aristocratic hand" on the door of Tasso's dungeon (23–24). After spending five months in Italy, Annenkov's persona brags he has learned to curse in Italian as well as Byron (36). In Switzerland, too, the *Letters* author follows in Byron's footsteps, seeking out the Diodatti villa near Geneva where the poet lived in 1816 (39).

Annenkov's gregarious narrator extends his affinity to Byron to English tourists. In riding across the Apennines into Italy, the Russian author bonds with an Englishman who shares his aesthetic delight in the scenery. Annenkov's persona further converges on the English as fellow "tourists" in Venice (fig. 2). He implicitly extends the term to himself as a gondola conducts him to the Piazza San Marco—a treasured site of "all tourists" (18–19). While living in Rome, he writes that after the age of antiquity and the age of Raphael and Michelangelo, the Eternal City has entered a "third age—the English one."

> There is, in fact, a multitude of English here, all with guidebooks. Even bespectacled ladies, hitching up their skirts with one hand from behind, and carrying site descriptions and little maps, climb

> up cupolas and columns. They gaze at the Ganymedes, the swans [Leda's seduction], and so forth. (27)

But "our tourists," too, are vacationing in Rome. Among those Russian visitors, Annenkov's narrator remarks officers and civil servants, all "having come here to have a good time" (*poguliat'*) and no doubt take in St. Peter's, the Vatican museum, and the Colosseum by moonlight (29–30). Good fellowship reigns among the Russian and English visitors, all happily doing and looking at the same things.

Discovering the Railroad

Touring the West in Annenkov's time swept Russians into the age of the railroad, the iron horse beyond Karamzin's imagination. During the early years of Nicholas I's reign, tsarist officials considered railroads a danger to "public morals" and sought to keep them out of Russia since "the mobility trains afforded could make people hard to control."[58] Russia had no railroad until 1838. A little over seventeen miles long, it ran from St. Petersburg to Pavlovsk through Tsarskoe Selo. In the view of a confidant of Nicholas I, the service was merely "a toy of the idle," going from the capital to the Vauxhall pub at the Pavlovsk station to drink, dance, and listen to gypsy music.[59] After nine years of planning and construction, a railroad between St. Petersburg and Moscow opened to the public in November 1851. Big crowds gathered to watch locomotives depart, but there was much "cynicism and doubt" about becoming a passenger. Russians assumed their railway personnel would be inexperienced, drunk, and rude.[60] Russian publications of the time also conjured train accidents as a common occurrence abroad, as in the case of a journalist imagining the ordeal of people claiming the bodies of loved ones killed in a wreck: the victim resembles "something like a red pancake," mutilated beyond recognition.[61]

---

58 Florinsky, *Russia*, 2:789; and Robin W. Winks and Joan Neuberger, *Europe and the Making of Modernity, 1815–1914* (New York and Oxford: Oxford University Press, 2005), 91.

59 Stephen L. Baehr, "The Troika and the Train: Dialogues between Tradition and Technology in Nineteenth-Century Russian Literature," in *Issues in Russian Literature before 1917. Selected Papers of the Third World Congress for Soviet and East European Studies*, ed. J. Douglas Clayton (Columbus: Slavica Publishers, 1989), 87, 98–99n18.

60 Haywood, *Russia Enters the Railway Age*, 150–70, 442–43.

61 N. P_ii, "Sovremennye izvestiia zagranichnye," *Moskvitianin* 5, no. 17 (1853), section 7: 70. See also O. P. Shishkina, *Zametki i vospominaniia russkoi puteshestvennitsy po Rossii*

The Russian tourist discovery of train travel abroad sowed a division similar to the one that emerged in Western Europe. By the 1840s railroad haters were speaking out in England and France. They experienced train travel as an "annihilation of space by time," a notion that Karl Marx formulated.[62] Railroad haters complained that speed reduced passing scenery to a blur and that mechanized movement degraded passengers to the level of parcels. Train travel also disrupted traditional travelers' relationship to one another. As Karamzin's *Letters* depicted, coach travel encouraged conversation among small groups of passengers. But on a train, complained the disgruntled, the rapid turnover of new faces discouraged efforts to strike up an acquaintance and prompted people to retreat into reading. While especially unhappy about being shot through scenery he found impossible to savor, an Englishman writing in 1844 articulated a more general yearning for the past: train travel meant that "all those exhilarating associations connected with 'the Road' are lost or changed."[63] But on the other hand, trains won partisans for their convenience, efficiency, and the new aesthetic experience they afforded. Nostalgia for pre-industrial travel thwarted the development of modes of perception appropriate to railroad journeys. However, "another kind of perception" began to develop, "one which did not try to fight the effects of the new technology of travel but, on the contrary, assimilated them." The new normal became mobile vision, a "panoramization of the world" that "turned the travelers' eyes outward and offered them the opulent nourishment of ever changing images."[64]

The Russian cultural assimilation of the railroad replicated those conflicting tendencies. After traveling abroad, Vyazemsky began opposing the train to the troika, a trope that proved very productive in Russian literature. One of his poems sets the train's "soul-killing modernity" (*dushegubnaia novizna*) against the good old days of the Road.[65] Annenkov's *Letters* provides an early example of the other side of the coin. While living in Brussels, the author profits from trains that make the capital a hub of cultural tourism: "I sit like a spider at the center of this web, and when I have barely

---

v 1845 godu, 2 vols. (St. Petersburg: Tipografiia Vtorogo Otdeleniia Sobstvennoi E. I. V. Kantseliarii, 1848), 1:11.

62 Alan Trachtenberg, "Forward," in Wolfgang Schivelbusch, *The Railway Journey: The Industrialization of Time and Space in the Nineteenth Century* (Oakland: University of California Press, 1986), xiv.

63 Schivelbusch, *The Railway Journey*, 53–58, 67, 73–76.

64 Ibid., 59–64.

65 Baehr, "Troika and Train," 94–95; and Billington, *The Icon and the Axe*, 383.

formed the idea of going to Ghent, Antwerp, or Bergen, I'm already there, while nonetheless miraculously finding myself back in my little room at the Grand Miroir hotel-tavern every night at ten." Thanks to the train, he sees "all the magnificent Gothic monuments" of Belgium. But in addition, Annenkov's narrator exercises panoramic perception to derive aesthetic pleasure from the passing scenery:

> The train races through sixty-some miles in three and a half hours, including stops. Fertile field of Flanders speed past your eyes; the colors of houses and roofs of happy Flemish villages leave an impression of a white ribbon bordered in red; countless smokestacks of Brabant factories run after one another like soldiers of a routed regiment; the sky overhead appears the only immobile thing in the world. (68–69)

Zhukova's *Sketches of Southern France and Nice* expresses equal enthusiasm for the train. In recollecting a railway trip from Berlin to Leipzig, the author shudders at the thought of travel in olden days: the "feudal baron" on a horse slogging along muddy roads; "our grandmothers and grandfathers" in cumbersome carriages creeping through monotonous terrain with no place to stay but "a wretched peasant hut;" and then the era of "the dreary diligence" that stole "twenty-four hours of one's life on trips that now take no more than four or five by railroad." True, the author continues, taking a train means giving up your freedom, once you settle into your compartment:

> The doors are locked, the engine gives a whistle, it has started to puff and will carry you along no matter how many thousands you would pay for it to stop, no matter how you may shriek, weep, or even faint. You are no freer than unfortunate animals transported by the dozen to market. But how many advantages you purchase with that momentary relinquishing of freedom!

The advantages go beyond efficiency, to encompass the pleasant aspects of coach travel. Far from annihilating space, Zhukova's train makes frequent stops, enjoyably immersing the passengers in a foreign culture. At every station the author remarks the "tidy, even swanky" hotels and coffee houses and describes boys streaming from inns with "trays of sweet pies, sandwiches, beer, and mulled wine" to sell to travelers. "Oh, the Germans, the nice Germans! How well they know how to enjoy life!" (25–28, 30).

Interaction with fellow travelers also preserves the sociability of coach journeys. The surprise is that the international mix of passengers in Zhukova's compartment—a Frenchman, a Dutchman, and a Georgian—converse in Russian.

Published the same year as Zhukova's book, Pogodin's *A Year in Foreign Lands, 1839: A Travel Diary* also participates in the pro-train tendency. The elevated railroad between London and Greenwich astonishes the author: "It's amazing, moving on a thousand arches like a Roman aqueduct, above buildings and roofs, at the level of tree-tops, smoke-stacks, and bell towers, over streets where carriages are moving and people walking [. . .]. What grandiose projects the English have!"[66] Aesthetics rather than engineering prowess impress him on a trip from Brussels to Antwerp:

> The steam engine's rooms [sic] are delightful—veritable boudoirs with soft seats, cushions, and mirrors. You can read, eat, snooze. What a pleasant, comfortable journey! You speed along as on a flying carpet. The route races through an unending garden, past lush pastures and fertile fields. Both sides of the tracks are even strewn with flowers here and there. A triumph of the artisan spirit![67]

Pogodin likewise found a train trip from Paris to St. Germain delightful.[68]

Trains had begun transforming the Russian tourist experience. But more far-reaching than that still limited exposure to the railroad was the belated Russian assimilation of the Byronic vacation mentality. By the 1820s English writers had fixed the public image of "the tourist" as a degraded traveler. On the other hand, Myatlev and Annenkov created a buoyant atmosphere of innocent Russian adventure, and that entailed making peace with English tourists as both competitors and mentors. For the Russian writers, the term *turist* remained neutral, connoting a travel-lover and serving both as self-description and as a designation of the lovable yokel other. By comparison to Karamzin's era, the traveling Russian population was more socially diversified (provincial Kurdyukova and her serfs, Annenkov's civil servants

---

66 M. P. Pogodin, *God v chuzhikh kraiakh, 1839. Dorozhnyi dnevnik*, 4 vols. (Moscow: Universitetskaia tipografiia, 1844), 3:206–7.

67 Ibid., 4:7–8.

68 Ibid., 3:165.

and officers), but no crowds of compatriots had begun amassing to spoil the stimulation and fun of being elsewhere.

These bellwether works of Myatlev and Annenkov displaced Karamzin's mixed model of discourse and sightseeing with an aestheticized, spectatorial style of travel. Italy came to the forefront, visual pleasures prevailed everywhere, and imagistic souvenirs reflected that emphasis. The tourist eye shook free of morality, especially with respect to the museum world, often viewed through a lens of gleeful eroticism akin to *Beppo*. For all its hilarity, *Sensations and Observations* had a serious, snobbish ramification in its presentation of art. Speaking over the head of his provincial heroine, the author upheld art as the highest form of cultural capital and ironically instructed readers on how to avoid yokel behavior in museums. At a time when there were no Russian-language tourist guidebooks for Western Europe, Myatlev marked a vast number of masterpieces *in situ*, thereby outlining a program of art appreciation. In his parodic handling, the Russian desire to satisfy a hunger for the fine arts frustrated at home became a self-justifying project to develop and display refined taste. Potential conflicts between aesthetic, political, and humanitarian agendas were ignored. Annenkov's *Letters* ignored them, too.

In giving such high priority to art, Myatlev and Annenkov adumbrated the split between aesthetic cosmopolitanism and domestic political concerns that would stir controversy for decades to come. Contesting imperial patriotism in a manner that intertwined Karamzin with the official ideology of Nicholas I's reign, *Sensations and Observations* rendered Kurdyukova's attachment to home as political cretinism, condoning serfdom as part of "our" way of life that West Europeans allegedly envy. In establishing that ironic perspective, the poem valorized travel as a broadening experience that sheds comparative light on the homeland and not always to the latter's advantage. If not so sardonically as *Childe Harold's Pilgrimage*, Annenkov's *Letters* cheered cosmopolitanism at the expense of demonstrative patriotism, underwrote the Byronic urge to escape home and the home-bound self, and represented Rome as a Byronic city of the soul: an enclave of eternal art offering refuge from the political turmoil of the modern world and Russia's problems in particular. Whatever its degree or manner of expression, the vacation mentality's disengagement from home would exacerbate anti-cosmopolitan Russian worries about the fragility of Russian national identity.

3

Nationalist Worries about Tourism: Pogodin, Belinsky, Zagoskin

When Pushkin's Onegin returns to Petersburg after his peregrinations in the southern borderlands of the empire, the high-society rumor-mill wonders what he might have become during that long absence:

> As what will he appear now? As a Melmoth?
> a cosmopolitan, a patriot?
> a [Childe] Harold? a Quaker? a bigot?
> Or will he sport some other mask? (8.8)[1]

Vladimir Nabokov glossed that literal translation of his as follows: "'Cosmopolitan,' a person at home in any country, but especially in Italy if an Englishman, and in France if a Russian."[2] Nabokov might have added that Onegin was already cosmopolitan before he left home. Pushkin etches the Russian capitals as places where cosmopolitan "contamination" exists in the streets for all to see: the German's bakery, the Finnish milkmaid, and Talon's French restaurant in Petersburg, and in Moscow the Central Asian Bukharans that Nabokov identified (in his casually authoritative manner) as "hawkers of Oriental wares, such as Samarkand rugs and robes."[3] The urban street scenes serve as backdrop and props on the stage where Pushkin's central characters perform cosmopolitanism in everyday life. In addition to his

1 Nabokov, translation in Pushkin, *Eugene Onegin*, 1:285.
2 Ibid., commentary 2.2:160.
3 Ibid., commentary 2.2:114.

dandyism, Bréguet watch, and Byron's poetry in his library, Onegin dines at Talon's, where he drinks champagne and eats rare roast beef, truffles, Limburg cheese, Strasbourg goose-liver pie, and pineapple, an emblematic luxury food in Russia. Lensky's appearance and abiding aesthetic and philosophical interests reflect his student experience in Germany. Tatyana knows French better than Russian, she reads (outdated) French novels, and her family's modest country estate has a modest English garden. The westernized Russian elite was living abroad at home.

But where was the "national man" that Julia Kristeva proposed we think about as the alter ego of "cosmopolitan man?"[4] Lotman's conception of theatricalization indicated why westernization made the Russian quest for an "authentic" national identity so fraught. The reforms of Peter the Great placed the Russian elite in the position of stage actors, performing roles they experienced as simultaneously foreign and their own.[5] With the rise of Russian national consciousness, increasing numbers of thinking Russian actors on that "stage" perceived cosmopolitan contamination not as joyous plenitude of the sort Rushdie celebrated but rather as a tortured condition closer to the existence of the main protagonist of V. S. Naipaul's *The Mimic Men*: the postcolonial Caribbean man's European education alienates him from "the bush" where he was born, but he cannot make a happy life in England either. He remains a stranger in the Europe he so long imagined as a promised homeland. In harmony with the historian Vasily Klyuchevsky's essay on Onegin's "ancestors," Herzen's remark on the condition of the westernized Russian elite closely anticipated Naipaul's vision: "Foreigners at home, foreigners abroad, idle spectators, spoilt for Russia by Western prejudices and for the West by Russian habits."[6]

Russian intellectuals and writers longing to see Russian "national man" clearly assert himself considered leisure travel abroad a counterproductive pursuit. This worry was already apparent in Karamzin's time. Writing in 1803, a critic of Sentimentalism argued that along with Gothic literature and Rousseau, the glamorization of travel in accounts such as Sterne's

---

4 See analysis of Byron from this angle in Wohlgemut, *Romantic Cosmopolitanism*, 95–96.

5 Schönle, *Authenticity and Fiction*, 78–79.

6 Quoted in the analysis of the "disaggregation of Russianness" in the multinational empire, in Yuri Slezkine, "Naturalists versus Nations: Eighteenth-Century Russian Scholars Confront Ethnic Diversity," in *Russia's Orient: Imperial Borderlands and Peoples, 1700–1917*, ed. Daniel R. Brower and Edward J. Lazzerini (Bloomington: Indiana University Press, 1997), 50, 57n111.

Sentimental Journey inflamed and seduced the mind: they sent Russians chasing "imaginary pleasures" abroad instead of staying home to cultivate responsible relationships to the fatherland, their families, and friends.[7] The Byronic Russian revolt from Karamzin's Enlightenment travel norms revived those old worries in the new context of Slavophile-Westernizer polemics concerning Russia's past and its destiny. Stimulated by the rise of Romantic literature and philosophy, the development of national consciousness and a search for the national past put Enlightenment cosmopolitanism on the defensive throughout Europe. In Russia, however, the nationalist turn posed an exceptionally acute dilemma. Peter the Great's modernization program had accustomed educated Russians to regard the Muscovite past as "brutal, uncivilized, and best forgotten." Yet, as of the 1830s, the practice of seeking guidance from West European authorities had acquainted Russians with the ideas of Herder, Schelling, and other thinkers who told them that in order to achieve European standards of nationhood, they had "to develop their Russianness."[8] They had to discover a usable *Volkgeist*.

The first of Pyotr Chaadaev's "Lettres sur la philosophie de l'histoire" (1836) turned this dilemma into a full-blown crisis. Chaadaev presented Russia as a gigantic but inconsequential empire, stretching "from the Bering Strait to the Oder" but having "given nothing to the world." Crippled by its Byzantine heritage, Russia had failed to develop the "high level of culture" and the drive for "self-improvement" that the West European "family of nations" had achieved. Russia was "a blank in the intellectual order," the "orphan among nations," a notion Isaiah Berlin paraphrased as "an abortion in the creative process" of civilization.[9] Chaadaev's Russians were feckless "nomads" living "in a narrow present, without a past as without a future, in the midst of a dead calm."[10] This assessment precipitated the Slavophile-Westernizer clash over national identity. On the one side stood religious nationalism and on the other—a universalistic secular vision of progress on European models. Chaadaev's esteem for Western civilization and the

7 Schönle, *Authenticity and Fiction*, 72–73.

8 Abbott Gleason, "Westernism," in Terras, *Handbook of Russian Literature*, 517; and Offord, *Journeys to a Graveyard*, 115–17.

9 P. I. Chaadaev, "Letters on the Philosophy of History," trans. Valentine Snow in Raeff, *Russian Intellectual History*, 160–73; and Berlin, "Introduction," 8.

10 On symbolic ramifications of the nomad trope, see Kleespies, *A Nation Astray*, esp. 47–80.

unifying force of Catholicism in particular goaded the Slavophiles into proclaiming Russian Orthodoxy the foundation of a national culture spiritually superior to Western enlightenment. Casting the West as the home of soulless rationalism, they formulated a nativist myth of pre-Petrine Russia as a socially harmonious, "primitive patriarchal monarchy" where the ruler shouldered the burdens of state, to leave the people free to follow their instinctive path to Christian righteousness.[11] "Slavophile" was a patriotic self-definition that implied "a monopoly of the true national spirit." Having staked that preemptive claim, the Slavophiles tagged their enemies "Westernizers" (or "Westerners"), a term suggesting "a foreign and unnatural origin."[12] The addressees, however, defiantly accepted the label and made it signify their Eurocentric vision of Russian national development within a Hegelian framework of the worldwide march of civilization. While sharing the Slavophiles' ardent concern with the welfare of the *narod*, Westernizers looked not to religion but rather the pragmatic processes of sociopolitical reform and the emulation of Western civic institutions and the rule of law.

An emblematic literary expression of the tensions between religious nationalism and the cosmopolitan receptiveness to the foreign was Alexei Khomyakov's (1804–60) lyric poem "A Dream" (1835), republished in 1843 under the title "West and East." Like his Slavophile peers, Khomyakov belonged to the middle-range landed gentry and was educated in the European tradition. In the mid-1820s, he spent two years abroad and studied art while living mainly in Paris. Khomyakov also visited northern Italy, Switzerland, and Austria. After returning to Russia, he published "Winter" (1831), a lyric featuring recollections of Italy as the "land of wonders," its art and scenery. But in accord with Khomyakov's deepening Orthodox faith, "A Dream" represented Western Europe as "the land of holy wonders" that are fading into darkness as dimming stars. Gloom is covering those lights of civilization that Russians were taught to revere. Concluding lines proclaim the East ascendant, destined to eclipse the dying West.[13] In the Slavophile

11 Quote from Raeff's commentary on the comprehensive exposition of Slavophile political ideology in Konstatin Aksakov's 1855 memorandum to Alexander II, "On the Internal State of Russia" ("O vnutrennem sostoianii Rossii"), trans. Valentine Snow in Raeff, *Russian Intellectual History*, 231–51. Inapplicable to the needs of a modern state, the memorandum was not published until 1881.

12 Malia, *Alexander Herzen and the Birth of Russian Socialism*, 207, 278–312 (quotes 281); and Gleason, "Westernism," 518.

13 A. S. Khomiakov, "Zima" and "Mechta," in his *Stikhotvoreniia i dramy*, 2nd ed. (Leningrad: Sovetskii pisatel', 1969), 88–89, 103.

view, Westernizers remained so blinded by the "holy wonders" of foreign arts, sciences, philosophy, and social thought that they failed to recognize cosmopolitanism as an engulfing tide endangering Russia's self-realization as a nation.[14]

How was tourism drawn into the philosophical controversy about cosmopolitanism? And to what extent did the issue fall along the Slavophile-Westernizer divide? The first focus of my investigation is the conservative nationalist Pogodin's *A Year in Foreign Lands*, the four-volume diary of his trip to Poland, the Czech lands, Italy, France, Germany, Switzerland, Austria, England, and Belgium. Lacking literary merit, this narrative makes no novelistic effort at persona-construction and is packed with trivia including the prices of many commodities and services. But despite its pedestrian character, *A Year in Foreign Lands* commands attention for its twofold censure of Byronic cosmopolitanism. First, Pogodin makes a show of resisting hedonism. He enacts the gratifications of cultural consumption but seeks to legitimize them by framing his journey as an officially authorized research trip rather than a vacation. The work theme revives the Enlightenment practice of travel as discourse in emulation of Karamzin whom Pogodin revered as a historian and travel writer. Secondly, *A Year in Foreign Lands* fashions "cosmopolitan" Russian tourists as an alien breed who serve as foils to the author's xenophobic patriotism.

My other selected vehicles of anti-cosmopolitan worries about tourism are Belinsky's critical essays and Zagoskin's feuilleton collection *Moscow and the Muscovites* (1842–50). Unlike the conservative nationalists Pogodin and Zagoskin, Belinsky expressed some favorable notions of cosmopolitanism. In 1845, for instance, he termed objectivity in the writing of history a "cosmopolitan" quality, as opposed to the Slavophiles' patriotic "nonsense" about the Russian past.[15] In 1847, however, Belinsky took a nationalist stand against cosmopolitanism in response to an essay by Valerian Maikov (1821–47), a child prodigy from an aristocratic family long prominent in the arts (his brother was the poet Apollon, their grandfather was the director of the imperial theater, and their father—a famous painter).[16] Belinsky's clash with Maikov has received attention but deserves a fresh look in the context

14 Andrzej Walicki, *The Slavophile Controversy: History of a Conservative Utopia in Nineteenth-Century Russia*, trans. Hilda Andrews-Rusiecka (Oxford: Clarendon Press, 1975), 343.

15 Belinskii, "Slavianskii sbornik," *PSS*, 9:190, 198, 200.

16 Billington, *The Icon and the Axe*, 375.

of tourism.[17] Maikov advocated "sensible cosmopolitanism" (the adjective is *razumnyi*, connoting common sense). What did this mean? And how did his disagreement with Belinsky intersect Russian leisure travel? In different but compatible ways, the writings of Pogodin, Belinsky, and Zagoskin show how the growth of Russian pleasure trips to the West contributed to anxieties about cosmopolitanism as a threat to Russian national development.

Legitimating Travel as Work

Although Pogodin would make the decadence of the West a signature theme of his writings, he displayed a social underdog's drive to acquire the status symbol of having "done" Europe on a Grand Tour. He was the son of a literate serf employed in an aristocratic household and manumitted along with his wife and children in 1806. Pogodin's educational development entailed a deep immersion in German Romanticism through participation in the Society of Wisdom Lovers (1823–25), a discussion circle devoted to the study of Schelling, Kant, Fichte, Spinoza, and other German philosophers. After receiving his professorship in history at Moscow University in 1826, Pogodin became friendly with his colleague Stepan Shevyryov (1806–64), a poet and teacher of philology. In 1831 Shevyryov accepted Princess Volkonskaya's offer to come live in Rome as her son's tutor. Craving a similar opportunity, Pogodin started studying Italian. He yearned to visit Italy and entertained hopes of becoming the curator of the museum of West European art that the princess proposed establishing in Moscow in affiliation with the university.[18]

German Romanticism and the allure of Italy find expression in Pogodin's *Adele* (1832), a novella based on his experience as a tutor of Prince Ivan Dmitryevich Trubetskoy's daughter Alexandra. The hero of the story is twenty-five-year-old Dmitry, a struggling scholar and novelist who fancies himself one of the "elect" towering over the "crowd." A self-styled humanist, he "loves people" and life's "highest pleasures," the pleasures of the mind.

17 Victor Terras, *Belinskij and Russian Literary Criticism: The Heritage of Organic Aesthetics* (Madison: University of Wisconsin Press, 1974), 16–17, 66–67, 97–99; Walicki, *The Slavophile Controversy*, 426–31; Iurii Mann, *Russkaia filosofskaia estetika, 1820–1830-e gody* (Moscow: Iskusstvo, 1969), 267–75; and Joseph Frank, *Dostoevsky: The Years of Ordeal, 1850–1859* (London: Robson Books, 1983), 227–29.

18 Blakesley, "Art, Nationhood, and Display": 916–17, 929. See excerpts from Volkonskaya's proposal with commentary in N. P. Barsukov, *Zhizn' i trudy M. P. Pogodina*, 22 vols. (St. Petersburg: Tipografiia M. M. Stasiulevicha, 1888–1910), 3:176–82.

He falls in love with his seventeen-year-old tutee Adele whose heart he seeks to win before her parents put her on the high-society marriage market. Dmitry daydreams of their wedding and a honeymoon abroad:

> We will luxuriate in everything beautiful and bow to everything lofty—the tombs of Schiller and Herder, Schelling's lectures, Raphael's Madonna, that hoary old man German Literature, French palaces, shows, *la rue de la Victoire*, the London harbor, Walter Scott's tranquil abode, railroads, the Lancaster schools, sunrises in the Alps, evening strolls along Lake Geneva, the *Venus de' Medici*, eternal Rome, Etna, Virgil's laurel wreath, Leonardo da Vinci's *Last Supper*, the ruins of Pompeii, St. Peter's cathedral—so many, many sweet sensations![19]

Nor will they miss the haunts of Shakespeare, Rousseau's attic, or the dungeons that held Luther, Galileo, and Dante. After returning to Russia, Dmitry and Adele will live in a village along the Volga. He will pursue his research and continue educating his wife while she raises their children. To adorn their home they will amass a collection of "the finest portraits of great people and copies of the masterpieces of Sculpture and Painting so that everything meeting the eye in our house shall foster good taste and elevate the soul!" (220). None of this comes to pass. A sudden illness kills Adele, and Dmitry dies at her funeral after leaping into her coffin to kiss her at last.

A Year in Foreign Lands enacts much of the cultural tourist program of *Adele* but only after Pogodin frames his trip as a research project. The work theme prevails in volume one which maps linear movement from Moscow to Petersburg, Warsaw, Prague, Vienna, and Trieste, to end in a four-day visit of Venice. An early chapter thanks the Russian ministry of education for taking a kindly interest in the author's journey.[20] At the beginning of 1838 Pogodin requested a leave of absence from the university in order to go abroad to recuperate from overwork and renew contacts with Slavic philologists he had met in Prague, Germany, and Vienna on a research trip in the summer of 1835. After approving Pogodin's request, Uvarov summoned him to discuss his travel plans. Uvarov said he expected Pogodin to study antiquity in "world history and especially Russian history." Pogodin replied

19 M. P. Pogodin, *Povesti*, 3 vols. (Moscow: Tipografiia S. Selivanovskogo, 1832), 3:212.

20 Pogodin, *God v chuzhikh kraiakh*, 1:32 (hereafter cited parenthetically).

that he had in fact already drawn up a three-point plan: first, to look for traces of the Normans in Italy, northern France, and England; secondly, to try to determine whether the original inhabitants of the Vendée were Slavs; and thirdly, to research Peter the Great's activities abroad.[21]

Uvarov's official send-off in January 1839 probably helps account for the fact that *A Year in Foreign Lands* rarely shows Pogodin having a good time. He performs professorial labors most assiduously in Poland and the Czech lands. During the month he spends in those regions, he confers with scholars to discuss language, history, religion, and culture as unifiers of the Slavic peoples. He visits libraries and manuscript collections and acquires materials for his archive. The research theme thins out but persists in Western Europe. Pogodin studies Slavic manuscripts in the Vatican museum, he seeks out professors at the university of Bonn, works in the Bibliothèque nationale in Paris, and confers there with Guizot, whom he scolds for neglecting the Slavs in his histories (1:49–57, 78–87, 96–97, 112–24, 2:83, 3:104–6, 112–13, 4:34–35). Even when Pogodin is not pursuing research, he conveys the conviction that tourism is a politically serious business, making it incumbent on Russian travelers to create a positive impression of their homeland. In shocked reaction to hearing English tourists criticize England, Pogodin asserts that speaking ill of Russia before foreigners is utterly unthinkable (4:98–99).

Xenophobic Aestheticism

The research theme of *A Year in Foreign Lands* belongs to a larger, deluded authorial effort to live up to Karamzin's *Letters*. In addition to reinstating the Enlightenment pursuit of travel as discourse, Pogodin extends Karamzin's critique of the West, to deplore bourgeois materialism, poverty, and other social ills.[22] Pogodin likewise takes pride in frugality. He styles himself "an enemy of all luxury" in Paris and expresses "patriotic" disapproval of "Russian spendthrifts" who cannot resist the deluxe Hotêl de la Russie in Frankfurt (3:6, 4:50). Despite these parallels, Pogodin never notices how horribly he lacks the virtues Karamzin displayed as a "citizen of the world." Along with his savoir faire, command of languages, and knowledge of the arts and sciences of Western Europe, Karamzin's narrator displays tolerance

21 Barsukov, *Zhizn' i trudy*, 5:198–206.

22 Offord, *Journeys to a Graveyard*, 125–134. Offord also covers Pogodin's xenophobia and patriotism and remarks his self-projection as a frantic, "neurotic tourist," 123.

and courtesy toward foreigners, including Jews in the Frankfurt ghetto. He visits their slum-like quarter, talks to them, and sympathizes with their recent success in getting a local theater to discontinue a production of Shakespeare's *The Merchant of Venice* which they found anti-Semitic.[23]

A Year in Foreign Lands represents the xenophobic antithesis. Pogodin raves about art and architecture, including the cathedrals of Milan, Paris, Cologne, Rome, and London. What "amazing monuments Western man has left," he exclaims in Milan (4:175). On the other hand, he demeans most West European people outside his professorial milieu. The treatment of Italy most fully represents his xenophobic aestheticism. Pogodin and his wife (whose name he never mentions) arrive in Venice unable to speak Italian. This creates insecurity that leads the author to stereotype Italians as born thieves, cheating tourists at every turn. The Russian visitors enjoy sightseeing, gondola rides, and a performance of music and ballet at the Feniche. But the Italian spectators at the theater are "scum" (*svolochi*). A theme of denigrating the lower classes of Western Europe similarly assumes most virulent form in Pogodin's Italian section which labels the Naples *lazzaroni* and other people of the South "savages" and "vermin." *A Year in Foreign Lands* degrades Catholicism everywhere but especially during Holy Week in Rome. In short, the ancient monuments of Rome and the art throughout Italy make that country the aesthetic peak of the author's tour, but he finds most Italians unbearable (1:187, 193–95, 2:96–97, 158, 163, 171).

If Italy suffers most, other countries fall into the pattern of combining xenophobia and aesthetic tourist pleasure. In France, for instance, Pogodin genuflects at the Louvre, Versailles, and other standard attractions. In Paris street life, too, he perceives an appealing stylishness and gaiety, as in the sight of beautifully dressed children playing by fountains in the Tuileries as parents and nannies watch over them. Yet, none of the positive impressions outweigh his recoil from the French stereotyped as a materialist people incapable of serious thought. London has impressive Gothic architecture, but the English are wholly devoted to amassing money. Switzerland is beautiful, but the local shepherds appear "barely human" (3:5–8, 95–6, 176, 4:149). Jews come in for special abuse in Rome, Belgium, Switzerland, and Poland (2:33–34, 4:19, 129, 225). In addition to maligning local people, Pogodin expresses hostility to other tourists: a Pole ("an evil spirit") in St. Peter's, a garrulous Spaniard on the Rhine, a German "pedant" reading a

23 Karamzin, *Pis'ma russkogo puteshestvennika*, 89.

guidebook to his family in the Piacenza cathedral, and "most obnoxious" of all—the English overrunning Italy (2:50, 102, 4:31, 195).

But no foreign tourists disturb Pogodin as profoundly as the "cosmopolitan" Russians he encounters at the Baths of Caracalla in Rome. Amidst the international crowd of visitors, he hears somebody speaking Russian. To trumpet his national origins, he raises his voice in conversation with his companion (the unnamed Shevyryov). He expects a warm response from the Russian strangers, but they walk by without even looking at him (and no doubt had their own story to tell about this incident). Pogodin attributes the aloofness to a "cosmopolitan" sensibility that has destroyed "national and even human feeling." Cosmopolitanism has sown division where national solidarity should have prevailed and has even eradicated a "human" instinct. In the author's view, those strangers have ceased to be Russian. They strike him as "foreigners," alien to his "homespun patriotism" (*kvasnoi patriotizm*).[24] This encounter reduces Russianness to the urge to blubber an unknown compatriot's neck at the sound of the native tongue. Refraining from such behavior makes one "cosmopolitan"—unpatriotic and downright inhuman.

Disconnected from receptiveness to foreign art and monuments, this primitive form of the "cosmopolitan" gives patriotic Pogodin license to pursue his self-aggrandizing project of amassing cultural capital. In churches, cathedrals, the Louvre, the Uffizi, and numerous other museums, he seeks out dozens of art works, including all those mentioned in *Adele.* Sweet Guido Reni is his favorite, but he seems determined to cover the whole field of West European art. He goes to Antwerp, for instance, for the sole purpose of seeing Rubens's *Descent from the Cross.* Claiming that words cannot express art's impact on the viewer, Pogodin substitutes ineffable sentiment for connoisseurship and measures authorial sensitivity by the clock (a standard way of asserting superiority to the tourist crowd).[25] Zhukovsky wrote that he contemplated the *Sistine Madonna* for an hour. Pogodin outdoes that, to claim two hours looking at Tintoretto's *Crucifixion* in the Scuola Grande di San Rocco in Venice. In the Vatican museum,

---

24 *God v chuzhikh kraiakh*, 2:51. Vyazemsky coined the term *kvasnoi patriotizm* in 1827 as an equivalent of *patriotisme d'antichambre* (lackey patriotism): see commentary in Belinskii, *PSS*, 9:739n201. The standard translation of *kvasnoi patriotizm* is "jingoism," whereas my alternative seeks to capture the homey connotation of kvas.

25 On the travelogue trope of prolonged gazing at art works, see Buzard, *The Beaten Track*, 106.

writes Pogodin, he spent an hour looking at *Laocoön* and lingered so long before the *Apollo Belvedere* that his friends had to drag him away (1:197, 2:32, 67, 71).

A Year in Foreign Lands suggests the author finds cultural tourism enriching. In the end, though, he seems to recall the trope of the dying West. As he and his wife prepare to go home, they mentally erase the travel experience: "The images of everything we saw have vanished from our minds—Florence, Rome, St. Peter's, the Alps, tunnels, palaces, the Palais Royal, Raphael, plays and shows—it is as though none of it ever existed" (4:216). The author has done his Grand Tour but makes a show of holding it cheap. As his coach approaches Moscow, the sight of church domes elates him. This concluding symbol of "Mother Russia" closes the religious frame set at the outset with Pogodin's rapturous description of the cross crowning the angel atop the Alexander column in Petersburg's Winter Palace Square (1:4, 4:228). From start to finish, *A Year in Foreign Lands* upholds Orthodoxy as the bedrock of Russian spirituality, conceived as a national treasure that trumps the cultural wealth and everything else in the West. Rather than opening Pogodin's mind, his trip merely reinforced his homespun patriotism.

His actual homecoming brought him new rewards from the state, along with the shock of the advancement of the Westernizers. In his absence treasured colleagues in the history department had died or resigned, and the charismatic Timofey Granovsky (1813–55) was among their replacements. This state of affairs made Pogodin take to reading the *Apocalypse* as a guide to modern times, as he sought to re-anchor himself in the good graces of the authorities. Soon after his return, he submitted to Uvarov a report on his trip. Its sole aim was to represent foreign Slavs (especially in the Austro-Hungarian Empire) as oppressed minorities looking to Russia for liberation just as "the Magi had looked upon the star in the East." Uvarov commended Pogodin and passed the report on to Nicholas I who bestowed upon "our traveler" a gift of two-thousand rubles.[26] To reposition himself in the political environment at home, Pogodin also established and assumed editorship of *Moskvitianin* (*Muscovite*, 1841–56), a journal that enjoyed official support. The first issue featured an article by Shevyryov. A romantic poet in his youth, Shevyryov had evolved into a reactionary who

26 Barsukov, *Zhizn' i trudy*, 5:330–45.

now symbolized Western Europe as a putrefying corpse, a metaphor that became prominent in the repertoire of conservative Russian nationalism.[27]

Under these circumstances *A Year in Foreign Lands* met a poor reception when it finally appeared as a book after serialization in the *Muscovite*. For one thing, the account's penny-pinching cost-accounting provoked parodies from Herzen, the poet Nikolai Nekrasov (1821–78), and others.[28] But in the view of the liberal-minded, the larger problem was the travel account's reductive opposition between Russia and Europe. Melgunov sent Pogodin a letter telling him that *A Year in Foreign Lands* was not worthy of a serious historian. On the one hand, wrote Melgunov, we have "your juvenile, unrestrained enthusiasm for everything Slavic" and on the other—"your hypochondriac, peevish view of all the rest of the West" (the "sickness" of Western Europe was another theme of official nationalism that Shevyryov and others advanced in the early 1840s). "Damn it all," continued Melgunov, "can we really not love what's ours without taking a hostile view of the foreign?" Aligning Pogodin with rigid Slavophilism, Melgunov claimed they were all "unjust to the West and do not want to understand it." He concluded that Russia needs a "third, higher cohort" that can reconcile necessary elements of both the Slavophile and Westernizer tendencies so as to recognize that "Russia is Russia but at the same time a *part* of *Europe* and that we should not oppose ourselves to it."[29] Melgunov's opposition to binary thinking would characterize Maikov's cosmopolitan outlook as well.

Belinsky's Shifting Conception of Cosmopolitanism

The sentimental patriotism of Pogodin's journal the *Muscovite* provoked scathing rejoinders from Belinsky, the Westernizers' most intransigent publicist of the time. In the early 1840s Belinsky differed from Pogodin in construing cosmopolitanism in a positive or at least neutral manner. Eventually, however, Belinsky reverted to an earlier idea of the cosmopolitan that converged on Pogodin's and similarly implicated Russian tourism. Belinsky's first anti-cosmopolitan outburst belongs to his aberrant phase of political quietism. Having acquired mainly from Bakunin an acquaintance with Hegel's argument that the real is rational and the rational is real, Belinsky embraced the regime of Nicholas I as a historical necessity. The

27 Riasanovsky, *Nicholas I and Official Nationality*, 155.

28 Offord, *Journeys to a Graveyard*, 122–23n53.

29 Barsukov, *Zhizn' i trudy*, 7:387–88.

critic's 1839 review of the officer Fyodor Glinka's (1786–1880) patriotic recollection of the Battle of Borodino asserts that "the state is the highest expression of social life," the formation that makes a people a nation rather than a tribe. The state, continues Belinsky, has a "holy" dimension because "the social bonds" it establishes derive not merely from a "contract for the common good" but from "the flesh and blood" of its people—the natural "substance" that produces in the population a physical "family resemblance" which, along with language, religion, form of government, a shared cast of mind, and way of life, makes a nation distinct.[30] This glorification of the state leads to the following conclusion:

> That is why the cosmopolitan is a false, ambiguous, strange, and incomprehensible phenomenon, a pale, hazy specter rather than a vivid, living reality. That is why, for instance, that a Russian who passes his early childhood in Paris and acquires his first vital impressions of life in a country alien to his native essence (substance) appears an amphibian—freakish and repellent as all amphibians. That is why the person who believes *ubi bene ibi patria* is an immoral, heartless creature unworthy of being called by the holy name of man. That is why, finally, that a traitor to his fatherland, a turncoat to his homeland is a villain, the sight of which makes the human heart shudder, a villain from whom humanity recoils in revulsion. (331–32)

This passage of the review of Glinka's memoir targets Faddei Bulgarin (1789–1859), a Pole who fought with Napoleon and then switched sides, to make a comfortable life for himself in Russia as an unscrupulous journalist and government informer. Bulgarin's vicious public hounding of Pushkin led the latter to strike back in writings that applied to the turncoat Pole the Latin phrase Belinsky repeats: "wherever life's good is my homeland." With Bulgarin in mind, Belinsky makes "cosmopolitan" incompatible with loyalty to the state to which one belongs by "substance," regardless of birthplace.

But after renouncing his Hegelian perceptions of the "rationality" of Nicholas I's regime, Belinsky drew both Pushkin and the practice of literary criticism into the "cosmopolitan" sphere. Writing in 1843, Belinsky maintained that despite Pushkin's "deeply national motifs," the "spirit of

30 Belinskii, "Ocherki Borodinskogo srazheniia," *PSS*, 3:330–32.

cosmopolitanism permeates his poetry." Pushkin is "a cosmopolitan in the world, the phenomena of which are equally beautiful, equally interesting in his eyes [. . .] he loves everything and clings to nothing, he hates nothing and denies nothing."[31] In conformity to *homo sum: humani nil a me alienum puto*, "cosmopolitanism" signifies here the receptiveness of versatile genius freely ranging across space and time for inspiration. On the same wavelength, an article Belinsky published the following month articulated a supranational vision of Russian cultural progress. Clearly alluding to Slavophiles, he was polemizing against "sham patriots" who want to erect "Chinese walls" between the homeland and the West. Belinsky declared that "the grand idea of national individuality" has now yielded to "the even greater idea of mankind": "Nations [*narody*] are beginning to recognize that they are members of the great family of man and share among themselves as brothers the spiritual treasures of their own nationality." Thanks to translation, Russian literature, too, has entered into this supranational cross-fertilization that "can only enrich each literature."[32]

Belinsky's fifth essay on Pushkin (1844) complicates the relationship between the native and the supranational but strikingly retains "cosmopolitan" as self-description. Being a good critic, declares Belinsky, requires approaching literature in the manner of an open-minded traveler in a foreign land. In order to judge the foreign fairly, "you must temporarily forget you are a citizen of your own country and become a complete cosmopolitan." So, too, must the critic bring to every literary work a detached and open "cosmopolitan" mind.[33] Pushkin's versatility, however, soon acquires strong nativist inflections in Belinsky's interpeting him as Russia's preeminent "national poet," the writer who captures the totality of "Russian reality" and ennobles even its prosaic aspects through artistry. The transformative power of poetry extends to the natural environment that previous authors found bleak: "Pushkin did not need to go to Italy to find pictures of beautiful nature. Beautiful nature was here at hand in Russia, in its flat, monotonous steppes, beneath its

---

31 Belinskii, "Bibliograficheskie i zhurnal'nye izvestiia," *PSS*, 7:35–36.

32 Belinskii, review of *Istoriia Malorossii*, by Nikolai Markevich, *PSS*, 7:45, 49.

33 Belinskii, "Stat'i o Pushkine. Stat'ia piataia," *PSS*, 7:306–7. A similar idea of the "cosmopolitan critic" emerged in Victorian England: see Tanya Agathocleous, *Urban Realism and the Cosmopolitan Imagination: Visible City, Invisible World* (Cambridge: Cambridge University Press, 2011), 54.

perpetually gray sky, in its sad villages, and in its cities, rich and poor." This assertion challenges the romantic cult of Italy and the country's allure to Russian tourists. Instead of flocking to Italy, implies Belinsky, Russians should learn to see their homeland through Pushkin's eyes. As though to reconcile the "cosmopolitan" and "national," the essay closes with a long quote from Gogol's well-known interpretation of Pushkin's "artistic multi-sidedness" taken as the evidence of a putative Russian national gift for assimilating other cultures.[34] The very versatility Belinsky had marked as "cosmopolitan" in 1843 thus becomes a distinct Russian "national" capacity.

Belinsky's eighth Pushkin essay (1844), devoted to *Eugene Onegin*, extends this argument, while portraying leisure travel as a practice counterproductive to Russian national self-definition. Belinsky observes that westernization has produced elite Russian "crazes" for foreign lands, especially England, France, and Germany. At home the Russian Anglophile lives in comfort, eats roast beef and pudding, drinks port, lays out an English garden, and has English horses, jockeys, and grooms. The Russian Francophile is a haughty fashion plate, speaks French like a Parisian, and prizes wit. The aficionado of Germany fixates on art and dreamy philosophizing. Whatever the Russian's craze may be, his home life prepares him for passing as a local abroad. Belinsky maintains that Germans and Englishmen radiate national identity wherever they go. On the other hand, Russians (as well as Frenchmen) travel without having "a passport stamped on their foreheads." While requiring an adequate command of a foreign language, being taken for a local is mainly a matter of the visual impression one creates "on the street, in a tavern, at a ball, in a coach." Belinsky favorably interprets the Russian tourist masquerade. Paraphrasing the Russian army captain Maxim Maximych speaking of Caucasian mountaineers in Lermontov's *A Hero of Our Time*, Belinsky argues that Russian tourists' success in passing for natives in the West represents a distinct national strength: the "amazing ability" to assimilate the ways of foreign peoples.[35] But is the theatricality of tourism not open to a different interpretation? Given the history of Russia's cultural and intellectual tutelage to the West, might tourist experimentation with identities not betray a national inferiority complex and alienation from the native realm?

34 Belinskii, "Stat'i o Pushkine. Stat'ia piataia," 331–32, 333–36. On the passage mentioning Italy, cf. Ely, *This Meager Nature*, 90.

35 Belinskii, *PSS*, 7:437–38.

The Polemic with Maikov and Its Aftermath

Incertitude about Russian "multi-sidedness" would emerge in Belinsky's clash with Maikov on the question of cosmopolitanism. A narrative poem by Valerian's brother Apollon appears to have influenced Belinsky's thinking. Apollon's *Two Fates* (1845) is set in Italy and centers on the Russian tourist Vladimir, whose social standing and profession are left to the reader's imagination. Vladimir has loved discovering scenery and art in Tuscany, Venice, and Rome. He has also enjoyed mingling with British, German, and French tourists. Russians, too, are thronging Italy, but Vladimir has learned to give them the cold shoulder. In an ironic reminiscence of the Baths of Caracalla incident in Pogodin's *A Year in Foreign Lands*, the *Two Fates* poet remarks:

> Как сладко нам среди чужих наречий
> Вдруг русское словечко услыхать!
> Так рад! Готов, как друга, ты обнять
> Всю Русь святую в незнакомой встрече!

> How sweet it is for us to hear amidst foreign tongues
> The unexpected sound of dear Russian!
> What joy! You are ready to embrace as a friend
> All Holy Russia in a stranger![36]

In actuality, Vladimir has usually found his traveling compatriots repugnant: they are either boorish nationalists who hate everything foreign or else pretentious Italy-lovers who disparage Russia and Russian culture. Having come abroad in a fit of Byronic spleen, Vladimir initially found Italy so enthralling that he thought about emigrating. After a long sojourn, however, he yearns to go home, as he explains to an acquaintance he encounters in Rome, a Russian count traveling with a Russian civil servant. The count calls himself a "cosmopolitan," the only option, he argues, for "a truly sensible person." Cosmopolitanism means acquiring new knowledge and honing refinement through an acculturation tour of Western Europe, especially Italy. The count idealizes Byron whom he regrets is no longer around to conduct him to the Colosseum at night. Cosmopolitanism also means identifying with human progress wherever it occurs in the struggle

36 A. N. Maikov, *Dve sud'by*, in *Polnoe sobranie sochinenii*, 4 vols. (St. Petersburg: A. F. Marks, 1914), 3:39.

against backwardness. Challenging the count, Vladimir and the civil servant speak up for home. Both are glad to have seen Italy but now love Russia all the more. They articulate a romantic nationalist view of the homeland as a realm of vast grainfields, wide rivers, primeval forests, and the idyllic country estate with its peasant huts. The count is appalled: "Having no sooner left their barbaric land, / The Scythians weep over the separation!" How strange, he adds, that the homesick Russians have not discarded their "patriotic ideas abroad" (49). Defiantly accepting "Scythian" as self-definition, Vladimir parts with the count and continues savoring his own inner vision of Russia. Belinsky's review of *Two Fates* found fault with the versification but wholeheartedly recommended the poem for dramatizing a question vital to Russian development: Who is right? The cosmopolitan or the nationalist?[37]

Belinsky's polemic against Valerian Maikov's "sensible cosmopolitanism" would opt for nationalism. Maikov's targeted essay concerns Alexei Koltsov (1809–42), a poet whose portrayals of Russian rural life reflected his upbringing and labors in a peasant family of cattle traders in the Voronezh region. As a confirmed disciple of Herder and Schelling, Belinsky subscribed to the Right Hegelian conviction that "the progress of humanity is realized in national endeavors and not in any cosmopolitan movement." Maikov, on the other hand, regarded nationalism as an "inertial force" inhibiting a culture's development of geniuses of world significance. This position represented "the Left Hegelian idea that human progress lay with a gradual elimination of the *particular* (that is, national), and the emergence of *universal* traits of human civilization."[38] One should add, though, that in parallel to Chaadaev's belief in the "absolute universality" of Christianity, so too was Maikov's conception of "universality" Christian and Eurocentric.[39]

Maikov's meandering analysis of Koltsov's poetry begins with a protest against binary thinking. All too often, asserts Maikov, observers pin reductive labels on people with strong opinions and then pit them against their putative opposites. This yields rigid dichotomies such as "the blind patriot" versus "the listless cosmopolitan," "the dull industrialist versus the indecent romantic, and so on and so forth." Maikov ultimately aims to show that cosmopolitanism should be sensible rather than listless. National categories, he argues, create artificial divisions between people, box them in,

37 Belinskii, *PSS*, 8:635–42.

38 Terras, *Belinskij and Russian Literary Criticism*, 98.

39 Chaadaev, "Letters on the Philosophy of History," 171.

and stunt their development. Starting from the simple premise that people are first and foremost human beings, sensible cosmopolitans strive to develop an impartial attitude towards their own country and seek to spread Christianity's universal ideals. The sensible cosmopolitan experiences kinship with peoples, natural environments, and cultures outside his native realm. The foreign is different but not alien. To illustrate sensible cosmopolitanism in action, Maikov looks to Italy (a choice that may take a dig at Belinsky's argument about Pushkin's not needing to go there). Maikov imagines how a painting of Italian scenery could motivate a Petersburg viewer to take a trip to Italy in order to experience the reality that inspired the artist. The exhibition visitor wants to feel that sunshine and taste those grapes. The journey teaches the Russian that Italy is not alien because nature everywhere is kindred to human beings (a strange silence here about *Italians*).[40]

Belinsky alludes to Maikov's essay in an 1847 article that set up a dichotomy between "the fantastic national identity" that Slavophiles have manufactured and "a fantastic cosmopolitanism in the name of humanity." In Belinsky's view, Slavophiles' "mythical" vision of pre-Petrine Russia prevents them from understanding the West and finding solutions to real problems. Slavophiles have nonetheless made their opponents start wondering whether Russian "multi-sidedness" is actually a weakness, demonstrating that westernization has deprived "us of a clearly manifested national character of the sort that, to their credit, distinguishes nearly all the peoples of Europe." In contrast to his earlier description of Russian travelers confidently passing for locals in the West, Belinsky now observes that trips abroad can rudely impress upon Russians their "lack of [national] character." Being abroad, argues Belinsky, makes Russians see that "our Russian Europeanism has absurdly, pathetically put our minds at rest concerning our Russian deficiencies, powdered and rouged but not eliminated." The travelers abroad learn they are neither genuinely European nor distinctly Russian. They go home confused, "not knowing who to take as a model, despite their sincere wish to become Russian."[41]

Maikov's advocacy of cosmopolitanism adds urgency to the task of solving what Belinsky terms "the enigma of Russian national spirit." Up to now, contends Belinsky, military triumphs and the formation of a strong state have demonstrated Russia's claim to be a great nation, a

40 V. N. Maikov, "Stikhotvoreniia Kol'tsova," *Otechestvennye zapiski* 49, nos. 11–12 (1846), section 5:1, 18–19, 39–43, 55.

41 Belinskii, "Vzgliad na russkuiu literaturu 1846 goda," *PSS*, 10:17–20, 25.

people above the historically insignificant level of a tribe. In intellectual and cultural spheres, too, Russia "will say its word" and has even begun to do so in literature. But the exact nature of the "word" remains a mystery that "our grandsons and great-grandsons" will solve. The revelation lies ahead but is bound to come, to assure Russia its rightful place in the family of nations. Belinsky then reversed his position on the relative importance of what he had so recently termed the "great idea of mankind," as opposed to the nation:

> What *individuality* is to the *concept* of a person, *nationality* is to the *concept* of humanity. [. . .] Without nationalities humanity would be a lifeless logical abstraction, a word with no content, a meaningless sound. In this respect, I am more ready to side with the Slavophiles than to stand with the humanist cosmopolitans because if the former make mistakes, they do so as people, as living beings, whereas the latter make even the truth sound like an abstract promulgation. But fortunately I hope to remain where I am, without taking either side.[42]

Belinsky's concluding remark expresses "a plague on both houses" attitude that would intensify in both directions. His mounting intolerance for Slavophile "myths" would explode in his "Letter to Gogol" (1847), provoked by the author's *Selected Passages from Correspondence with Friends* (1846), a book that incurred official disfavor as well because of its reactionary, mystical rejection of the Petrine foundations of political and economic development under Nicholas I. On the other hand, just a month after his polemic with Maikov, Belinsky resumed his attack on cosmopolitanism in response to an article by E. I. Guber, a feuilleton writer, poet, and Russia's first translator of Goethe's *Faust*. Guber thought Russian literature was overdoing the lowly clerk theme as exemplified in Dostoevsky's *Poor Folk* (1845), the novel Belinsky had hailed as a masterpiece of the new naturalist tendency exposing the misery of Russia's urban underdogs and the peasantry. In Guber's opinion, art should be "cosmopolitan" to avoid falling into a rut: poor clerks are an important topic, but Russian writers need to follow imagination wherever it leads them, to ancient Rome or Greece, for instance. Guber's proposition resembled the "cosmopolitan spirit" Belinsky once attributed to Pushkin's œuvre. Now, however, Belinsky asserted that

42 Ibid., 21, 28–29.

great art is never cosmopolitan: it is "not a vagrant without a home or country" but rather the product and developer of a national culture.[43]

In speaking of art, Belinsky nearly always meant literature, and he had next to no interest in painting or sculpture. While abroad seeking treatment for tuberculosis in 1847, Belinsky, who spoke no foreign languages, visited the Dresden picture gallery with Annenkov as his guide. The last article Belinsky published shortly before his death in 1848 contained his belligerent response to the *Sistine Madonna*. Although granting the work's "nobility" and "elegance," he interpreted the painting through a lens of social inequality:

> This is not the mother of the Christian God. This is an aristocratic woman, a tsar's daughter, the *ideal sublime du comme il faut* [the fashionable world's ideal of the sublime] [. . .]. She looks at us with cold indulgence, simultaneously wary of being sullied by our gaze and of offending us plebes, should she turn away from us. The infant she holds in her arms is more frank. There is a trace of arrogance in his curled lower lip, and his whole mouth exudes contempt for us *racaille* [riffraff]. In his eyes one sees not the future God of love, peace, forgiveness, and salvation, but rather the ancient, Old Testament God of wrath, fury, punishment, and retribution.[44]

A déclassé intellectual whose father was a naval doctor, Belinsky writes here with a chip on his shoulder, to render the museum not as a storehouse of treasures of universal value, but rather as a socially divisive institution shoring up barriers between the cultural elite and commoners such as himself—"us plebes." Whereas Pogodin's lowly social origins appear to have helped fire his passion for West European art as cultural capital, Belinsky veered in the opposite direction, to construe art appreciation as an upper-class luxury, irrelevant to Russia's political struggle. As he put it in responding to a friend's 1846 article about visiting the art collection in Vienna's Belvedere palace, publications on "such topics as painting" make

43 Belinskii, "Sovremennye zametki" (1), *PSS*, 10:92.

44 Belinskii, "Vzgliad na russkuiu literaturu 1847 g.," *PSS*, 10:308–9. A similar response figures in the letter to V. P. Botkin, July 7/19, 1847, ibid., 12:384. See discussion in Irene Pearson, "Raphael as Seen by Russian Writers from Zhukovsky to Turgenev," *Slavonic and East European Review* 59, no. 3 (July 1981): 359–60.

"strange" reading these days, and "many people think so," he combatively added.[45]

After returning to Russia, Belinsky reasserted his objections to cosmopolitanism in philosophical rather than aesthetic terms in a letter to the liberal journalist and Moscow University professor of jurisprudence Konstantin Kavelin (1818–85):

> I can't stand exalted patriots with their constant effusions about kvas and kasha. Embittered sceptics I find a thousand times better because hatred is often just a peculiar form of love. I confess, though, that I find pathetic and disagreeable the serene sceptics, abstract people, passport-less vagrants roaming humanity [*brodiagi v chelovechestve*]. However hard they try to convince themselves that they're living in accord with the interests of some country or other which, in their opinion, represents humanity, I don't believe in those interests. [. . .] For me, Peter [the Great] is my philosophy, my religion, my revelation of everything concerning Russia. He sets the example for everybody, large and small, who wants to do something and be useful.[46]

In upholding Peter as inspiration for the abolition of serfdom and other reforms he wished for Russia, Belinsky allied himself to Enlightenment cosmopolitanism as opposed to the aesthetic, Byronic brand. The most likely exemplar of the "passport-less vagrants" was Bakunin. Their friendship had soured, and Belinsky had just encountered him in Paris in 1847. Doomed never to achieve anything useful, Bakunin was on the move, eloquently professing in several languages "his love for humanity in the abstract," yet ready to "wade through seas of blood" for the revolution.[47]

---

45 Belinskii, Letter to P. N. Kudriavtsev, May 15, 1846, *PSS*, 12:281. See also P. N. Kudriavtsev [Nestroev, pseud.], "Bel'veder (iz putevykh zapisok Russkogo)," *Otechestvennye zapiski* 45, no. 3 (1846), section 8: 14–42.

46 Belinskii, *PSS*, 12:433. My digital search of Belinsky's collected works established that, contrary to a widely spread assumption, he did not coin the phrase *bezrodnyi kosmopolit* ("homeless" or "rootless cosmopolitan"). It is true, however, that this formulation about nationally adrift, "passport-less vagrants" was harnessed to increasingly anti-Semitic Soviet anti-cosmopolitanism under Stalin as of the late 1940s, as in Z. Papernyi, "Perechityvaia Belinskogo," *Literaturnaia gazeta*, June 5, 1948, 1. My thanks to Jan Adamczyk for this reference.

47 Berlin, "Herzen and Bakunin on Individual Liberty," in *Russian Thinkers*, 113.

Turgenev's historical novel *Rudin* (1856) points to the Bakunin connection in Belinsky's late thinking about cosmopolitanism. The story is set in the 1840s, the time of Russia's Hegel craze and "superfluous men," the frustrated nobles who find no constructive outlet for their talents under Nicholas I. Bakunin is generally regarded as the prototype of Rudin, a superfluous man type and philosophizing sponger.[48] At the outset of the novel Rudin arrives at the country estate of an acquaintance. At social gatherings his rhetorical flights on the soul, love, and humanity awe most of his listeners. Natalya, the young lady of the manor, falls in love with him but privately puts him to shame when she sees he is a phony, not prepared to make her his life partner in glorious, progressive endeavors. After Rudin flees the estate, the character Lezhnyov (who will marry Natalya) denounces him as a spineless windbag. Lezhnyov asserts that Rudin's "bitter, hard fate" is that he does not know Russia: "Cosmopolitanism is rubbish, the cosmopolitan is a zero, worse than a zero. Outside nationality there's neither art nor truth nor life nor anything."[49] The observation paraphrases Belinsky's polemic against Maikov. The collapse of Belinsky's friendship with Bakunin also resonates in the *Rudin* passage (Turgenev knew them both well, saw them in Paris in 1847, and was privy to their disagreements). Lezhnyov's anti-cosmopolitan position is not the author's but rather a reminiscence of the intellectual ferocity of Belinsky, a departed friend who Turgenev would affectionately recall in a memoir.

Cherchez la femme

Belinsky's epistolary denunciation of "abstract" servants of humanity closely resembles the attack on cosmopolitanism in Zagoskin's *Moscow and the Muscovites*, the collected feuilletons narrated by the fictional nobleman Bogdan Ulich Belsky, a patriot up in arms against Russian subservience to Western Europe. A section of the book that Belinsky favorably reviewed consists of Belsky's harangue against the introduction of

48 Marshall Shatz, "Bakunin, Turgenev and *Rudin*," in *The Golden Age of Russian Literature and Thought: Selected Papers from the Fourth World Congress for Soviet and East European Studies, Harrogate, 1990*, ed. Derek Offord (Basingstoke: Macmillan Press, 1992), 103–14; and John Randolph, *The House in the Garden: The Bakunin Family and the Romance of Russian Idealism* (Ithaca: Cornell University Press, 2008), 277

49 I. S. Turgenev, *Sobranie sochinenii*, 12 vols. (Moscow: Khudozhestvennaia literatura, 1975–79), 2:110.

foreign words into Russian.[50] Belsky's pet peeves include "civilization." We do not need this French import, he argues, because Russian *prosveshchenie* (enlightenment) means the same thing. Not merely redundant, this linguistic borrowing allegedly encourages Russians to construe "civilization" as something exclusively foreign. Are we to assume, asks Belsky, that "Russians will become an enlightened people" only when they have turned into Frenchmen, Germans, or Englishmen, only when they "sling mud at everything native and speak a corrupted, half-Russian tongue? [. . .] If cultivated Germans, Englishmen, Dutch, Swedes, and even Italians (who bear no resemblance whatsoever to the French) all have the right to call themselves enlightened people, then why do Russians not merit the same title, without losing their national particularity?"[51] The word "cosmopolitanism" strikes Belsky as another import undermining Russian national pride. Self-defined "citizens of the world," he asserts, "seek to justify their strange citizenry through love of all humanity. Therefore, the *gumannyi* [humane] person and the cosmopolitan are nearly the same thing." It turns out, however, that there are two kinds of cosmopolitanism, or "humaneness" (*gumannost'*):

> One is expressed not in words but in deeds, while the other is mostly limited to eloquent verbiage. The one has its foundation in God's law, the other rests on enlightenment and philosophy, and experience has repeatedly shown us how fragile and unreliable those foundations are. The love of all humanity—what a beautiful, lofty sentiment! In the spiritual sense, it is perhaps the primary Christian virtue, but let as look at it in practice in our mundane, sinner's world. The cosmopolitan, that is—the "humane" person, does not want to hear about restricted love; he loves everybody in equal fashion and concerns himself not with the personal welfare of one generation or people, but rather with the universal, general welfare of the whole human race. It is not enough for him to be the good father of a family, a useful member of society, a loyal son of his homeland—he is above all that. He is a citizen of the world. All this is beautiful, sonorous, and magnificent on paper but in actuality almost always serves

50 Belinskii, "Sovremennye zametki" (2), *PSS*, 10:186–87.

51 M. N. Zagoskin, "Pervoe maia" (1848), *Sochineniia*, 2 vols. (Moscow: Khudozhestvennaia literatura, 1987), 2:422–24.

> as a specious pretext for not thinking about anybody's welfare except one's own.[52]

Knowingly or not, the cranky Gallophobe Belsky is articulating a Rousseauesque view of the cosmopolitan as a shirker of personal, family, and national responsibilities.[53]

While not given to philosophizing, the Russian noblewomen of *Moscow and the Muscovites* actively participate in cosmopolitanism's "unconditional love of the *West*" (Zagoskin's italics serving here to convey emotional intensity).[54] Belsky is a bachelor who failed to find a soul mate. He wanted to marry somebody who "did not resemble a French mademoiselle, a German Fraulein, or English miss, but was simply an educated, cultivated young *Russian* noblewoman who loved her fatherland [*otechestvo*], its language, and even its customs." In Moscow society he met young ladies who shared his interest in music, painting, literature, and history. They read "all the French poets," *La Nouvelle Héloïse*, Edward Gibbon, Jacques Bénigne Bossuet, and Barthélémy's *Voyage du jeune Anacharsis en Grèce*. But alas, those women refused to speak Russian, sneered at Russian artists, and took no interest in "our native writers" or Russian history. They pined for Italy and Paris, and when they actually had such a trip in store, they become "utterly unapproachable." This experience led Belsky to seek a wife among the provincial nobility. The typical provincial miss "praises everything that is ours." She loves dashing through snow in a troika, she follows folkways of fortune-telling, and she curses the French, especially Napoleon. She speaks Russian, to be sure, but just because she knows no other language. She does not beg to go to Italy, but only because she has never heard of it. After a couple of meetings, such gaps in her education make her unbearably boring.[55]

As though to take revenge for his disappointments with women, Belsky stereotypes Russia's fair sex as status-seeking spendthrifts crazy about foreign travel, the fashionable world's version of cosmopolitanism in action. Evidently annoyed by Myatlev's mockery of his novel *Homesickness*, Zagoskin blames "the famous Madame Kurdyukova" for creating a tourist epidemic among provincial Russian women as well as those of the capitals.

---

52 Ibid., 424–25.

53 On Rousseau, consult Sophia Rosenfeld, "Citizens of Nowhere in Particular," in Delanty and Inglis, *Cosmopolitanism*, 1:192

54 M. N. Zagoskin, "Literaturnyi vecher" (1847), *Moskva i moskvichi. Zapiski Bogdana Il'icha Bel'skogo* (Moscow: Moskovskii rabochii, 1988), 185.

55 Zagoskin, "Moskovskii starozhil" (1842), *Sochineniia*, 2:382–83.

Responsible husbands rule the roost, to keep their wives at home. But just let the man die, and "the inconsolable widow" rushes abroad to assuage her grief, first along the Rhine and then in Paris where she will stay two years.[56] In the process, of course, the fool exhausts the inheritance that her hard-working spouse left her.

A similar scenario plays out in Zagoskin's *A Trip Abroad* (1850), a drama set in Moscow. The main protagonist is Marya Oborkina whose feminine wiles wear down her much older husband Andrei's resistance to her travel plans. Inspired by George Sand's novels, Marya protests against the tyranny of her spouse and whines that if he really loves her, he will take her abroad. When all else fails, she persuades a doctor to certify she has signs of tuberculosis requiring treatment in Western Europe. Marya conceives her trip as social climbing. She wants to see her name in the newspaper on the list of worthies going abroad, and she wants to make her "friends" green with envy.[57] Even before Andrei has made his decision about the journey, she outlines the itinerary to an acquaintance who has just announced her own upcoming trip to Carlsbad, Dresden, and Switzerland. Marya first dismisses Carlsbad as a place now so "vulgar" (*poshlyi*) that one might as well go to Lipetsk (an old spa long past its heyday), Voronezh, or Penza (two stereotypical provincial towns). Andrei and I, she continues, will be spending at least three years abroad, beginning on the Rhine, then Switzerland, northern Italy, Lago Maggiore, Venice, winter in Rome, several months in Paris, then back to Italy, to Naples, and perhaps Palermo. Fooled by the medical certificate, Andrei agrees to the tour. Marya packs the minimum because she plans to buy a new wardrobe in the West. She will not be caught hauling "Russian rags and trash abroad." She and Andrei tour Europe for two years, during which time his worst financial nightmares come true. Their estate goes to rack and ruin, and the conniving wife returns home genuinely sick with dropsy.[58] In their reckless pursuit of the status symbol of a Grand Tour, the Russian women of *Moscow and the Muscovites* enact a craze for the West that makes them the feminine counterparts of the philosophizing cosmopolitan men that Zagoskin subjects to scornful Rousseauesque critique.

56 Zagoskin, "Literaturnyi vecher," 185–86.

57 On the newspaper practice of publishing the names of Russians going abroad, see Dostoevskii, "Peterburgskaia letopis'" (1847), *PSS*, 18:21.

58 Zagoskin, *Poezdka za granitsu*, in *Moskva i moskvichi*, 319–42, esp. 319–20, 324–26.

Dismal Years Ahead

As a companion to the preceding discussion of Byron's rise as a romantic travel authority in Russia, this chapter has underscored Karamzin's 1818 misjudgment of the cosmopolitan as a species too "metaphysical" to discuss, either "to praise or condemn." To the contrary, cosmopolitanism aroused new controversy as Russian leisure travel expanded in the era of official nationalism. On the one hand, Myatlev's *Sensations and Observations* and Annenkov's *Letters from Abroad* mapped cultural tourism as a cosmopolitan adventure on the elitist Byronic template, a project that entailed depreciating and in Myatlev's case lampooning demonstrative Russian patriotism. Annenkov alone explicitly embraced cosmopolitanism as a tourist experience, but Myatlev's yokel story endorsed the same sensibility through parody. Maikov articulated compatible values in imagining a Russian traveler's delighted discovery of Italy, presented as an expression of "sensible cosmopolitanism," an outlook he pitted against parochial nationalism.

On the other hand, anti-cosmopolitans perceived Russian tourism as a threat to national solidarity. The anxiety was primarily a conservative nationalist phenomenon, but as Belinsky's contributions illustrate, the discussion did not fall neatly along the Slavophile-Westernizer divide. In fashioning himself as a patriot at the Baths of Caracalla, Pogodin identified "cosmopolitan" Russian visitors as "foreigners," "strangers" whose emotional attachment to their country had snapped or perhaps never existed. Belinsky, for his part, conjured the Grand Tour circuit as a stage, where the westernized Russian elite, accustomed to living abroad at home, grew more incapable than ever of defining and joining the "authentic" Russian fold. Rather than contemplating the process, Zagoskin simply laid down the law, to declare the Russian mania for foreign travel conclusive evidence of cosmopolitan contempt for Russian culture. In performing anti-Byron travel norms of patriotism and the pursuit of Enlightenment type work, Pogodin most explicitly contested hedonism. However, Belinsky and Zagoskin, too, both disapproved the pleasure principle. Unable to accept leisure travel as a politically innocent adventure, they shared Pogodin's perception of vacationing as idle flight from the responsibilities of Russian nation-building, either in opposition to the West for Pogodin and Zagoskin, or in Belinsky's case, on the foundation of Peter the Great's achievements.

A controversy about tourism as a cosmopolitan pursuit had arisen in Russian print but was squelched during the last seven "dismal years" of

Nicholas's reign. The 1848–49 revolutions in Paris, northern Europe, and Piedmont panicked the Russian state into intensifying tyranny at home. Between 1848 and 1855 "censorship terror" paralyzed intellectual life to such an extent that public discussion of politics, philosophy, and religion became virtually impossible. Journals retreated into compiling bibliographies and airing safe, trivial subjects.[59] Excluded from serious discussion, cosmopolitanism was trivialized. An anonymous feuilleton series in *Syn otechestva* (*Son of the Fatherland*) featured Frants Frantsovich Cosmopolitsky (his first name and patronymic are based on *Frantsiia*, the Russian for France). Cosmopolitsky is a self-satisfied Petersburg mediocrity who speaks some French and enjoys intimidating naïve Russian provincials newly arrived in the capital.[60] "Cosmopolitan" was also the name of a French poodle in *Fantaziia* (1851), a frothy little play by the popular fictional poet Kozma Prutkov, the invention of the poet and prose writer Alexei Tolstoy (1817–75) and his cousins the Zhemchuzhnikov brothers Alexander and Vladimir.[61]

Many major writers fell afoul of the authorities during these dismal years. In 1849 Dostoevsky was arrested, subjected to mock execution, and sent to a Siberian prison labor camp for his participation in the Petrashevky political discussion circle. In 1852 Turgenev was arrested, locked in a guardhouse, and banished to his country estate for over a year for having published an obituary of Gogol. Evoking Boccaccio's *Decameron*, Turgenev would recall the period as a plague for educated Russian society. As Bruce Lincoln underlined, "Turgenev had a point." Beginning in 1848, "men who had been (and would again become) some of Russia's most sensitive and creative writers, abandoned their nightly discussions of social problems relating to Russia's past, present, and future, and instead devoted the evenings to reading and writing erotic verse, to drinking, cards, and prostitutes." There was likewise a surge in the popularity of pornographic literature among the intelligentsia.[62]

The revolutionary disturbances of the late 1840s also led the Russian state to impose draconian foreign travel restrictions that lasted throughout the Crimean War. Many Russians, including Khomyakov, yearned to

59 This discussion of the "dismal years" closely follows Lincoln, *Nicholas I*, 296, 316–24.

60 Anon., "Pokhozhdeniia provintsiala v stolitse," *Syn otechestva* 37, no. 5 (May 1850), section 7: 50–63.

61 Koz'ma Prutkov, *Sochineniia* (Moscow: Khudozhestvennaia literatura, 1960), 175.

62 Lincoln, *Nicholas I*, 323.

go to the 1851 Great Exhibition in London, particularly to see Joseph Paxton's glass and iron Crystal Palace.[63] Instead, the vast majority of Russians had to content themselves with newspaper reports (often based on foreign sources) about the exhibition and its architectural symbol of modernity. Nevertheless, as Zagoskin's conniving Marya Oborkina illustrated, Russians could beat the system if they obtained a medical attestation or simply had good connections. An erstwhile participant in the Russian army's assault on Chechnya, the Russian officer Boris Gersevanov published in 1848 an account of his recent rambles in the Swiss Alps and northern Italy. Declining to follow the implicitly Karamzinian practices of the "learned traveler" (*obrazovannyi puteshestvennik*), Gersevanov contentedly identified himself as a "mere tourist" (*prostoi turist*), a *flâneur* seeking "pleasure every step of the way."[64] Other escapees were Annenkov, back in Paris in 1848; Turgenev who spent most of 1849–50 in the West; and Vyazemsky who went to Carlsbad for a cure and visited Switzerland, France, and Italy with his wife in 1853-54. They stayed four months in Venice in a rented apartment with a garden and a balcony on the Grand Canal. Vyazemsky said his wife spent most of the time sitting on the balcony, while he merrily plunged into *le métier de touriste* (the business of being a tourist), with the aid of French and Italian guidebooks.[65] Pogodin, too, went abroad in 1853–54, the fifth of his six trips to Western Europe. The travel letters he published at the time even advertised what a pleasant time he was having. In continuity with his glowing representations of train travel in England, France, and Belgium in *A Year in Foreign Lands*, Pogodin's travel reports of the early 1850s feature him dining with champagne on an "enchanting" train from Strasbourg to Paris. In the French capital he eats at the "celebrated Provençal Brothers" restaurant and goes to the Louvre to see the *Venus de Milo* and paintings by Correggio,

63 On Russian newspaper coverage of the exhibition and the Crystal Palace as symbol of modernity, see Dianina, *When Art Makes News*, 40–57; and Sarah J. Young, "The Crystal Palace," in Martinsen and Maiorova, *Dostoevsky in Context*, 176–84.

64 B. N. Gersevanov, "Iz putevykh vpechatlenii turista. Pereezd cherez Simplon," *Otechestvennye zapiski* 57 (1848), section 8: 5 and 10. Gersevanov published several other accounts of his tourist adventures in Western Europe.

65 P. A. Viazemskii, *Staraia zapisnaia knizhka*, in *Polnoe sobranie sochinenii*, 11 vols. (St. Petersburg: M. M. Stasiulevich, 1878–87), 10:5–44 (quote 32). Among Vyazemsky's guides was A. Quadri, *Huit jours à Venise* (1821, 1831, 1842): see Nina Kauchtschischwili, *L'Italia nella vita e nell'opera P. A. Vjàzemskij* (Milan: Società editrice vita e pensiero, 1964), 118–20.

Raphael, Murillo, and Leonardo. He roams the Jardin des plantes and visits Versailles, a symbol of prerevolutionary France for which he has warm words: "I love Versailles with its magnificent galleries, wide hallways, endless passages, gala staircases, big terraces, trimmed walkways, diverse copses, hermitages, ponds, and fountains, the Versailles of Louis XIV who seems still alive, gazing out a little window." In glorifying the French palace and gardens, Pogodin endorsed an *ancien régime* similar to the one still existing in Russia. His letter concluded with an ebullient account of visiting Fontainebleau and Bayeux to see the tapestry commemorating the Battle of Hastings.[66]

Pogodin's celebration of tourist pleasures in France was strangely out of synch with the onset of the Crimean War. In 1844 Nicholas I had dubbed the author of *A Year in Foreign Lands* "our traveler." Were the authorities of the early 1850s letting Pogodin assume the public role of the homeland's token tourist abroad, as though to suggest Russians were free to go wherever they liked? One can only wonder. But the well-known truth behind any such a charade was that the authorities shut the gates to the West for most Russians in 1848. The restrictions hit teachers and students particularly hard, and officials even began summoning home Russians already abroad.[67] Having left Russia with his family in 1847, Herzen received such an order in Paris but took the cue to emigrate, get his fortune transferred from Russia with the assistance of Baron James Rothschild, and pay to become a naturalized citizen of the Swiss canton of Fribourg.[68] In clamping down on trips to Western Europe, the state gave a fillip to domestic tourism, particularly in the North Caucasus, a territory that Romantic literature had made so enticingly exotic to Russian readers.

66 Pogodin [M. P.], "Otryvki iz zagranichnykh pisem," *Moskvitianin*, no. 1 (January 1854), section 7: 9–11 and 13–18 (quote 13).

67 Riasanovsky, *Nicholas I and Official Nationality*, 218; and Lincoln, *Nicholas I*, 300.

68 Malia, *Alexander Herzen and the Birth of Russian Socialism*, 389–90.

4

Vacationing in the Caucasus: Authenticity and the Sophisticate/Provincial Divide

The tourist allure of the West provoked the defensive opinion that Russians should see their own country first. Catherine the Great purportedly articulated that view, and it gained momentum with the growth of Russian national consciousness. By the early nineteenth century a Russian yearning for cultural self-sufficiency, a concern with the national balance of payments, and a growing ethnographic interest in the peasantry prompted efforts to promote "internal touring as a patriotic alternative to travel abroad."[1] An example was the preface to Fyodor Glinka's *Letters of a Russian Officer* (1815–16). An eight-volume work, the *Letters* concludes with an often touristic account of the author's discovery of German lands and France during the campaigns that culminated in the occupation of Paris in 1814. In these sections of Glinka's saga, campaigning takes on the air of a poor man's Grand Tour.[2] Containing maps and illustrations, volume eight is entirely devoted to Paris, where the demobilized author wears fashionable French clothing and represents himself sightseeing, looking at art, trying a fancy restaurant, and enjoying his well-appointed hotel near the Palais Royal. The domestic counterweight to the poor man's Grand Tour occupies the second and third volumes of the *Letters*. There Glinka narrates his explorations of rural areas around Moscow, Kyev, Tver, and his

1 Dickinson, *Breaking Ground*, 69, 105–7.

2 Ibid., 147–55; and Susan Layton, "Russian Military Tourism: The Crisis of the Crimean War Period," in Gorsuch and Koenker, *Turizm*, 45–48.

native Smolensk during 1810–11. In roaming the heartland, he wrote in 1815, Russians will not find "pictures, marble statues, and other artworks," but they will get to know "our" natural environment and "the customs of our own folk."[3] Glinka thus sought to encourage a purposeful, non-touristic form of domestic travel that would build national consciousness through contact with peasants and the land.

Russians seeking alternatives to vacationing abroad favored the empire's non-Slavic edges, where scenery could compete with Western Europe's. Soon after Catherine annexed the Crimea in 1783, she began conceptualizing the new possession as the empire's "paradisial garden," a trope elaborated in poetry and travel writing.[4] As Pyotr Sumarokov's account put it in 1803, the Crimea was simply a "piece of paradise," surpassing Italy and Switzerland.[5] His contemporary, the sentimental journalist, poet, and prose writer Pyotr Shalikov, similarly designated the Carpathians "our local Alps" and proclaimed Poltava alone superior to Italy, Switzerland, "happy France," the upper Rhine, and the Swabian plains.[6] Similar claims proliferated in the Romantic era. Much imitated by other poets and travel writers, the mountain landscape of Pushkin's *The Prisoner of the Caucasus* originated a Russian concept of the "Caucasian Alps."[7] In an 1822 account of his trip to the North Caucasus, viewed through the lens of Pushkin's captivity tale, the artillery officer and botanist Ilya Radozhitsky used a Russian equivalent of *paese felice* to declare "the happy land" (*schastlivyi krai*) of Georgia "Russia's Italy."[8] As a final example of the trend, the minor poet Vasily Dmitriev (1777–1820) touted southern and western Siberia as a region offering Russian travelers the equivalent of diverse scenic marvels abroad: "Pyrenees valleys, Tivoli cascades, Swiss beauty, and Alpine mornings."[9] To the standard Russian way of thinking, the whole empire was "our fatherland," and the scenery was as good, if not better than Western Europe's.

---

3 Dickinson, *Breaking Ground*, 204.

4 On the paradise trope and its political functions under Catherine, see Andrei Zorin, "Krym v istorii russkogo samosoznaniia," *Novoe literaturnoe obozrenie* 31 (1998): 130–37; and Andreas Schönle, "Garden of the Empire: Catherine's Appropriation of the Crimea," *Slavic Review* 60, no. 1 (Spring 2001): 1–23 (quote 3).

5 Andrei Mal'gin, *Russkaia Riv'era: Kurorty, turizm i otdykh v Krymu v epokhu Imperii, konets XVIII—nachalo XX v.* (Simferopol': SONAT, 2004), 41–42.

6 Dickinson, *Breaking Ground*, 129.

7 Layton, *Russian Literature and Empire*, 36–53.

8 I. T. Radozhitskii, "Doroga ot reki Dona do Georgievska na prostranstve 500 verst," *Otechestvennye zapiski* 15, no. 41 (1823): 343–44.

9 Ely, *This Meager Nature*, 53–54.

But what about the cultural dimensions and the transactions between Russian vacationers and native peoples of the non-Slavic borderlands? The question holds special interest for the North Caucasus. Russian spa culture arose there during the military conquest of Chechnya and Daghestan that lasted from 1817 to 1859 when Imam Shamil surrendered after twenty-five years of leading the Islamic resistance movement. Russians generally regarded expansion into the Caucasus as a European civilizing mission, even though there was ambivalence about violence as the means to that end. In identifying with empire-building in Asia as a European pursuit, Russians extended the hegemonic discourse of Western orientalism. As of Pushkin's time, however, the ideology of the civilizing mission coexisted with Romanticism's glamorization of the Orient as part of the Russian self and Russian culture, historically and geographically located between East and West.[10] Romanticism conspired in this respect with the officially propagated metaphor of the empire as a family, a notion we have already remarked in Madame Kurdyukova's proud claim to Kirghiz ancestry. As that detail from Myatlev's poem illustrates, Romanticism encouraged Russians to visualize the empire not as a coercive order but rather as a convivial Euro-Asian assembly under their rightful authority.[11]

The case of music demonstrates that the romance of the Orient could evolve from fantasy into a matter of genuine cosmopolitan engagement. Pushkin's Caucasian captivity poem features a fabricated "Circassian Song" that the Russian captive hears native maidens singing as they do a folk dance. The "Bela" section of Lermontov's *A Hero of Our Time* similarly faked native culture by transferring a Russian folk song to Chechnya. On the

---

10 On orientalized people orientalizing others, see Milica Bakić-Hayden, "Nesting Orientalisms: The Case of Former Yugoslavia," *Slavic Review* 54, no. 4 (Winter 1995): 917–31; Maria Todorova, *Imagining the Balkans* (New York: Oxford University Press, 1997), 17–20, 58–59; and the notion of "reverse Orientalism" in Dickinson, *Breaking Ground*, 164. For representative analyses of Russian orientalism vis-à-vis imperial borderland peoples, see Marc Bassin, "Russia between Europe and Asia," *Slavic Review* 50, no. 1 (Spring 1991): 1–17; Monika Greenleaf, *Pushkin and Romantic Fashion: Fragment, Elegy, Orient, Irony* (Stanford: Stanford University Press, 1994), 108–55; Layton, *Russian Literature and Empire*, 72–88; Offord, *Journeys to a Graveyard*, 19–22; Geraci, *Window on the East*, 2–8, 358; and Katya Hokanson, "Russian Women Travelers in Central Asia and India," *Russian Review* 70, no. 1 (January 2011): 6–7.

11 On "ornamental" dimensions of "visualizing" an empire, see David Cannadine, *Ornamentalism: How the British Saw Their Empire* (London: Allen Lane, The Penguin Press, 2001), esp. 64–69. On cosmopolitan "conviviality" versus a "coercive order," consult Walter D. Mignolo, "The Many Faces of Cosmo-polis: Border Thinking and Critical Cosmopolitanism," in Delanty and Inglis, *Cosmopolitanism*, 2:58–60.

other hand, Alexander Bestuzhev-Marlinsky (1797–1837), the most popular Russian prose writer of the 1830s, learned Azeri and included in his story *Ammalat-Bek* (1832) his translations of Daghestani battle songs, an element of the work that reflected his more general study of languages and cultures of the Caucasus.[12] In 1834 the Siberian-born Russian composer Alexander Alyabev (1787–1851) published the songs he gathered during fieldwork among Circassians, Georgians, and other Caucasian peoples. Like Alyabev, later composers, including Mily Balakirev (1837–1910), also drew upon music of the Caucasus, to create culturally hybrid works reflecting their conviction that Asia held the key to a "special destiny" for Russian music.[13]

But whereas individual composer-travelers built cosmopolitan bridges to the foreign, Russian tourist transactions with natives of the Caucasus were typically limited to consuming spectacles and buying souvenirs. Romantic literature provided preparation for this tourist regime by creating theatrical, ornamental images of the *gortsy* (mountaineers, highlanders), the elastic term Russians applied to North Caucasian peoples, primarily Circassians, Chechens, and Lezgins.[14] Pushkin's *Prisoner of the Caucasus* set the spectatorial perspective. Unconscious and bleeding from a head wound, the Russian captive is dragged into the village in chains. After a rapid recovery assured by the attentions of the Circassian woman who immediately falls in love with him, the unfettered prisoner spends much time observing the villagers' way of life. He perceives the freedom-loving, weapon-bedecked warriors as paragons of masculine grace, vigor, and horsemanship. He watches their war games that may entail beheading a slave, and he "admires the beauty / of their simple, martial clothing."[15] A similar theatricality figured in Russian accounts of the 1826 coronation of Nicholas I. As reported in the press, the festivities in the Winter Palace afforded a colorful tableau of Circassians, Kabardinians, Georgians, Armenians, Kirghiz,

12 A Decembrist exiled to Siberia, Bestuzhev spent five years there before being transferred into the Russian army in the Caucasus, where he died in battle. After his arrest he was allowed to publish under the pseudonym Marlinsky.

13 Malcolm Hamrick Brown, "Native Song and National Consciousness in Nineteenth-Century Russian Music," in Stavrou, *Art and Culture in Nineteenth-Century Russia*, 67–70; and Orlando Figes, *Natasha's Dance. A Cultural History of Russia* (London: Penguin Books, 2003), 389–91.

14 King, *The Ghost of Freedom*, xiii; and Bruce Grant, *The Captive and the Gift: Cultural Histories of Sovereignty in Russia and the Caucasus* (Ithaca and London: Cornell University Press, 2009), 21, 134–36. Used as a generic ethnonym, *cherkesy* (Circassians) figured interchangeably with *gortsy* in Russian writing.

15 Pushkin, *SS*, 3:99.

3. This picture from the *Peoples of Russia* (1862) album extended the ornamentalist traditions of Russian print culture's treatments of North Caucasian mountaineers in the Romantic era. Placed in their homeland mountain milieu, the Chechen (left side) and the Lezgin have a static stately demeanor, finely crafted weapons, and splendid clothing that all bring to mind the staging of an ethnographic exhibition.

and Kalmyks in indigenous military dress. Nicholas independently had a contingent of smartly outfitted North Caucasians, the so-called Circassian guards.[16] Marlinsky elaborated the ornamental mode, compounding classical and Homeric references. His Circassian warriors are "majestic," the

16 Wortman, *Scenarios of Power*, 1:290–91, 293. On military pageantry in Bestuzhev-Marlinsky's *Ammalat-Bek*, see also Katya Hokanson, *Writing at Russia's Border* (Toronto, Buffalo, London: University of Toronto Press, 2008), 180–81.

very "model of Ajax or Achilles," and the Daghestani Ammalat-Bek is a man of "uncommon beauty, svelte as the *Apollo Belvedere*."[17] Lermontov's *A Hero of Our Time* qualified this tradition. The Circassian Bela, who sings and dances for Pechorin, conforms to ornamental type, but the novel's star mountaineer, Kazbich, is a slovenly little man. He has finely crafted weapons, however. More importantly, Pechorin becomes a surrogate of native masculine beauty when he achieves "dandy" elegance in the "Circassian outfit" he wears for horseback riding in the resort region.[18] Pechorin thus set a theatrical example for Russian men vacationers.

In creating a glorious foreign landscape and ornamentalizing the mountaineers, Romantic writings packaged the North Caucasus for ethnic Russian tourist consumption just as Walter Scott's novels played a major role "in 'packaging' Scotland for touristic consumption by the dominant English."[19] The notion of packaging suggests a theme park, an advanced form of the tourist industry's curating the natural environment and "staging" local culture for profit.[20] As formulated in the first edition of MacCannell's *The Tourist* (1976), "staged authenticity" referred to the kind of tourist attraction that purports to give visitors an inside view of some aspect of the local way of life, such as revealing the "inner workings" of a business.[21] The commodification of "culture" now more typically concerns arts, crafts,

17 A. A. Bestuzhev-Marlinskii, "Pis'mo k doktoru Ermanu," *Sochineniia*, 2 vols. (Moscow: Gosudarstvennoe izdatel'stvo khudozhestvennoi literatury, 1958), 1:300; and *Ammalat Bek*, in *Povesti i rasskazy* (Moscow: *Izdatel'stvo "Sovetskaia Rossiia,"* 1976), 196-97.

18 Lermontov, *SS*, 4:81 (subsequent citations appear parenthetically).

19 Buzard, "The Grand Tour," 44.

20 Representative discussions are Buzard, *The Beaten Track*, 175–77, including citation of Cohen, "Authenticity and Commoditization," 376–78; Chhabra, "Staged Authenticity," 702–19; Berghoff, "From Privilege to Commodity?," 168-80; and John K. Walton, "British Tourism between Industrialization and Globalization: An Overview," in Berghoff and Korte, *The Making of Modern Tourism*, 123.

21 MacCannell, *The Tourist*, 98–102. On the torrent of criticism of the concept, especially from anthropologists, see Dean MacCannell, "Staged Authenticity Today," in his *The Ethics of Sightseeing* (Berkeley, Los Angeles, London: University of California Press, 2011), 14. Particularly severe was Edward M. Bruner, *Culture on Tour: Ethnographies of Travel* (Chicago and London: University of Chicago Press, 2005), a book that called "staged authenticity" a product of the 1970s' "deep structuralism" trading in "limited binaries" such as front/back, real/show, authentic/inauthentic. In Bruner's view, today's paid performances of Maasai tribal dances, for instance, are not simulacra but rather "new culture," largely precipitated by "international mass tourism" and specifically meant "for a tourist audience", 5, 191. But for the counter-argument that the commodification of culture is bound to diminish its meaningfulness to the local community, see the anthropologist Davydd J. Greenwood, "Culture by the Pound:

reenactments of historical events, heritage festivals, and the like. Along with the theme park, the cultural festival, featuring music and dances, is most relevant to the imperial Russian past. The illustrative writings I examine here express differing tourist assessments of authenticity. Are the cultural exotica really real or merely put on for tourists and therefore unworthy of a sophisticate's attention? A related question is whether tourism's own transmuting impact on local culture as well as the environment represents "capitalist improvement" and progress of the civilizing mission or rather creeping destruction of a place where the different-from-home kind of visitor can feel truly "elsewhere."[22] In sowing division on those issues, the resort experience generated "hostility rather than community" among Russian vacationers, despite the nationally coalescing power of the imperial idea.[23]

My exploration of these issues observes a fault line between Petersburg and the provinces. On the different-from-home side stands Alexander Druzhinin (1824–64), a prominent Petersburg man of letters who published several Caucasian spa stories after taking a cure for tuberculosis in Pyatigorsk, Kislovodsk, and Essentuki in the summer and early autumn of 1851.[24] Druzhinin was a novelist, a leading literary critic of the art for art's sake persuasion, and a translator of *King Lear* and other Shakespeare plays.[25] A friend of the painter Pavel Fedotov (1815–52), Druzhinin also followed developments in the Russian visual arts, collected art, and published art criticism.[26] On the home-plus side stands Evgeny Verderevsky (1825–67?), a journalist and poet from the provinces. Born in Saratov, he was a prominent commentator on the Caucasus in his time. Between 1847 and 1853 he held a civil service job in Perm, a town widely regarded in Russia as

---

An Anthropological Perspective on Tourism as Cultural Commoditization," in Smith, *Hosts and Guests*, 171–85.

22 European travelers invariably consider themselves agents of "capitalist improvement" in pre-capitalist foreign lands, in Mary Louise Pratt, *Imperial Eyes: Travel Writing and Transculturation* (London and New York: Routledge, 1992), 61.

23 Quote from Culler, "Semiotics of Tourism," 158.

24 A. V. Druzhinin, *Povesti. Dnevnik*, ed. B. F. Egorov and V. A. Zhdanov (Moscow: Nauka, 1986), 467.

25 Offord, *Portraits of Early Russian Liberals*, 156–68; and Michael R. Katz, "Introduction" to *Polinka Saks* and *The Story of Aleksei Dmitrich* [trans. Michael R. Katz], by Aleksandr Druzhinin (Evanston: Northwestern University Press, 1992), 1–13.

26 Known as the Russian Hogarth, Fedotov (1815–52) was an amateur painter who enjoyed success at Petersburg exhibitions in 1849 and 1850. His close connections to the Petrashevsky political circle to which Dostoevsky belonged made him a target of government persecution which apparently hastened his early death in a mental clinic.

an emblematic backwater, still evoked as such by Chekhov.[27] Verderevsky then went to Tiflis to work in the chancellery of the Caucasian Viceroy Mikhail Vorontsov (1782–1818). Verderevsky became the literary editor of the newspaper *Kavkaz* (*Caucasus*), and he published *Plen u Shamilia* (*Shamil's Captives*, 1856), an account of the captivity of the Georgian princesses Anna Chavchavadze and Varvara Orbeliani, recently ransomed after being held hostage for eight months in Shamil's household in Daghestan.[28] Based on Verderevsky's exclusive interviews with the princesses in Tiflis, the book became a best-seller in Russia. This success very likely encouraged Verderevsky to publish *From Trans-Uralia to Transcaucasia: Sentimental, Humorous, and Practical Travel Letters* (1857), a narrative written and partly serialized prior to the Crimean War.[29] More than half of Vederevsky's book concerns his ten-week vacation in the North Caucasus and Georgia in 1853.

"Transcaucasian" journeys were not new at the time. After Russia annexed Georgia in the early nineteenth century, Russian periodicals and newspapers ran dozens of descriptions of south Caucasus regions and the Georgian Military Road from Vladikavkaz to Tiflis. According to Charles King's estimates, at least seventy-eight relevant accounts appeared in the 1830s, seventy-one in the 1840s, and 103 in the 1850s. Although many individual efforts were no more than "notes on road conditions, post houses, and distances," the size of the whole corpus no doubt contributed to Russian imperial consciousness of Georgia and the way to get there.[30] But if Verderevsky had many predecessors, he seems to have been the first to introduce Russian readers to the Borjomi spa. While the Georgian section of his book must remain secondary here, it will serve to underscore the

27 For Herzen's and Chekhov's visions of Perm, see Lounsbery, *Life Is Elsewhere*, 65–66, 209. Positive as well as negative dimensions of Perm's public image are treated in Vladimir Abashev, *Perm' kak tekst: Perm' v russkoi kul'ture i literature XX veka* (Perm': Izdatel'stvo Permskogo universiteta, 2000), 19–20, 31–38, 56, 86–94.

28 For a textual analysis of this captivity narrative, see Susan Layton, "Imagining a Chechen Military Aristocracy: The Story of the Georgian Princesses Held Hostage by Shamil," *Central Asian Survey* 23, no. 2 (June 2004): 183–203.

29 E. A. Verderevskii, *Ot Zaural'ia do Zakavkaz'ia. Iumoristicheskie, sentimental'nye i prakticheskie pis'ma s dorogi* (Moscow: V. Gautier, 1857). Designating Siberia, "Trans-Uralia" (*Zaural'e*) is Verderevsky's invented term to parallel "Transcaucasia" (*Zakavkaz'e*), the Russocentric designation of lands "beyond" the Caucasus mountain range.

30 King, *The Ghost of Freedom*, 122.

imperialist role of the spa as Russian home-plus in Georgia no less than in the North Caucasus.

At Home in Asia

In 1803 the Russian army initiated the development of Caucasian spa culture with the establishment of cure facilities for sick and wounded officers in Pyatigorsk, known at the time as Hot Springs. General Alexei Ermolov's brutal campaigns in Chechnya and Daghestan in 1817–1827 then made curative tourism feasible for civilians. Military operations necessitated amelioration of the Georgian Military Road, the laying of new roads, and the building of bridges. The new infrastructure would facilitate tourist journeys. As interpreted in Pushkin's Caucasian captivity poem, Ermolov's initial victories seem to have secured a new territory for Russian leisure travel. After declaring that "the Russian sword has conquered all," the poem's militaristic epilogue ends with an image of a horseman riding alone on a mountain road where outsiders once feared to venture. Thanks to the Russian army, implies the poet, the route is now open to anybody who cares to come.

Although Pushkin's captivity tale did not depict the spas, it stirred the imagination of readers in such a way that turned the Caucasus into a tourist attraction. The Caucasian spas of the Romantic era catered to tourists seeking home-plus. As Timothy Phillips has underlined, Russia's curative resorts were institutions imported from the West, and in the Caucasus they formed little pockets of Western languages (especially French), architectural styles, fashion, and the practice of the cure as established abroad. Russians even nicknamed their resorts after famous foreign ones. Fashionable in the early nineteenth century, Lipetsk in the Tambov province, for example, was christened "Russia's Spa, Russia's Baden."[31] For the home-plus tourist, a Caucasian spa provided an exotic natural setting in which to engage in familiar practices such as speaking French. As that alone should suggest, Russian aristocrats and well-to-do lesser nobles comprised the Caucasian resorts' dominant clientele in Pushkin's time.[32] Such was the impression of Radozhitsky, writing from Georgievsk in 1823. He reported seeing carriages

---

31 Timothy S. J. Phillips, "Health and Sociability: The Cultural Significance of the Resort in Nineteenth-Century Russia" (DPhil diss., Oxford University, 2005), 42–44, 302–4, 338–44.

32 Ibid., 217, 227–35.

of Russian families from "the best society" en route to the Caucasian spas. The girls were pretty and clearly not sick.[33] Those people were on vacation.

As enclaves of Europeanized Russian culture, the spas largely segregated visitors from local people. In Pushkin's era, some curious tourists visited nearby villages of "pacified" mountaineers.[34] Belinsky, though, was not among the venturesome. While taking a cure in Pyatigorsk in 1837, he reported seeing "many Circassian men" in the vicinity, but his fear of being kidnapped and tortured kept him glued to the spa.[35] Segregation was the order of the day in the Romantic era, as illustrated in the classic representation of the North Caucasus tourist scene in the "Princess Mary" section of Lermontov's *A Hero of Our Time.* Other parts of the novel portray or mention Circassians, Kabardinians, Chechens, Georgians, Ossetians, and Shapsug. In "Princess Mary," however, no Russian spa client ever encounters a native.

While Pushkin put the Caucasus on the Russian leisure travel map, he apparently was also the first to perceive spa development as despoliation rather than progress. He had discovered the Pyatigorsk region in 1820 as a camper with members of General Nikolai Raevsky's family. Nine years later Pushkin found that tourism had made big inroads, as recounted in *A Journey to Arzrum* (1836), the narrative of his unauthorized trip to observe the Russian army at war in Turkey. "In my time," recalled Pushkin, "the baths were housed in hastily built huts." The campers used "scoops made of bark or the bottom of a broken bottle" to drink from mineral springs bubbling out of the mountains. But now, continued Pushkin, there are shops, "magnificent bath houses," bridges and pavilions, "spotless footpaths," "rectilinear flower-beds," guard-rails along steep paths, and police injunctions posted on bathhouse walls. The site had become orderly, clean, and pretty, but in the process had lost the charm of its "wild state" (5:417–18). Expressing here a travel pioneer's pride in roughing it, Pushkin lamented the triumph of tourist comfort, commercialization, and security at the expense of nature's authenticity and the spirit of adventure.

---

33 I. T. Radozhitskii, "Progulka k kavkazskim mineral'nym vodam," *Otechestvennye zapiski* 16, no. 44 (1823): 387–88.

34 S. D. Nechaev, "Otryvki iz putevykh zapisok o iugo-vostochnoi Rossii," *Moskovskii telegraf* 7, no. 1 (1826), section 1: 32; and P. S. Sumarokov [P. S.], "Pis'ma s Kavkaza," *Moskovskii telegraf* 33, no. 11 (June 1830): 320–22.

35 Belinskii, letter to M. A. Bakunin, June 28, 1837, *PSS*, 11:138.

By the mid-1830s the Caucasian spas had grown sufficiently popular to have their high-season bustle and off-season calm. Hotels, more bathing facilities, pavilions, and shops awaited the visitor. In Pyatigorsk tourists could buy Persian rugs, shawls, cloth and silks, embroidered "Circassian slippers" for ladies, nielloed jewelry, wine, tea, books, equestrian paraphernalia, fleecy lambskins, tall sheepskin hats, and *burki* (singular *burka*)—cloaks of felt or short curly lamb wool, worn by some Russian partisans during the Napoleonic Wars. Other popular souvenirs were daggers, swords, and *cherkeski*—the frontier-style tunics with cartridge pockets sewn across the chest.[36] According to Lermontov, merchants in Stavropol turned a nice profit by passing off shoddy daggers to gullible Russian tourists keen to obtain an authentic Caucasian weapon.[37] Whatever souvenir the tourist chose, on the road or at the spa, those commodities functioned at home as "booty of empire," strengthening the imperial consciousness and travel urges of Russians who saw them.[38]

The Taint of Provinciality

As tourism expanded, the spa scene recreated the sociocultural hierarchies of home that set the Russian capitals against the provinces. "Princess Mary" accordingly frames the resort as a hotbed of "power games."[39] The action unfolds as the journal of Pechorin, the high-society dandy who styles himself a Byronic loner, consumed by ennui and despising the common herd. On his first stroll through Pyatigorsk, he spots Russian provincials by their clothing: the men's "worn-out, old-fashioned frock coats" and the "dressy attire of their wives and daughters." Pechorin assumes these people are

36 A. E. Rozen, *Zapiski dekabrista* (Leipzig: Duncker and Humbolt, 1870), 373, 383–84; V. A. Manuilov, ed., *Lermontovskaia entsiklopediia* (Moscow: Izdatel'stvo Sovetskoi entsiklopedii, 1981), 221, 459; Lermontov, "Kavkazets," *SS*, 4:145; and M. P. Pogodin, ed., *Aleksei Petrovich Ermolov. Materialy dlia ego biografii*, 4 vols. (Moscow: Katkov and Co., 1864), 4:293–94.

37 Lermontov, "Kavkazets," 143.

38 Quote from Turner and Ash, *Golden Hordes*, 19. On souvenirs as circulating "visual evidence" of empire, consult also Willard Sunderland, "Shop Signs, Monuments, Souvenirs: Views of the Empire in Everyday Life," in *Picturing Russia: Explorations in Visual Culture*, ed. Valerie A. Kivelson and Joan Neuberger (New Haven and London: Yale University Press), 104–8.

39 Joe Andrew, "Another Time, Another Place: Gender and Chronotope in the Society Tale," in *The Society Tale in Russian Literature from Odoevskii to Tolstoi*, ed. Neil Cornwell (Amsterdam and Atlanta: Rodopi, 1998), 127–32.

"families of landowners from the steppes," and he thinks with satisfaction of the "Petersburg cut" of his own military coat.[40] In sartorial terms he has fenced off the provincials to whom he will never deign to speak. Steppe landowners had actually begun frequenting the spas by this time. As one of them put it in the mid-1830s, "our kind" feels obliged to visit the Caucasus at least once, just as a Turk makes a pilgrimage to Mecca.[41]

Although Pechorin locates the "steppe" people within a "multilingual crowd," other visitors never prevent his solitary enjoyment of nature. Alone in his rented house in Pyatigorsk, he gazes out the window at mountain vistas and the flowering cherry trees in his garden. In solitude he strolls along "viny paths" near the spa spring. Alone on horseback in his Circassian garb, he roams the environs and refreshes himself in the wilderness. His dominant aesthetic register at the spa is the "picturesque," a concept that marginalizes the sublime to produce a milieu hospitable to tourist pursuits such as picnics, promenades, and horseback riding (4:62, 64, 67). Including children with nannies, the vacationers can look at Mt. Elbrus through a telescope on a viewing platform, but they are never exposed to blizzards, avalanches, or other hazards of rugged mountain country, so prominently featured in the novel's "Bela" episode. As seen through Pechorin's eyes, the spa acquires the character of a vast, secure garden for his private enjoyment.

What Lermontov never revealed was that some of his Russian contemporaries found the Caucasian resorts boring and provincial. In the opinion of the Petersburg civil servant Turovsky, writing in 1841, anybody who has stayed in Pyatigorsk knows "how many dissatisfactions and inconveniences a visitor has to endure." With the exception of the "luxurious" facility named for Nicholas I, "the bathing facilities are poor, even bad," and the food is bad. Pyatigorsk is dusty, the air smells of sulfur, and the balls—the main form of entertainment—become monotonous. As on a previous visit, Turovsky hiked up Mt. Mashuk and scrawled his signature on a monument alongside other Russian graffiti. He took a perfunctory look at the view but said the most enjoyable resort activity for men was to peek into the women's bathhouse. Kislovodsk was more to Turovsky's liking for the

---

40 Lermontov, *Geroi nashego vremeni*, in *SS*, 4:63 (subsequent citations appear parenthetically).

41 Ia. Saburov, "Poezdka v Saratov, Astrakhan' i na Kavkaz," *Moskovskii nabliudatel'*, no. 2 (May 1835): 176.

air quality as well as the cuisine that included trout, pheasant, and fruit.[42] Two Petersburg brothers in Ekaterina Lachinova's novel *Escapades in the Caucasus* (1844) regard Pyatigorsk even more harshly as a shabby, typically provincial Russian town. Army officers at leisure, the snooty brothers perceive spa clients cavorting at a ball as a backwoods freak-show, complete with the standard "provincial ladies' man."[43]

The taint of provinciality was the kiss of death for a Russian vacation venue, and such assessments may have helped prompt the official decision to make the resorts more attractive.[44] In practice, this meant more conformity to European norms. Following his 1844 appointment as viceroy, Vorontsov established a new spa directorate with medical, aesthetic, and managerial divisions. The reorganization led to changes that initated in the early 1850s a public debate never resolved: Are "our" resorts as good as Europe's or not? In 1853 the spa doctor Ivan Drozdov assured prospective visitors that Pyatigorsk had "become much better" since 1846: there were new buildings in Italian and Moorish style, a stone theater, and an artificial waterfall (a common feature of aristocratic gardens in Europe and Russia).[45] Singing the same hymn of progress, Konstantin Zelenetsky, a lycée teacher from Odessa, declared that anybody who has visited both Western Europe and the Caucasus knows that Pyatigorsk is "in no respect inferior to Carlsbad, Marienbad, or any other spa" abroad.[46] Vsevolod Melnitsky, a steamboat

---

42 N. G. Turovskii, "Dnevnik poezdki po Rossii v 1841 godu," ed. N. F. Golombievskii, *Russkaia starina* 155 (September 1913): 499–501, 503.

43 E. Lachinova [E. Khamar-Dabanov, pseud.], *Progulki na Kavkaze*, 2 vols. (St. Petersburg: K. Zhernakov, 1844), 2:117; and Phillips, "Health and Sociability," 316–17n38. On the trope of "the provincial ball" as a showcase "for failures of taste" in central Russia, see Lounsbery, *Life Is Elsewhere*, 67–68. By contrast to these negative views of Pyatigorsk, the spa appears as an elegant retreat of the aristocracy in Adèle and Xavier Hommaire de Hell, "Les Steppes de la mer Caspienne, le Caucase, la Crimée et la Russie méridionale" (1843–45), in *Le Voyage en Russie. Anthologie des voyageurs français aux XVIIIe et XIXe siècles*, ed. Claude de Grève (Paris: Robert Laffont, 1990), 766. One should keep in mind, though, that Xavier was traveling on an official Russian mission to investigate prospects for scientific and industrial development.

44 Although I quote some of them differently, most of the sources, as well as the facts in this paragraph come from Phillips, "Health and Sociability," 48, 313–14, 329–30. My biographical information comes from P. S. Zaionchkovskii, ed., *Istoriia dorevoliutsionnoi Rossii v dnevnikakh i vospominaniiakh*, 5 vols. in 13 parts (Moscow: Kniga, 1976–89), 2.2:38, 62.

45 I. I. Drozdov, "Kavkazskie mineral'nye vody," *Sovremennik* 39, no. 5 (May 1853), section 2: 3–5.

46 K. P. Zelenetskii, "Kavkazskie mineral'nye vody v 1852," *Moskvitianin* 2, no. 6 (1853), section 7: 47.

captain and editor of the *Morskoi sbornik* (*Naval Miscellany*), asserted: "I know nothing better, nothing more reliable in Russia than Pyatigorsk in summertime. It is a southern belle, it is Italy, where I have never been, the Italy I picture in my dreams." Moreover, traveling to the Caucasus had been marvelous on an inexpensive but luxurious Volga steamer ordered as usual for Russian companies from Holland or Sweden.[47] In 1860 a Russian army doctor laid down the patriotic law on the international spa competition: Russian spas are better because they are Russian. True, he granted, European resorts offer more pleasures, but those "pleasures are foreign, not ours." Quoting Alexander Griboedov's (1795–1829) satirical verse drama *Woe from Wit* (first published in 1833 with censor's cuts), the doctor added: "And surely you have not forgotten the words of the poet: 'Even the smoke of the fatherland is sweet and agreeable to us.'"[48]

Seeking Authenticity

By the 1850s the Caucasus presented a new high-season tourist scene characterized by overcrowding, herd behavior, and Russian clients' increasing contact with mountaineers. In high season Russian visitors quickly filled all available lodgings, as Tolstoy discovered in 1852 when he failed to find a room in Pyatigorsk. Tourists thronged local markets and trooped to festivals in nearby Kabardinian villages, where they sampled local food and watched cultural spectacles: folk dancing, *dzhigitovka* (acrobatic horsemanship), and displays of marksmanship.[49] But in addition, the spas themselves were becoming contact zones where tourists encountered mountaineers, working or simply hanging around. The porous boundary between the resort-home and the foreign reflected the fact that combat in Chechnya and Daghestan was winding down. As more villages came under Russian control (at least ostensibly), war-weary local people were opting for various forms of accommodation and cooperation with imperialism.

---

47 G. I. Mel'nitskii, "Pereezdy po Rossii v 1852 godu," *Sovremennik* 39, no. 6 (June 1853), section 6: 134; ibid. 40, no. 7 (July 1853), section 6: 17 and 20; and ibid. 40, no. 8 (August 1853), section 6: 138.

48 Degaev [sic], *Rukovodstvo dlia edushchikh v Piatigorsk i Tiflis i piatigorskie mineral'nye vody, s prilozheniiem marshruta* (Moscow: Tipografiia Lazarevskogo Instituta vostochnykh iazykov, 1860), 83–84.

49 A. Opul'skii, *L. N. Tolstoi na Kavkaze. Literaturno-kraevedcheskii ocherk* (Ordzhonikidze: Severo-Ossetinskoe kniznoe izdatel'stvo, 1960), 162; and I. I. Drozdov, "Zapiski kavkaztsa," *Russkii arkhiv* 3, no. 10 (1896): 221.

Druzhinin's stories represent the frustrated tourist quest for authentic foreign culture at the North Caucasian resorts of his time. "The Singer" (1851) announces this theme. The heroine is the "charming tourist" Sophie, a Petersburg belle of French origin married to a Russian.[50] She entrances Russian men at the resort, including the unnamed narrator (who reports at the end that he married her after her husband's sudden death). At the outset of "The Singer" the time has come to return to Petersburg. On the journey, the narrator tells readers that his men traveling companions hope savages will attack the coach to give them a chance to show Sophie their courage. But alas, the roads are perfectly safe. Druzhinin's adventure-hungry tourists are too late to encounter Caucasian highlanders up in arms. Way too late, in fact, to judge by "The Legend of the Sulfur Springs" (1855), set in a somewhat earlier phase of the conquest. That story features Russian tourists at Essentuki trotting around wearing daggers. Keen to pass as heroes in the eyes of visiting ladies, some of those men even hire law-abiding natives to stage an "attack" (2:25). From this tale's vacation perspective, Caucasian warrior culture has shrunk to a staged, paid performance.

"The Singer" satisfies the tourist yearning for martial authenticity in the hybrid figure of the Terek Cossack. At the nature-loving narrator's suggestion, he and Sophie get out of the coach on their homeward journey to take a parting look at the scenery. She is not impressed. "I've seen all the mountains of Europe," she sniffs. "I've never been as bored as I was this summer. My cure didn't work, I had no fun and don't feel rested" (503). Suddenly the tourists hear gunshots and catch sight of distant horsemen in *burki* and sheepskin hats. "*Circassiens!*" gasps Sophie (echoing Princess Mary's reaction to Pechorin's sudden appearance on horseback in native dress). Druzhinin's tourists hide in the bushes. As the horsemen approach, their singing in Russian becomes audible to reveal they are Cossacks.[51] Their magnificent voices, the melody, and the rosy light of the sunset make a stunning spectacle for the tourists, watching but unseen (505–6).

The vacationers take the Cossack song home as booty of empire. Sophie plays it on the piano at a Petersburg soirée after a visiting French woman (the story's disagreeable eponymous heroine) sings Bellini and Verdi arias.

---

50 A. V. Druzhinin, "Pevitsa," *Sobranie sochinenii*, 8 vols. (St. Petersburg: Akademiia nauk, 1865-67), 1:496 (subsequent citations to this edition appear parenthetically).

51 For literary representation of the shared frontier culture of Terek Cossacks and mountaineers, see Druzhinin's story "Mlle. Jeanette" (1852), ibid., 546–79 and, most famously, Tolstoy's novella *The Cossacks* (1863).

The singer proposes to vocalize the Cossack song, too, but the narrator requests instead an aria from *Lucia di Lammermoor*.[52] Cossack and West European music co-exist for the pleasure of the cosmopolitan Petersburg audience. "The Singer" thus exemplifies the Russian elite's hegemonic enjoyment of the empire imagined as a storehouse of exotic cultures to consume selectively along with West European arts. The Russian vacationers had no contact with mountaineers, but the colonized Terek Cossacks stand in as exemplars of frontier authenticity.

Caucasian Conmen and the Tourist Crowd

As depicted in "The Singer," the Cossacks preserve their culture even as tourism expands in their native realm. On the other hand, Druzhinin's highlanders abandon their traditional way of life to chase tourist rubles by means fair or foul. "A Russian Circassian" (1855) centers on the figure of the colonized North Caucasian conman. The tale's central protagonist is the bizarrely named Matvei Kuzmich Makhmetov, a henpecked, ethnic Russian squire who seeks adventure in the Caucasus following a chance encounter with Aslan Makhmetov, a fat and homely but sly mountaineer who successfully performs clichés of Russian Romanticism. Aslan introduces himself to the squire as a Circassian prince. But the reader learns he is actually an itinerant "oriental" peddler of the sort to be found in Petersburg selling "chibouks, rose water, and nielloed silver knickknacks engraved 'Caucasus, such-and-such a year.'"[53] Aslan finagles hospitality for the night at Matvei Kuzmich's country home. He arouses the squire's imagination with yarns about the wild frontier, convinces him they must be distant relatives, and invites him to visit Pyatigorsk. Before departing, Aslan proposes a ritual gift exchange to consecrate their friendship. He goes away with a valuable silver snuff box in return for a heap of rubbishy Caucasian garb. In the ensuing months Matvei Kuzmich reinvents himself as a Circassian, dressing in the clothes received from his *kunak* (consecrated friend). After reading Marlinsky, who further enflames his imagination, he goes to Pyatigorsk, where his Circassian regalia, complete with a dagger his hotel proprietor

52 On Druzhinin's love of Italian opera, consult N. B. Aldonina, "Obraz Rima v publitsistike A. V. Druzhinina," in *Obraz Rima v russkoi literature. Mezhdunarodnyi sbornik nauchnykh rabot*, ed. Rita Giuliani and V. I. Nemtsev (Rome and Samara: Izdatel'stvo OOO NTTs, 2001), 165.

53 Druzhinin, "Russkii cherkes," *SS*, 2:184.

advises him not to wear, makes him the laughingstock of more worldly Russian vacationers.

The story's narrator looks down on the whole tourist crowd. In language full of reminiscences of *A Hero of Our Time*, he stresses the "poetic sensations" of vacationing after the summertime "flocks" have gone: "If only lovers of the spas' high season knew how delightful and fetching Pyatigorsk is in the autumn [. . .] then perhaps many of them would not rush back to Russia [in August]." Lacking the narrator's sensitivity to the environment, the awful tourists include picnickers who toss empty bottles over a cliff. The author also ridicules the specimen Russian vacationer heading home in a "modest travel cart" loaded with cheesy-sounding souvenirs: "a dagger, a stick inscribed 'Caucasus,' a full set of Circassian vestments, a piece of coarse silk, or a cherry-wood chibouk" (206–7, 212, 216). The "modest," springless vehicle and the banal commodities suits the likes of Matvei Kuzmich, the specifically provincial Russian tourist.

Druzhinin's storyteller maligns Caucasian natives as well as the tourist crowd. Aslan is unmasked when Matvei Kuzmich discovers him selling "pomade, chibouks and fat cigarettes" in a shabby Pyatigorsk market stall. Aslan also repairs daggers and swords but has the reputation of being a "coward and cheat" and gets caught substituting an inferior blade on a Cossack customer's weapon (216, 218).[54] Complementing Aslan is Izmail, a native ne'er-do-well whose interaction with Russians at the spa has led to his addiction to vodka and gambling. With the intention of robbing Matvei Kuzmich, Izmail proposes conducting him into the mountains. The tourist becomes suspicious and does not go, but Izmail gets away with the money advanced for guide services. Druzhinin's other stories feature North Caucasian people honorably seeking tourist rubles. Using a common misnomer for a mountaineer, "The Singer" mentions an old "Tatar" woman working as a spa attendant (493). "The Legend of the Sulfur Springs" features tourists tossing coins to mountaineers dancing the *lezginka* (an athletic Lezgin folk dance) in the street (46). Unlike Druzhinin's depiction of the Cossack chorus unaware of their Petersburg audience in "The Singer," the dancers at the sulfur springs spa have a pathetic, money-grubbing air. The author renders the spectacle as something ersatz, staged merely for tourists.

---

54 Native peoples of the North Caucasus dominated metalworking in the frontier economy: see Thomas M. Barrett, "Crossing Boundaries: The Trading Frontiers of the Terek Cossacks," in Brower and Lazzerini, *Russia's Orient*, 240.

Druzhinin's spa stories express a politically confused authorial discontent with the tourist situation Russian imperialism has produced. His novel *The Story of Alexei Dmitrich* (1848) treats Russia's ongoing war in Chechnya as a civilizing mission. Yet, his spa vacation stories betray the worry that Russian tourism is damaging rather than improving the North Caucasus's natural environment and native population. Druzhinin's anti-tourist authorial personas articulate nostalgia for the lost milieu of Pechorin's "Princess Mary": pristine wilderness to enjoy in solitude, no run-ins with the natives, tasteful souvenirs (Pechorin buys a Persian rug in Pyatigorsk), pleasant lodgings, and plenty of champagne.

Self-Definition as a Modest Tourist

The anti-tourist values of Druzhinin's resort stories stand in mutually illuminating contrast to the unqualified enthusiasm for high-season vacationing in Verderevsky's *From Trans-Uralia to Transcaucasia*. Verderevsky presents himself as a "tourist" in the spectatorial mold. From start to finish, he privileges the eye and assumes that seeing is knowing.[55] "I will describe only what I see," he promises at the start (4). However, in the course of the narrative his visual priorities change, along with his sense of being a tourist. At first, he uses "tourist" as a plain synonym for traveler. The term covers Alexander von Humboldt and accommodates the Russian author's own "statistical" agenda of investigating economic activity in "boundless Russia," region by region (27, 33).[56] In May 1853 Verderevsky leaves Perm to visit the Irbit trade fair in Siberia. He then comes back across the Urals, lingers in Perm, and heads for the Caucasus. But well before reaching his ultimate destination, he scraps his knowledge-gathering agenda. He showcases his subjectivity instead and occasionally switches from prose to verse. Once he reaches the Caucasus, he fully incarnates the "tourist" as a vacationer with a nascent leisure-travel industry to serve him. He regrets there are no "Russian guidebooks" for his itinerary, but in their absence he relies on Pushkin, Marlinsky, and Lermontov (4).

---

55 On this issue, see Kristi Siegel and Toni B. Wulff, "Travel as Spectacle: The Illusion of Knowledge and Sight," in *Issues in Travel Writing: Empire, Spectacle, and Displacement*, ed. Kristi Siegel (New York: Peter Lang, 2002), 109–16.

56 On the prior history of such projects, see Naily Tagirova, "Mapping the Empire's Economic Regions from the Nineteenth to the Early Twentieth Century," in Burbank et al., *Russian Empire*, 125–29.

Verderevsky represents the Russian Empire as a culturally cosmopolitan Euro-Asian state. He uses some French and displays familiarity with French and English literature, Rembrandt, and Italian opera (48, 110–11, 123, 161, 223). In parallel to this self-presentation, the book seeks to shore up all Russia's European credentials. Russian merchants in Irbit drink Clicquot and Roederer. Youths in Ekaterinburg sport the latest London fashions, brought home from the "universal exhibition," an apparent reference to the 1851 Great Exposition in London. Cossacks play "airs from *Norma* and *Semiramide*" along with their folk music at a festival in Novocherkassk (25–27, 42, 110–11).

The author's insistent themes of national and personal inclusion in Europe reflect his touchiness about provinciality. Fondly recalling his life in Perm, he bristles at the town's reputation as a "dismal backwater" (*glush' pechal'naia*), a phrase borrowed from Pushkin's playful narrative poem *Count Nulin* (1827).[57] True, admits Verderevsky, many Perm people do nothing but play cards and gossip. But on the other hand, there are cultivated residents and literary salons, where a gracious hostess will oversee passionate discussion of Pushkin, Lermontov, Turgenev, and other writers. She then provides a lovely dinner with fine wines. (46–50). As Verderevsky continues his trip, he fashions himself as an increasingly assertive provincial. Pitted against his homey ideal of Perm, Kazan proves abrasive. A coachman deposits the author before the Rozanov Rooms, the "best hotel" in town. The doorman intimidates the traveler on sight. The overpriced room is dusty, the expensive dinner is bad, and a theater cashier displays "vulgar snobbery" toward the author. The provincial traveler lashes back, castigating Kazan as a pretentious city trying to match Moscow. How much better are unassuming little towns such as Perm, Vyatka, or Tomsk—places where you can find a "modest hotel" with no "useless doorman." Your room will be clean, and you will have a simple but good meal instead of "scraps called dinner" (74–76).

By contrast to the Kazan experience, the provincial tourist gains a sense of control and obtains value for his money in the Caucasus. The land symbolically welcomes him. First glimpsed from a distance, "indescribably

57 A comic parody of Shakespeare's *The Rape of Lucrece*, Pushkin's poem recounts a bored wife's nearly committing adultery not in Perm but rather on a remote estate in a generic "dismal backward" of rural Russia, a place of "mud, bad weather" and the "howling of wolves," in *SS*, 3:181–83. On the Shakespeare connection, see A. D. P. Briggs, *Alexander Pushkin: A Critical Study* (Beckenham, Kent: Croom Helm, 1983), 102–8.

beautiful Elbrus" is a "host" rising up to meet him— a "guest," "come from afar." In Pyatigorsk the guest finds an "excellent" hotel and then rents an apartment with a balcony and view of the mountains. He will find his accommodations in Kislovodsk and Zheleznovodsk equally satisfying (116–17, 121–22, 135, 140). Adopting a guidebook idiom, he offers insider's tips on shopping: "If you come to Pyatigorsk, you must not fail to buy a bridle and saddle of local manufacture." You might patronize an established artisan such as "the Armenian Tumanov." But you should also look for bargains in little workshops on side streets where you can observe "two or three dirty Tatars" working metal, to produce nielloed snuff boxes and other "truly elegant items" (166–67). This passage exhibits a characteristically modern tourist impulse to track down an authentic "work display" in a foreign land.[58] Verderevsky's narrator also purchases a collection of insects, including scorpions and "Caucasian scarabs." Along with the gemstones and seals he bought in Ekaterinburg as a "souvenir of the Urals" and the "white topazes" he will buy from Ossetian urchins along the Georgian Military Road, the Pyatigorsk acquisitions comprise the tourist's booty of empire (36, 168, 189–90).

On the whole, the author thrives at the spas. The sole interference comes from Russian "snobs." They are recognizable on sight. Apparently Petersburg people, they sport fancy vacation wardrobes, and one of them ostentatiously wields a silver pencil in public, no doubt writing a travel journal. In this context the author positions himself among the "modest" tourists, eavesdropping on snobbish conversations at the table d'hôte in a Pyatigorsk hotel. Some snobs maintain the Caucasian mineral waters are not as pure as they used to be, while other snobs tout the superior price-quality ratio of Western resorts they have visited (123–25, 145–46). In either case, Verderevsky's persona has missed the boat. According to some snobs, he is too late for the peak spa experience at home. According to others, the problem is not time but space: he has come to a domestic watering hole, inferior to what the West has to offer. Pretty soon, predicts one snob, nobody will bother coming to the Caucasus.

58 MacCannell, *The Tourist*, 62–63.

Provincial Imperialist

The socially challenged author finds compensation in identifying with the empire's ideology of the civilizing mission. Under the supervision of a "Tatar" guide, he joins a Russian tourist group to ride mules to the top of Mashuk to admire the cross erected there by the Russian army. Verderevsky reads the cross as a symbol of Russian Orthodoxy's victory over Islam throughout the Caucasus. Never mind that Russian tourists have scrawled graffiti all over the cross. Far from deploring the defacement of a religious monument, the author savors the graffiti as evidence of Russians' power to impose their presence far from home (150–52). Verderevsky identifies so thoroughly with the Russian state's claim to be improving the Caucasus that he tries to play *Kulturträger* on an individual level. In Pyatigorsk he invites two Kabardinians to his house to show them marvels of civilization. The guests especially like his telescope, and on a subsequent visit they steal it. Later, one of the men turns up at night heavily armed and asks the author for a "loan." The tourist hands over the money. He has learned his lesson and passes it on: visitors in the Caucasus "should refrain from close contact with mountaineers and not let them into their house under any circumstances" (155–56). The thieving, potentially violent highlander is but one negative stereotype in Verderevsky's xenophobic repertoire. The flip-side of Kabardinian violence is Kabardinian laziness, set against the industriousness of German settlers near Kislovodsk. A Georgian cadet in national dress figures as a moron spouting "broken Russian" in a Pyatigorsk restaurant. Ossetians are mercenary, overcharging the author for cheese and wine along the Georgian Military Road. In Dusheti on the outskirts of Tiflis, "noisy" local women knitting on their porches give the author the unpleasant impression of "some little Jewish town in our Belorussia" (139, 147, 204, 212).

Positioning himself among uncongenial Asian strangers, Verderevsky's persona perceives all inroads of Russian tourism as progress. He marvels at the "gigantic," English-style cure facility under construction in Kislovodsk and judges the recent stone structures of Pyatigorsk superior to the wooden ones of Lermontov's time. From the tourist standpoint of the 1850s, new hotels, bathhouses, shops, omnibus services between the spas, coaches and drivers for hire, flower peddlers, regular concerts by Russian military bands, dance classes, an English-style sculpture garden, and a new shooting gallery in Pyatigorsk all enhance the leisure travel experience. Furthermore, mountaineers are cooperating in the development of tourism, not only by serving as guides but by providing folkloric spectacles. The author joins "crowds of strollers" on the boulevard in Pyatigorsk to watch Kabardinians from Russian-allied villages racing and performing *dzhigitovka*. Pleased with the grassroots initiative to commodify culture, Verderevsky reports that "generous spectators" showered the natives with money (124, 128, 132–33, 135–36, 157).

The performance of local culture for a tourist audience climaxes in the author's account of a group excursion to a Kabardinian village. The Russian men and women tourists arrive in the morning on horseback and in carriages. Upon entering the village, they see "around two-hundred" mountaineers "fully armed" and dressed in "festive clothing," standing or mounted on horses in front of the house of their host – the village headman, a Russian-speaking prince in Russian service. Inside the house the visitors find an assembly of local people so good-looking that they assume they must be the "flower of the village aristocracy." About to go to Petersburg for service in an "Islamic squadron," two particularly handsome youths "evidently do not yet have any idea about the ways of civilization." In timid silence they gaze at the Russian guests "as though paying us for the interest with which we have studied the wild, beautiful figures they cut." Here the author strikes the pose of a colonizer-king on vacation condescending to visit overawed natives. Local girls serve a "Kabardinian dinner," next come *dzhigitovka* and folk dances, one of which reminds the author of the *lezginka* in Mikhail Glinka's opera *Ruslan and Lyudmila* (a redoubling of theatricality, from "them" to "us" and back again). In conclusion, he writes that the most "picturesque of all the spectacles we saw that day" was a "semi-savage convoy" of galloping mountaineers, whooping and shooting as they escorted the tourists out of the village. Never mentioning the money that must have exchanged hands, Verderevsky represents the festivities as a testimony to the mountaineers' reverence for the colonizers, and he treats the experience as something no Russian tourist should miss (157–65).

Tourist Georgia

In pressing on to Georgia, Verderevsky entered a territory long under Russian rule. Alexander I annexed Georgia after the last Georgian king, Giorgi XII, appealed to the Russian state to establish a protectorate over his country weakened by in-fighting and vulnerable to attack by the Persians who had sacked Tiflis in 1795. Russia first took control of Giorgi's eastern kingdom of Kartli-Kakheti (1801) and then the western principality of Mengrelia (1803), the kingdom of Imereti (1804), the Ossetia region (1806), and the principality of Abkhazia (1810). The incorporation of Georgia into the empire entailed violence and the establishment of a complicated modus vivendi with the Georgian elite. Many Georgians nobles served in the imperial administration and the Russian army. They fought in the Napoleonic Wars and in campaigns against North Caucasian highlanders. Members of the Georgian nobility nonetheless harbored nationalist aspirations made manifest in an 1832 conspiracy that Russian authorities thwarted. Assuring the security of Georgia was a gradual process that brought Russian civil servants, Russian and foreign adventurers, diplomats, merchants, and tourists into the territory. As already indicated, many of those travelers had taken pen in hand by Verderevsky's time but without displaying memorable, literary flair.

The aestheticizing leap came in 1830 when Pushkin published "The Georgian Military Road," a preview of *A Journey to Arzrum*.[59] The narrative established a symbolic topography representing Georgia as a realm of pastoral beauty in opposition to the dread sublimity of the Military Road.[60] Rugged and subject to avalanches, the route runs through the Darial gorge, where the "roaring" Terek River "flings murky waves" against the rock walls. A verdant panorama awaits the author at the mountain crest: "The sudden transition from the dreadful Caucasus to pretty Georgia is delightful." Leaving behind "bare cliffs," "gloomy gorges," and the "dreadful" Terek, the traveler beholds sunny valleys, the sparkly, "merry" Aragvi River, green hills, gardens, and fruit trees (422–23, 426–27). Pushkin's symbolism generated extravagant extensions. His own lyric "The Caucasus" described the Terek at Darial as a ferocious, caged animal, a metaphor amplified in *Ammalat-Bek*. Marlinsky's "wildly beautiful, bellowing Terek" is a Promethean "genius [. . .] struggling against nature," "a ferocious beast;" and in a flash of lightening at night, its "murky, tumultuous waves" resemble thrashing "spirits of hell [. . .] pierced by the archangels' sword."[61]

59 See commentary in Pushkin, *SS*, 5: 648.

60 On the contrasting "aesthetics of infrastructure" in Russian and Georgian writings about the Road, consult Paul Manning, *Strangers in a Strange Land: Occidentalist Publics and Orientalist Geographies in Nineteenth-Century Georgian Imaginaries* (Boston: Academic Studies Press, 2012), 59–80.

61 Bestuzhev-Marlinskii, *Povesti i rasskazy*, 181.

At the opposite topographical pole, Pushkin's "pretty" Georgia blossomed into a gendering of the territory as an oriental woman.[62] The extravagant epitome figured in an unfinished novel by the minor writer Alexander Shishkov. After an arduous trek across the mountains, Shishkov's traveling Russian hero perceives sunny Georgia as "a beautiful woman, lying luxuriously on a multicolored carpet, her head resting on the snowy Caucasus as on a white pillow, while the fragrant roses of Gulistan bloom at her feet!" The air is "sweet as the kiss of a bride-to-be for her fiancé, home from a campaign."[63] In a reminiscence of the *paese felice* idea of Italy, Lermontov's narrative poem *The Demon* depicted Georgia as "a happy, splendid land" (*schastlivyi pyshnyi krai*)—a country of luxuriant valleys spreading out like a carpet, a land of roses, nightingales, timid deer, and beautiful women sweetly singing. Functioning as an emblem of her homeland, the heroine Princess Tamara first appears in a chador as she goes to the Aragvi to fetch water (2:46–47). Simultaneously a Russian Italy and a Russian Orient, Lermontov's Georgia stands in his œuvre as the antithesis of the Georgian Military Road. The journey from Tiflis to Vladikavkaz that opens *A Hero of Our Time* runs the symbolic gamut from the idyllic Koyshaur valley to the life-threatening terror of a blizzard at the Pass of the Cross on the mountain crest. Including "Gifts from the Terek," Lermontov lyrics accentuate the dread of the Road, especially at Darial, the site of the castle of the wicked queen of "Tamara," who lures travelers to her orgies and then has them murdered each morning and cast into the river.

In addition to establishing the symbolic topography, Pushkin's *Journey* contains an account of a roughly two-week stay in Tiflis. Under Stalin in the 1930s, Pushkin's visit became part of an official narrative of Russian-Georgian friendship.[64] In fact, the *Journey* expresses a surface-skimming tourist view and underwrites Russia's presumption to be civilizing Georgia. Accentuating exoticism, the author depicts a multiethnic street scene of Georgians, Armenians, Circassians, Persians, and Russian civil servants moving among basket-laden donkeys and ox carts. He conducts readers to

62 Layton, *Russian Literature and Empire*, 192–211.

63 A. A. Shishkov, ["Otryvki iz nezakonchennogo romana o 'kakhetinskom bunte' 1812 goda,"] in *Russkie pisateli o Gruzii*, ed. Vano Shaduri (Tbilisi: Zaria vostoka, 1948), 426, 432.

64 Giorgi Maisuradze and Franziska Thun-Hohenstein, *Sonniges Georgien. Figuren des Nationalen im Sowjetimperium* (Berlin: Kulturverlag Kadmos, 2015), 207–12, including a reproduction of Ekaterina Bagdavaje's painting of a handsome, swarthy Pushkin promenading with Georgian women and men in national costume.

4. *Jardin aux environs de Tiflis.* Prince Grigory Gagarin's picture of music-making Georgians and a dancer exemplifies the exquisite attention he gave to nationally distinct clothing and objects (such as these musical instruments) throughout his resplendent, oversized album *Le Caucase pittoresque*, now a collector's item.

the old-town bazaar to see Georgian, Turkish, and Persian goods for sale, and he designates outstanding monuments: "Asiatic" versus European-style architecture in the old and new quarters of the city, the "luxurious" public baths, aqueducts, and the monastery on Mt. Kazbek. He regards the aqueducts as evidence of a high level of "civilization" (*obrazovannost'*) and yet deems Georgians' "intellectual capacities" underdeveloped. On the other hand, Georgians are praised for hospitality and artistic flair: they are a "merry, gregarious people" with excellent Kakhetian wines and distinct musical traditions (427, 429–31). After watching war in Arzrum, the author back in Tiflis enjoys a culturally hybrid show staged in his honor by a Russian acquaintance. A Russian eyewitness recalled a program of "varied music, singers, dancers, bayadères, and troubadours from all the Asian peoples then in Georgia," including Georgians, of course. Recitations of Byron's poetry figured as well.[65]

65 Untitled recollections of K. I. Savost'ianov, in G. I. Dolgobanov, ed. *Letopis' zhizni i tvorchestva Pushkina*, 3 vols. (Moscow: IMLI RAN, Nasledie, 2001, 2009, 2016), 1.2:103–4.

Pushkin's *Journey* and its offshoots set the tropes of Russian travel writing into Verderevsky's era and beyond. For the spectacle of nature, Marlinsky's Darial, Shishkov's Georgia as woman, and the Koyshaur valley segment of *A Hero of Our Time* proved particularly popular. Travelers also sought out the monuments and old-town bazaar that Pushkin had marked.[66] By the 1840s, Tiflis shopping opportunities had expanded, thanks to "fashionable shops with goods straight from Paris." Russian and Western newspapers were also available, adding to the cosmopolitan atmosphere.[67] The religious writer Andrei Muravyov highlighted cross-cultural hospitality in recounting his interaction with Georgian clergymen who assisted his pilgrimage to their monasteries and churches.[68]

If unable to achieve host-guest contact beyond the cash nexus, other Russian travelers certainly relished Georgia's foreignness. Instead of recounting the trip from Russia, an anonymous account from 1850 immediately plunged the reader into Tiflis, framed as an impressive architectural ensemble in a glorious landscape:

> Before our eyes, lies picturesque Tiflis with its palaces, cupolas, towers, and semi-subterranean *sakli* (earthen huts) built into hillsides and surrounded by shady gardens where the most delicate fruit trees grow, such as peach, fig, pomegranate, lotus, mulberry, grapes, and quince. Winding through town like a snake, the Kura vanishes in the distance beyond grassy hills, and the deep blue Georgian sky rises as a vault high overhead.[69]

Mingling with Georgians in a Tiflis tavern, the author enjoys the cheap, delicious local wine and food, and the performances of Georgian patrons spontaneously improvising poetry (8–9). This self-fashioning expresses pride in getting off the beaten path, to be in this case the only tourist in a foreign eating-place not meant for tourists. However, in representing

66 A-v., "Zametki na puti iz Moskvy v Zakavkazskii krai," 13–18; Dmitri Bakradze, "Stseny iz gruzinskoi zhizni," *Sovremennik* 25, no. 1 (January 1851), section 6: 2–5; A. N. Murav'ev, *Gruziia i Armeniia*, 2 vols. (St. Petersburg: Tipografiia III Otdeleniia, 1848), 1:27–38; Mel'nitskii, "Pereezd po Rossii v 1852 godu," 149–60; A. L. Zisserman, "Desiat' let na Kavkaze," *Sovremennik* 47, no. 9 (September 1854), section 5: 6–12 and 22–23; and D. Lomachevskii, "Novaia Voenno-Gruzinskaia Doroga," *Russkii vestnik* (July 1857), section 1: 171–72.

67 A-v., "Zametki na puti iz Moskvy v Zakavkazskii krai," 18.

68 Murav'ev, *Gruziia i Armeniia*, 1:80–295.

69 Anon., "Iz zapisok o Gruzii," *Syn otechestva* 52, no. 4 (1850), section 7: 5.

a clash between "Europe and Asia," the happy consumer of Tiflis exotica left no doubt about his imperialist values. Across from the "large chancellery buildings, constructeded in the latest style," stretch "dilapidated, flat-roofed" homes of Armenians and the "semi-subterranean *sakli* of Georgians, Persians, and Tatars." In the street "Europe and Asia walk hand in hand. [. . .] The wife of a Russian civil servant walks next to a Georgian woman wrapped in a chador, a Don Cossack rides beside a wild Ararat Kurd; and raggedy Imeretian, Ossetian, and Lezgin *mushaty* (porters) jostle against Muscovite entrepreneurs."[70] Juxtaposed to the unprepossessing Asian abodes and the local menial laborers, the chancellery quarters and the Muscovite capitalists signify European progress. This imagery reflected Vorontsov's rebuilding program that was transforming Tiflis from a "garrison town to [an] urban imperial outpost" with wide thoroughfares, new residential areas on the Kura, theaters (one for Georgian plays, the other for Russian), and an opera house with an Italian company.[71]

Verderevsky's account extends the travel literature's symbolic topography and the trope of Tiflis as half European, half Asian. At the Darial gorge our provincial tourist quotes *A Journey to Arzrum* and *Ammalat-Bek*, and upon crossing the mountains he writes that the first glimpse of Georgia's valleys simulates "the embrace of a captivating woman, tenderly greeting a traveler upon his exit from the infernal desolation," the explicitly Dantesque Hell of the Military Road (191–97, 207).[72] Never interacting with Georgians, Verderevsky's persona consumes Tiflis as theater, typically viewed through an orientalizing prism. In the old-town bazaar, for instance, he gawks at Armenian tradesmen taking siestas in their shops. They lie sprawled on the floor with their heads on a pile of watermelons or a wine sack, an image of "Asian" sloth pitted against the "European" habits and architecture of the new town with which the author identifies (219–25).

---

70 Ibid., 6. For a similar vision of capitalist progress, with a touch of regret for the loss of the "picturesque," see Ia. P. Polonskii, "Tiflisskie sakli" ("Tiflis Huts"), *Sovremennik* 40, no. 7 (July 1853), section 1: 5–7, reprinted in *Polnoe sobranie sochinenii*, 4 vols. (St. Petersburg: Tipografiia R. Golike, 1885–86), 3:410–86. The work reflects the writer's service in Voronstov's chancellery.

71 King, *The Ghost of Freedom*, 86.

72 In 1868 a Russian traveler would find a copy of Dante's "Divina comedia" [sic] at the Vladikavkaz posting station: see I. Kretovich, "Don, Kavkaz i Krym. (Iz putevykh vospominanii)," *Vestnik Evropy* (June 1868): 726.

The most original part of Verderevsky's Georgia segment concerns Borjomi, another spa Russians first developed to serve the military.[73] Russian troops were stationed there during the Russo-Turkish war that Pushkin observed in Arzrum. After an ailing regimental commander found the mineral waters effective, he ordered constructions to facilitate their use. In 1837 an army doctor sent to Petersburg and Moscow his analysis of the waters' composition and therapeutic effects, and the healing powers of the Borjomi springs were publicized in Russia. Impressed by the results of his daughter Catherine's Borjomi cure in the early 1840s, the reigning Caucasian Viceroy, General Evgeny Golovin, named a spring after her and another after himself. The facilities that Verderevsky visited opened in 1850, and four years later the authorities commissioned the building of the first bottling plant.

Verderevsky's Borjomi functions as a Europeanized Russian home in Asia. The resort's lush environment satisfies the tourist author's preference for "picturesque" nature, as opposed to the "savage" Georgian Military Road. He loves Borjomi's European architecture, including his hotel in the style of an "English abbey" (born in Britain and educated at Cambridge, Vorontsov was an Anglophile; and Nicholas I affected an English gentleman's style of life at his Peterhof "cottage" designed by an English architect).[74] Underscoring Russian spa culture's connection to the military, Verderevsky calls attention to a monument commemorating Golovin's establishment of the "Catherine Springs" and the visit of Nicholas I and his family in 1842. These concluding passages also feature Vorontsov presiding over a table of guests who dance after dinner to the music of an orchestra (230–31, 242). It is not clear whether Verderevsky had a seat at the table. But on the printed page, he coexists with the Russian power elite: the Caucasian Viceroy, the reigning tsar, and the specters of Nicholas I and General Golovin. Excluding Georgians, Verderevsky's Borjomi is a Russianized place epitomizing Russian domination of all Georgia.

In relation to other writings this chapter has considered, Verderevsky's travelogue stands out for its unqualified approval of domestic tourism as an arm of empire-building. Both in the North Caucasus and Georgia, the

73 D. G. Dzhanshiev, *Perl Kavkaza: Borzhom. Abastuman. Vpechatleniia i mysli turista*, 4th ed. (Moscow: I. N. Kushner, 1900), 39–42; McReynolds, *Russia at Play*, 173; and "Borjomi (water)," Wikipedia.

74 On Vorontsov and Nicholas, respectively, see King, *The Ghost of Freedom*, 84; and Wortman, *Scenarios of Power*, 1:338–39.

author minimizes contact with local people and seeks no insight into their cultures. Having regretted his effort to induct mountaineers into "civilization" in Pyatigorsk, he avoids mingling with Georgians in Georgia, never venturing into a Tiflis tavern, for instance. Although we can only wonder how Druzhinin might have described Georgia, his spa stories coincide with Verderevsky's attitudes toward North Caucasian peoples. Both writers suggest that the military conquest created prickly relations between Russian spa guests and the locals. By the early 1850s, a Russian resort client might apparently encounter a native working as a guide, a bath house attendant, or souvenir salesperson. Both authors express anxiety about the interethnic transactions: Are these people really "pacified" or not? Expressing doubts on that score, Druzhinin and Verderevsky stereotype mountaineers as sluggards, hustlers, cheats, and thieves ready to prey on tourists.

The idea of the empire as a convivial multiethnic family rather than a coercive order fails to survive the tourist test, and the local people take all the blame. The unvarnished truth of coercion would emerge in "Black Sea Letters" (1857), a travel account envisioning tourist development of the region between the Kuban River and the sea coast of Abkhazia. Writing from Ekaterinodar (now Krasnodar), the author laments that the current war blocks Russian civilian access to this "beautiful corner of the world." The only way to see it now is "to participate in dangerous expeditions. [. . .] But it is not appealing to get acquainted with a new place under a hail of bullets, and so that is not likely to entice a tourist." In the meantime and at a safe distance, the tourist might consume military conquest as a sound and light show:

> Sometimes you will hear from across the Kuban an exchange of gun shots and thunderous cannon. That means one of our detachments has gone into action: taking a village, burning it down, and firing grape-shot into the forest bushes under which mountaineers are lying. Sometimes at night you will be struck by the unique splendor of burning forests, reeds, and steppe grass across the Kuban. On our side—darkness, but over there fiery streams twist along hills and dales, huge tongues of fire waver, in some spots a blazing flame will flare up with a black column of smoke before it, and the murky Kuban glitters then like steel reflecting the sky's crimson color. For the time being, this is all the non-combatant observer along the Kuban can boast about.

> Venturing any further is difficult and dangerous. [. . .] In a word, our Kuban River boundary sets a limit for the tourist and his inquisitiveness.[75]

If only this "thankless war" would end!—concludes the author. Then "civilization" could penetrate the region and teach the natives to "exchange poverty and pillage for amenities of life and a plentiful peace" (92). If not specified in those concluding remarks, securing the region for Russian tourism is clearly part of the agenda, advanced with hair-raising disregard for the victims of imperialist violence.

The writings of Druzhinin and Verderevsky show that even during ongoing military conquest in the Caucasus, snobbery fostered division among vacationers at the spas. In the represented resort milieu of the 1850s, Lermontov's gulf between the Petersburg elite and the "steppe" people still obtains. But whereas Pechorin's journal silenced the provincials, Verderevsky's travel account gives them a voice. Having situated himself among the "modest" tourists in Pyatigorsk, Verderevsky's persona asserts his right to vacation in the Caucasus as he sees fit, no matter what Petersburg "snobs" may say or do. He is happy with accommodations, amenities, group activities, and the grassroots commodification of culture. Conscious of his lowly status at home, he finds compensatory gratification in playing lord of the empire, an experience no tour of Western Europe could match.

Speaking from the other side of the Petersburg-provincial divide, Druzhinin's anti-tourist stories suggest that for the more culturally demanding Russian visitor, the privilege of being top dog in the imperial power structure could not offset the drawbacks of a Caucasus spa vacation. In his writings, tourism has led to overcrowding, the stupidities of provincial visitors in particular, a perceived loss of "authentic" culture, moral corruption of mountaineers (vodka-addiction, gambling, thievery), and despoliation of nature. The minimal entertainment bores the Petersburg belle, and she ranks the scenery beneath Western Europe's. Despite valuing the Cossack song as tourist booty, she certainly lacks the ethnographic way of seeing that might have prompted her to return to the North Caucasus instead of going abroad for European comfort and high culture. Russians so disposed

---

75 N. I. Voronov, "Chernomorskie pis'ma," *Russkii vestnik*, no. 2 (March 1857), section 3: 91–92. Educated at Kharkov University, Nikolai Voronov (1832–1888) was a journalist and ethnographer who served for many years as president of the Caucasus branch of the Imperial Russian Geographical Society as of the early 1860s.

would get their opportunity after the Crimean War, and Druzhinin was among them. In the spring of 1857, shortly before his sole trip to the West, he recorded in his diary that he had dreamed of "something like Venice and distant mountains." Anticipating the "joy of traveling," he added: "I'm going, going, going! [. . .] I'm going to Europe, not in my dreams but for real!"[76] He would have lots of company headed that way.

76 Druzhinin, diary, April 9, 1857, in *Povesti. Dnevnik*, 403.

Part Two
Shocks of Modernization

5

Inundating the West after the Crimean War

The memoirs of the radical intellectual Nikolai Shelgunov (1824–91) recollected the loss of the Crimean War as a rude jolt that awakened Russians from a deep sleep. He wrote that official nationalism under Nicholas I lulled people into imagining that "Russia was the biggest, richest, mightiest country and could serve as the 'bread basket' of Europe or else, if desired, could leave Europe without food and in the extreme, if necessary, could subjugate all other nations."[1] The fall of Sevastopol destroyed that illusion of invincibility. Complacency vanished, to reveal Russia's military and economic backwardness, the country's stagnation, and the corruption of the government and army in the closing phase of Nicholas I's reign. The truth was plain to educated Russians all along the political spectrum. Wounded national pride was endemic but offset by the widespread belief that winning the war might have been even worse than losing, since victory would have vindicated the Nicholas system.[2] Now facing an uncertain future, Russians had to "find ways of re-establishing national self-respect, either by rapid modernization on a Western model or by formulation of a new mission inspired by distinctive Russian, or Slavonic, values."[3]

The Great Reforms were meant to bring Russia into line with the industrialized, democratizing nations of northwestern Europe. In April 1856, just a month after signing the Treaty of Paris ending the war, Alexander II

---

1 N. V. Shelgunov et al., *Vospominaniia*, 2 vols., ed. E. Vilenskaia, E. Ol'khovskii, and L. Roitberg (Moscow: Khudozhestvennaia literatura, 1967), 1:196.

2 Sh. M. Levin, *Ocherki po istorii russkoi obshchestvennoi mysli: Vtoraia polovina XIX–nachalo XX v.* (Leningrad: Nauka, 1974), 316–30.

3 Offord, *Journeys to a Graveyard*, 199.

informed the nobility that he intended to abolish serfdom, the barbarism the historian Sergei Solovyov called "this sore, this shame" that excluded Russia "from the society of European, civilized peoples."[4] The emancipation of the serfs in 1861 was preceded by other momentous changes. In 1857 Alexander II obtained financing from French and English banks for the expansion of Russia's railroad network, the limited extent of which had contributed to the loss of the war.[5] The state also liberalized trade, beginning with the tariff act of 1857 that removed several import prohibitions and more generally abandoned protectionist policies prevalent under Nicholas I.[6] The relaxation of censorship and modifications of the educational system revitalized intellectual life and literature. Public discussion of substantive issues resumed with the appearance of new journals representing a range of political positions. Literature reawakened with the publication of many works of enduring renown, including Tolstoy's *Sevastopol Stories*, Turgenev's *Fathers and Children*, and Goncharov's novel *Oblomov*. Translations of foreign literature proliferated anew, and universities again offered courses in Western and world literature.[7] Other formerly proscribed subjects were reintroduced into the curriculum, admission to universities became easier, and prohibitions on the importation of scholarly materials were relaxed.[8] Reforms of 1864 transformed the judicial system, introduced trial by jury, and established the *zemstva* (singular *zemstvo*), the rural councils that oversaw infrastructure, public health, and education.

A boom in foreign travel was another consequence of the regime's cosmopolitan turn. Soon after Alexander II ascended the throne, new regulations made it easier and cheaper to obtain a passport, and the authorized period of staying abroad increased from two to five years, with extensions readily granted.[9] Most Russians had been stuck at home since 1848, a situation that surely intensified the frustration, which Herzen in 1854 characterized as the norm in a country where foreign travel had always been a

---

4 Wortman, *Scenarios of Power*, 2:59.

5 Haywood, *Russia Enters the Railway Age*, 590.

6 Florinsky, *Russia*, 2:940.

7 N. A. Dobroliubov, *Sobranie sochinenii*, 9 vols. (Moscow and Leningrad: Gosudarstvennoe izdatel'stvo khudozhestvennoi literatury, 1961–64), 3:257.

8 Abbott Gleason, *Young Russia: The Genesis of Russian Radicalism in the 1860s* (New York: Viking Press, 1980), 118.

9 Anon., "Prodolzhenie letnei emigratsii za granitsu i novyi zakon, oblegchaiushchii prebyvanie za granitseiu," *Otechestvennye zapiski* 113 (July 1857), section 3: 22; and Wortman, *Scenarios of Power*, 2:24.

privilege rather than a right. In the worldly-wise manner of an exile in the West since 1847, Herzen wrote that a Russian's eyes "are constantly turned toward the door, locked by the tsar and opened only slightly and rarely. Going abroad is the dream of every decent person. We strive to see, to touch the world we know from study, the splendid and majestic facade of which, as it was formed by the centuries, has amazed us since our childhood [. . .]. The Russian rushes across the border in a kind of intoxication—heart wide open, tongue unbound."[10] Representing the Parisian bourgeoisie as the vulgar, materialistic emblem of the capitalist West, Herzen warned Russian travelers that their journeys would prove disillusioning. But for Russians locked up since 1848, the traditional vision of the "splendid and majestic façade" continued to exert an irresistible attraction. Unfortunately for the state, many people rushing out the "door" were Russian nobles terrified that the end of serfdom would ruin them financially. The independently wealthy and those newly flush with cash from selling their estates "transferred capital outside Russia and took frequent foreign holidays. Between 1856 and 1860 alone, the number of travelers leaving Russia climbed from 17,000 to 275,000, and their lengthy sojourns outside the country put a heavy strain on the nation's reserves of foreign currency."[11] Understandably, however, that situation was not publicized at the time.

Prior to the emancipation, the dominant public image of foreign travel was rather that of a rite of passage variously motivated. The vacation mentality found new release, as Russians leaped at the chance to visit the West. Along with the resumption of "pure" tourism came other types of journeys that blended purpose and pleasure. Journalists integrated sightseeing and entertainment into their assignments abroad. Another mobile group, adept at mixing work and leisure, were Russian students, permitted once again to further their education in the West, particularly at German universities and the Sorbonne. By the late 1850s thousands of Russian youths were taking advantage of those opportunities.[12] Many went in search of scientific knowledge, such as Alexander Borodin (1833–87), the future chemist and

10 Herzen, *Pis'ma iz Frantsii i Italii*, appendix: "Pis'mo k Sh. Riberoliu," *SS*, 5:218.

11 Dietrich Geyer, *Russian Imperialism: The Interaction of Domestic and Foreign Policy 1860–1914*, trans. Bruce Little (Leamington Spa, Hamburg, and New York: Berg, 1987), 24.

12 Dobroliubov, *SS*, 3:257.

composer.[13] Some aspiring Russian artists, too, resumed studying abroad. True, a public exhortation for Russian painting to go national appeared in 1856, when the soon to be famous art and music critic Vladimir Stasov (1824–1906) argued that "the future of Russian art lay in genre painting" dealing with contemporary native subjects "rather than in imitations of classicism or the latest trends in West European art."[14] Nonetheless, the Romantic period's tradition of seeking training and inspiration abroad found a few emulators, as illustrated by the case of Nikolai Ge (1831–94), who produced pictures of local scenery while living in Italy between 1857 and 1870.[15] Trips abroad for the purpose of buying art also resumed but were risky for amateurs.[16] Druzhinin had interestingly issued a warning in an 1852 story recounting the Russian narrator's five-hour visit of the art collection at the luxurious country estate of a compatriot friend. Purchased in Rome and now kept aside in a locked rotunda is "an excellent forgery in the style of Guercino," a portrait of St. Paul that is an aesthetic marvel but an embarrassing one, serving as a reminder of the need to be vigilant when shopping for art in the West, especially in Italy.[17]

In the turbulent atmosphere of new freedom, *turizm* acquired the public profile of a distinctly modern practice that had just come Russia's way. As my introduction indicated, Melgunov and other observers of the time perceived changes that were similar to those that had marked a break with Karamzin in writings of the Romantic era: more travelers, increased social diversity, more women on the move, and industrialized transport supplanting the coach. In the post-Crimean War period, however, the modernization of Russian tourism seemed to have advanced from creep to leap, to produce a qualitatively new phenomenon. Viewed as both a symptom and engine of modernization, tourism became intertwined with the reform era's larger questions of social change, moral values, aesthetics, and national and political ideologies. Old questions of travel norms and priorities now

---

13 I. S. Aksakov, *Ivan Sergeevich Aksakov v ego pis'makh*, 4 vols. (Moscow: Tipografiia M. G. Volchaninova, 1886–87), 3:360

14 John O. Norman, "Pavel Tretiakov and Merchant Art Patronage, 1850–1900," in Clowes et al., *Between Tsar and People*, 95.

15 Lo Gatto, *Russi in Italia*, 144–45, 151–52.

16 Fearing that several oil paintings he purchased in 1854 were forgeries, the textile merchant Pavel Tretyakov (1832–98) turned to amassing his munificent collection of Russian art, whose authenticity he could verify: Norman, "Pavel Tretiakov and Merchant Art Patronage," 95.

17 Druzhinin, "Istoriia odnoi kartiny" (1852), *SS*, 1:524–45, quote 530.

acquired fresh urgency. Did Byronic cosmopolitanism remain a viable travel regime? Or to the contrary, should vacationing now yield to journeys aimed at acquiring knowledge vital to Russia's future? A journalist of 1857 believed travelers could combine the two agendas. He asserted that "all these trips" to the West could "yield useful benefits along with pleasure," if Russian visitors would compare "the foreign with the native" and think about how "to accept and apply what is good, after rejecting the bad."[18] This author was writing tourists into a collective project of assessing Western Europe with a mind to determining the ideal balance between tradition and change for Russia's development.

But how feasible was this amalgamation of pleasure and purpose? The difficulty of combining cultural tourism with a political fact-finding mission will be weighed here in a clash between the radical plebeian critic Chernyshevsky and the conservative, elitist poet Fet, well-established by 1850 for his lyrics of tender love, the Greek anthology repertoire, wondrous gardens, and other beauties of nature. Beginning with his master thesis *Aesthetic Relations of Art to Reality* (1855), Chernyshevsky disparaged the museum world of the West and maligned the typically upper-class Russian tourists it attracted. Fet, on the other hand, was a vehement advocate of art for art's sake and enacted an art pilgrimage in his *From Abroad*, a linear narrative that begins with the author's departure from Kronstadt on a Prussian steamship and then treats each stop on his itinerary: Stettin, Berlin, Dresden, Carlsbad, Frankfurt, Strasbourg, Paris, and Marseille.[19] Chernyshevsky's polemical stance on the art pilgrimage and Fet's counter-polemic announced dim prospects for efforts to fold political purpose into a regime of highbrow cultural consumption.

The Advent of *Turizm*

In re-opening gateways to the West, the reformist orientation of Alexander II's regime made fashionable again a cosmopolitan gusto for travel. Annenkov's *Letters from Abroad* had represented spa tourism as a cosmopolitan opportunity for Russians to mingle with the sociocultural elite of Western Europe. Fet's *From Abroad* and an 1857 travel account by the writer

18 Anon., "Prodolzhenie letnei emigratsii," 22–23.

19 A. A. Fet, *Iz-za granitsy. Putevye vpechatleniia*, *Sovremennik* 60, no. 11 (November 1856), section 1: 71–117; ibid., 61, no. 2 (February 1857), section 1: 237–71; and ibid., 64, no. 8 (August 1857), section 1: 81–128.

Dmitry Grigorovich (1822–99) also found stimulating the "cosmopolitan," multinational throng of visitors in Carlsbad and Nice, respectively.[20] Melgunov elaborated the perspective in hailing *turizm* as a reflection and promoter of modern-day internationalism. While regarding Karamzin as a Russia's first tourist, Melgunov emphasized that most Russians of bygone times were stay-at-homes, partly because travel had been difficult but mainly for psychological reasons: "attachment to the hearth, a lack of curiosity, fear of the new, hostility to the foreign," and the belief that everything "ours" is better just because it is "ours." But now, continued Melgunov, "we live in an era of rapprochement between people and nations." The paved roads, steamships, and locomotives that make travel easier are merely the technological component of a larger, sociocultural conjuncture: "Everywhere, in all spheres, distance is shrinking, intercourse is expanding, borders are vanishing, security and customs barriers are being relaxed, people are ceasing to be foreign to one another, and former stay-at-homes are visiting neighbors near and far."[21] In this rosy vision of a cosmopolitan community of nations, Russians are second only to the English in their zest for tourism, and that exploratory urge remains rooted in the old, nagging perception of central Russia's lack of scenic and cultural appeal.

Melgunov ventured to outline a typical postwar Russian tourist itinerary. He said Russian trips began in Germany, particularly at spas with gambling casinos. Next came Paris, the one place no Russian would have dreamed of missing. If tourists had enough money after the casinos and Paris, they went to Italy, Melgunov's own evident favorite (6). Corroborating the Russian tourist patronage of Italy was the existence of a Hôtel de Russie in Naples as early as 1855.[22] England was an afterthought in Melgunov's itinerary. He said Russians considered it edifying but boring, especially after Paris (11). That comment was misleading, as we will see shortly. In addition, Melgunov oddly omitted Switzerland, a country that kept attracting Russians, largely for the authenticity-of-nature quest that Karamzin would have found familiar.[23] As another active observer of the late 1850s put it,

20 Fet, *Iz-za granitsy* (November 1856), 110; and D. V. Grigorovich, *Korabl' "Retvizan" (God v Evrope i na evropeiskikh moriakh)*, in *Polnoe sobranie sochinenii*, 3rd ed. (St. Petersburg: A. F. Marks, 1896), 9:326.

21 Mel'gunov, "Turisty voobshche," 1.

22 Vladimir Iakovlev, *Italiia. Pis'ma iz Venetsii, Rima i Neapolia* (St. Petersburg: Tipografiia Koroleva, 1855), 64.

23 For a biographical, literary study of Russian attraction to Switzerland from the eighteenth into the twentieth century, see Mikhail Shishkin, *Russkaia Shveitsariia.*

Russians were going to Switzerland in search of "picturesque" scenery, for the sake of which "they scorn comforts and are prepared to spend the night in a solitary hut on a mountain."[24]

Far from arousing a sense of tedious obligation, London strongly attracted postwar Russian visitors for two reasons. First, it acquired a new positive image as a cultural tourist destination less expensive than Paris. Botkin made the budgetary point in an article about his two-week holiday in London in 1859. He also praised London's musical offerings, the parks, and the fine weather from April to September. He wrote of hotels and furnished apartments to suit every pocketbook and standard of "comfort and elegance." Botkin advised Russian travelers to learn some English before their trips in order to patronize hotels run by Englishmen rather than the inevitably "dirty, bad" foreign-run, mainly francophone establishments.[25] Herzen, too, would add to London's cultural appeal in publicizing well-attended concerts of Russian music held there in the summer of 1860. Offering balm to the Crimean War wounds of Russian national esteem, Herzen presented these events as musical revanchism, "an aesthetic invasion of Russian sound, advancing from one victory to the next."[26]

More historically important than postwar London's cultural appeal was the interest it aroused in Russians wondering what industrialization would mean for their country. Britain emerged from the war as the world's one remaining global power, and London came to epitomize for Russians the pros and cons of capitalist development. Beginning with Chernyshevsky prior to the war, young radicals valorized Western and specifically British achievements of industry and engineering, including suspension bridges and the Crystal Palace reconstructed in Sydenham in 1854. In an unsigned article published that year, Chernyshevsky, on the basis of his readings, hailed the Palace as a universally recognized "miracle of art, beauty, and splendor."[27] The plebeian revolutionary, poet, translator, and journalist Mikhail Mikhailov (1829–65) extended the tradition in his travel cycle *London Notes* (1859), which argued that the benefits of industrialization

---

Literaturno-istoricheskii putevoditel' (Moscow: Vagrius, 2006).

24 A. Ia. Panaeva [N. Stanitskii, pseud.], *Russkie v Italii*, *Sovremennik* 67, no. 2 (February 1858), section 1: 298–99.

25 V. P. Botkin, "Dve nedeli v Londone 1859 goda," in *"Ia bereg pokidal tumannyi Al'biona": Russkie pisateli ob Anglii, 1646–1945*, ed. O. A. Kaznina and A. N. Nikoliukin (Moscow: ROSSPEN, 2001), 281–82.

26 Herzen, *SS*, 14:291.

27 Young, "The Crystal Palace," in Martinsen and Maiorova, *Dostoevsky in Context*, 179.

outweighed its negative impact on workers and the natural environment.[28] Other publications in Russia took the opposite view. Serialized in 1854 in *Sovremennik* (*Contemporary*), the leading journal of progressive opinion, Charles Dickens's *Bleak House* abounded in images of the toxic Thames, filthy air, piteous working children, and the insalubrious coops of the poor. Thoroughly consonant with Dickens's vision of London was Dostoevsky's anti-capitalist travel narrative *Winter Notes on Summer Impressions*. In between the two extremes fell the London section of Goncharov's *The Frigate Pallas*. Goncharov represented air pollution, the dirty Thames, and dehumanizing utilitarianism at work in the English capital. But on the other hand, he was impressed by the British Museum, could not resist shopping for some useless trinkets, and was awed by the kaleidoscopic sight of a steam locomotive rushing above a "glittering ocean" of brightly lit streets.[29] Conflicting Russian travel accounts thus played a prominent role in making London a focus of debate about industrialization, material progress, and consumerism.

Pent-up desire to visit Western Europe crossed the Slavophile-Westernizer divide. A patriotic contingent now considered Russian pleasure trips to the West unseemly and even harmful. Such was the view of the Kharkov University professor K. P. Paulovich, whose book *Observations on Italy* (1856) presented Italy as an overrated tourist destination bound to corrupt Russians' hearts and minds.[30] Evidently referring to such attitudes, Turgenev's essay "From Abroad" (1858) observed that some Russians now regarded the West's cultural richness as a strictly European rather than universal patrimony. With the air of expressing a "patriotic sentiment," those people have become fond of intoning that "there is no profit to be had from foreign wealth."[31] For his part, the Slavophiles' chief ideologue Konstantin Aksakov donned an Old Testament mantle to rebuke hedonists of every stripe:

28 See my ch. 7 for close reading of Mikhailov's travel writings. Some of the latter appear in volume 3 of his *Sochineniia*, 3 vols. (Moscow: Gosudarstvennoe izdatel'stvo khudozhestvennoi literatury, 1958).

29 I. A. Goncharov, *Polnoe sobranie sochinenii i pisem*, 20 vols. (St. Petersburg: Nauka, 1997–2017), 2:39–40, 46; and commentaries in Offord, *Journeys to a Graveyard*, 200–1; and Bojanowska, *World of Empires*, 25–31. Bojanowska accentuates Goncharov's positive responses to London, including its "consumer culture" (26).

30 Commentary in Dobroliubov, *SS*, 7:626.

31 Turgenev, "Iz-za granitsy," in *SS*, 12:184.

> Work is man's duty, his normal condition on Earth. Work alone gives one the right to enjoy life [. . .]. "In the sweat of thy face shalt thou eat bread, till thou return unto the ground." Such is man's fate and duty. Life is not pleasure, as some people think.[32]

This killjoy attitude implicated the new craze for tourism, if only implicitly. The vacation mentality also offended the Slavophile sentiments of Anna Tyutcheva, daughter of the poet Fyodor Tyutchev (1803–73) who had published patriotic religious verse during the war. A lady in waiting at the Russian court, Anna recorded in her diary for January 1856 that the minister of foreign affairs Count Nesselrode and his entire social set took a celebratory attitude toward the fall of Sevastopol: they cared about nothing else except getting back to Paris and going to Italy to eat pasta and hobnob with local aristocrats.[33] All the same, not every Slavophile found tourist trips to the West an affront to national mourning. Tyutcheva's future husband Ivan Aksakov (1823–86), the younger brother of Konstantin, toured Germany, France, Italy, and Switzerland for several months in 1857. Letters to his parents made clear that he did this with their blessing and that he found it exhilarating to see newly accessible countries he knew only from books.[34]

Whatever their motivations, itinerary, or politics, Russians were purportedly streaming to the West in record numbers. In July 1857 *Annals of the Fatherland* remarked that thanks to the "new law" favoring travel, Russians were effecting a summer "emigration" abroad.[35] A glut of travel narratives in the periodical press contributed to the impression of mass mobility. In the fall of 1857 the *Contemporary* conjured a legion of traveling Russians with pens in hand: "This past summer, with few exceptions, all literary Petersburg went abroad – poets, journalists, *feuilletonistes*, critics, and so forth, so that all Petersburg's newspapers and journals are flooded with letters, travel notes written on the run (*à vol d'oiseau*), travel impressions, *récits de voyage*, recollections of Europe, etcetera, etcetera."[36] As the minor poet Boris Almazov put it in his verse "The Tourist" (1859): "Seduced by

---

32 K. S. Aksakov, *Estetika i literaturnaia kritika* (Moscow: Iskusstvo, 1995), 393.

33 A. F. Tiutcheva, *Pri dvore dvukh imperatorov. Vospominaniia, dnevnik, 1853–1882*, 2 vols. (1928–29; repr. Cambridge, MA: Oriental Research Partners, 1975), 2:98.

34 Aksakov, *Aksakov v ego pis'makh*, 3:313–45.

35 Anon., "Prodolzhenie letnei emigratsii," 22.

36 Anon., "Peterburgskaia zhizn'. Zametki novogo poeta," *Sovremennik* (September 1857), section 5: 82.

the cut-rate passport fee, / Everybody is rushing to leave the dear fatherland."[37] Writing travelers reinforced the big numbers theme in observing vast amounts of Russian graffiti on walls in Carlsbad and on monuments in Italy.[38] One such traveler also reported that beggars in Rome were adding Russian words to their vocabularies, another sign of growing numbers of Russians touring Italy.[39] According to Melgunov's undocumented figure, 43,000 Russians received passports to visit Western Europe in 1858 alone. He found the number astounding and added that less than 3,000 were related to medical needs.[40] In 1857 Ivan Aksakov encountered "swarms" of Russians in Paris, and on his second trip abroad three years later he found "a multitude of Russians everywhere in Germany" and several fully booked pensions catering to Russians on Lake Geneva.[41] No doubt remembering those occasions, he would assert in 1863 that Russian tourists "rushed to Europe and flooded it" right after the Crimean War.[42] As captured by the trope of an inundation, postwar Russians perceived a big spike in the number of travelers as the principal sign of a revolution in their national practice of tourism.

The social diversity of Russian holidaymakers was greater than ever. Turgenev represented the aristocratic old guard. After languishing at home for three years when foreign travel was prohibited, he hastened back to France and Italy in 1856. Alongside the wealthy nobles came an influx of members of the plebeian intelligentsia. With the emancipation looming, the nobility of Turgenev's generation faced the prospect of diminished economic and social status (and hence the sale of estates and the flight of capital abroad). Their power in the cultural sphere was challenged as well by the increasingly radical *raznochintsy* journalists Chernyshevsky, Dobrolyubov, and their followers at the *Contemporary*. Work as reporters abroad enabled some members of that middle-class cohort to make journeys to Western Europe otherwise beyond their financial means. Besides Mikhailov, who publicized his experiences in

37 B. N. Almazov, "Turist," in *Poety 1860-kh godov*, ed. I. G. Iampol'skii (Leningrad: Sovetskii pisatel', 1968), 607.

38 Fet, *Iz-za granitsy* (November 1856), 112; and G. P. Danilevskii, *Pis'ma iz-za granitsy*, in *Polnoe sobranie sochinenii*, 24 vols., 8th ed. (St. Petersburg: Tipografiia A. F. Marks, 1901), letter from April 22/10, 1860, 24:140.

39 Danilevskii, *Pis'ma iz-za granitsy*, 139.

40 Mel'gunov, "Turisty voobshche," 6.

41 Aksakov, *Aksakov v ego pis'makh*, 3:314, 364, 490.

42 I. S. Aksakov, "Iz Parizha," *Polnoe sobranie sochinenii*, 7 vols. (Moscow: Tipografiia M. G. Volchaninova, 1886–87), 2:98.

England and France, Dobrolyubov, too, extensively toured Western Europe during 1860–61, when he went abroad to seek medical treatment for tuberculosis and to write at-the-scene reports on the Risorgimento.

In addition to reflecting the emerging clash between aristocratic liberals of Turgenev's generation (soon dubbed "the men of the 1840s") and the young radicals ("the men of the 1860s"), the new tourist scene bore witness to Russian women's liberation. Women were rebelling against traditional restrictions, entering universities, traveling abroad, and defying established sexual morality.[43] Among the women now embarking on the tourist circuit was the writer Avdotiya Panayeva (1819 or 1820–93). From a family of actors, she specialized in stories of women's liberation and used the masculine pseudonym N. Sitnitsky. Panayeva first went abroad in 1844 with her husband Ivan Panayev, a *Contemporary* editor. By the late 1840s she had divorced and become the common-law wife of the poet and senior *Contemporary* editor Nekrasov with whom she visited Italy and France in 1856. A product of that trip was Panayeva's *Russians in Italy* (1858), a hybrid work on the boundary between travelogue and the literary genre of the society tale. Two other liberated women of the time left travel diaries, published only in the Soviet era: Lyudmila Shelgunova who formed a *menage à trois* with her husband Nikolai and Mikhailov and visited England and France with one or both of them during 1858–59; and Dostoevsky's temperamental lover, the former serf Apollinaria Suslova, who traveled with him in the West in 1863.[44]

The collapse of social barriers to leisure travel made newly problematic the element of masquerade that Belinsky had identified in writing about Russian tourists successfully passing for foreigners abroad. While clothing could announce a person's wealth or its absence, the freedom to dress as one liked when traveling eradicated the vestimentary code of uniforms that operated at home. Beginning in the reign of Nicholas I, civil servants and students as well as military men had uniforms with distinguishing features. Russian tourists lost their anchor in that semiotic system, and that made them fear mingling with compatriots of lesser rank. In a satirical vignette

---

43 Richard Stites, *The Women's Liberation Movement in Russia: Feminism, Nihilism, and Bolshevism, 1860–1930* (Princeton: Princeton University Press, 1978), 29–63.

44 L. P. Shelgunova, *Iz dalekogo proshlogo. Perepiska N. V. Shelgunova s zhenoi,* in N. V Shelgunov et al., *Vospominaniia*, 2:74–95 (the diary mainly concerns trips of 1856 and 1858 with her husband); and Polina Suslova, *Diary*, in F. M. Dostoevsky, *The Gambler*, trans. Victor Terras, ed. Edward Wasiolek (Chicago: Chicago University Press, 1972), 201–302.

by the art critic and memoirist Pavel Kovalevsky (1823–1907), the mighty Table of Ranks helps explain the anxiety attack of a Russian tourist on the threshold of a train compartment. Before daring to take a seat, the newcomer wonders who are the passengers already installed: "Among his fellow travelers is there a cobbler, a person with no official rank, a Jew, or some Russian who cannot count a general among his uncles?"[45] The "cobbler" most likely referred to a middle-class Englishman, a category of tourists whose numbers began exploding in 1855 when Cook offered his first foreign tour, taking in Germany, Belgium, and France. Cook's democratization of Continental tourism made upper-class English travelers nervous about rubbing elbows with social inferiors abroad.[46] In Kovalevsky's estimation, though, Russians fretted about this risk more than any other national group, a peculiarity reflecting the all-important role of rank in Russian society.

The Railroad

While Kovalevsky used train travel to address the obsession with rank, other writings of the period shed a broader light on the cultural history of the railroad in Russia. As of 1855 Russia had only around 654 miles of railway, a paltry figure that made the state's plan to expand the network a topic of tremendous public interest. A flurry of articles appeared in 1857: where would the new railroads run, what purposes would they serve, what social impact would they have, and how would they alter the personal experience of travel?[47] The network reached nearly 2,050 miles by 1862, the year the Petersburg-Warsaw line opened after eleven years of construction. Originally undertaken with the aim of strengthening military control over Poland, the railroad to Warsaw made it possible for the first time for Russians to travel by train from home and transfer to lines into Western Europe.[48] As in the Romantic period, then, most Russians of the immediate post-Crimean War period discovered the "iron horse" abroad.

---

45 P. M. Kovalevskii, "Putevye vpechatleniia ipokhondrika," *Sovremennik* 77, no. 10 (October 1859), section 1: 363.

46 Buzard, *The Beaten Track*, 55–59.

47 K. S. Aksakov, *Estetika i literaturnaia kritika*, 390–92. See also Anon., "Znachenie predpolozhennoi seti zheleznykh dorog dlia raznykh kraev Rossii," *Otechestvennye zapiski* 111 (March 1857), section 3: 21–22; and Anon., "Prodolzhenie letnei emigratsii," 22–23.

48 Haywood, *Russia Enters the Railway Age*, 536–44, 589–90, 594n33.

The old divide between haters and lovers of trains persisted in travel accounts of the 1850s. Published in London, the second (1858) edition of Herzen's *Letters from France and Italy* carried romantic nostalgia for the Road into the post-Crimean War period.[49] The account fashions the author as "a revolutionary tourist."[50] The main narrative line treats Herzen's development as a political thinker convinced that the Russian peasant commune provided a basis for a Russian form of socialism that would avoid the ravages of capitalism. The *Letters* nonetheless gives much attention to tourism and even begins on that note. Written in Paris, the first paragraph of Herzen's narrative asserts that "an idle person can have no better life than that of a tourist; there are countless things to do, you have to see everything, rush everywhere, and you think you are accomplishing something."[51] The self-irony is evident, and yet the *Letters* does in fact dramatize the author's indulgence of tourist curiosity. He explores the whole cultural spectrum ranging from the exquisite pleasure of seeing "Racine at home" at the Comédie française, to observing the vulgar "sensuality" of crowds at the Bal Mabille and other Paris *cafés chantants* (32, 50–52). But before recounting any tourist activities, Herzen deplores train travel, as discovered in Germany. Expressing the idea of annihilating space, he declares that railroads were invented to make Europe disappear and eventually to efface it from human memory. For the train traveler, the signified of "Europe" is dwindling down to a sequence of railroad station cafes, "a number of points illuminated by lanterns." On this view, even coach transport in France operates as a dehumanizing "machine" and lacks "all the poetry" of the Russian road. Travelers in France purportedly do not talk to one another, and there are no Russian-style adventures of negotiating with station masters for fresh horses (16, 69).

By contrast to Herzen's nostalgia for the Road, other Russian accounts of the 1850s share the positive reactions to the railroad that Annenkov, Zhukova, and Pogodin had expressed. Destined to make a fortune as an industrial entrepreneur, Vasily Kokorev (1817–89) left an account of a train trip from Berlin to Cologne in 1859. He was amazed by the smoothness of the ride and the excellent, efficiently served food at stations along the

---

49 On the previous publication, see my ch. 2, n. 42.

50 *Journeys to a Graveyard*, 178. Offord's reading of Herzen focuses on the political dimension, 174–95.

51 Herzen, *Pis'ma iz Frantsii i Italii*, 15.

route. He admired the Rhine and the "famous cathedral" in Cologne.[52] Fet's *From Abroad* praised the convenience, comfort, and efficiency of trains and prayed for the expansion of railroads in Russia. En route from Berlin to Dresden, the author exclaims: "What a marvelous thing the rail network is! Wherever you think of going is close. In five years or so, God willing, for us Russians, too, going from Petersburg to Odessa will be closer than going from Pskov to Novgorod or from Voronezh to Tambov." The compression of space brings Odessa "closer" to the Russian capital, whereas lesser towns, including the exemplary backwater Tambov, remain relatively "far" from one another, on the assumption that the provinces will long lack train service. Fet's persona also perceives Western Europe unfolding alongside the train as clearly as a stage performance. Never reduced to a blur, the passing scenery and towns rivet his gaze, as though he were a spectator at a "new theater," watching a "new play" he previously knew only from hearsay.[53]

With a similar adjustment to panoramic vision, Panayeva's *Russians in Italy* represents a railroad journey from Laibach (now Ljublyana) to Trieste as a thrilling triumph of engineering. The author asserts that the train ride makes the passenger feel "buoyant and free," as though "recovering from a serious illness." During the trip's entire ten hours, it is impossible to "tear one's eyes from the window, even for a minute."[54] The infrastructure and the locomotive's conquest of awesome terrain amaze the traveler:

> The puffing engine plunges into a hollow between mountains that rise higher and higher, and as the cars fearlessly thunder along, the track narrows. Winding like a snake, the engine climbs but no sooner reaches the top of a cliff, when another juts up, obstructing the railway now tapering down to a little path on the edge of a precipice at the base of a huge wall of yellow marble. [. . .] Eventually, granite massifs completely block the way. . . . The train screeched piteously as a frightened mouse, darted into a narrow aperture in a mountain and, thrashing in the dark, filled the vault with an extraordinary rumble and hissing. In a moment the cars exit into the light, to dash across a long bridge over an abyss and disappear again into the dark of a tunnel. Seeping springs wet

52 V. A. Kokorev, "Iz putevykh zametok. Pis'mo k S. A. Khrulevu," in *Utro. Literaturnyi sbornik*, ed. M. P. Pogodin and colleagues at *Moskvitianin* (Moscow: Tipografiia Barfknekhta i Ko., 1859), 208–10.

53 Fet, *Iz-za granitsy* (November 1856), 71, 93.

54 Panaeva, *Russkie v Italii*, 308.

> the granite cliffs, as though shedding a stream of tears over man's disfiguring them. (309)

The assault on nature gives pause but does not cancel the passenger's excitement over the feats of technology. Aesthetic pleasure intensifies with the "magical transformation" of scenery as the train descends to Trieste. "A native of the North," now seeing the Mediterranean for the first time, the author is swept away by "the grandeur of this picture": the deep blue and greenish tones of the ocean, sparkling sun on the water, the vividness of foliage, and snowy mountain peaks gleaming like mother-of-pearl (310). In parallel to Fet's metaphor of train travel as a theatrical show, Panayeva's detailed visualization of topography and scenery manifests the modern capacity to develop a new aesthetics of perception appropriate to a railroad journey. These travelers assimilate industrialized transport instead of fighting it as Herzen did.

Art: Universal or National?

A good half of the Slavophile Khomyakov's idiosyncratic article "A Letter to Petersburg about the Railroad" (1845) dealt with the fine arts set in opposition to trains. A polymath who invented a steam engine and patented it in England, Khomyakov attributed universal value to technology and science and argued that Russia should take full advantage of Western achievements in those areas. "In the realms of pure and applied sciences," he wrote, "every people uses the discoveries and inventions of others, without lowering self-respect or losing rights to independent development." Khomyakov's "Letter" advocated the expansion of Russian railroads in particular, primarily on strategic grounds. In order to be able to concentrate military forces as needed, he presciently argued, Russia required a rail network centered in Moscow with lines connecting the Baltic to the Black and Caspian Seas.[55] Khomyakov spoke up again in favor of Russian railroad expansion in 1857, this time emphasizing commercial rather than military benefits.[56] But while never questioning the universal value of technology and science, he urged Russia to protect its cultural distinctiveness from

---

55 A. S. Khomiakov, "Pis'mo v Peterburg po povodu zheleznoi dorogi," *Sochineniia*, 3 vols., 2nd ed. (Moscow: Tipografiia Lebedeva and Universitetskaia tipografiia M. Katkova, 1878–82), 1:452–53, 457, 463.

56 Khomiakov, "Pis'mo k izdateliu *Russkoi besedy*," ibid., 607–11.

Western contamination (an idea conforming to his ironic trope of the "holy wonders" in "A Dream"). Khomyakov conceded that Russia had little to show in the fine arts but believed his homeland would overtake the West in those domains. Russian painters, he contended, would attain the supreme religiosity of the icon and thereby surpass Raphael's Madonnas, the works Khomyakov considered the spiritual pinnacle of Western art. On this view, Raphael remained a fundamentally national figure, rooted in Italian culture and Italian ways of life, just as the paintings of the Flemish and the Dutch bore traces of their national origins.[57]

In conceiving culture as a nationally bounded domain, Khomyakov disputed the cosmopolitan assumption that had always underwritten art-oriented tourism: art is an expression of human creativity that can matter to anybody anywhere no matter when or where it was produced.[58] In Goethe's classic formulation, the well-prepared cultural tourist was a pilgrim: "Pilgrime sind wir alle, die wir Italien suchen." The art pilgrim sought Italy above all but was drawn to the display of art everywhere, including the Dresden picture gallery which Goethe proclaimed a "temple" receiving devotés from far and wide.[59] A strong Russian echo of Goethe's outlook was Zhukovsky's essay on the *Sistine Madonna*. Based on a trip of 1822, the essay represented the visit of the Dresden gallery as a quasi-religious cultural ritual that elevates the author to a plane of existence above the milling crowd of ordinary visitors, inattentive and incapable of appreciating Raphael's spiritually profound achievement.[60] Without designating museums "temples," the travel narratives of Pogodin, Annenkov, and Myatlev—despite their political differences—all portrayed art housed in Western Europe as objects of universal value. Closer in spirit to Goethe was Apollon Maikov's "After Visiting the Vatican Museum" (1845), one of twenty-nine lyrics comprising his travel cycle *Sketches of Rome*. The poem opens with a Byronic reminiscence of the collection's two most famous statues: "I still hear the howling and roar of Laocoön, / Apollo's bowstring reverberates in my ears." The poet is "a pilgrim from afar," now wandering "among sacred things," a sequence of ancient statues. Ending on a note of Russian cultural intimidation by the Vatican, he imagines himself as one of the "savage

57 Khomiakov, "Pis'mo v Peterburg," 458–66.

58 Appiah, *Cosmopolitanism*, 118–35.

59 Germain Bazin, *The Museum Age* (New York: Basic Books, 1967), 159, 164.

60 Schönle, *Authenticity and Fiction*, 98–105. Schönle's analysis provides the base for my discussion of parallels between Fet and Zhukovsky, below.

Scythians" come from "across the Dneiper," to stand wonderstruck at the grandeur of Byzantium.[61]

The art pilgrim trope in Russian travel writing culminates in Herzen's *Letters from France and Italy*. While still living in Russia, Herzen spent many pleasurable hours visiting the collection of the Hermitage, especially Raphael paintings and the copies of his Vatican palace *stanze*.[62] The experience evidently whetted his appetite for museum visiting abroad. Hailing Goethe as the pathbreaker of art-oriented journeys to Italy and Rome in particular, Herzen's *Letters* quotes and expands upon the phrase "Pilgrime sind wir alle". All Rome, writes Herzen, overwhelms the visitor with its "wealth of exquisite art works, that finality of genius, that eternal beauty before which a person halts with reverence and a tear—moved and shaken to the depths of his soul, purified by what he has seen and reconciled to many things." Traveling from all over the globe, pilgrims come here to "pay homage to art"—literally, to perform an act of "worship" (*poklonenie*). Having recounted his flight from Paris in despair at the suppression of the 1848 revolution, the author attributes to the Vatican collection the capacity to restore one's faith in humanity. In art—these "great guarantors of human power"—the dispirited visitor "finds again something to bless in life."[63] Art comprises a "universal" (*obshchechelovecheskoe*) rather than national patrimony, a proposition Herzen accuses Slavophiles of denying because they consider "everything universal foreign, everything civilized alien" (24). While sarcastically exaggerating the Slavophile position, Herzen's outlook on universality represents the dominant highbrow Russian tourist outlook of the time. Another particularly striking example is the literary historian and enlightened censor Alexander Nikitenko's (1804–77) 1857 essay on seeing the *Sistine Madonna*. Raphael's painting, he wrote, is "a panhuman creation" (*vsechelovecheskoe sozdanie*), a work of "world-wide significance," "an annunciation of divine truths precious to the interests of all mankind."[64]

61 Maikov, "Posle poseshcheniia vatikanskogo museia," *PSS*, 1:66–67.

62 Herzen, letters from 1839–40 to various correspondents, *SS*, 22:22–23, 65–67, 88.

63 Herzen, *Pis'ma iz Frantsii i Italii*, 86, 102. Goncharov's *The Frigate Pallas* drolly engages the pilgrim tradition in writing of tourists and locals trooping "from temple to temple, from museum to museum" in London. The author is among them, however, and values the learning experience: *PSSiP*, 2:46.

64 A. V. Nikitenko, "Rafaeleva sistinskaia madonna," *Russkii vestnik* (October 1857), section 1: 589, 593.

Chernyshevsky's Attacks on the Art Pilgrimage

In continuity with Belinsky's perception of the *Sistine Madonna* as an exclusionary, aristocratic painting, Chernyshevsky's *Aesthetic Relations of Art to Reality* denied the universality of art and the meaningfulness of pilgrimages to see it. Whereas Khomyakov's disagreement with aesthetic cosmopolitanism was nationalist, Chernyshevsky's was that of an evolving socialist revolutionary who regarded the fine arts as upper-class fetishes, unnecessary to the mass of humanity. Chernyshevsky was well versed in imaginative literature, and his thesis made clear that he had read widely in aesthetic theory from Plato to Lessing. Disputing the elite Russian consensus of the period, his primary aim was to overturn Kant's doctrine of the autonomy and specificity of art and aesthetic experience. His efforts to do this fully justify Victor Terras's observation that the *Aesthetic Relations of Art to Reality* reads as the work of an author "who is not only uninterested in art but who actually has never bothered to find out what art is. It is the argument of a man unfamiliar with aesthetic experience."[65]

As made evident in a preparatory essay from 1854, Chernyshevsky was itching for a fight with an envisioned audience of the cultural elite:

> Go admire your *Apollo Belvedere*, you lofty minds for whom everything earthly is base and lacks the worth of a dead statue on a marble pedestal! For people who have not attainted your disinterested heights, a young housewife delighted as a child at having neatly furnished her modest little three- or four-room apartment is more poetic than all the Venuses of the Medicis and the Louvre; and we dare deem her sentiment more poetic than yours if all you do in life is try to explain Greek statues and *Iphigenia in Taurus*. "Vulgarity, triviality!" you will say again and again. No! A human being is not vulgarity, and in your cold idols there is less poetry than in [Gogol's] Akaky Akakyevich delighted with his overcoat.[66]

---

65 Terras, *Belinskij and Russian Literary Criticism*, 237. See also E. Lampert, *Sons against Fathers: Studies in Russian Radicalism and Revolution* (Oxford: Oxford University Press, 1965), 213–221; and William F. Woehrlin, *Chernyshevskii: The Man and the Journalist* (Cambridge, MA: Harvard University Press, 1971), 153–59.

66 N. G. Chernyshevskii, "Kriticheskii vzgliad na sovremennye esteticheskie poniatiia," *Polnoe sobranie sochinenii*, 16 vols. (Moscow: Gosudarstvennoe izdatel'stvo khudozhestvennoi literatury, 1939–53), 2:155

The polemical intent to stir up a fuss no doubt explains many of the inanities that the *Aesthetic Relations of Art to Reality* articulates about painting and sculpture, the art forms Chernyshevsky ranks as the poorest, while defining utilitarian, engagé literature as the most valuable. Chernyshevsky contends that "beauty is life; the most lofty, genuine beauty is beauty a person encounters in reality, not beauty created by art."[67] This materialist assumption restricts art to functioning as "a textbook of life." A seascape, for instance, can broaden a landlocked viewer's knowledge of the world. On a personal note, Chernyshevsky assigns to portraits the function of serving as the "surrogate" of an absent person (14, 77–78). Compared to "nature's work," however, paintings and sculptures are "extremely crude, awkward, and ungainly." No painted apple has a real fruit's vivacity; painters cannot capture the true tone of flesh or represent objects in bright sunlight; and "even Raphael" displays "ignorance of anatomy." The "deadness" and "immobility" of statues makes them inferior to life, and no sculpture can be as beautiful as "the multitude of living people" one may encounter any day on a city street (38–39, 51, 55, 57, 59). Although Chernyshevsky grants that some statues are better than others, he expresses a "seen one, seen them all" attitude. True, he grants, the *Venus de' Medici*, *Venus de Milo*, and *Apollo Belvedere* are not available in Russia, but the Hermitage has works by Canova that give us the "courage to judge" how good sculpture can be.[68] No need to go abroad, then, if you can see a Canova in Petersburg. Chernyshevsky further insists that no art is "immortal," as artists and their admirers like to pretend. Paintings and statues deteriorate, they get lost, or go out of fashion. They are ephemeral objects that have no eternal, sacred significance (42, 47–52, 55, 84).

In contesting the rationale of an art pilgrimage, Chernyshevsky anticipates Pierre Bourdieu's critique of the museum as an institution naturalizing invidious social distinctions.[69] Chernyshevsky's arbiter of aesthetic value is the common man, "the aesthetically healthy person" content with the beauty of life at his place of residence, wherever that may be (while conceding that the Crimea and Switzerland have exceptionally glorious scenery, Chernyshevsky, as did Belinsky before him, urges readers to

67 Chernyshevskii, *Esteticheskie otnosheniia iskusstva k deistvitel'nosti*, in *PSS*, 2:14.

68 The Hermitage collection included Canova's *Three Graces*.

69 Pierre Bourdieu and Alain Darbel, with Dominique Schapper, *The Love of Art: European Art Museums and Their Public*, trans. Caroline Beattie and Nick Merriman (Cambridge, UK and Malden, MA: Polity Press, 1997), 14–107.

learn to take visual pleasure in central Russia's natural environment).[70] Chernyshevsky pits his ideal of aesthetic health against "worshippers of art," the upper-class sophisticates whose study of the "latest aesthetics" have given them "artificial tastes" and left them too "corrupt" to perceive "the riches reality offers" (11, 40, 42, 45, 59, 74). Even if the aesthetically healthy person wanted to visit a museum, he might not have the leisure or the means to do so (the Hermitage, we should recall, still had a dress code requiring frock coats). In view of the socioeconomic disparities, Chernshevsky argues that the convenience and accessibility of reproductions make them more valuable than actual art works: "Only people who go to a gallery," he writes, can admire a painting hanging there, whereas reproductions make the image available "throughout the whole world. Anybody can admire it when he pleases, without leaving his room, getting up from the couch and changing from his dressing gown [*khalat*]" into street clothes (77). In shunning class-marked museums, the "aesthetically healthy" homebody begins to resemble Goncharov's monumentally inert Oblomov, who spends much of the day lying on his couch in his dressing gown.[71] Advanced in the name of democratizing culture, Chernyshevsky's preference for facsimiles makes evident his total lack of sensitivity to what Walter Benjamin's essay on art in the age of mechanical reproduction called the "aura" of an original. A copy of the *Mona Lisa* may be indistinguishable from the original, but "only the original has the aura," "the connection with the hand of Leonardo."[72]

In reviewing *Italy: Letters from Venice, Rome, and Naples* (1855) by the Petersburg journalist Vladimir Yakovlev (1817–84), Chernyshevsky reiterates the argument about reproductions, impugns the moral character of cultural tourists, and advocates a political consciousness-raising alternative to their agendas. Yakovlev's book extends the traditional tourist romance of Italy. Byron mediates the *Italy* vision of Venice as the poet's beloved "ocean queen," no less dear in the present than in her glory days (*Childe Harold's Pilgrimage*, 4.17–18). In the Russian account, "shc" is "a faded beauty," "a faded coquette" but still the splendid "Palmyra of the seas."[73] Yakovlev begins his tour with the Piazza San Marco presented

70 Ely, *This Meager Nature*, 148–49.

71 Goncharov published the first chapter of his novel in 1849 under the title "Oblomov's Dream."

72 Appiah, *Cosmopolitanism*, 134.

73 Iakovlev, *Italiia*, 3–4, 26, 56, 60.

as a site where artists from Raphael to Canova, writers from Petrarch to Goethe, and potentates from Peter the Great to Napoleon have stood. Instead of taking the traveling tsar as a prompter of Russian national pride, however, Yakovlev elevates Italian culture over that of his homeland. Addressing the reader in guidebook style, the author writes that the mosaics of the San Marco basilica "might make you think that you were in the Kremlin or Kyev, if so many of the mosaics did not manifest the masterly designs of Titian, Tintoretto, and Veronese" (5–7). As Yakovlev continues his visit, he views the architecture, canals, gondolas, and the play of light through the lens of Canaletto, whose paintings he assumes his Russian audience knows well. More generally, Yakovlev exalts Venice as an incomparable "realm of the arts." With an enchanted eye for "poetic" dilapidation, he perceives the city as "an entire museum of divine palaces, cathedrals, and semi-ruins" (22, 26, 31, 33).[74] He descends from the heights of art to warn readers about tacky souvenirs such as replicas of gondolas made "only for tourists;" and he recommends the hotel where he stayed on the Grand Canal: two well furnished rooms, an attentive staff, and at a price only half of the cost of "the alarming lodgings" one finds in the Russian provinces (51, 63).

While rating Yakovlev an engaging author, Chernyshevsky's review discredits his whole program of traveling to see and write about art. The critic contends that the Italian-journey genre generally fails to justify the aesthete's putative need to leave home to see Raphael paintings, "broken columns," and other "objects that everybody else finds a crushing bore." The trips are pointless because reproductions allow anybody to become acquainted with such things "perfectly well without ever leaving his study." Raising the moral stakes, Chernyshevsky goes on to malign the cultural pilgrimage to Italy as an undertaking not merely unnecessary but downright inhumane. Scribbling art lovers, he asserts, display indifference to people. If Russians cannot refrain from going to Italy, they should at least do it for the purpose of living among the residents in order to study them, their customs, and way of life.[75]

74 On the romantic Russian attitude to ruins, see Andreas Schönle, "Broken History and Crumbling Stones: The Romantic Conception of Architectural Preservation," *Slavic Review* 71, no. 4 (Winter 2012): 745–65.

75 Chernyshevskii, *PSS*, 2:628–30.

Fet's Polemical Defence

Aesthetic Relations of Art to Reality succeeded in riling the cultural elite (particularly incensed was Turgenev who called Chernyshevsky a "stinking cockroach" who should be flayed).[76] Among the indignant aesthetes, Fet stands out as the first to defend the art tour. *From Abroad* does not depict crowds of Russian tourists in the West but signals modern times in other ways: first, through the prominent theme of the railroad and secondly, by featuring new things to see on the Continent. Those include the ongoing reconstruction of Paris under the direction of Baron Haussmann, a separate room for the *Sistine Madonn*a following a renovation of the Dresden gallery completed in 1856, and the new Pinakothek in Berlin with a fresco cycle in progress designed by the painter Wilhelm von Kaulbach (1805–74), illustrating the history of mankind.[77]

In defiance of Chernyshevsky's antagonism to the cultural elite, *From Abroad* sculpts the author as an apolitical gentleman fully at leisure to pursue art-oriented tourism. The divergence from Fet's actual condition merits underlining. He was christened Afanasy Afansyevich Shenshin, the son of a Russian nobleman and his German companion Charlotte Foeth. But in 1835, after noticing the boy was born out of wedlock, Russian ecclesiastical authorities stripped him of noble status, his father's name, his right to inheritance, and even his Russian citizenship, in view of his mother's foreign origin. Fet's precarious socioeconomic position weighed heavily upon him until his 1860 marriage of convenience to the tea-fortune heiress Maria Botkin, Vasily's sister. *From Abroad* suppresses all those real-life problems. Foregrounding the aesthetic enjoyment of tourism, Fet uses the travel genre to fashion a more enviable self, a person who does not "exist at home."[78]

After establishing museums as his top priority in Berlin, the *From Abroad* author fully emerges as an art pilgrim in Dresden and Paris, the installments that mount a polemic against Chernyshevsky. While paralleling Zhukovsky's *Sistine Madonna* essay in many respects, Fet's treatment of Raphael's painting as a sacred object reads in context as a riposte

76 Frank, *Dostoevsky: The Years of Ordeal*, 247.

77 Fet, *Iz-za granitsy* (November 1856), 80–83, 97; and ibid. (August 1857), 117.

78 Quote borrowed from the discussion of British women travelers in Susan Bassnett, "Travel Writing and Gender," in Hulme and Youngs, *The Cambridge Companion to Travel Writing*, 234.

to Chernyshevsky's sneering at the elitist "worship" of art. Fet's persona sets out to see the Dresden gallery's whole collection, a standard tourist ambition producing museum fatigue. His impressions blur, his eyes grow tired, and he becomes "semiconscious as a sleepwalker," wandering from room to room. Then comes the claim of a sudden awakening: "Under the impact of Murillo, Rubens, Correggio, Rembrandt, Holbein, and others, I had completely forgotten Raphael's *Madonna*." This presentation of the *Sistine Madonna* as an afterthought heightens the spell it will cast on the viewer. He assumes the demeanor of a pilgrim. With humbly downcast eyes, he crosses the "cherished threshold" of the *Sistine Madonna* room as though entering a temple. Slowly he raises his eyes, to become transfixed by the painting. Fet's narrator insists that no reproductions of this "immortal work" can do it justice. All copies, he declares, "not only fail to remind" one of the original but "crudely distort" it.[79] This formulation speaks back to Chernyshevsky's insensitivity to the aura of an original art work, his denial of art's immortality, and his notion of portraits as surrogate reminders of the absent subject.

As the *Sistine Madonna* draws in Fet's persona, he claims to perceive more than artistry in the painting. Again like his fellow poet Zhukovsky, he arrogates for himself a special kinship to Raphael's "divine vision." Fet's narrator finds himself in the company of other visitors sitting on "little sofas" as they silently admire the painting. But in a state of "imperturbable bliss," the author elevates himself over the others, to assert that "God has contrived to make me an accessory to the vision." Known at home as a poet, this viewer defiantly enacts what Chernyshevsky configured as the conceit of artists fancying themselves superior to ordinary people. After interpreting the *Madonna* as a sacred object, Fet offers technical analysis with special attention to the sight-lines of Raphael's figures (yet another extension of Zhukovsky's essay). In conclusion, the *From Abroad* art pilgrim asserts he spent two hours looking at the *Sistine Madonna*, an hour longer than Zhukovsky claimed (102–3). Pogodin's *A Year in Foreign Lands* has already illustrated this tactic for suggesting exceptional sensitivity to cultural objects. But again in the immediate Russian context, Fet's specification of time appears to take a dig at

79 Fet, *Iz-za granitsy* (November 1856), 100–2.

Chernyshevsky's contention that few people can look at a painting for more than fifteen minutes.[80]

The Paris section of *From Abroad* embraces the erotic as part of the "sacred" treasure trove of the West. Represented as a complement to the *Sistine Madonna*, the second major object on Fet's art agenda is the *Venus de Milo*, one of the statues Chernyshevsky declared superfluous for Russians who could see a Canova work in Petersburg. This installment of *From Abroad* devotes eighteen of its thirty-four pages to visiting the Louvre described as the most important tourist site in Paris. "Lofty aesthetic enjoyment awaits you," Fet's narrator promises in guidebook style, as he conducts readers to the *Venus de Milo* after pausing to look at the *Discobolus* and other works in the museum's Greek sculpture division.[81] *From Abroad* underscores the erotic appeal of the *Venus de Milo*, represented as a maiden goddess newly born (a perception more elegantly conveyed in the lyric poem Fet published on the statue while still in Paris).[82] In the eyes of the museum visitor, the marble simulates "velvety" skin with the resilience of a flower that has just opened, as yet untouched by human breathe or the dawn's "joyful tear" of dew. "The goddess is not playing the coquette, not trying to please," he asserts. But please she does. The description highlights the tantalizing drapery that the "exquisite twist of the body has made fall to the thighs." Fet's tour-guide persona urges the reader to enjoy Venus from every angle: "Circle all around her and breathlessly admire the inexpressible freshness of the figure, the austere luxuriance of the maiden bosom" (260–66).

Ultimately, though, the narrator enacts disinterestedness, to idealize the statue as a timeless artefact independent of its creator and as "sacred" as the *Sistine Madonna* "The artist does not exist," declares Fet's persona, "he has turned into the goddess." The art pilgrim sternly lectures his envisioned Chernyshevskian antagonist: "Everything the marble unwittingly sings to you is the goddess speaking, not the artist. Only such art is pure and sacred, all else is its profanation." In conclusion, the narrator throws down a challenge: "If after sitting for an hour before the *Venus de Milo* or the Dresden *Madonna*, you are not convinced of this eternal verity, then courageously say: 'Lead me away! I am blind, blind from birth'" (266). The

80 Chernyshevskii, *Esteticheskie otnosheniia*, 50.

81 Fet, *Iz-za granitsy* (February 1857), 253 (subsequent citations appear parenthetically).

82 A. A. Fet, "Venera Milosskaia," *Stikhotvoreniia i poemy, 1839–1863*, vol. 1 of *Sochineniia i pis'ma*, 20 vols. (St. Petersburg: Akademicheskii proekt, 2002–), 275.

metaphor of blindness evokes Chernyshevsky's insensitivity to the visual arts. More specifically, Fet's phrase "courageously say" (*govorite smelo*) double-voices Chernyshevsky's claiming the "courage to judge" (*smelost' sudit'*) sculpture without having seen the *Venus de Milo* or any other statue in West European collections.

Shortly before closing time forces Fet's narrator to leave the Louvre, he exalts the capacity of the "true artist" to soar above photography. Speaking of Paul Potter's cows, sheep, and goats, *From Abroad* declares that a great painter "lends you his artistic eyes, revealing delights you would not have noticed. That is art's triumph over the daguerreotype. A daguerreotype is a mirror of nature. A painting, too, may mirror nature but refracts it through the magical interpretive prism of the human spirit" (270). In the immediate Russian cultural arena this utterance is a forceful, parting jab at Chernyshevsky for his evident failure to understand that painting can achieve something different from photography.

If not portraying the author exclusively as an art pilgrim, *From Abroad* unapologetically luxuriates in aestheticism as a gentleman's pleasure. Whatever the object, the authorial gaze is that of the sociocultural elite, a perspective underlined in an episode near the narrative's end. The author recounts there his stay at the chateau of friends in Rozoy-en-Brie near Paris (the unnamed hosts were Pauline Viardot and her husband Louis, and Turgenev was also there).[83] A high-society retreat, the chateau has an elegant drawing room, a piano, and magnificent grounds. The author highlights a dinner party and features himself enjoying cigars at the chateau and then on his train ride back to Paris. A weighty detail, the enjoyment of cigars had become a symbol of Russian "gentry hedonism," thanks in large part to Nekrasov's poetry and public persona as an author articulating social compassion and yet guiltily fond of creature comforts, including gourmet feasts and luxury hotels.[84] The cigar-smoking gentleman author of *From Abroad* adamantly declines to politicize travel. Symbolized by grand boulevards, opulent carriages, and diamond earrings on sale, the glittery Paris of the haute bourgeoisie ultimately strikes him as "a fancy Viennese

---

83 On visiting the Viardots, see A. A. Fet, *Vospominaniia*, 3 vols. (1890, 1893; repr. Moscow: Kul'tura, 1992), 1:149–63.

84 Konstantine Klioutchkine, "Between Sacrifice and Indulgence: Nikolai Nekrasov as a Model for the Intelligentsia," *Slavic Review* 66, no. 1 (Spring 2007): 48. On the expensive hotels, discovered abroad, see N. N. Skatov, *Nekrasov* (Moscow: Molodaia gvardiia, 1994), 184.

cake left standing for a month in the window of a patisserie."[85] This is about as political as *From Abroad* ever gets, and the thought behind the metaphor of a stale confection is itself a stale replication of an anti-Paris Russian tradition dating back to Fonvizin.[86]

From Abroad concludes with a lyrical flourish, to reassert the author's primary identity as a poet rather than a social butterfly. After returning to Paris, he travels by train to Marseille to catch a boat to Italy. On the leg of the journey from Dijon, "the steam engine's hexameter" lulls him to sleep, to round off his travelogue's theme of an art-loving poet assimilating the railroad. Viewed from the deck of the boat leaving Marseille, the buildings lining the bay resemble a "herd descended from the hills to drink from an azure lake." As the vista widens, the traveler contemplates the "savage glory" of ruddy cliffs. Changing from "a translucent azure" close to the shore, the waves "gradually acquire the lustre of green glass, and masses of them, rolling heavier and heavier in harmonious rows, run to crash against the base of the cliffs. Farewell, France. Tomorrow I will be in Italy!"[87]

Fet would in fact find Italy disillusioning during the roughly three months he spent there in the fall of 1856 with his ill sister, who was seeking medical treatment. A lyric published after his return to Russia represents the poet visiting the ruins of "Caesar's palace," apparently in Rome. The first thing that catches his eye there is a pile of rubbish. Extending the *hic transit gloria* sentiment, a companion poem begins: "Italy, you lied to my heart." It conjures a decaying realm teeming with beggars who interfere with the tourist's aesthetic pleasure.[88] A lack of compassion for the poor would persist in Fet's 1881 memoirs. He wrote that poetry had led him to imagine Italy as a "beautiful queen," but the real Italy he encountered was teeming with "dirty, greedy beggars."[89] We will never know if Fet had something more generous to say in the narratives about Genoa, Pisa, Rome, and Naples that he wrote during his trip. He sent those to the *Contemporary*,

85 Fet, *Iz-za granitsy* (August 1857), 107–116, quote 116.

86 Anti-French tendencies of Russian travel writing are privileged as "constant themes" in Roman Jakobson, "Le mythe de la France en Russie," in *Russie, folie, poésie*, ed. Tzvetan Todorov, trans. Nancy Huston et al. (Paris: Seuil, 1986), 159–65. For a corrective, see the anthology of favorable Russian travel impressions of France in *Les Russes découvrent la France au XVIIIe et XIX siécle*, ed. Vera Miltchina and Alexander Ospovat, trans. Camille Lambert et al. (Paris: Éditions du Globe, 1989).

87 Fet, *Iz-za granitsy* (August 1857), 127–28.

88 Fet, "Na razvalinakh Tsezarskikh palat" and "Italiia," *Sochineniia i pis'ma*, 1:243–45.

89 Fet, *Vospominaniia*, 1:170.

but they never appeared. Fet's memoirs suggested that those writings were lost or misplaced.[90] It seems more plausible, however, that Chernyshevsky vetoed them, in accord with his opinion that *From Abroad* was largely "empty—tawdry at times, if sometimes not entirely stupid."[91] That hostile attitude toward a leading proponent of art for art's sake soon became de rigueur at the *Contemporary*, whose utilitarian attacks on Fet would drive his poetry out of the press for twenty-some years.

As exemplified in the clash between Fet and Chernyshevky, the post-war persistence of aesthetic cosmopolitanism met new resistance from the rising cohort of Russian socialist revolutionaries. Fet's *From Abroad* fashioned the author as an art-loving gentlemen, declining to shoulder the tasks of sociopolitical reportage. At the opposite extreme of this disinterested absorption in art and monuments, Chernyshevsky articulated a new kind of political and moral disaffection to the wealthy nobility's style of cultural tourism and travel writing. On his view, a cultivated appreciation of art was not an attainment but rather a socially marked moral deficiency, a testimony to valuing "dead" statues and paintings more than people. Disapproving the art pilgrimage, Chernyshevsky advocated trips and reportage that would add to Russians' store of information about socioeconomic and political conditions abroad.

While the tone was distinctly strident, this radical call for knowledge-building journeys belonged to a broader new tendency to demand civic benefits from foreign travel. Alexander II had flung open the long-locked "door" to the West, allowing variously motivated Russians to go abroad in record numbers. Idle consumers, landowners fearing the end of serfdom, students and professors eager for knowledge, journalists, and people with medical needs moved alongside the well-prepared cultural tourists, men and women alike. The multitude of travelers, their social diversity, and the increasing reliance on industrialized transport comprised the advent of modern *turizm* in the public eye. But whatever their aims in going abroad, Russians were now under new pressure to assess the West as a model for their country's future. The political stakes of foreign travel had begun to rise, a collective agenda of national reconstruction was taking shape, an agenda to further if only through a relatively casual, touristic gathering of impressions of what was good and what was bad about Western civilization.

90 Ibid., 169.
91 Chernyshevskii, letter to N. A. Nekrasov, February 7, 1857, *PSS*, 14:340.

Travelers throughout history have always compared their homelands to foreign countries they are visiting. But for post-Crimean War Russians, facing the prospects of modernization on a society-wide scale, there was a compelling new need to sort out the benefits and costs of Western-style material progress. The first postwar Russian literary classic to grapple with that issue was Tolstoy's vacation story *Lucerne*.

6

Tourist Angst: Aesthetics, Moral Imagination, and Politics in Tolstoy's *Lucerne*

Tolstoy's story *From Prince Nekhlyudov's Notes: Lucerne* bears many signs of the times. It participates in the fashionable production of travel *zapiski* (notes, memoirs), it dramatizes the post-Crimean War condition of wounded Russian national pride, and it seeks to negotiate the competing claims of pleasure and purpose: aesthetic self-expansion, on the one hand, and on the other, sociopolitical reportage on a foreign land. The narrator Prince Nekhlyudov arrives in Lucerne eager to experience the authenticity of nature. The scenery instantly fills him with a joyous sense of connection to his fellow men. In short order, however, the idyllic alpine realm becomes the site of callous human behavior. A crowd of English visitors give no money to a busker whose performance they have enjoyed. The incident turns Nekhlyudov's holiday into a moral and political crisis, leading him to denounce the Swiss republic and Western civilization in general. *Lucerne* treats the West's parliaments, constitutions, and professed dedication to freedom, justice, and equality as façades that hide a dog-eat-dog world ruled by money. Under such circumstances, the romantic notion of tourism as an innocent multiplication of events is also an illusion. In Tolstoy's telling, vacationing in the capitalist West turns out to be a morally compromising, materialistic pursuit made to order for the English and best avoided by anybody with humanitarian values.

More dominant than ever on the Continental tourist circuit, the English were salient targets for Tolstoy. In 1850 *Fraser's Magazine* remarked that "the English are to be found in every nook and cranny that contains a fraction of a civilized or uncivilized population." Our "wandering countrymen" are everywhere—in museums, cathedrals, mountains, ruins, and any other tourist attraction "you" readers may choose to visit. Greatly contributing to this impression of omnipresence were the expanding support sectors that English travelers had been establishing in local economies on the Continent ever since the 1820s. Hotels were named and organized to attract English customers. English doctors, pharmacies, grocers, bookshops, and other conveniences clustered to serve the same clientele. William Thackeray's novel *The Newcomes* (1855) termed these aggregates of tourist services "little Englands." Finding a little England at every turn dismayed English travelers who went abroad in search of the "authentically foreign in the foreign place."[1] English tourists disliked English tourists. *Lucerne* articulates the counterpart of tourist antagonism across a national divide, as Karamzin, Pogodin, and Rostopchina had done in earlier times.

Tolstoy's story was based on his experience in Lucerne in the summer of 1857 on his first trip abroad. His diary entry of July 7 recounts that when returning to his hotel from a brothel that evening, he came upon a talented busker playing a guitar and singing Tyrolean songs:

> I gave him something and proposed he go sing in front of the Schweizerhof [hotel]. But nothing happened, he ashamedly walked away, muttering, while the crowd followed laughing. And before, the crowd had gathered there on the street and on a balcony to listen silently. I caught up with him and invited him to the Schweizerhof for a drink. They led us into a side room. This artist is a mediocrity but touching. As we drank, the waiters started laughing, and the doorman sat down with us. That made me explode. I cursed them out and become extremely upset. The night is a marvel. What do I want, what do I so passionately desire? I don't know, but it's not a good of this world. How can one not believe in the immortality of the soul!—when you feel such grandeur in your soul.[2]

This diary entry outlines the plot of *Lucerne* and adumbrates the major character traits of its narrator, Prince Nekhlyudov: sensitivity to nature and

1 Buzard, *The Beaten Track*, 88–91.

2 L. N. Tolstoi, *Polnoe sobranie sochinenii*, 90 vols. (Moscow: Gosudarstvennoe izdatel'stvo khudozhestvennoi literatury, 1928–58), 47:140–41.

music, spontaneous altruism toward a downtrodden stranger, and a yearning for religious truth.

The creative process that produced *Lucerne* altered the diary in two ways that burnished Prince Nekhlyudov's moral image. First, the brothel disappeared. Secondly, and of much greater import, the "crowd" became English. Tolstoy encountered many English tourists on his trip, but he perceived them through the lens of Thackeray, the writer who mediated the transformation of the diary's generic "crowd" into the English of *Lucerne*. In a letter to Botkin dated July 23/August 4, 1857, Turgenev dismissed Tolstoy's story as "a mixture of Rousseau, Thackeray, and the Orthodox Catechism."[3] Although Turgenev's remark is well known, the Thackeray connection deserves a closer look as part of the cultural history of Russian tourism. Commentary on *Lucerne* has also neglected the resonance of English tourists in Karamzin's *Letters of a Russian Traveler*. Boris Eikhenbaum likened *Lucerne* to Alexander Radishchev's *A Journey from St. Petersburg to Moscow*, a classic Russian example of the travelogue genre used to expose social injustice.[4] The general point is well taken, but parallels with Karamzin are stronger. As in Karamzin's *Letters*, Tolstoy's Nekhlyudov is visiting Western Europe, and he squares off against the English cast as an insensitive, materialistic horde that represents an evil empire.[5]

While *Lucerne* makes plain the author's moral vision, the opposition that Nekhlyudov draws between himself and the English proves unstable in the post-Crimean War context of wounded Russian national pride. The boundary between "me" and "them" grows porous as the Russian prince falls prey to "tourist angst"—the self-disgust arising from a tourist hater's recognition that he, too, is a tourist.[6] The self-implicating thrust of Nekhlyudov's assault on the English has remained obscure in the critical literature. In describing *Lucerne* as a Russian's encounter with "English tourists," commentators imply that Nekhlyudov is a man above tourism. True, his reflections on beauty, music, and Christian brotherhood reveal

3 Turgenev, *PSSiP*, *Pis'ma*, 3:138.

4 Boris Eikhenbaum, *Lev Tolstoi*, 2 vols. (1928–31; repr. Munich: Wilhelm Fink Verlag, 1968), 1:313.

5 An inversion of Karamzin also occurs in *Lucerne*. English travelers in the *Letters* lavishly reward the Italian street musician that they ask to sing for them in a Bern tavern. The Russian author listens, too, but gives the singer nothing.

6 Alan Brien, "Tourist Angst," *The Spectator*, no. 6840 (July 31, 1959): 133; and Paul Fussell, *Abroad: British Literary Traveling between the Wars* (Oxford: Oxford University Press, 1980), 49.

an exceptional mind. Prefiguring the lonely God-seeker type of Tolstoy's mature writings, Nekhlyudov yearns for universal brotherhood. In Richard Gustafson's interpretation of Tolstoy's œuvre, Nekhlyudov represents the "stranger" type seeking to become a "resident" in a spiritual abode of human fellowship and communion with God.[7] But as Gustafson and others have observed, self-dissatisfaction creeps into Nekhlyudov's antagonism to the English. In my view, the negative emotion betrays the Russian prince's realization of his complicity in the capitalist tourist system he vilifies. His effort to inculpate the English alone collapses, bespeaking authorial uncertainty about Russian national identity and the moral legitimacy of Russian vacationing abroad in the aftermath of the Crimean War.

Aesthetic and Moral Preparation

The aesthetic values that *Lucerne* seeks to create as exclusively Russian fully emerged during Tolstoy's postwar alliance with writers advocating art for art's sake. During the siege of Sevastopol Tolstoy served as an artillery officer and wrote his first two *Sevastopol Stories*. Lionized at home as a war hero and literary genius, he was welcomed into the circle of Petersburg's preeminent men of letters. There he became party to debates over the role literature should play in Russian society. On the one side stood the cohort of "aesthetic critics" Druzhinin, Botkin, and Annenkov along with their allies Turgenev and Fet.[8] Influenced by Schiller's philosophy of art, the aesthetic camp exalted the artist as a member of the elect operating in an eternal realm. On the other side of the conflict stood Chernyshevsky with his utilitarian program of harnessing art to education and progressive sociopolitical causes. After the publication of his thesis *Aesthetic Relations of Art to Reality*, Chernyshevsky elaborated his views in essays declaring Pushkin inferior to Gogol reductively interpreted as a muckraker exposing

7 Richard F. Gustafson, *Leo Tolstoy: Resident and Stranger. A Study in Fiction and Theology* (Princeton: Princeton University Press, 1986), 20–27. See also E. B. Greenwood, *Tolstoy: The Comprehensive Vision* (London: J. M. Dent, 1975), 53–56; Liza Knapp, "The Development of Style and Theme in Tolstoy," in *The Cambridge Companion to Tolstoy*, ed. Donna Tussing Orwin (Cambridge: Cambridge University Press, 2002), 171–72; and Justin Weir, *Leo Tolstoy and the Alibi of Narrative* (New Haven: Yale University Press, 2011), 128–32.

8 Belinsky applied the term "aesthetic critics" to commentators focusing on formal properties rather than the politics of a literary work.

the misery of Russia's urban poor. Druzhinin counter-attacked to defend Pushkin as the superior writer because he demonstrated the power and glory of "free art" beyond didacticism and tendentiousness. The young Tolstoy preferred the "pure" art position, in part because of his aristocratic disdain for Chernyshevsky. In opposition to Chernyshevsky, *Lucerne* proclaimed unfettered art the world's highest good, and Tolstoy would develop that theme in *Albert* (1858), a story about a talented but socially marginal, alcoholic violinist.

Prior to his alliance with the "pure" art camp, however, Tolstoy expressed a powerful moral imagination, specifically suspicious of tourism. A diary entry written in 1855, during his combat duty in the Crimea, formulated the project of founding a global religion:

> Yesterday a conversation about faith and the divine suggested to me a grand, enormous idea, for the realization of which I feel I could devote my life. That idea is the founding of a religion suitable for the development of humanity, the religion of Christ, but cleansed of faith and mystery, a practical religion, not promising future bliss, but giving bliss on earth. [. . .] To act *consciously* to unite people by means of religion—here is the foundation of a thought that will, I hope, captivate me.[9]

The representation of war in the *Sevastopol Stories* conforms to that universalistic religious outlook which anticipated the old Tolstoy's Christian pacifism. The moral fervor of the young Tolstoy, envisioning the founding a religion for the good of all mankind, prompted an Augustinian attack on the lust of the eyes in his first Crimean War narrative, "Sevastopol in December." Highly prized by Alexander II, the story was a rousing tribute to the courage and patriotism of the Russian commanders, the troops, and the people of Sevastopol during the siege. But precisely because Tolstoy recognized patriotism as the willingness to die for one's country, he condemned

9 Tostoi, *PSS*, 47:36–37, largely as translated in Orwin, "Chronology," in Orwin, *The Cambridge Companion to Tolstoy*, 4. On the global reach of Tolstoy's concept of brotherhood, see Galina Galgan, "The Idea of World Renewal," in *Tolstoy and the Concept of Brotherhood*, ed. Andrew Donskov and John Woodsworth (New York: Legas, 1996), 23–38; and Gary R. Jahn, "Brother or Other: Tolstoy's Equivocal Surrender to the Concept of Brotherhood," in ibid., 71–88. Consider also Appiah's remark on the old Tolstoy as one of those "friends of cosmopolitanism who make me nervous," in Tolstoy's case because of his inveighing against patriotism, as though the nation has "no rightful claim" on a person's conscience: *Cosmopolitanism*, xiv.

the urge to consume war as a spectacle. Imitating the style of a guidebook, "Sevastopol in December" places the reader ("you") in the position of a visitor taking a walking tour of the city. The author-guide leads you around and urges you to *look* at the sights.[10] The excursion starts at sunrise on Sapun Hill, where you have a lovely view of the town and the bay. You then go to the harbor and see an animated marketplace. The next site is the military hospital, where the strong-nerved may venture into the amputation room to see "war in its real guise—blood, suffering, and death."[11] The last stop on the tour is a front-line bastion where you see a Russian peasant soldier fatally wounded. Inculpating the reader as a "tourist of death," "Sevastopol in December" aims to make its audience realize the "singularly inappropriate" nature of sightseeing in a military arena.[12]

The guidebook idiom of *Lucerne* polemically engages two previous texts. The first was an article that a Russian officer named Komarnitsky published in the *Odessa Herald* in April 1854. Addressing the reader in tour-guide fashion, Komarnitsky staged a visit of Sevastopol but featured only picturesque sights.[13] The second text was the first installment of Druzhinin's "Observations of a Petersburg Tourist," a long-running feuilleton he published during the Crimean War. Configuring tourism as the pleasure of *looking*, Druzhinin's persona uses *turist* as a self-descriptive synonym of the *flâneur*. He defines every creative writer as a "consummate tourist" (*turist iz turistov*), a "traveling observer," moving about in search of inspiration, at home or abroad, in the city or country. The author's sophisticated friends tell him he must go to Paris to practice *flânerie* because there is not enough to see in Russia, not even in the capital. Setting out to prove those skeptics wrong, the "Petersburg tourist" finds plenty to ogle in the city streets, especially on the main thoroughfare, Nevsky Prospekt. He likens tourism-*flânerie* to attending an "opera" whose chorus is no less visually captivating than the "diva."[14] The metaphors evoke picturesque urban spectacles with a cast representing the whole social spectrum.

---

10 Gary Saul Morson, "The Reader as Voyeur: Tolstoi and the Poetics of Didactic Fiction," *Canadian-American Slavic Studies* 12, no. 4 (Winter 1978): 469–80.

11 Tolstoi, "Sevastopol' v dekabre," *PSS*, 4:9.

12 Morson, "The Reader as Voyeur," 470.

13 Iu. V. Lebedev, "Tolstoi na puti k 'Voine i miru' (Sevastopol' i 'Sevastopol'skie rasskazy')," *Russkaia literatura*, no. 4 (1976): 70–71.

14 Druzhinin, *SS*, 8:111–17, 644–45. On Russia's emergent "culture of spectacle," see Katia Dianina, "Passage to Europe: Dostoevskii in the St. Petersburg Arcade," *Slavic Review* 62, no. 2 (Summer 2003): 239–44.

Lucerne takes issue with Druzhinin's presentation of tourism as the amoral indulgence of the lust of the eyes. Moving beyond the Sevastopol arena, where the spectacles include the suffering and death of soldiers, *Lucerne* suggests that a relish for visual pleasure may be bad even in a vacation setting. Focused on the Alps and Lake Lucerne, the story's aestheticism becomes morally tainted in relation to the homeless busker, eking out a living in the streets. Tolstoy thus introduced into Russian literature (and perhaps world literature) the problem of "the non-human tourist," a type E. M. Forster scathingly identified in 1948 as the authorial position of John Ruskin's autobiography *Praeterita* (1907). In recalling his travels abroad, Ruskin wrote that he and his family, in ignorance of foreign languages, had moved about in isolation from local people. They had traveled not "for adventures, nor for company" but rather "to see with our eyes and to measure with our hearts."[15] *Lucerne*, by contrast, advances the proposition that the aesthetic priority on *looking* may inhibit such "hearts" from developing a proper sense of Christian brotherhood toward strangers of the host country.

Tolstoy the Traveler

When the young Count Tolstoy set off on his first journey to Western Europe, he was not consciously concerned with the ethical issues of sightseeing. In conformity to the outlook of his most high-minded aristocratic peers, he conceived his tour as a necessary acculturation program. A revealing document, written at Yasnaya Polyana in November 1856, was a letter to Valeriya Arsenyeva, a neighbor Tolstoy was assessing as a potential wife. Tolstoy told her that his matrimonial ideal was to settle in the country but spend at least five months each year abroad or in Petersburg in order not to "fall behind the times or become provincialized." Tolstoy's letters to Arsenyeva are one of the most literary cycles of his personal correspondence. He explored here the role of the husband as a stern educator of a young wife. One may therefore quibble about the sincerity of his stated opinion. However, at this stage he had not completely shed the values and practices of the Russian aristocracy. High society regarded the Grand Tour as a rite of passage, and Tolstoy wanted to experience it—not to learn "what hats and waistcoats are being worn," he added to Arsenyeva, but rather to keep abreast of West

15 Buzard, *The Beaten Track*, 331.

European cultural and intellectual life. Such trips, he suggested, could lift a person above the mediocrity and turpitude of Russian serf-owners preoccupied with "the buying and selling of people, or collecting quitrent from the peasants, when every student already knows it's shameful."[16] In Tolstoy's formulation, Russian provinciality signified not only cultural backwardness and the banality of everyday life but also the evil of serfdom—a wide range of deficiencies to be counteracted through edifying trips abroad.

Tolstoy pursued cultural tourism most assiduously in Paris which he reached on February 21, having traveled from Moscow on the coach route via Warsaw. After living in the French capital for nearly two months, Tolstoy described himself as a "complete ignoramus" profiting from the city's boundless cultural riches. A letter to a Russian friend enumerated the attractions he foresaw keeping him in Paris for weeks to come: "Enjoyment of the arts, the Louvre, Versailles, the Conservatoire, the quartets, the theaters, the lectures at the Collège de France and the Sorbonne, and above all the social freedom of which I had no conception whatsoever in Russia." Another letter reiterated the invigorating discovery of freedom observable in street life. Even Russian officers "fooling around with the whores" struck Tolstoy as accumulators of travel experience of potential benefit to their homeland. There is not a man among those officers, he asserted, who is "such a blockhead as not to be affected by this feeling of social freedom which constitutes the main charm of life here, and which is impossible to judge without experiencing it." Contact with Western Europe "completely changes" a Russian's outlook, Tolstoy concluded, with the assumption that this would energize the move to reforms at home.[17]

As recounted in a letter to Botkin, what Tolstoy called the "stupid and callous" mistake of watching a public execution by guillotine ended the cultural enchantment and propelled him from Paris to Geneva.[18] He would remain in Switzerland three and a half months, a prolonged stay that indicates how thoroughly he enjoyed active leisure in the Alps: an eleven-day walking tour, swimming, rowing on lakes, and excursions to popular sites such as the Rigi and the Pissevache, a waterfall Myatlev's Madame Kurdyukova had gone to see (at the Rigi, however, Tolstoy complained of

16 Tolstoi, *PSS*, 60:116–17.

17 Tolstoi, letters to D. Ia. Kolbasin, March 24/April 5, and to Botkin, March 24–25/April 5–6, *PSS*, 60:165–67. Translation in *Tolstoy's Letters*, ed. and trans. R. F. Christian, 2 vols. (New York: Charles Scribner's Sons, 1978), 1:94–95 (hereafter, *Letters*).

18 Ibid., to Botkin, 167 (*Letters*, 1:95).

"the same stupid view of nature and people," meaning English tourists with "their blankets, Murray guides, and maps").[19] Even before the public execution, the streetwalkers and the unmarried couples cohabiting in hotels had led Tolstoy to regard Paris as a city where "the debauchery is terrible!"[20] Mechanized murder and sexual depravity blended with distaste for the railroad in a remarkably literary letter he wrote but never sent to Turgenev. This composition portrayed Paris as the corrupting modern city in opposition to purifying, bucolic Switzerland. In describing his "vile railroad journey" to Geneva, Tolstoy famously wrote that "the railway is to traveling what the brothel is to love—just as convenient, but just as inhumanly mechanical and deadly monotonous." With the public execution still fresh in his mind, he summed up his Paris stay as "one and a half months in Sodom," where "filth" filled his soul: "Two whores, one guillotine, idleness, and vulgarity." Switzerland then emerged as deliverance from the metropolis he had initially embraced as a cultural capital of the West. Tolstoy recalled sitting next to the driver of a stagecoach as they rode through the "wonderful spring moonlit night." Upon reaching Geneva, Tolstoy found the Gospels in his hotel room and proceeded to read. That self-image grandly concluded the mapping of a journey of purification. In flight from Paris, debauchery, mechanized killing, and the train, the author in Switzerland restores contact with a pre-modern world of horse-drawn transport, resplendent nature, and Christianity.[21]

Literary Lenses

This epistolary account of spiritual renewal in Switzerland anticipates the Rousseauesque vision of nature in *Lucerne*. In old age Tolstoy declared Rousseau and the Gospels the two major influences on his life. He shared Rousseau's faith in the goodness of nature, extended his critique of civilization, and held the same views on childhood, education, language, and other matters.[22] "The Profession of Faith of the Savoyard Vicar" in Rousseau's

---

19 Diary from July 2/14, *PSS*, 47:140–41.

20 Diary from March 20/April 1, 1857, ibid., 120.

21 Tolstoy's closing words call attention to the letter's literariness: "Please don't try to make what I'm writing to you now fit into the general idea you've formed about my personality," ibid., 60:169–70 (*Letters*, 1:97–98).

22 Milan I. Markovitch, *Jean-Jacques Rousseau et Tolstoï* (Geneva: Slatkine Reprints, 1975 [1928]), 175–229, 314–38, 355–63; Isaiah Berlin, "Tolstoy and Enlightenment," in his

Emile was particularly important for *Lucerne*. Tolstoy read it in the early 1850s and "adopted the Vicar's resolution to eschew all metaphysics except what was necessary to objectively ground moral freedom and law. He abandoned forever any attempt to directly connect with God or metaphysical truth, and became instead a kind of transcendentalist." Written at "the peak of Tolstoy's Rousseauesque fervor," *Lucerne* represents Swiss scenery in a quasi-religious mode. When Nekhlyudov first gazes upon the mountains and the lake, he perceives nature as a glorious, harmonious whole. In nature rather than culture he locates humanitarian social instincts, the "principle of love."[23] On a more mundane level, Rousseau helped set Tolstoy's tourist itinerary, and Byron may also have had an impact. Like many other travelers before him, Tolstoy sought out Clarens and other haunts featured in Rousseau's writings. Karamzin had enacted this pilgrimage. But thanks to the English guidebook industry, Byron was the more salient seeker of Rousseau shrines in the 1850s. As we will see, Nekhlyudov in *Lucerne* uses John Murray's handbook for Switzerland. Tolstoy presumably did, too, and may have well followed the 1854 edition which incorporated extensive quotes from Byron's poetry and travel journal recounting his impressions of the routes of *Julie, ou la nouvelle Héloïse*.[24]

Whereas Rousseau and perhaps Byron mediated the idyllic landscape of *Lucerne*, Thackeray inspired the English despoilers. Tolstoy's diary indicates that he enjoyed socializing with English tourists in Switzerland at the very time he was writing the story. Those occasions included an evening at a hotel in Sarnen where he "chatted and made music" with a group of English people.[25] In populating *Lucerne*, however, Tolstoy extended Thackeray. Tolstoy's Crimean War service no doubt heightened his receptiveness to the theme of English tourist arrogance in Thackeray's *The Book of Snobs* (1848). Tolstoy read that work in June 1855 and alluded to it in "Sevastopol in May."

Russian Thinkers, 240, 246, 251; and Knapp, "Development of Style and Theme," 164.

23 Donna Tussing Orwin, *Tolstoy's Art and Thought, 1847–1880* (Princeton: Princeton University Press, 1993), 68–79, 89–90; Orwin, "The Return to Nature: Tolstoyan Echoes in *The Idiot*," *Russian Review* 58, no. 1 (January 1999): 92–93; and Orwin, "Introduction," in Orwin, *The Cambridge Companion to Tolstoy*, 55.

24 John Murray, III, *Handbook for Travellers in Switzerland and the Alps of Savoy and Piedmont*, 6th ed. (London: John Murray, 1854), 120–21, 128, 151, 158–60.

25 Tolstoi, diary, June 29/July 11, *PSS*, 47:142. On Tolstoy's interaction with British culture throughout his career, see W. Gareth Jones, "Introduction," in *Tolstoi and Britain*, ed. W. Gareth Jones (Oxford: Berg, 1995), 1–26.

Thackeray's typology of snobs differentiates the British from the French. "There is always something uneasy" in the Frenchman's patriotic conceit:

> He brags with so much fury, shrieking, and gesticulation [. . .] that one can't but see the poor fellow has a lurking doubt [. . .]. About the British Snob, on the contrary, there is commonly no noise, no bluster, but the calmness of profound conviction. We are better than all the world; we don't question the opinion at all; it's an axiom. [. . .] WE are the first chop of the world: we know the fact so well in our secret hearts that a claim set up elsewhere is simply ludicrous.[26]

Nationalist conceit characterizes all social classes, from the "lord in his travelling-carriage" to his servants—"John in the rumble." This "amazing and indomitable insular pride" dictates the Englishman's treatment of foreigners "he condescends to visit." Those lesser beings include the Russians of St. Petersburg. In the author's view, it is this indomitable arrogance that "makes us so magnificently hated throughout Europe as we are" (61). The sketch's concluding image of the snob adds military inflections and extends the cliché of the insensitive English traveling by the book:

> That brutal, ignorant, peevish bully of an Englishman is showing himself in every city of Europe. One of the dullest creatures under heaven, he goes travelling Europe under foot, shouldering his way into galleries and cathedrals, and bustling into palaces with his buck-ram uniform. At church or theatre, gala or picture-gallery, HIS face never varies. A thousand delightful sights pass before his bloodshot eyes, and don't affect him. Countless brilliant scenes of life and manners are shown him, but never move him. He goes to church, and calls the practices there degrading and superstitious: as if HIS altar was the only one that was acceptable. He goes to picture-galleries, and is more ignorant about Art than a French shoeblack. Art, Nature pass, and there is no dot of admiration in his stupid eyes. (62–63)

This portrait of the phlegmatic "bully" no doubt struck a satisfying chord in Tolstoy and other post-Crimean War Russian readers.

---

26 William Makepeace Thackeray, *The Book of Snobs* (1848; repr. Teddington: Echo Library, 2006), 61.

In addition to *The Book of Snobs*, Tolstoy knew Thackeray's *The Kickleburys on the Rhine* (1851), a comic novella that represents English tourism as colonization of the Continent.[27] A Russian translation appeared in 1851 under the title *English Tourists*, but it deleted all of Thackeray's Russian characters (perhaps to imply that Russian tourists never stoop to idle consumerism).[28] Thackeray's story centers on the mess the English matron Mrs. Kicklebury creates while vacationing with her husband, daughter, and son-in-law in Rougetnoirburg, a Baden-Baden type spa. Mrs. Kicklebury catches the gambling bug and loses much of the family bank account after observing the phenomenal winning streaks of a Russian boy, a "young Calmuck" [sic], at the roulette table with his mother, "a Russian princess" with a "very high complexion." "Russian princes" are among the "pleasure-seekers from every nation in the world" who come to the casino to try their luck, and the sole "frantic" gambler, "gnashing [his] teeth," is "Count de Mustacheff, a Russian of enormous wealth." The big crowd is multinational, but the English dominate, unique in their power to create home away from home. "We flock the one after the other, we faithful English folk," proclaims the author. "We can buy Harvey Sauce and Cayenne Pepper and Morison's Pills in every city in the world. We carry our nation everywhere with us and are in our island wherever we go."[29] This representation of English tourist hegemony anticipates the "little Englands" conceit of *The Newcomes*, another novel Tolstoy read before going abroad.[30]

---

27 William Makepeace Thackeray [Michael Angelo Titmarch, pseud.], *The Kickleburys on the Rhine*, 3rd ed. (London: Smith, Elder, 1851). In the epistolary draft of *Lucerne* addressed to Botkin, June 26/July 8, 1857, Tolstoy reported seeing "lots of Kickleburys" (*PSS*, 60:201).

28 William Makepeace Thackeray, *Angliiskie turisty*, trans. A. Butakov [Russian translation of *The Kickleburys on the Rhine*], *Otechestvennye zapiski* 76, no. 6 (1851), section 8: 106–44. Thackeray's "young Calmuck" and his mother became a "lady" and her "boy;" "Russian princes" disappeared from the casino, and Count de Mustacheff became a "rich man." Compare Thackeray [Titmarsh], *Kickleburys on the Rhine*, 46–51, 76; and *Angliiskie turisty*, 127–29, 139.

29 Thackeray [Titmarsh], *The Kickleburys on the Rhine*, 34; and *Angliiskie turisty*, 122.

30 The *Contemporary* serialized *The Newcomes* in 1855–56. Tolstoy read the novel in July 1856, at least partly in English: see *PSS*, 47:82–84, 346.

Tourist Angst

Tolstoy heeded Thackeray's satires of English tourists but had not mastered his art of isolating the writing self from the other. The opening paragraph of *Lucerne* exposes the problem. Nekhlyudov introduces himself as an upmarket Russian vacationer under English authority: "Yesterday evening," he begins, "I arrived in Lucerne and have lodged in the Schweizerhof, the best hotel in town."[31] The prince then quotes his Murray handbook description of Lucerne. Nekhlyudov adds that all the guidebooks recommend the town as one of the "most romantic places in Switzerland." Yet, this Russian visitor's chosen vade mecum is precisely Murray.[32] Nekhlyudov further explains that guidebooks have drawn "masses" of "travelers of all nations" here, "especially the English" (3). The English overrun Lucerne, but Nekhlyudov is there too, on the beaten path to the "romantic" Swiss town, on the beaten path to the Schweizerhof, the hotel that tops Murray's list of recommended accommodations, with special mention of the views the rooms afford of the lake and the mountains.[33] Nekhlyudov thus enters the story as a dupe of fashion, doing as the English do, as prompted by their leading guidebook.

Prince Nekhlyudov is the only aristocrat in town, and his title hints at inner nobility—the aesthetic and moral sensitivity he presumes to monopolize. Badly needing to differentiate himself from the apparently bourgeois English crowd, the prince falls into a long-standing anti-tourist pattern of English travel writing. Although Tolstoy was no avid reader of Byron, *Lucerne* replicates the role-distancing strategy of world literature's prototypal anti-tourist Childe Harold. In harmony with Byron's letters describing English tourists as "a pack of staring boobies," Harold fashions himself as the heroic wanderer, participating in tourism but not in the ordinary manner. The "crowd" cannot claim him as one of their own: he stood "among them but not of them; in a shroud / Of thoughts which were not their thoughts" (3.113). Emulating the Byronic pose, English writers engaged in "denunciation, evasion, and putative transcendence of merely 'touristic' purposes and behaviour." They presumed to possess "saving inner qualities"—exceptional

31 Tolstoi, *Iz zapisok kniazia D. Nekhliudova. Liutsern*, in *PSS*, 5:3 (subsequent citations appear parenthetically).

32 Begun in the 1830s, the Murray guidebook enterprise covered more than sixty countries by 1848: Buzard, *The Beaten Track*, 65–67, 72.

33 Murray, *Handbook for Travellers in Switzerland*, 36.

sensitivity, superior knowledge, and good breeding: the "aristocracy of inner feeling" that displaced wealth as the signification of the traveling elite.[34]

Lucerne establishes a similar clash. Tolstoy's thinking, however, had native roots in the controversy that surrounded Peter I's introduction of the Table of Ranks: the story attributes to Nekhlyudov the nobility of character and mind, as opposed to the putative English fixation on power and money. Tolstoy's prince introduces the English as imperialist despoilers, overjoyed by their power to remake foreign places to suit them. Now installed in his room and apparently relying on his Murray guidebook, Nekhlyudov explains that "in olden days," there was a "covered, sinuous wooden bridge" where the hotel now stands. But "thanks to the huge influx of the English—their requirements, their tastes, and their money," the old bridge was torn down and replaced by a rectilinear quay, rectangular five-story buildings, straight rows of lime trees, and little green benches. "Rejoicing in what they have produced," English visitors stroll along the modernized lakeside which Nekhlyudov finds "vulgar." He, on the other hand, relishes the pleasure that "one experiences from solitary contemplation of the beauty of nature" (3–5). The prince manages to obtain this Karamzinian experience in the stillness of his room, where he positions a chair by a window at an angle excluding the crowd from his field of vision. The epistolary draft of *Lucerne* accused the English of insensitivity to art as well as untrammeled scenery: Nekhlyudov likens the tourist transformation of the lakeside to gluing "gilt edging onto the chin of Raphael's [*Sistine*] *Madonna*," a painting Tolstoy had not yet seen.[35]

The dinner episode of *Lucerne* contains the first sign of Nekhlyudov's self-dissatisfied consciousness of mirroring the English. Mentioning again that English visitors dominate Swiss tourist sites, he dehumanizes them through Gogolian synecdoche. At the Schweizerhof table d'hôte, set for some one hundred guests, the English are reduced to an assemblage of white lace and collars, white faces, white teeth—real and false, and white ring-bedecked hands. During the meal the English speak little and only in whispers. Without indicating the language he tried, Nekhlyudov recalls his efforts to strike up a conversation, but all in vain. The English are "dead faces," dead souls, the Gogolian synecdoche would imply. Yet the hegemonic

34 Buzard, *The Beaten Track*, 80–81, 121–22, 153–54.

35 Tolstoi, *PSS*, 60:201.

English exert an "irresistible influence" on the Russian prince: "I became just as dead" (6).

The anglicizing of Nekhlyudov is underlined through his recollection of the unpretentious Paris pension where he experienced vacation as cosmopolitan community. He attributes the warm, animated tourist atmosphere there to "French sociability," a national trait of the visited place, as opposed to the English chill of the luxury Swiss hotel (6). Under a French rather than English influence, the twenty pension guests with whom Nekhlyudov regularly interacted represented many countries, including a Spanish countess, an Italian abbot who recited Dante, and an American doctor. The small scale and shared capacity to communicate in French, however faultily, created a sense of home away from home. Removing Nekhlyudov in space and time from the Schweizerhof table d'hôte, the recollected pension stands in for the family, affirming social values of a pre-capitalist world.

The Tyrolean busker's performance in front of the Schweizerhof momentarily renews Nekhlyudov's tourist experience of fellowship with a nationally mixed group of strangers. Depressed and lonely, he takes a walk after dinner. At first he perceives moonlit Lucerne as a sordid, dirty little town. The distant sound of music suddenly lifts his spirits, and he hastens to join the assembly of tourists and hotel employees listening to a shabby-looking little singer with a guitar. The audience's rapt attention illustrates the unifying potential that MacCannell attributed to the modern-day tourist attraction as "the locus of a human relationship between un-like-minded individuals, the locus of an urgent desire to share" and establish a connection to strangers.[36] The singer, of course, has no official standing, but he is performing for money, and as he later tells Nekhlyudov, his finest song—"L'air du Rigi"—was expressly written to entertain "the travelers" (15). His performance is a grassroots commodification of culture, local but shaped to satisfy the assumed tastes and expectations of tourists.

The magic moment of sharing among strangers ends as soon as the Tyrolean stops singing and holds out his cap for tips. His timid manner and "nearly incomprehensible" French spoken with "a half-Italian, half-German accent" provoke laughter from the crowd. Feeling sorry for the busker, Nekhlyudov alone gives him "some centimes," as listeners in the street disperse and those on the hotel balconies turn away (10). The Russian prince has distinguished himself from the stingy crowd. However, in mentally

36 MacCannel, *The Tourist*, 203.

registering the singer's poor French, he signals his own, aristocratic Russian command of that language.[37] Briefly disguised by the unifying magic of the music, Prince Nekhlyudov's class consciousness presages the failure he will experience in his effort to befriend the busker.

Instead of passively pitying the singer, Nekhlyudov takes action that displays the cosmopolitan sentiment of feeling responsible for each and every human being. This occurs in a symbolic scenario of Crimean War hostilities. Upon reentering the Schweizerhof, he deliberately elbows an Englishman exiting with his family. The prince then tracks down the singer and offers him a drink. The musician seeks to steer him to a modest pub, but Nekhlyudov insists on returning to the deluxe hotel for champagne. He wants the English and Schweizerhof employees to *see* his act of charity.[38] Relegated to a side room where a maid is washing dishes, Nekhlyudov and his guest attract the overly familiar attention of waiters and the hotel doorman. Under those derisive eyes, the Russian prince's effort to raise the singer's consciousness of social injustice goes all wrong. The busker bears no rancor and grows ill at ease, as Nekhlyudov's fury mounts. Since no English are available, the prince vents his anger against the hotel personnel who stare at him blankly, apparently finding his German incomprehensible. The thought of bludgeoning an English woman tourist's head crosses Nekhlyudov's mind, and he imagines himself charging into an English trench in Sevastopol, "to slash and hack" (19). The military scenario ends in a rout. Finally yielding to Nekhlyudov's demands, the waiter ushers them into the dining room where the prince seats himself and his mortified guest next to an English couple who flee and complain to the management.

Yet even as Nekhlyudov makes a spectacle of his extraordinary, interventionist ethics in relation to a stranger, a motif of "aristocratic" clothing suggests he is imprisoned in a predetermined role in the Swiss theater of tourism, a role way too "English" for his self-respect. At the outset of the musical interlude, he asks the Schweizerhof's "aristocratic lackey and cook" for information about the busker (9). The adjective implies the two hotel employees are wearing fancy uniforms. But to all indications, the story's one actual aristocrat, the Russian prince, sports trappings of wealth as much as the English. Whether in the dining room, in the streets or on the balconies of their rooms, the English of *Lucerne* wear "sumptuous clothing."

---

37 Weir, *Tolstoy and the Alibi of Narrative*, 131.

38 Ibid., 128.

This includes a chubby-kneed little girl in a dress of "fine lace" (8, 9, 12). However, Nekhlyudov's wardrobe evidently marks him, too, as a rich tourist. In the champagne drinking scene, he explodes at the waiters for discriminating against the "poorly dressed" Tyrolean. You are polite to me, yells the prince, just because "I am wearing good clothes." The semiotics of clothing binds Nekhlyudov to the English, his fellow residents in the hotel that "houses the rich" (18, 20). Clothes signify the money that wins deference in the capitalist tourist milieu, and the Russian prince has fallen into the pattern. Moreover, he later admits to himself that he has conformed to the materialist logic: being better dressed than the hotel lackey gave him the license to insult that employee "with impunity" (24).

No Russian Alternative

To counteract tourist angst after parting with the singer, Nekhlyudov seeks to reassert his difference from the English. In his hotel room and out on the quay, he reviews what has happened and elaborates an opposition between money and "poetry": English materialism, on the one hand, and on the other—his own recognition of art as the higher good. In his estimation, ninety-nine percent of the clients of the Schweizerhof believe money is "the greatest blessing in the world" (21). Tourist materialism becomes intertwined with British empire-building. In Nekhlyudov's mind, the crowd's disregard for the busker is even more momentous than colonial oppression and violence regularly covered in newspapers. In the name of civilization, the English ride roughshod over India and kill Chinese (the Opium Wars), while the French massacre Algerians. The draft of *Lucerne* added that Russians are killing Chechens.[39] Victims of Russia's ongoing colonial war in the Caucasus thus crossed Tolstoy's mind. But as etched in *Lucerne*, the global network of imperial power and wealth must exclude Russia's military campaigns in the North Caucasus. The Russian Empire needs innocence to shore up Nekhlyudov's personal claim to moral superiority over the English vacationers.

Tolstoy historicizes the moral issue in a manner that finds no Russian alternative to the Western civilization he denounces. Nekhlyudov thinks that what happened to the singer does not represent "the eternally evil side of human nature" but rather the moral, emotional costs of civilization in

39 Tolstoi, *PSS*, 60:210.

the form it has reached in the modern time and place he has discovered as a tourist. In the "lousy" Swiss republic, states Nekhlyudov, "civilization, liberty, and equality have been brought to the highest point, and the most civilized travelers from the most civilized nations congregate here." But while the British and French publicly profess concern with bettering the whole human race, they appear incapable of kindness toward an individual human being. "Can it be," asks Nekhlyudov, "that the spread of the sensible and selfish association of people that is called civilization destroys and contradicts the need for instinctive, loving association?" Coming from Russia, where processes of capitalist modernization were barely underway, he idealizes the family and the village as traditional, nurturing structures whose values are violated in the Swiss luxury-tourist milieu. His mental interrogation of the English renders leisure travel a senseless uprooting: "Why have you all abandoned your fatherland, your relatives, your occupations, and finances to flock to the little Swiss town Lucerne?" On this view, a vacation is not an innocent adventure but rather the evasion of work and the responsibilities of raising a family. Similarly, Nekhlyudov asserts that the English crowd's display of inhumanity toward the busker would be "impossible in any village—German, French, or Italian" (19, 21, 23, 24). That list of countries imagined as sites of brotherly rural communities conspicuously omits Russia where the peasant masses remained enslaved. Nekhlyudov's native land provides no support for his anti-capitalist ideal of supranational brotherhood.

His ultimate retreat into Rousseauesque transcendentalism similarly isolates him from Russian life. After condemning Western civilization, he yields again to the aesthetic power of nature, to end the story in a state of quietism. Gazing at the moonlit sky, Nekhlyudov paraphrases the abbey Savoyard to posit an inscrutable harmony governing the world. The prince now feels like a "paltry worm," unable to see and judge the grand designs of Providence (26). Perhaps the busker is happy and the English not as ghastly as they seem. Everybody is innocent, including Nekhlyudov who is now forgiving himself for having fallen into the trap of luxury tourism. He humbly relocates himself in a heavenly mansion, a spiritual "residence," in Gustafson's terms. Yet, no semblance of a Russian home exists in *Lucerne*. Throughout the story the prince resides in the Schweizerhof. The reader learns nothing about his life in Russia, except for the indications of military service in Sevastopol. In denying Nekhlyudov a homecoming, Tolstoy excludes an opportunity for national affirmation. Karamzin's *Letters* set the round trip template ending with the Russian traveler's joyful return to his empire (Kronstadt rather than his home in Moscow). Pogodin's *A Year in Foreign Lands* also enacted a joyful return, specifically to Moscow. Whatever the points of exit and reentry, the pattern of jubilant homecoming valorized Russia over the foreign.[40] *Lucerne*, on the other hand, leaves Nekhlyudov displaced and looking for answers in the Swiss sky. In opposition to the capitalist realm discovered as a tourist, he idealizes village life, the family, brotherhood, and communion with God. This antimodern cluster veers toward classic Slavophilism but never finds native ground. The symbolic sites of Nekhlyudov's ideal are all outside Russia: Swiss scenery, the Paris pension, German, French, and Italian villages.

The humiliations of the Crimean War and the persistence of serfdom in Russia made the easy national affirmations of previous Russian travel accounts untenable for Tolstoy. Karamzin, Fonvizin, Pogodin, and other Russian writers posited materialism, ruthless individualism, rationalism, and rigid legalism as the dominant features of Western civilization and set them against the warm, expansive Russian soul and the Russian *potential* to forge a more humane society. Most recently, Herzen's *Letters from France and Italy* indicted the Parisian bourgeoisie in similar terms and formulated a socialist conception of Russian national character rooted in the "instinctive allegiance to community" that he perceived in the Russian peasant commune.[41] *Lucerne* parallels those earlier critiques of the bourgeois capitalist West but finds no Russian answer to the problem.

Tolstoy's Path to National Affirmation

Unlike Prince Nekhlyudov left in limbo in the Schweizerhof, Tolstoy fled the luxury tourist milieu the very day after he began drafting *Lucerne* as a letter to Botkin. Tolstoy rented two attic rooms in what he described as a "simple, rural, and terribly nice" family pension on the lakeshore. "What a charming place Lucerne is," he now told Botkin, "how wonderful I find everything here." Having extricated himself from the guilt-inducing hotel that "houses the rich," Tolstoy finished writing *Lucerne* at the pension and

40 On this general dynamic in travel narratives, see Robert Reid, "Ethnotope and Imperium in Nineteenth-Century Russian Literature," in *Why Europe? Problems of Culture and Identity*, vol. 1: *Political and Historical Dimensions*, ed. Joe Andrew et al. (London: Palgrave Macmillan, 2000), 199.

41 Offord, *Journeys to a Graveyard*, 251.

enjoyed the unceremonious, human scale of interacting (largely through sign language) with his hosts—the elderly, nearly deaf proprietor and her pretty seventeen-year-old daughter. Here, too, Tolstoy again took delight in the beauty of nature viewed from his room: the tendrils of vine leaves at his window, black poplars in the distance, and "through their leaves the still lake in a moonlit glow."[42]

Tolstoy's aesthetic sensibility remained intense upon his return to Russia. After six months in Western Europe, he found the homeland "disgusting." "I just don't like *it*," his diary added—the "coarse, deceitful life surrounding me on all *sides*" (August 6 and 8, 1857). Later that month he confided those feelings to his cousin Alexandra, a lady in waiting at the Russian court and a cherished friend so much older than Tolstoy that he nicknamed her "Granny." He called Russia a "backwoods" realm of "patriarchal barbarism, thieving, lawlessness" and endemic brutality such the sight of a noblewoman thrashing her serf girl in the streets of Moscow. Tolstoy added that he found consolation in the privacy of his country manor, reading Homer and playing Beethoven piano sonatas with "tears running down my cheeks."[43] Despite Tolstoy's rejection of the bourgeois materialism he had witnessed abroad, his trip opened his eyes to a world free of serfdom. In that optic, Russia appeared to him more backward than ever, and an immediate response was to retreat into art of foreign origins.

Tolstoy's position in the literary arena paralleled the epistolary self-image of aesthetic retreat from Russian reality. As suggested by Turgenev's comparison of *Lucerne* to "the Orthodox Catechism," nobody in the "pure" art camp liked the story. Tolstoy himself, however, passed the harshest judgment in a letter to Nekrasov, editor in chief of the *Contemporary* where *Lucerne* appeared in September 1857. Tolstoy called the story "a commonplace and loathsome thing [. . .]. I've made a complete fool of myself and you too, it seems." [44] Apparently feeling he had naively ventured into treacherous political waters, Tolstoy was again drawn to art for art's sake. Writing to Botkin in January 1858, he expressed weariness with the public bickering about if, when, and how the serfs should be liberated. In Tolstoy's opinion, only art's unique "power of goodness" might reconcile the warring camps. He accordingly wanted to establish a new, internationalist journal devoted to defending "the independence and eternity of art" above the political fray.

42 Tolstoi, letter to Botkin, June 27/July 9, *PSS*, 60:214–15 (*Letters*, 1:101).
43 Tolstoi, letter to A. A. Tolstaya, August 18, 1857, *PSS*, 60:222–23 (*Letters*, 1:106–7).
44 Tolstoi, letter from October 11, 1857, *PSS*, 60:225.

Seeking to "attract everything that is purely artistic" in Russian and foreign literature, the journal would be a bulwark against "the sordid stream of politics" now "seeking to engulf everything and, if not to destroy art, at least to sully it."[45] *Albert* similarly proclaimed art the "highest manifestation of human power," granted only to a "select few."[46] Tolstoy took this position again in an 1859 speech he delivered upon his induction into the Society of Lovers of Russian Literature. Identifying himself as "a one-sided lover of belles-lettres," he deplored the "muckraking" tendency as a "political tide threatening to swamp all literature."[47] In the audience was the Slavophile Khomyakov, a founder of the society who published his responses to the talks of new members. Defending the need for muckraking when social problems were as grave as Russia's, Khomyakov tactfully argued that Tolstoy himself had in fact already used his brilliant literary gift to expose the suffering of Russian peasants in the story "Three Deaths" (1859).[48]

By the end of 1859 Tolstoy abandoned the "pure" art position. He broke away from the St. Petersburg literati, set up a school for serf children on his estate, and began devoting more time than ever to managing his property. In 1860 he also started producing a pedagogical journal to publicize his ideas on education. Tolstoy's rapprochement with the peasants, his absorption in country life, and his turn away from creative writing to what he termed "useful" pursuits marked the first of his recurrent, active efforts to make the Russian peasantry the beneficiaries of those ideals of Christian brotherhood that he had articulated in his 1855 plan for establishing a global religion. A definitive statement of his new, anti-tourist outlook appeared in a letter from March 1, 1860 to a friend at the time Boris Chicherin (1828–1904), a lawyer, historian, and liberal politician. Chicherin had written Tolstoy a letter that imagined him vegetating in the country in a *khalat*, lazy Oblomov's favorite garment. Tolstoy's response was furious:

> You give me advice in a nonchalant and kindly way about how *an artist should develop, what a beneficial effect Italy has on one with its monuments and its sky, and other banal commonplaces—how harmful inactivity in the country is—life in a khalat—and how I should get married and write nice stories, etc.* [. . .] I will merely say

45 Tolstoi, letter to Botkin, January 4, 1858, *PSS*, 60:246–49 (*Letters*, 1:115–16).
46 Tolstoi, *PSS*, 5:50.
47 Ibid., 271–72.
48 A. S. Khomiakov, *O starom i novom: Stat'i i ocherki* (Moscow: Sovremennik, 1988), 308–9.

> in reply to your advice that at our age and with our means, roaming around away from home is, in my opinion, just as bad and just as unseemly as writing stories which are pleasant to read. At our age, when you have reached [. . .] an awareness of the uselessness and impossibility of seeking enjoyment; when you feel that what seemed like torture has become the only true substance of life—work and toil—then searchings, anguish, dissatisfaction with yourself, remorse, etc.—the attributes of youth—are inappropriate and impossible. I won't say that it is necessary to, but rather that it is impossible not to do the sort of work of which the fruits are the ability to see sufficiently far ahead to be able to give yourself up entirely to that work—one man to plow the land, another to teach young people to be honest etc. But the self-delusion of so-called artists [. . .] is the meanest baseness and falsehood on the part of the person who yields to it. To do nothing all one's life and to exploit the labor and all the good things of other people in order to reproduce them, badly and perhaps worthlessly [. . .] is a monstrous, foul act.[49]

Tolstoy did not reveal that he was teaching. He merely said he was doing satisfying work that often filled him with "sinful pride" vis-à-vis his peers. In conclusion, he remarked that he intended to stay put at Yasnaya Polyana, and he urged Chicherin to come see what he was doing there, after the latter finished his travels.

This letter shows how radically Tolstoy's perspective on travel had changed since 1857 when he described regular journeys to the West and long sojourns in Petersburg as necessary defensive measures against Russian provinciality. The response to Chicherin re-evaluates the native Russian realm in relation to aesthetic wonderlands abroad, especially Italy. For Tolstoy, teaching the peasants had become the greater good, superior to tourist "roaming" in the West. Moreover, the peasant masses were rising in his estimation, as expressed in his pedagogical article "Who Should Learn to Write from Whom—the Peasant Children from Us or We from the Peasant Children?" (1862).[50] Based on his classroom experience, Tolstoy's article endows the folk with intelligence humanly superior to the book learning of the westernized Russian elite. While the letter to Chicherin foresaw a

49 Tolstoi, *PSS*, 60:327–28 (*Letters*, 1:133–34).

50 Tolstoi, *PSS*, 8:301–24.

long continuation of living and teaching at Yasnaya Polyana, Tolstoy would in fact spend nearly a year abroad during 1860–61, but he undertook the journey for the purpose of learning about pedagogical theory and practice in the West, particularly in Germany and England.

Like thousands of other Russians of the postwar period, Tolstoy in 1857 had seized the opportunity to tour the West. A professed "ignoramus" in Paris, he evidently approached travel with the hope of filling gaps in his education. However, he had embarked on the project without clearly establishing his aims and priorities. What should motivate Russian trips abroad: pleasure or purpose, acculturation or political knowledge-building? And if Russian travelers chose to privilege the latter regime, what might they bring home: blueprints for modernizing their country on the capitalist model or reactive inspirations for building instead upon native institutions, customs, and values? Tolstoy's journey clarified his position on those questions that he had not even posed to himself before going abroad. His trip itself furthered his progress along a path of liberation from the dominant values of Russian high society and turned him into a permanent enemy of pleasure trips, no matter how exalted the traveler's professed aim.

7

Cosmopolitans, the Crowd, and Radical Killjoys: Turgenev, Other Writers, and the Critics

Although Tolstoy encountered multitudes of Russian tourists in Paris and Switzerland, *Lucerne* opted for a structure of national difference, pitting the lone, sensitive hero against English hordes. In the same period, however, Turgenev and several other writers directed tourist phobia against the post-Crimean War wave of Russians "flooding" the West. As in England right after the Napoleonic Wars, so too in the early years of the reign of Alexander II did thousands of Russians seem to be going abroad and cropping up everywhere as unwelcome reminders of home. In experiencing this situation for the first time in Russian history, Turgenev and like-minded writers laid claim to the inner nobility of character and intelligence, pitted against the compatriot tourist crowd in thrall to rank and money. The target was no longer English, but the self/other dynamics matched those of *Lucerne*.

The post-Crimean War gallery of degraded Russian tourists had antecedents in Myatlev's *Sensations and Observations of Madame Kurdyukova Abroad*. Apparently hoping to profit from the renewed Russian rage for visiting Western Europe, a publisher re-issued Myatlev's poem in 1857. This proved a miscalculation. The poem met a poor reception, as exemplified by Chernyshevsky's describing it as "tedious chatter."[1] Kurdyukova's provinciality, her espousal of official nationalism, and her contentment with serfdom had evidently become too painful for the new audience to find funny. Nevertheless, there was thematic continuity between Myatlev's yokel

1 Chernyshevskii, *PSS*, 3:590.

and portrayals of the post-Crimean War Russian tourist crowd. Traveling with serfs, Kurdyukova likes passing for a countess abroad. She looks down on *les prolétaires*, including "peasant" chambermaids in little towns. She loves shopping, attaches much importance to clothes, and cannot make any sense of art. The trope of the Russian as a provincial in the West and the themes of telltale clothing, shopping, and incapacity to appreciate high culture reappear in the postwar articulation of tourist phobia. The motif of social pretentiousness returns to mask national insecurity about the West, and traveling serfs reappear but with very new connotations.

By contrast to the mirth of the Kurdyukova saga, post-Crimean War tourist phobia took the form of corrosive satire reflecting new sociopolitical turbulence in Russia. In the Romantic era Kurdyukova functioned as a comic instrument affirming the poet's cosmopolitan values. The Tambov innocent abroad enjoys her trip and strives to meet highbrow norms of sightseeing, museum visits, fine dining, and active leisure (the donkey ride to Pompeii). On the other hand, the run-of-the-mill tourists of the postwar boom are not merely culturally unprepared but xenophobic, spiteful, and reactionary. They appear seriously afflicted by a national inferiority complex that they attempt to mask through lying, snobbery, displays of wealth, and expressions of contempt for foreign people and places. They are obsessed with rank and oblivious to the shame of serfdom. The vulgar rich prevail. But whatever the Russians' financial resources, they prove incapable of deriving any benefit from their trips and blame the West for their disappointments. This syndrome bespeaks resentment toward Western Europe, a perennial phenomenon but one that the humiliations of the Crimean War had aggravated.

Turgenev was the outstanding formulator of the elitist, politically liberal anti-tourist discourse. Among the best educated Russians of his generation, he began roaming around Europe while studying philosophy in Germany during 1839–40. At that time he came to adore Italian art, especially Raphael; he later collected art and enjoyed the reputation of a connoisseur. This chapter treats his celebrated story *Asya* and his lesser known essay "From Abroad," both written in Rome in 1857 and published in Russia the following year. The principal ephemera that will complement Turgenev's writings are Melgunov's "Tourists in General and the Russians in Particular" and Druzhinin's feuilleton collections *Our People Abroad* (1857) and *Russians Abroad* (1860). The discussion will also take into account two stories by prominent women writers of the time: Panayeva's *Russians in Italy* and *On the Verge* (1858) by

Evgeniya Tur (1815–92), the *nom de plume* of Elizaveta Salhias de Tournemir. As in the cases of Turgenev and Panayeva, all the other writings reflect first-hand impressions. Melgunov was married to a German and lived abroad for extended periods as of the late 1830s. Druzhinin traveled in the West in 1857. Tur's parents whisked her abroad in her youth to thwart her romance with a social inferior. In France in 1838 she married the French Count Tournemir. After they settled in Russia in the 1840s, he was expelled for dueling. This left her alone to earn a living as a writer and journalist, most successfully as a critic and contributor to children's literature. Tur also published accounts of her trips to the Caucasus and the Crimea.

In valorizing tourism, while maligning the Russian crowd, Turgenev and the other liberal sophisticates comprised a united front. There was, however, a porous boundary between them and the *raznochintsy* intelligentsia. Practicing what he preached, Chernyshevsky never indulged in leisure travel, and he apparently loathed the little he saw of Western Europe during the short trip he made to London in 1859 to discuss political differences with Herzen.[2] On the other hand, the impecunious Dobrolyubov and Mikhailov, on journalistic assignments abroad, profited from tourism as a democratizing opportunity. This was a distinctly Russian phenomenon, recorded for posterity in the travel accounts *Paris Letters* (1858–59) and *London Notes* by Mikhailov, whose revolutionary activities in Russia would lead to his arrest and deportation to Siberia in 1861. But while Mikhailov fashioned a self-image as a progressive enjoying his discovery of the West, the radical killjoy in chief Chernyshevsky propagated an ascetic work ethic and redoubled his promotion of travel for strictly political or practical purposes. Chernyshevsky's attitudes found an ally in Pisarev, the caustic, dandyish critic of noble origins best known today for his essay calling for the "abolition of aesthetics" and the transformation of literature into a strictly utilitarian, educational endeavor. The popularization of science, hoped Pisarev, would eventually supplant "useless" creative writing entirely.[3] In opposition to the anti-aesthetic anti-tourism that Chernyshevsky

2 An enlightened priest, his father found the letter (now lost) puzzling: "If being abroad is really not entertaining," he asked, "then what is it that attracts the throngs of our compatriots going over there?" See N. M. Chernyshevskaia, *Letopis' zhizni i deiatel'nosti N. G. Chernyshevskogo* (Moscow: Gosudarstvennoe izdatel'stvo khudozhestvennoi literatury, 1953), 172.

3 D. I. Pisarev, "Realisty" (1864) and "Razrushenie estetiki" (1865), *Polnoe sobranie sochinenii i pisem*, 12 vols. (Moscow: Nauka, 2000–2013), 6:226–28, 274–77, 283–85, 339–48; and 7:365–69.

originated, Melgunov and Druzhinin defended the art pilgrimage as a personally enriching and even nationally beneficial undertaking. Ranging from Turgenev's aristocratic disdain for the crowd to Chernyshevskian intolerance for the very practice of vacationing, these clashes over travel regimes and individual comportment demonstrated how tourism was contributing to cultural, sociopolitical, and moral agitation in Russia in the years leading up to the emancipation.

Self and Others in *Asya*

In a letter sent from Rome to a friend in Russia in 1857, Turgenev asserted: "Among any fifty Russians abroad, forty-nine are best avoided."[4] That conviction resonates in *Asya* and "From Abroad." The unnamed nobleman who narrates *Asya* is a bachelor recollecting a trip he made along the Rhine in his youth. At the outset of the story, set in a German town inspired by Sinzig, he is rapidly recovering from a spa romance with a young widow who abandoned him for a Bavarian lieutenant. The narrator cultivates the air of a Byronic individualist, free of financial worries, and roaming "without a plan." The very sight of a tour guide makes him shudder, and he ignores standard tourist attractions: monuments, "unusual mountains, crags, waterfalls."[5] His atypical, mind-numbing foray into Dresden's Grünes Gewölbe (which had delighted Karamzin) merely reinforced his determination to avoid the tourist rut.[6] The plot of Turgenev's story is set in motion when the narrator meets his compatriots Gagin and Asya, who turns out to be Gagin's half-sister born of their father's liaison with a serf. After several weeks of meetings between these three, Asya, under the spell of the stand-offish, class-conscious narrator's wary attraction to her, summons him to a tryst. The storyteller listens to her declaration of love and then sternly rebuffs her. Looking back twenty years later, he still regrets having let social proprieties repress his feelings for this capricious but charming semi-serf maiden.

While resembling certain Turgenev love stories set in Russia, *Asya* provides an occasion for the author's representation of gauche compatriots

4 Turgenev, letter to E. E. Lambert, November 3/15, 1857, in *PSSiP*, *Pis'ma*, 3:163.

5 Turgenev, *SS*, 6:199 (subsequent parenthetical citations refer to this edition).

6 Fet also dismissed the Grünes Gewölbe as a tourist trap that art lovers should avoid: *Iz-za granitsy* (November 1856), 97–98.

abroad.[7] The narrator first encounters Gagin and the heroine while watching German students singing and drinking on the sidewalk terrace of a hotel during a festival. Upon hearing another spectator in the crowd speaking Russian, the storyteller turns to look at him, feels an immediate affinity, and initiates a conversation. But in an aside to the reader, he explains that he usually shuns Russian tourists:

> To tell the truth, I was reluctant to strike up acquaintances with Russians abroad. Even from a distance I would recognize them by their walk, the cut of their clothes, and mainly by the look on their faces. An expression of self-satisfaction and contempt, even imperiousness would suddenly change to caution and timidity. In a flash, the person was on guard, anxiously looking around. "Good lord! Have I done something wrong? Are people laughing at me?" that fleeting expression seemed to say. An instant later, the majestic physiognomy would be restored, alternating now and then with blank perplexity. Yes, I avoided Russians. But I had liked Gagin right away. (202–3)

On the basis of Gagin's face, his clothes, and demeanor, the narrator has sized him up, to decide he is worth speaking to. A few moments of conversation confirm that Gagin is indeed the storyteller's kind of person—another young Russian gentleman traveling for pleasure and enthralled by the bucolic scenery masterfully depicted throughout the tale. Gagin also shares the storyteller's passion for the arts and even dabbles in painting. From the first encounter of these Russians, leisure travel shapes up as a gentlemen's club.

Being in a foreign land stirs the *Asya* narrator's national consciousness but produces no coherent picture of what it means to be Russian. The first evocation of Russian national character occurs at the song fest. Gagin remarks that even at play, the Germans seem tame by contrast to "our" students who by now would have been smashing windows and chairs. Asya has a boisterous streak that intersects this concept of the unbridled Russian. On a walk with her brother and the narrator, she breaks a branch from a tree and carries it on her shoulder like a rifle. This attracts the disapproving

7 On thematic similarities between *Asya* and other Turgenev stories, see Eva Kagan-Kans, *Hamlet and Don Quixote: Turgenev's Ambivalent Vision* (The Hague: Mouton, 1975), 41–51; and Frank Friedeberg Seeley, *Turgenev: A Reading of his Fiction* (Cambridge: Cambridge University Press, 1991), 153–55.

stares of a "prim" English family. Asya defiantly gazes back at them and bursts out singing. In both cases, the story marks rowdiness as a Russian trait by contrast to West European restraint. Russia also exists in the narrator's mind as the idyllic country estate. The smell of hemp by a roadside reminds him of home and makes him wonder what he is doing in foreign lands, among foreign people. On the day after the encounter with the stuffy English, Asya similarly arouses his nostalgia for genteel Russian life in the country. Wearing an old-fashioned dress as she sits quietly doing embroidery, she strikes him as "a perfectly Russian girl," unassuming as a "housemaid" (212). Encountered abroad, Asya exhibits conflicting traits. Moreover, she sharply differs from Gagin and the narrator, two well-educated youths with a predilection for philosophizing about questions such as the "significance of the artist in our time" (213). This fondness for empty talk is also marked as Russian but divides the men from Asya, as well as "our" imagined students. Russians are not alike, and travel brings out their differences.

Asya's gender, her "cameleon" personality, and half-serf origins make her the subordinate among the three tourists. Convinced her illegitimacy precludes her making a good marriage in Russia despite her considerable education, Gagin foresees her living abroad. A cultural initiation is underway.[8] Yet, Asya rarely articulates a response to the foreign land. She even becomes part of the scenery in the narrator's eyes when the trio visits picturesque ruins. Arriving after Asya, the men find her mounting the towering ruins (a symbolic image of the "bounder on ancient ledges").[9] Asya merges with the tourist attraction that the narrative technique never allows her to focus through her own eyes. At another point in the tale she reminds the narrator of Raphael's Galatea in the Villa Farnesina in Rome. The comparison advertises his aesthetic sensitivity and knowledge of art, while preserving Asya's status as an object of the male gaze. Denied a subject position, Asya assumes the guise of a surface-skimming tourist unprepared to make her trip edifying.

8 Joe Andrew, "Introduction," in Turgenev, *Asya*, ed. with notes and vocabulary F. G. Gregory (London: Bradda, 1992), 15–22. On the resonance of the contemporary Russian woman question in *Asya*, consult also Victor Ripp, *Turgenev's Russia: From* Notes of a Hunter *to* Fathers and Sons (Ithaca and London: Cornell University Press, 1980), 162–72, esp. 169–70.

9 Peterson, *The Clement Vision*, 66.

A matter of good breeding, art connoisseurship, and a more general aristocratic education, the narrator's "cosmopolitan snobbery" ultimately cannot accommodate Asya's "nativist ignorance."[10] Turgenev's storyteller feels so secure in his social identity, sophistication, and sensitivity that he does not have to pretend tourism is something beneath him. Near the ruins a German woman runs a stand selling "beer, gingerbread, and seltzer water to tourists" (208). While Asya romps, the narrator and Gagin drink steins of beer. By including the central protagonist among the beer consumers, Turgenev conveys his awareness that in an increasingly commercialized world, all leisure travelers are tourists (in this respect, he anticipates Henry James).[11] The point is to travel in a cultivated, knowing manner—thinking and feeling differently from the tourist flocks. Turgenev's two art-loving gentlemen calmly enjoying a beer fit this pattern of being alone in the crowd, whereas Asya is violating etiquette she would observe at home. At no other time does the narrator find her so irritating, so impossible to imagine as his wife.

Provinciality

Most alien to the *Asya* narrator are the specimen tourists—those strangers whose lack of refinement he measures in their ill-cut clothing and anxious demeanor. Other authors etch a more specifically provincial syndrome. Panayeva's *Russians in Italy* opens with an authorial interpretation of post-Crimean War Russian tourism as a "vanity" project to combat the national feeling of cultural marginality with regard to the West. The trip becomes a status symbol to wield at home.[12] Druzhinin, for his part, devotes much of *Our People Abroad* to establishing a Russian tourist provinciality complex. The work begins with a parody of two conventional types of narrator: the faux naïf, chatterbox *feuilletoniste* and the insecure, belated travel writer.[13] Druzhinin's persona has promised to send his editor a travelogue, but he fears he cannot match the competition. So many Russians, he writes, are now publishing reports of their trips abroad. And how sophisticated and

10 Ibid., 64, drawing parallels with Henry James's *Daisy Miller.*

11 On James, see Buzard, *The Beaten Track*, 252–53.

12 Panaeva, *Russkie v Italii*, 291.

13 On the chatterbox feuilleton convention, see Joseph Frank, *Dostoevsky: The Seeds of Revolt, 1821–1849* (Princeton: Princeton University Press, 1976), 219; and Dianina, *When Art Makes News*, 98–101.

informative they are! They brag about never using guidebooks, they report that Paris has gas lights, Italy is disappointing, and Berlin has a street called Unterlinden. "After all that," cries Druzhinin's narrator, "how dare I raise my voice?"[14]

He soon shows his true colors. Dropping the mask of a "simple soul," he asserts his cultural and moral superiority to Russia's "exemplary tourist." Typically from Petersburg, the exemplary tourist is terrified of appearing provincial abroad. In Paris, London, and other Western cities bigger than the Russian capital, the visitor is courteous and delighted with everything. All the while, though, he harbors the fear that the local Europeans look down on him because he is Russian. What a transformation occurs when he visits a small town! There he loudly complains about its tedium and rusticity. He is rude to chambermaids and the rest of the working class (521–22). The Petersburg tourist's double standard of behavior manifests the national anxiety about provinciality as the condition of the whole Russian Empire. Hailing from the city reputed to be Russia's most cosmopolitan, the Petersburg tourist lords it over the residents of lesser towns abroad, just as he does at home. On the other hand, Paris, London, and other cultural capitals intimidate and remind him of all Russia's cultural marginality in Western eyes.

Melgunov's essay "Tourists in General and the Russians in Particular" adds further nuances to the trope of provinciality in an international framework. At one point the author politely pretends the category is inclusive. "We Russians," he writes, tend to act like "provincials" abroad, amazed at everything we see. The inclusive "we" proves disingenuous, however. The article proceeds to gradually distinguish the hard-core Russian tourist as a provincial. As in *Asya*, telltale clothing helps set the author apart from the putative majority. All too often, claims Melgunov, Russians arrive abroad in attire that makes West Europeans take them for Samoyed (a nomadic ethnic group native to northwestern Siberia). Predictably worse than the bumpkin clothes are the typical Russian tourist's personality and superficiality. In olden days, muses Melgunov, solitary adventurers such as Montaigne, Sterne, and Karamzin performed a "feat" when they undertook their journeys for the pleasure of "seeing the world, getting to know foreign peoples, and enriching themselves with impressions." Old-time travel rested on solid preparation; it was pursued at a slow pace and entailed hardships and inconveniences. On the other hand, today's typical Russian tourists

14 Druzhinin, *Nashi za granitseiu*, in *SS*, 8:518.

are multitudinous, ill-educated, and content to get a quick look at foreign lands. They have no serious interests, no heroic spirit of adventure, and they crave steady comfort. In sum, they "travel not with the aim of learning something beneficial but rather for the same reason that country people go to Moscow or Petersburg when they have no occupation to keep them at home. Such is the majority of vagabonding Russians who comprise the contingent of tourists in the strict sense of the word."[15] This evocation of the Russian capitals' allure for Russians from the boondocks underscores Melgunov's central idea of the tourist other as a mirror of provinciality widespread at home.

At the same time, Melgunov suggests that the tourist clout of the English can make any Russian feel marginal abroad. Reprising Thackeray's "little Englands" theme, the article imagines leisure travel as a sphere of international competition where Russians lag far behind the English. Wherever you go, writes Melgunov, you will find a British Hotel, English breakfast and other English foods, meal times to suit the English, English churches, English newspapers, reading rooms, bookstores, English pharmacies, and the omnipresent sign "English spoken." Why, asks Melgunov, do Russians not transport Russia abroad? Why do menus not feature kvas, kasha, cabbage soup, or *kulebiaka* (usually salmon in puff pastry)? Why do hotels not provide samovars? Why no "Russian spoken" signs? Despite the outcome of the Crimean War, the British empire's wealth and power fail to answer those questions for Melgunov. He defensively asserts that "we" are as rich as "they" are (2–3, 10). But this lame protest simply proves that money alone cannot buy the tourist authority of the English. Melgunov declines to face the fact that the post-Crimean War persistence of "little Englands" abroad mirrors the British Empire's status as the preeminent global power. Russian visitors of the time are not merely belated but also politically disadvantaged arrivals on the tourist scene. Citizens of a humiliated empire, they find themselves outdone at every turn by the English, the undisputed masters of the Continental leisure-travel circuit.

Lying, Materialism, and the Shame of Serfdom

The sidelined Russian tourists retaliate in various ways. Some invent self-aggrandizing personas. In Druzhinin's *Our People Abroad* a Russian visitor at

15 Mel'gunov, "Turisty voobshche," 1, 5, 10.

a cafe on the Piazza San Marco brags to everybody within earshot that his Petersburg palace is a replica of the Ca' Foscari. He claims he belongs to the tsar's inner circle, holds a high post in the ministry of foreign affairs, and commanded sixteen assaults on enemy trenches during the siege of Sevastopol. A Russian tourist on a cruise along the Rhine declares that he often goes hunting with Napoleon III. After listening to those compatriots, Druzhinin's narrator speaks, revealing that he, too, is Russian. He calls the liars' bluff and embarrasses them before West Europeans. The tourist on the Rhine reproaches the author: "Russians should support each other in foreign countries," he whines (519–21). To his way of thinking, national origins alone should create solidarity among Russians abroad. It is a matter of "us" against "them."

More prevalent than lying are the tactics of ostentatious dressing and spending. "Ultra-fashionable," "sumptuously dressed" Russian tourists (especially women) represent the norm in *Our People Abroad* and Panayeva's *Russians in Italy*. In an evident effort to project aristocratic status, the tourists fall into excess, and the practice boomerangs to mark them as Russians trying too hard. Their national origins are as glaring as those of their bumpkin cousins—the ill-dressed tourists of *Asya* and Melgunov's hapless "Samoyed." The specimen Russian tourists of the 1850s strive to win respect in the West by "lavishly throwing money around." This behavior buys the deference of hotel personnel who address them by aristocratic titles, a form of flattery whose mercenary motive the self-important tourists fail to recognize. They do not realize it is impossible to buy the "friendly disposition of civilized people." Along with big spending in hotels, a "passion" for shopping empties the pocketbooks of Russian women, while "secret expenses" (obviously of an erotic nature) deplete the funds of Russian men. Gambling, too, ruins many a Russian man on tour. According to Melgunov, only one in a thousand Russians can resist the casinos or at least gamble in the moderate manner of the English, French, Italians, and Germans. For Russians, the risk of addiction appears so great that Melgunov advises them to avoid casinos altogether. Gambling proves especially costly. But whatever form of material consumption the Russian tourists favor, they pass the time on steamers and trains boasting about how much money they are spending.[16] The cash outflow sets the whole "value" of their trip.

---

16 Druzhinin, *Nashi za granitseiu*, 523, 525–27; Panaeva, *Russkie v Italii*, 296–98; and Mel'gunov, "Turisty voobshche," 7–8.

Some Russian visitors take their wealth as a license for callous behavior toward West Europeans. The cruellest incident occurs in Rome at carnival time. From the balconies of their posh rented apartments, Russian tourists dangle trinkets to tempt revelers in the streets, but when Italians jump at the bait, the Russians throw quicklime rather than flour into their faces.[17] Druzhinin's *Our People Abroad* contains a longer anecdote about nasty tourists in Cologne. After admiring the cathedral and purchasing the requisite bottle of eau de Cologne, the author joins other guests for dinner in his hotel restaurant. He enjoys practicing his German in conversation with traveling German students, and he shares the assembled company's pleasure in listening to the music of two humble old musicians playing a fiddle and a harmonium. The author admits the music would sound schmaltzy anyplace else. Here, though, it contributes to the cultural authenticity effect. The convivial atmosphere changes with the arrival of a party of Russian fashion plates. Speaking Russian, the latecomers ridicule the rustic decor, the food, the harmonium, and the "disgusting music." Switching to French, they talk about "the dukes, princes, and counts" they socialized with in Baden-Baden. After the last song, the harmonium player walks around with his cap, and the appreciative diners respond generously. One of the snooty Russian women ostentatiously offers a gold coin, only to learn that German dignity is not for sale. The musician politely refuses the money: "You did not like our music. Then why pay for it?" The shamed Russians flee the dining room, while the remaining diners invite the musicians to join them for a bottle of wine.[18] Acting as a foil to "our" despicable people abroad, Druzhinin's narrator is the only Russian to enact cosmopolitan receptiveness to foreign culture and people: the Cologne cathedral, the music, the local wine, the German musicians, and German diners at his hotel (522–25).

Tourist xenophobia combines with anxiety about social democracy. After visiting several Italian cities, Panayeva's wealthy General Karazin is so "sick of Italians" that he vows to stop patronizing the Italian opera in Petersburg. He explains that he has come abroad to confirm his belief that "Europe is rotten." This he has done, and he now believes he has the authority to spread the word to comrades at home: "I know, old buddy, I've seen it,

17 Panaeva, *Russkie v Italii*, 302.

18 Pitting humble local musicians against ill-mannered rich Russians, the episode may allude to Tolstoy's *Lucerne*, where Prince Nekhlyudov alone gives money and drink to the busker.

I've been there."[19] The general has found democracy even worse than anticipated. He observes that there is indeed more freedom in the West than in Russia, but in his view this simply gives "riffraff" access to upper-class turf:

> You see peasants in the galleries, they're worming their way into museums! What business do they have doing that? [. . .] And what about those common tables in the taverns. Is it really permissible to seat decent people next to some cobbler and what's more to serve *him* while making *you* wait? I admit—at first I thought these redheads were lords, I deferred to them, but then it turns out they're brewers or salesmen. Back home such rascals wouldn't dare sit down before I did! And people keep shrieking—it's the English aristocracy. (319)

So speaks a Russian reactionary keen to keep the serfs down at home, and in satirizing him the author endorses the opposing impetus to reform. For Panayeva's oafish general, the observation of democracy in action shores up his allegiance to the status quo in Russia. The middle-class English tourists he has mistaken for lords represent the threatening potential of destabilizing Russia's *ancien régime* social structure and the sacrosanct Table of Ranks.

Panayeva's authorial persona drives home this theme by casting Russian serfs as "piteous, involuntary travelers," forced to accompany their masters on tour. Knowing no foreign languages, the serfs suffer the lot of the "deaf and dumb":

> [They] languish in noisy hotels and garrets. They somnolently pass the Alps, seas, and cities. All Italy's natural beauty counts for nothing compared to a warmly heated bathhouse; so, too, the pleasures of European life compared to a samovar in the company of fellow serfs in a dirty kitchen, where they show off their perspicuity and make jokes about their owners. (293–94)

The author seeks to enter into the traveling serfs' "piteous" situation but ends up emphasizing the cultural and emotional gulf between them and their masters. Observed on tour in the West, the uncouth serf appears ominously alien, resentful, and duplicitous in relation to the Russian elite.

19 Panaeva, *Russkie v Italii*, 322, 343.

Apollon Maikov's sketch "A Stroll around Rome with My Acquaintances" suggests that encounters with social democracy abroad might sow seeds of rebellion in a traveling serf.[20] Abroad with his master Andrei Blagochiny, the serf Vanka finds the Italian capital dilapidated compared to home: so many crumbling buildings in Rome, so many statues without noses![21] Vanka lacks the acquired romantic taste for ruins but surely needs no special training to judge the sight of Italian commoners humiliating his master and another Russian nobleman, Chuprunov. Accustomed to deference from underlings at home, Chuprunov tries to shove his way to the head of a line of Italian artisans and workers waiting to buy theater tickets. When a local man resists, the pushy Russian seeks to pull rank in the manner of the Very Important Personage in Gogol's *The Overcoat*, the big bureaucrat whose ferocious tongue-lashing brings on Akaky Akakyevich's fatal illness. "Do you know whom you are dealing with, you boor?!" shouts Chuprunov. In fact, the Italians have no idea. They take the tourist for the pompous ass that he is and laugh at him. He brandishes his walking stick, the threatened Italian pulls a knife, but yet another Russian traveler steps in to prevent violence (and tosses the offended man some coins). Chuprunov slinks away fuming, the object of Italian ridicule. The equally abashed Blagochiny follows, after giving Vanka a kick and yelling at him: "Shove off, idiot! What're you staring at so bug-eyed?!"[22] The serf remains silent but has had the potentially transformative experience of witnessing the free workers of Italy assert equal rights vis-à-vis a Russian of high standing at home.

"The Majestic Tourist" section of Druzhinin's *Russians Abroad* also represents a repulsive Russian mistreating his serf in the West.[23] The main character is an arrogant Russian count traveling with his equally stuck-up Russian wife, their niece (a baroness), and a decrepit serf in livery, "a walking incarnation of servility" (567). After the feuilleton narrator bails out a sassy Russian friend who lost all his money at roulette in Baden-Baden, the two of them encounter the aristocrats on a train going from Hamburg to Berlin. At a rest stop the count is rude to the gambler. When the train

---

20 Maikov, "Progulka po Rimu s moimi znakomymi," *PSS*, 4:189–215. This edition dates the work 1848 but does not indicate when it first appeared in print.

21 On Maikov's humorous treatment of Russian disregard for Roman ruins, see Schönle, "Broken History and Crumbling Stones," 764. The ironically applied adjective *blagochinnyi* means "decent, decorous."

22 Maikov, "Progulka po Rimu s moimi znakomymi," 204–6, 214–15.

23 Druzhinin, "Turist velichavyi," *SS*, 8:565–73.

moves on, the count retires to his compartment and closes the curtains for a siesta. While "the majestic tourist" sleeps, the abused gambler takes revenge by telling German passengers at each stop that the serf in livery is a Russian Barnum transporting freaks to exhibit in Berlin. As the rumor spreads from station to station, growing crowds of curious Germans cluster at the count's curtained window. When he finally opens the curtains at a stop, the sight of the mob startles him. He sends his serf to make inquiries. After hearing the terrified lackey's report, the count blames him for the mess and gives him a brutal dressing-down on the platform. The vicious serf-owner thus makes a spectacle of himself under Western eyes. He exits the story as a genuine Russian freak, a freak of feudalism long extinct in the West.

Cultural Ineptitude

In addition to all their more appalling traits and practices, the Russian crowd is predictably too ignorant and unimaginative to make travel an edifying experience. Turgenev's "From Abroad" maintains that "traveling in a foreign country is like learning a foreign language; it enriches a person's inner life." Most of "our tourists," however, are too ill-prepared and insular to benefit from their trips. Wherever they go (with the exception of Paris), they appear bored, weary, and perplexed. They fail to realize that they create their own misery. "Nine out of ten Russians," asserts the author, are incapable of functioning outside a tourist bubble. They "see towns, buildings, faces, clothing, mountains, fields, and rivers, but they have no actual, living contact with the people among whom they are traveling." Whether from "inborn timidity, pride, laziness, or an inability to establish friendly relations with people," they rarely venture beyond "the milieu of hotels, waiters, long bills, bells, tables d'hôte, hired servants, hired carriages, hired donkeys, and guides." Their ignorance precludes the enormous pleasure to be derived from visiting a country whose language, culture, and history one knows well. For most Russian tourists, "the names of cities, historical personage, and events remain simply names." All they can report from their trip is that "Frankfurt is bigger than Nuremburg, and Berlin is even bigger" (12:182–83).

In the realm of cultural consumption, nothing so deeply separates cosmopolitans from the herd as sensitivity to art. Turgenev's writings of the 1840s had already begun treating Russians as ignoramuses in this sphere.

The eponymous hero of the story "A Hamlet of the Shchigrov District" (1849) tells acquaintances in Russia about his sightseeing during a two-year sojourn in the West: "I was in Italy, I stood before [Raphael's] *Transfiguration* in Rome, and I stood before the *Venus* [*de' Medici*] in Florence" (1:261). He stood and stood, but nothing happened. The ironic perspective is strictly the author's, creating the meaning that no character in the story ever realizes: art cannot speak to the deaf. The elitist postwar narratives by Turgenev and his contemporaries elaborate this theme of Russian tourist ignorance of art. The specimen tourists of Panayeva's *Russians in Italy* know that cultivated people set store in seeing paintings and sculptures in the West, and so they make arduous journeys to Italy with children and serfs in tow. However, once they get there, they discover that they do not actually like art. To avoid such costly mistakes, the author advises readers to make a trial run at the Hermitage. While some Russians prove unreceptive to art, others make a show of being "art lovers." In Rome they broadcast their visits of the Vatican, artists' studios, galleries, and palaces. This "love" of art is evidently more about being *seen* rather than seeing. Women are the worst offenders (so depicted by a woman writing under a masculine pseudonym). Two Russian ninnies blow all their money on clothes in Hamburg. That ends their trip. But in order to impress people back home, they "give ecstatic accounts of the 'Dresden Madonna,' Raphael, the delights of Italy," the Vatican, and an excursion to the top of Vesuvius.[24] Those vanity travelers had assimilated the norms of fashionable tourism but could not budget their funds.

Melgunov's essay and Turgenev's "From Abroad" more pointedly valorize art as the acid test that separates cosmopolitans from the crowd. Both mention the *Sistine Madonna* in particular. Melgunov portrays Russian women tourists chattering about their dresses while standing before Raphael's painting (and apparently blocking the view of sensitive men). The ladies come abroad to expand their wardrobes, a priority that makes them wild for Paris—"that pandemonium of shoppers' delights." Without differentiating men from women, Turgenev's "From Abroad" imagines the average Russian's confrontation with the *Sistine Madonna* as an excruciating ordeal. In a sheep-like manner, Russian tourists seek out art works that guidebooks label "famous," but they have no idea how to respond to them. In this regard, continues the author, "the little room in the Dresden gallery that displays the *Sistine Madonna* provides a particularly instructive

24 Panaeva, *Russkie v Italii*, 293, 304–6.

spectacle. For so many years, generation after generation of Russian travelers have sat and still sit there on a sofa that has repeatedly struck me as a site or instrument of spiritual torture, not inferior in that respect to instruments of physical torture, those remnants of the Dark Ages now displayed for curious visitors in ancient armories." The author adds that tourists baffled by art can always turn to nature since the appreciation of "a beautiful landscape, a waterfall, or a mountain" requires no special aesthetic training. Upon occasion, he continues, "I have even seen [Russian] generals of both the military and civil hierarchies" enraptured by such sights. "But nature alone cannot fully satisfy a person. One idyll resembles the next, and the general remains a general."[25] On this condescending view, "they" can have the scenery. Art is for "us," the connoisseurs. Turgenev's demotion of scenery by comparison to art again anticipates James, who in 1873 portrayed Switzerland as an over-developed, tourist "show country" that he was pleased to leave to "the mass of mankind."[26]

Turgenev's elitist outlook chimes with contemporary satires of the Russian tourist as a philistine, actively hostile to art and aesthetic experience in general. Panayeva's reactionary General Karazin nearly dies of boredom in Florence and believes no self-respecting Russian would want to become an artist. He also considers Venice nothing special. In reaction to his compatriots raving about the city on a moonlit gondola ride, he snorts that "Mother Russia," too, "has her Venices," places with "water all around" such as Gorokhovets, a monastery town on the Klyazma river around two hundred miles east of Moscow (324, 327–28). A more vociferous philistine features in Tur's *On the Verge*. The heroine Tatyana Istomina is a wealthy young widow who visited Europe several times with her husband to relieve Petersburg ennui. After his death she went to Italy again. In Rome she studied art, found a Papal Mass in St. Peter's square particularly moving, and more generally derived cultural satisfaction from her travels.[27]

Memories of her sojourn in the West have become a cosmopolitan component of her inner life without having altered her belief that the Russian nation has a claim on her conscience.[28] In hopes of doing her part for the commonweal, she has retired to her family's country estate to

25 Mel'gunov, "Turisty voobshche," 6, 8; Turgenev, "Iz-za granitsy," 183–84.

26 Buzard, *The Beaten Track*, 95.

27 Evgeniia Tur [E. V. Salhias de Tournemir, pseud.], *Na rubezhe*, *Russkii vestnik* (October 1857): 859; and (November 1857): 337.

28 See again Appiah on the old Tolstoy in *Cosmopolitanism*, xiv.

use her money to build a school for the peasants and renovate the local hospital. During a cholera epidemic, when she opens her manor house to the sick and assumes duties as a nurse, she falls in love with a rural Russian doctor. All Tatyana's values—her receptiveness to Western culture, her dedication to social service, and rejection of the high-society life her parents think she should be leading—set her against the Petersburg cousin who comes to court her in the country soon after taking his own Grand Tour. This youth has nothing good to say about the West. He dismisses Germany as a "petty bourgeois" country and calls Italy "a damned hole," a land of "thieves, bandits, and swindlers." He proclaims Michelangelo nothing but "a big name," an indication of his general disregard for art. "The time has come," brays the philistine, "for us to cease bowing to authorities and to form an independent point of view." Tatyana's cousin scorns ancient ruins, too: "What do I care if Caesar's palace or a temple of the Vestal Virgins once stood someplace?"[29]

Revolutionaries on Tour

While this anti-aestheticism and resentment of "authorities" smacks of the nihilism Turgenev would incarnate in Bazarov in *Fathers and Children*, several budding revolutionaries of the pre-emancipation years leaped at the chance to take part in tourism previously monopolized by the aristocracy. The familiar letters of those men show that a well-educated Russian's sense of belonging to an aristocracy of feeling was not entirely a function of socioeconomic status or political orientation. Sent to the West in 1858 to develop his expertise in forestry management, Shelgunov spent time at a German spa. A letter to his wife conveyed his impressions of the Russian visitors: "The educational disparities within our so-called educated middle class are most evident abroad. That's the reason, no doubt, that respectable Russians over here do not like to socialize with strangers from home." Like the narrator of *Asya*, Shelgunov identified Russians on sight and shunned them: "I've run into a lot of Russians here. I think they're Russians because only Austrian aristocrats and Russian generals and their ladies could behave so stupidly."[30] Dobrolyubov expressed a similar tourist phobia in a letter from Dresden in June 1860. He wrote that the number of Russian

29 Tur, *Na rubezhe*, (November): 335–36.
30 Letters from June 21 and 27, 1858, in Shelgunov et al., *Vospominaniia*, 2:90, 91.

visitors there had reached the point of "indecency." When inane Russian loudmouths surrounded him at a theater, he sought to hide his nationality. Using German, he asked the man seated next to him a question, only to discover that this stranger, too, was a Russian holding himself aloof from the tourist crowd.[31] As his acquaintance Nikolai Novitskii (1833–1906) would put it, Dobrolyubov was "an aristocrat in heart and soul."[32]

Although Dobrolyubov never made public his personal experiences as a tourist, his blossoming in the West was extraordinary and moving. Abroad for a little over a year, he visited Germany, France, Switzerland, Italy, and Sicily. He went hiking in the Alps, extensively toured Italy (Florence, Venice, Rome, Naples, Pompeii, Genoa and Milan), and brought home lots of Italian and Swiss postcards.[33] Expressing an open mind toward the West and its residents, his letters contain no critiques of capitalism or the French bourgeoisie. He liked trains and made at least one railway journey in France the occasion for conversing with a young French fellow traveler, to learn about his life; and Dobrolyubov was struck by how *well* the Frenchman lived on much less money than tourism was costing.[34] Writing from a pension in Paris, Dobrolyubov told a Russian friend that his hosts were treating him as one of the family. Their kindness had made him stop worrying about his poor French, and the convivial parties at the pension were allowing him to realize his life-long dream of learning to dance. In short, he was "coming to resemble a human being," a person with the "right to live and enjoy life and not be exclusively on call to exercise my talents for the benefit of mankind. Here nobody sees me as the acidic critic, nobody expects from me an outpouring of venom."[35] Later in his wanderings, Dobrolyubov fell in love with an Italian woman in Messina and proposed to her. With self-deprecating irony, he then informed Chernyshevsky that he "had decided to give up performing any further feats for the field of Russian letters." He looked forward instead to learning a "new trade" and leading a modest family life in a secluded "corner of Italy."[36]

31 Dobroliubov, letter to M. I. Shemanovskii, June 11/23, 1860, *SS*, 9:422.

32 S. A. Reiser and V. V. Zhdanov, eds., *Dobroliubov v vospominaniiakh sovremennikov* (Leningrad: Gosudartsvennoe izdatel'stvo khudozhestvennoi literatury, 1961), 254.

33 Dobroliubov, letters to various correspondents, *SS*, 9:427, 462–71; and L. N. Samsonov, in Reiser and Zhdanov, *Dobroliubov v vospominaniiakh*, 329–30.

34 Dobroliubov, letters to various correspondents, *SS*, 9:427, 462–71.

35 Dobroliubov, letter to A. F. Kavelina, November 14/26, 1860, ibid., 454.

36 Dobroliubov, letters from May 28/June 9 and June 12/24, 1861, ibid., 471–74.

Interaction with West Europeans had liberated Dobrolyubov from his Russian public persona as the "acidic critic," but that self-transformation remained hidden from the Russian readership. In view of his precarious health, his Italian sweetheart's parents forbade the marriage. After returning to Russia in July 1861, Dobrolyubov resumed his job as a *Contemporary* gadfly up to his death in November. Chernyshevsky drafted an obituary revealing Dobrolyubov's desire to "remain in Italy forever" but deleted that passage for publication.[37] In step with that editorial decision, other public commemorations from the radical camp would represent Dobrolyubov as a single-minded political actor who had "worked constantly in Italy," derived no personal benefit from his trip, and grew even sicker in the West than he had been at home.[38]

A New Revolutionary Tourist

Among the *raznochintsy* revolutionaries of the early reform era, Mikhailov stands out for his public performance of tourism as a democratizing opportunity in *Paris Letters* and *London Notes*.[39] His paternal grandfather was a serf, owned by a member of the Aksakov family and beaten to death for protesting when his master broke his promise to free him. The manumitted son became a civil servant in the mining industry and married a Kirghiz (Kazakh) princess in the Orenburg region where Mikhailov was born. Raised in the little mining town of Uletskaya Zashchita, he received his early education from tutors at home. In 1845 he published his first poems and then attended Petersburg University (1846–48) where he impressed

37 Chernyshevskii, in Reiser and Zhdanov, *Dobroliubov v vospominaniiakh*, 382.

38 M. A. Antonovich, ibid., 242; and A. M. Skabichevskii, *N. A. Dobroliubov. Ego zhizn' i literaturnaia deiatel'nost'* (St. Petersburg: Tipografiia tovarishchestva "Obshchestvennaia pol'za," 1894), 68–71.

39 The sparse commentary on *Paris Letters* and *London Notes* consists mainly of Stalin-era Soviet interpretations that categorize the works as muckraking "political journalism" (*publitsistika*) and ignore the author's enjoyment of tourism. For representative examples, see P. S. Fateev, "Mikhail Ilarionovich Mikhailov (1830–65)," in M. I. Mikhailov, *Stikhotvoreniia* (Leningrad: Sovetskii pisatel', 1950), 24–25; and A. V. Kushakov, "M. L. Mikhailov," in M. L. Mikhailov, *Sobranie stikhotvorenii* (Leningrad: Sovetskii pisatel', 1953), 16–19. Some of Mikhailov's positive responses to France and England are noted in Iu. D. Levin, "Poeziia Mikhailova," in M. I. Mikhailov, *Sobranie stikhotvorenii* (Leningrad: Sovetskii pisatel', 1969), 11–12. ("Ilarionovich" was a variant of Mikhailov's patronymic "Larionovich").

Shelgunov as a "walking bibliography of foreign literature."[40] In addition to his command of Russian and European literature, he knew French, German, English, and Latin—all signaled, if only marginally, in his travel writing. Although penury forced Mikhailov to take a temporary civil service job in Nizhnii Novgorod, he continued pursuing his literary ambitions and eventually returned to Petersburg, where he enjoyed recognition for his poetry and stories, his translations of Heine and others, and his journalism on women's liberation.[41]

In sending Mikhailov abroad to write reports for the *Contemporary*, Nekrasov advised him to "touch lightly on politics."[42] This word to the wise licensed the creation of a complicated authorial persona with antecedents in Herzen's *Letters from France and Italy*. Mikhailov's narrator is a revolutionary tourist of new times, a champion of modernization who blends progressive reportage with an acculturation regime typical for the Russian elite. Briefly but significantly, he participates in the tendency to malign the postwar tourist crowd. While visiting the palace of Versailles, he encounters a traveling Russian family admiring paintings of French royalty, Crimean War victories, and other French military triumphs. The author calls the family's attention to the absence of images concerning the French revolution, but they do not get the point. He alone perceives the "sword of reaction" overhanging the Second Empire of Napoleon III.[43]

Mikhailov's anti-bourgeois political perspicuity marks him as a protégé of Herzen (who hosted him and the Shelgunovs in London). *Paris Letters* represents the local bourgeoisie as vulgar materialists addicted to vaudevilles that put their lives on stage. Even death is grist for Mikhailov's anti-bourgeois mill: he drolly reports that during his stay in Trouville, "an estimable bourgeois" from Paris drowned while attempting a swimming feat in the sea.[44] As an enemy of bourgeois morality, Mikhailov gives much attention to the condition of women abroad. His Paris hotel proprietor is a feminist living with her lover, and one entire installment of *Paris Letters*

40 Shelgunov et al., *Vospominaniia*, 1:239.

41 On Mikhailov and feminism, see Richard Stites, "M. L. Mikhailov and the Emergence of the Woman Question," *Canadian-American Slavic Studies* 3, no. 2 (January 1969): 178–99; and Jennifer Lonergan, "M. L. Mikhailov and Russian Radical Ideas about Women, 1847–1865" (PhD diss., University of Bristol, 1995), 188–93.

42 Lonergan, "Mikhailov," 166.

43 *Parizhskie pis'ma*, *Sovremennik* 72, no. 10 (October 1858), section 1: 469.

44 Ibid. (November 1858), section 1: 190–91.

is devoted to Proudhon on the woman question.[45] In London's Haymarket slums the author observes prostitutes, viewed as victims of a hypocritical social order. Along with bourgeois hypocrisy, *London Notes* elaborates the themes of materialism and avarice. With an epigraph from Schiller, the first installment introduces the English capital as the realm of Mammon: "und es herrscht der Erde Gott—das Geld."[46] Finally but importantly, Mikhailov's accounts also expose poverty, although with a light touch, as we will see.

While progressive reportage is extensive in both travel cycles, it jostles against a scenario of self-expansion on the part of a social underdog relishing aristocratic-style tourist pleasures on a low budget. Recounted in the opening installment of *Paris Letters*, Mikhailov's big coup in this regard is to have found the Hôtel Molière, a charming but inexpensive place "not in Reichard's famous guide." On a side street off the Boulevard des Italiens and close to a mouth-watering bakery, the hotel has a courtyard where one can sit after dinner to enjoy a cup of coffee and smoke a cigar.[47] Already remarked in Fet's *From Abroad* as a symbol of gentry hedonism, the cigar alone proclaims that Mikhailov's persona has stormed old social barricades to leisure travel. His culinary adventures and his concern with good, affordable lodgings become leitmotifs of *Paris Letters* and *London Notes*; and when it comes to food, there is no contest between the French and the English.

The Hungry Tourist Eye

Mikhailov's central paradigm of tourism is the pleasure of looking. The author identifies himself as "an inveterate tourist" and *flâneur* with a John Murray guidebook.[48] The Christmas shopping season in Paris reveals to him the intensity of his visual appetite. Shopping is a spectacle, beginning with expensive stores likened to museums with "exhibitions of presents."[49]

---

45 Ibid., no. 9 (September 1858): 270–73; ibid. 73, no. 1 (January 1859), section 1: 163–82.

46 *Londonskie zametki*, *Sovremennik* 77, no. 9 (September 1859), section 3: 147–48; ibid. 76, no. 6 (June 1859), section 3: 219.

47 *Parizhskie pis'ma* (September 1858), 269–71. On the guidebook of Heinrich Reichard (1751–1828), see Dostoevskii, *PSS*, 28.2:26, 377n5.

48 *Parizhskie pis'ma* (November 1858), 182, 199, 215. Mikhailov identifies the guide as his "little red book," an allusion to the nickname that Thackeray gave the Murray series. On the latter point, see Buzard, *The Beaten Track*, 75.

49 For general discussion of a parallel between shopping and visiting museums in Haussmannian Paris, see Andrea Witcomb, *Re-Imagining the Museum: Beyond the Mausoleum* (London and New York: Routledge, 2003), 19–21.

The open doors invite you to "walk right in, stroll through the shining salons, and admire the luxury, even if you do not buy anything." In this "varied exhibit," however, you will almost certainly see something you cannot resist. Shopping thus teaches the author how right Balzac was to call the human eye an "avid, spoiled" organ making "boundless demands." And besides, unlike the dour clerks of London, Parisian sales personnel make shopping a pleasure: "Go into a big store or little shop and buy something if only for a franc or just a few centimes. The salesman is so courteous, so attentive that you even start feeling a little embarrassed. You would think you were doing him a great service" by purchasing something.[50]

Boundless demands of the authorial eye produce an eclectic sightseeing program. Mikhailov's narrator visits many standard, aesthetically pleasing tourist sites besides Versailles: the Louvre and other art museums, the Bois de Boulogne, Notre Dame, cathedrals in Normandy, St. Paul's, Westminster Abbey, and the Tower of London. He conducts readers on a Paris walking tour from the Madeleine to the Bastille, with a long stop to ogle the "orgy" of a public ball at the Opera.[51] Precisely because his *flâneur* eye is so avid, Mikhailov's persona tends to dislike trains. He finds them too fast to allow seeing the passing sights properly. On an express from Berlin to Paris, for instance, he barely catches a glimpse of "the huge mass of the Cologne cathedral," flashing past in the distance. French trains induce claustrophobia as well: "they keep you locked in your compartment like a wild beast in a cage."[52] Apart from the railroad, however, achievements of industry and engineering entrance him. He praises suspension bridges and modern architecture, especially the Crystal Palace (which he does not see) and the Palais d'Industrie on the Champs-Elysées. Expressing no regrets about the destruction of old Paris, he approves the tree-lined new boulevards of Haussmann's project of urban renewal. In London Mikhailov visits a tunnel under construction beneath the Thames and in Paris predicts that a tunnel beneath the Channel will someday connect England to France.[53] Man's technological ingenuity appears unlimited.

50 *Parizhskie pis'ma*, *Sovremennik* 73, no. 2 (February 1859), section 3: 233–34; *Londonskie zametki* (September 1859): 145–46.

51 *Parizhskie pis'ma* (February 1859), 215–32, quote 225.

52 Ibid. (September 1858), 289; ibid. (November 1858), 182.

53 *Parizhskie pis'ma* (December 1858), section 1: 616–18; *Londonskie zametki* (June 1859), 231.

In the tourist's visionary perspective, the benefits of industrialization outweigh the environmental and human costs. In London he observes "filth and slops" in the Thames and smoke belching from chimneys. Coal dust and soot impair breathing and dim the sun, casting an eerie crimson light on "monstrous omnibuses" and the "frightful" crush of crowds of pedestrians. And yet, "for the unaccustomed observer, there is something make-believe, fantastical" in this eye-riveting urban scene, so different from anything existing in Russia, now merely on the threshold of industrialization. London, writes Mikhailov, would be more beautiful "if smooth stone banked the river, and the shores were lined with trees, dotted with flower beds, and built up with elegant palaces and so forth [. . .] But there is no shortage of such beautiful scenes in the world outside London, and none of them has imparted to a single city the durable grandeur that London's morose, grim ambiance has given to it."[54] From this standpoint, the author demotes even Venice, imagined with perpetually clean waters and air, both better for human health.

Whether standard tourist sites or displays of industry and technology, the visual seductions blur the author's perceptions of poverty. Having taken in the "exhibits" of luxury goods in Paris shops, he observes that "all these riches are laid out for the wealthy. The poor person can only look at them and envy people with lots of money. Yet, even the poor man is not forgotten." The shopping spectacle continues in the streets, where customers of modest means find affordable items at sidewalk stalls. Here, for instance, are cheap toys delighting the children. This comprehensive tableau of consumer satisfaction minimizes socioeconomic disparities in Paris. Mikhailov's persona is similarly reluctant to linger on poverty during a tour of Normandy focused on scenery and Gothic cathedrals. On his way to Jumièges, he comes upon a row of hovels and asks his coachman to stop. Finding the door of a hut ajar, the author enters and walks around the empty home. He observes "pitiful bedding made of pauper's rags" and other signs of a wretched existence, but the naturalistic details are left to speak for themselves. As for the tourist, back in his coach, he happily announces that "the new, clean houses of Duclair finally shielded me from those walls" of sooty hovels that had lined the road.[55]

London Notes erects a similar aesthetic barrier between the author and the local working class. In the London Bridge area, where his hotel is

54 *Londonskie zametki* (June 1859), 230, 235–37.

55 *Parizhskie pis'ma* (November 1858), 221; ibid. (February 1859), 233–34.

located, he remarks "stinking side-streets" and "dismal dens of poverty and debauchery." But by then, the docks, depots, ships, and commercial activity along the Thames have overawed him. He asserts that all this impressive evidence of Britain's "unprecedented wealth" makes "you forget the equally unprecedented poverty" of London's toilers. The *London Notes* epitome of working-class misery is not a real person but rather the heroine of a poem Mikhailov translated: Thomas Hood's "Song of a Shirt" about a London seamstress living in a garret and too poor to afford a rug. The poetic language aestheticizes poverty. Eventually, *London Notes* completely abandons the working class, to take the reader on a walking tour of "monumental London." This begins with a long visit of St. Paul's and ends at Westminster Abbey, two monuments prompting the author to extol Gothic architecture, with a long citation from Louis Viardot's "excellent book on London museums."[56] Lingering in the Poets' Corner of the abbey, to reflect on the British authors buried or commemorated there, Mikhailov's persona liberates himself from political reportage, to accentuate his identity as a cultural tourist and a creative writer experiencing a cosmopolitan kinship with foreign authors and architects.

Whereas Herzen's *Letters from France and Italy* privileged politics over tourism (concentrated in Italy, particularly in Rome and the Vatican museum), *Paris Letters* and *London Notes* reverse the priorities. Although Mikhailov's narratives register misgivings about bourgeois capitalist societies, the critiques are muted. Whatever its off-putting aspects, Paris ultimately strikes the author as "a splendid city." The place seems to mesmerize, transcending socioeconomic injustice, tiresome frivolity, and the vapidity of much of the current theater and literature. Similarly, "the majestic side" of London leaves a lasting impression in Mikhailov's account, despite the city's pollution and poverty.[57] These value judgments summed up a commoner's irrepressible gratitude for the tourist experience his work as a journalist made possible. The aesthetics of the project did not wholly eclipse Mikhailov's humanitarian concerns but made clear that even for a politically progressive Russian, leisure travel abroad could be an exhilarating, cosmopolitan expansion of the homebound self.

Mikhailov had come a long way from Uletskaya Zashchita but was doomed to die back in Siberia, in circumstances that erased the enjoyment of

56 *Londonskie zametki* (June 1859), 237–38; ibid. (August 1859), section 3: 224–26, 233–34.

57 Ibid. (June 1859), 219–20.

tourism from his afterlife. The emancipation of the serfs precipitated a political crisis that radicalized him. The terms of the liberation favored landowners. Left-leaning, mainly young Russians were incensed, and peasant protests occurred, most notably in the village of Bezdna in the Kazan province, where government troops massacred several hundred demonstrators in the spring of 1861.[58] In June Mikhailov returned to London to have Herzen's Free Press print the proclamation "To the Young Generation." Written by Shelgunov, perhaps with Mikhailov's assistance, the proclamation marked the beginning of overt agitation aimed at overthrowing Alexander II and fomenting peasant revolt. Mikhailov smuggled home 600 copies of the proclamation, which fell primarily into the hands of students at Petersburg University. After an informer betrayed him, Mikhailov was arrested in September and assumed the full blame. The first person charged with a political crime under Alexander II, he was brought to trial in December and sentenced to six years hard labor in the Siberian mines.[59] After his death there in 1865, commemorative poems represented him as a Christ figure, "an exiled martyr," whose "beautiful heroic feat and holy sufferings" would forever inspire rebels against Russian despotism.[60] Erasing the gusto for tourism that Mikhailov's travel writings expressed, the posthumous fashioning whittled him down to a unidimensional secular saint and placed him in the catalogue of revolutionary martyrdom where Dobrolyubov had already been inscribed.

Aesthetics and the Killjoy Work Ethic

While Mikhailov represented his journey as a knowledge-building venture affording much pleasure, Chernyshevsky hardened his killjoy outlook. A new target was *Letters on Spain* (1857) based on a trip of 1845 and partly serialized in the late 1840s and 1851. While praising the book as one of the most notable, well-written Russian travel accounts of recent times, Chernyshevsky regretted Botkin's extensive attention to art, scenery, and alluring Andalusian women. Seeking to provide a corrective to the tourist eye, Chernyshevsky's review concentrated on Spain's sociopolitical and economic problems. The critique also disapproved leisure in principle.

58 Gleason, *Young Russia*, 160–61.

59 Joseph Frank, *Dostoevsky: The Stir of Liberation, 1860–1865* (Princeton: Princeton University Press, 1986) 134–38; and Gleason, *Young Russia*, 161.

60 N. P. Ogarev, "Mikhailovu," in Mikhailov, *Sobranie stikhotvorenii* (1969), 503–4; see also 501–2, 506–10 for poems by others.

Chernyshevsky asserted that "man by nature finds pleasure in working, has an innate need for work and languishes in boredom if he does not work, if his idleness is not merely rest after work—rest stimulating him to return to work with replenished forces."[61] In opposition to every aspect of Chernyshevsky's position, Druzhinin's review of *Letters on Spain* commended *Letters on Spain* as a sensitive "tourist" whose aesthetic priorities made absorbing reading.[62]

Chernyshevsky doubled down on his anti-tourist position in his review of *Asya*, "A Russian at a Rendezvous" (1858). The article portrays Turgenev's narrator as a typical representative of the cultivated high nobility, the class long considered Russia's "best people." Chernyshevsky argues that prior to the tryst with Asya, the storyteller appears sensitive and full of "lofty feelings." But in devastating the naïve semi-serf heroine, he exhibits the "senseless cruelty" of "a man more rotten than an out-and-out scoundrel." The essay goes on to indict Turgenev's entire generation of elite liberals as grandiloquent "superfluous men" unable to act. The story of failed romance thus becomes a parable of political paralysis. Chernyshevsky concludes that Russia's putatively "best people" are actually the worst, too absorbed in their erotic lives to ponder urgent questions of Russia's future: "administrative and judicial improvements, financial reforms, and the liberation of the peasants." Envisioning a progressive audience, Chernyshevsky insinuates that "we," unlike Turgenev, realize we should be working rather than vacationing, in order to "render our society some service."[63]

Chernyshevsky paved the way for Pisarev's contributions to the public discussion of tourism. Writing in 1859, Pisarev stated that one should not regard work as "a sad necessity but rather as a lofty pleasure," achieving perfect harmony between "duty and desire."[64] On this view, travel should be an extension of work, an argument Pisarev made in reviewing "Mr. Vodovozov's" epistolary travel articles about pedagogy in Germany.[65]

61 Chernyshevskii, *PSS*, 4:222–45, quote 230.

62 Druzhinin, *SS*, 7: 381–414.

63 Chernyshevskii, *PSS*, 5:156–59, 166–71.

64 Pisarev, "Vliianie iskusstva na vospitanie" (1859), *PSSiP*, 1:234.

65 The otherwise unidentified author must have been Vasily Vodovozov (1825–86), a St. Petersburg-born pedagogue, educational theorist, polyglot translator (ancient Greeks, Byron, Goethe, etc.), poet, and children's writer. His *Elementary School Reader* (1871) competed with Leo Tolstoy's primers: see Ben Hellman, *Fairy Tales and True Stories: The History of Russian Literature for Children and Young People (1574–2010)* (Leiden: the Netherlands, Loninlkijke Brill NV, 2013), 146–47, 454–55.

Pisarev argued that "intelligently undertaken" journeys are those that contribute to the traveler's professional development, as opposed to the wanderings of "a tourist" seeking "mere recreation." Whoever the traveler may be—a naturalist, historian, agronomist, politician, or pedagogue, his line of work should set his itinerary, sightseeing program, and the point of view from which he will "examine everything that merits his attention." In drawing poets and artists into the purview, Pisarev contended that their job was to observe nature and "typical individuals" in foreign lands, an assignment that apparently precluded looking at art. Whatever the traveler's specialty, continued Pisarev, an account produced from his professional point of view will have a "unity of thought" more useful and engaging than the customarily unfocused "impressions and travel notes of a tourist."[66] In sum, the "specialist" alone has a good reason to travel. Although this article reduced tourism to "mere recreation," Pisarev would soon grant that it was possible to be a "learned tourist" (*obrazovannyi turist*), out to see "everything worthy of attention."[67] The implied connotation of practical "worth," however, remained consistent with the killjoy notion of travel as work.

Pisarev's bias against "mere recreation" predisposed him against Italy, just as it had Chernyshevsky. Pisarev loved Kovalevsky's iconoclastic "Pictures of Italy" serialized in *Annals of the Fatherland* in 1858. Although Kovalevsky published art criticism, his Italian travel accounts say little about art and contest the aestheticizing tourist gaze through naturalism akin to the physiological-sketch genre's exposure of the seamy sides of cities. The "pictures" begin with Venice, presented as the first of many Italian cities that will disillusion the Russian visitor. First, Kovalevsky demystifies the Byronic trope of a boat traveler's first glimpse of Venetian palaces gloriously rising through the mist. From the deck of his ship heading into port, Kovalevsky's persona sees virtually nothing but "a panorama of black, ugly, dilapidated buildings," "clustering masts of commercial ships," "a grubby wharf and the start of an even grubbier quay filled with the frenzied cries of fish mongers." The prosaic, workaday scene resembles Amsterdam rather than Venice.[68] Kovalevsky proceeds to up-end Byron's "gendered

66 Pisarev, *PSSiP*, 1:229.

67 Pisarev, "Apollonii Tianskii" (1861), ibid., 3:127.

68 P. M. Kovalevskii, "Kartiny Italii: Venetsiia," *Otechestvennye zapiski* 116, no. 2 (February 1858), section 1: 442–45 (reprinted with related essays in his *Etiudy puteshestvennika: Italiia, Shveitsariia, puteshestvenniki i puteshestvie* [St. Petersburg: Tipografiia V. Golovina, 1864]). Serialized in *Annals of the Fatherland* in 1846–47, Charles Dickens's *Pictures from Italy* (1846) recoiled from "commonplace shops" and other signifiers of

geography" of Venice as a woman before an "enamoured male spectator from the North."[69] From the perspective of the apartment he rents on the Grand Canal, the Russian author finds the shimmering vistas of water and sky reminiscent of mirages in the Russian steppes, a native landscape he pronounces no less beguiling than Venice.[70] The Russian air is fresher, too. When the tides of the lagoon recede, especially along small canals, Venice becomes "a gigantic cesspit" full of garbage, rubbish, and dead rats (462–63, 470–73). Proud to believe he had penetrated the city's showy facade, Kovalevsky in his collected articles would personify Venice as a "beautiful slattern," most genuine when not attempting to amaze an admirer, that is, when not dolled up to suit the hoodwinked tourist crowd.[71] Pisarev considered Kovalevsky's 1858 article "a marvelous thing!"—a bracing dose of realism revealing the truth of everyday life in Venice.[72]

Pisarev's position on cultural tourism was still evolving, however. Just a month after reviewing Kovalevsky, Pisarev praised the style and content of Yakovlev's travel essay "The Eternal City," centered on St. Peter's. Pisarev specifically welcomed Yakovlev's concentration on subjects that the Rome installment of Kovalevsky's "pictures" cycle marginalized or ignored: antiquities, "historical monuments, immortal creations of architecture, sculpture, and painting," and the "great personages Bramante, Michelangelo, and Raphael."[73] In short order, however, Pisarev would endorse the outlook of Turgenev's Bazarov, the revolutionary and advocate of science who declares art in general and Raphael in particular useless. "I would rather be a Russian cobbler or a Russian baker rather than a Russian Raphael," proclaimed Pisarev on his way to fully articulating his ambition to abolish aesthetics.[74]

---

the workaday realm of Italian cities. On this tendency in Dickens and other British travel writers, see Buzard, *The Beaten Track*, 173–74. For a lovely poem that manages to renew the trope of Venice as queen of the seas, see K. K. Pavlova, "Venetsiia" (1858), *Polnoe sobranie stikhotvorenii* (Leningrad: Sovetskii pisatel', 1939), 117.

69 On this Byronic trope and its extension in Henry James, consult Buzard, *The Beaten Track*, 132–38.

70 On Gogol's trope of the steppe as a vast expanse of water, see Ely, *This Meager Nature*, 93.

71 Kovalevskii, *Etiudy puteshestvennika*, first page of unnumbered preface.

72 Pisarev, *PSSiP*, 1:26–27, 412.

73 Ibid., 1:106, 108.

74 "Kukol'naia tragediia s buketom grazhdanskoi skorboi," ibid., 6:218.

Defending Cultural Tourism

Even before radical utilitarianism peaked, Melgunov defended cultural tourism as a pursuit that could contribute to Russian national welfare. Well-prepared travelers, he maintained, need not pursue formal studies abroad in order to derive self-improving, civic benefits from their trips. They simply "absorb" Western Europe's "atmosphere of civilization," especially in Italy with its vitalizing climate and wealth of wonderful art, architecture, and scenery. Enriched and energized by European culture and scenery, these cosmopolitan tourists return home ready and eager to contribute to the postwar tasks of rebuilding the wounded Russian nation.[75] Exactly how such a travel regime would translate into effective work remained a mystery. Melgunov clearly wanted to legitimize highbrow tourism, but in the end he merely articulated the hope that it might be nationally beneficial. A few years later, though, he would publish an essay on the national economic need for Russians to overcome their endemic indolence and apathy. Remarkable for its time, this discussion anticipated some of Max Weber's ideas about the Protestant work ethic.[76] If offering no concrete proposals for harnessing tourism to national welfare, Melgunov was attempting to find an ideal, productive balance between work and leisure, aesthetic cosmopolitanism and civic conscience in an individual Russian citizen's life.

While Melgunov engaged the radicals' demand for practical benefits from travel, Druzhinin's sketch "The Steppe Tourist" (1860) stubbornly sought to shore up the Russian gentleman's authority as an art expert, the very competence Chernyshevsky had long declared useless.[77] Druzhinin's eponymous protagonist is a well-to-do, good-natured Russian provincial who has been in Florence a couple of weeks when the narrator checks into the same hotel. Identified as Jacques de Lopouchnikoff on his *carte de visite*, the bored steppe tourist seeks out the author because he believes they both relish food, drink, and Italian chambermaids. The hotel staff told him, after all, that upon arrival the narrator had lunch with wine in his room and then summoned a maid to bring up flowers and close the shutters for his siesta. The narrator explains that he has come to Florence for the art. The steppe

75 Mel'gunov, "Turisty voobshche," 11.

76 See translation of the essay in Philip Shasko, "Nikolai Alexandrovich Mel'gunov on the Reformation and the Work Ethic," *Comparative Studies in Society and History* 9, no. 3 (April 1967): 258–65.

77 Druzhinin, "Turist stepniak," in *SS*, 8:556–65.

tourist, on the other hand, has never set foot in a museum. Homesick and craving company, he nonetheless tags along with the narrator to go to the Pitti. In the gallery, Lopouchnikoff darts goggle-eyed from nude to nude. Standing before each painting, he elbows other visitors and howls out his approval. After making him sit down and be quiet, the narrator gives him pointers on art appreciation. A joyful new world opens for the yokel as he begins to fathom the marvels of painting. In walking around Florence with the narrator, Lopouchnikoff sees the city, too, with newly appreciative eyes. But alas, the acculturation program abruptly ends when the steppe tourist's wife and mother-in-law arrive from Rome and haul him off to Naples. The henpecked provincial has nothing good to learn from those harpies, who have come abroad just to brag about their trip when they get home, a pattern of behavior the author types as feminine.

Still contesting Chernyshevsky's revolt against the elitism of art appreciation and the exclusionary character of museums, Druzhinin's sketch offers a parable of the Russian nobleman's claims to cultural authority. In Florence the author assumes the position of the Russian provincial's teacher. He successfully sensitizes the yokel to the value of painting and inculcates the proper standards of museum behavior. To the radicals, however, this must have seemed a dying squeak from the Russian nobility no longer enjoying a monopoly on cultural discourse. As Chernyshevsky gained editorial command of the flourishing *Contemporary*, the reading public en masse deserted Druzhinin's rival journal *Biblioteka dlia chteniia* (*Library for Reading*, 1834–65). "The Steppe Tourist" was striving to arbitrate cultural tourism but was no match for the nihilism intensifying among the radicals. It was left to *Fathers and Children* to dramatize the streak of self-satisfied ignorance in the worldview of Bazarov, the doomed revolutionary who denigrates the Vatican museum, Raphael, Pushkin, and Schubert as representatives of the aristocratic culture he aspires to eradicate in Russia.

This chapter has focused on the emergence of a Byronic form of tourist phobia in the writings of Turgenev and other politically liberal aesthetic cosmopolitans in the aftermath of the Crimean War. A new phenomenon in the national culture, the alien Russian tourist crowd that these authors depicted bore all the sorry traits that English elitists had been attributing to traveling compatriots ever since Byron's era: insularity, superficiality, bad manners, ignorance, and insensitivity, especially with regard to art. But in addition, the Russian specimen tourists display drastic moral deficiencies related to the loss of the Crimean War and their country's long history of cultural marginality vis-à-vis the West. Smarting from this double wound, the Russian tourist crowd seizes travel as an occasion to take revenge on Western Europe through self-aggrandizing lies, ostentatious spending, and overdressing. Their materialist mind-set prompts them to take money as a measure of all things and to mistreat financially underprivileged natives abroad: service personnel in little towns, the Cologne restaurant singers, and Italians attempting to enjoy the Roman carnival. "Our tourists" have xenophobic tendencies; they are obsessed with rank and exhibit the barbaric mentality of the unrepentant serf-owner.

With Turgenev leading the way, the tourist phobia that emerged in the 1850s was largely the creation of politically liberal art-lovers of the nobility, aspiring to police the practice of leisure travel from within. But as the contributions of Mikhailov, Panayeva, and Tur have illustrated, the traditional, elitist norms were now spreading with the accelerated breakdown of class and gender barriers to tourism. As a noblewoman, Tur was the least sociologically significant. Panayeva, Mikhailov, and Dobrolyubov, however, show us the evolution of cultural tourism as a democratizing opportunity for middle-class Russians.

Notions of Russian national character flicker into view in the writings investigated here but are subordinate to the dominant picture of tourism as a nationally divisive phenomenon. Leisure travel accentuates differences between Russians by bringing them together in unfamiliar contexts that test their ability and preparation to make the journey rewarding. "We Russians should support each other," whines Druzhinin's playacting visitor on the Rhine. But automatic bonding on the basis of nationality is exactly what tourist phobia proclaims impossible. Instead of positing national solidarity, the narratives considered here represent Russians squaring off against Russians: refined cosmopolitans versus provincials, materialists, and philistines; the reform-minded against the reactionary; men against women stereotyped as shoppers, vanity travelers, and non-contenders in the appreciation of art; and in the case of Panayeva speaking under a masculine pseudonym—the new-age woman traveler, adventurous and discriminating, as opposed to brainless members of her sex. Over all, Russian *turistki* come in for harsher criticism than their men compatriots, and this may well reflect women's emancipation. Russian women were demanding new liberties and parity with men. But were women making good use of their travel opportunities? Tur's Tatyana alone passes the test in a manner that provides a limited parallel with Dickens's *Little Dorrit*, serialized in *Annals of the Fatherland* in 1857. Slipping away from her bossy, social-climbing chaperon Mrs. General, Dorrit explores Venice and Rome on her own, pursuing a thoughtful acculturation program in distinction from the shallow English crowd.[78] After her tour, little Dorrit's "cosmopolitan experiences go with her" when she marries in England and settles into what Dickens describes as "a modest life of usefulness and happiness."[79] Acculturation in Italy leaves Tur's Tatyana poised to follow a similar trajectory (she goes insane instead, when her high-society parents block her marriage to the country doctor). As a rule, though, Russian writers—men and women alike—would tend to represent the *turistka* as an idle consumer throughout the nineteenth century.[80]

More generally, the structure and themes of the phobic discourse that arose in the post-Crimean War period would persist in Russian print and personal documents throughout the imperial age. Cultivated travel lovers would continue to regard a journey to Western Europe as an edifying rite of passage, and many would attribute special importance to art appreciation. To burnish their self-images, the travel-lovers would extend the gallery of Russian others that emerged in the late 1850s: the materialist vanity traveler, the compulsive gambler, the boorish general, the woman spendthrift and clothes horse, and the self-satisfied ignoramus unable to appreciate art. In the epilogue of Turgenev's *Fathers and Children*, for example, "our tourists" in Dresden find nothing better to do than "run after" Pavel Kirsanov, the self-exiled, embittered, and wasted man of the 1840s excluded from the novel's final tableau of the family happiness that his brother Nikolai and nephew Arkady have achieved with their wives and babies on their country estate in Russia. "Elegantly dressed" as he strolls on the Brühl Terrace at "the most fashionable time," Pavel himself has become the main sight

78 Buzard, *The Beaten Track*, 108, 146–47.

79 James Buzard, "'The Country of the Plague': Anticulture and Autoethnography in Dickens's 1850s," *Victorian Literature and Culture* 38 (2010): 419.

80 For early twentieth-century literary efforts to redeem the Russian woman traveler, see Lidiia Veselitskaia [V. Mikulich, pseud.], "Vo Florentsii" ("In Florence"), in her *Dusia. Rasskazy* (Petrograd: Novoe vremia, 1915), 25–58; and "V Venetsii" ("In Venice"), in her *Rasskazy* (St. Petersburg: Izdanie A. S. Suvorina, 1911), 199–276. The Russian heroine of the former story is an artist working in a studio in Florence and sightseeing in her spare time. During a prenuptial vacation in Venice, the heroine of the second story discovers that her art-loving fiancé and foster mother are having an affair. The aesthetic cosmopolitans' immorality alienates her from tourism and motivates her quest for self-realization in work as a nurse.

that the Russian tourist crowd seeks to see, while implicitly ignoring the *Sistine Madonna* and all the rest of the paintings housed in the adjacent Gemäldegalerie (3:332).

Yet, even as "our tourists" became objects of satire in the late 1850s and early 1860s, Chernyshevsky and Pisarev turned the tables, to malign the traveling sophisticates and their aesthetic priorities. Stigmatizing leisure and hedonism, the radicals formulated a nationally distinct, maximalist brand of anti-tourism that denied the value of the cultural capital that cosmopolitan travelers maintained every intelligent, sensitive Russian tourist should be seeking. Adding insult to injury, Chernyshevsky assailed the refined traveler's moral character. In opposition to scenic and cultural tourist agendas, the radical killjoys sought to redirect foreign travel toward sociopolitical investigation and the acquisition of practical knowledge. Their anti-elite, anti-aesthetic disapproval of Russian holidays abroad would find a strange political bedfellow in the conservative nationalist Dostoevsky.

8

Dostoevsky's Anti-Cosmopolitan Animus toward Tourism

During Dostoevsky's career as a novelist and journalist, his tourists evolved from buffoons to something on the order of a fifth column menacing Russian nation building. In the 1840s he was already imagining the tourist as somebody else. Charting the mental breakdown of a Russian clerk tormented by his low rank in the civil service, Dostoevsky's novella *The Double* (1846) features two Englishmen who come to Petersburg for the sole purpose of seeing "fretwork in the Summer Garden."[1] Having done so, they scamper back home. Dostoevsky's feuilleton "Petersburg Chronicle" (1847) sarcastically accuses "an ultra-famous tourist" from France of spreading misconceptions about Russia after a hasty visit (18:24). The Frenchman in question was almost certainly the Marquis de Custine, the Russophobe author of *La Russie en 1839*, a banned book widely read in Russia. Surface-skimming links those two 1840s' versions of the tourist, to leave us wondering how Dostoevsky might have reacted to the Russian tourist "inundation" of the West immediately following the Crimean War. At that moment he was far from home, completing the ten-year punishment he incurred in the 1849 round-up of the Petrashevsky circle. Dostoevsky spent four years in a hard labor stockade in Omsk, an experience fictionalized in his *Memoirs from the House of the Dead* (1861). After the prison camp, he finished his sentence in military service in

1 Dostoevskii, *PSS*, 1:156 (subsequent citations to this edition appear parenthetically).

Semipalatinsk, a garrison town in the Kazakh steppe. He did not return to Petersburg until mid-December 1859.

While the young Dostoevsky conceived the tourist as somebody else, he yearned to go abroad, especially to Italy. In conjunction with Schiller's aesthetics of the sublime and beautiful, Russia's romantic passion for Italy captured Dostoevsky's imagination. As an unhappy student at the Russian military Academy of Engineering, he daydreamed of writing a novel of Venetian life. Shortly after his graduation in 1843, he translated George Sand's *La Dernière Aldini*, a Venetian novel representing the triumph of revolutionary social Christianity over the decadent Italian aristocracy during the struggle for national independence.[2] Dostoevsky also adored Rossini's operas, a taste his novella *White Nights* (1848) evoked with self-mockery and lingering affection (2:123–27). In 1846 Dostoevsky was planning a trip to Italy, France, and Vienna. Apparently feeling a need to justify the journey when funds were short (as usual), he told his brother Mikhail that he had to go abroad "not for fun but for medical treatment. Petersburg is hell for me. It's so hard, so hard to live here!" Dostoevsky pinned great hopes on Italy, not for consultations about his epilepsy but rather as a haven for artistic self-realization. There in Italy, he imagined, he would be "at leisure, at liberty," to write a novel "for myself." He would market it independently, to end his "enslavement and literary dependence" on Petersburg publishers.[3]

Dostoevsky revisited his youthful reveries in an 1861 evocation of his prison term and exile as an unplanned journey that took him to Siberia instead of Italy. By that time he had re-established his literary reputation with *Memoirs from the House of the Dead*, the novel *The Insulted and Injured* (1861), and other prose fiction, all serialized in *Vremia* (*Time*, 1861–63), the journal he launched in January 1861 and edited with Mikhail. In July that year, Dostoevsky recollected his yearning for the "magic" of Italy in a letter to the poet Yakov Polonsky (1819–98), an acquaintance traveling in Austria at the time:

> Italy is right next to you. Is it possible you're not tempted to visit it? Lucky man! How many times, from earliest childhood, have I dreamed of going to Italy. Ever since the novels of Anne Radcliffe, which I read when I was eight, all sorts of Catarinas, Alfonsos, and Lucias have been running around in my head. As for the Don

---

2 Frank, *Dostoevsky: The Seeds of Revolt*, 70, 129.

3 Dostoevskii, letter from October 7, 1846, *PSS*, 28.1:127–28.

> Pedros and Doña Claras, I still rave about them even now. Then it was the turn of Shakespeare—Verona, Romeo and Juliet—the devil only knows what magic was there! Italy! Italy! But instead of Italy I landed in Semipalatinsk and before that the House of the Dead. (28.2:19)

This self-presentation contains an overlooked reminiscence of *Childe Harold's Pilgrimage*, a work Dostoevsky no doubt read in his youth when he purportedly learned by heart *The Prisoner of Chillon* in Zhukovsky's translation.[4] In speaking of Venice as a woman, the *Childe Harold* poet declares:

> I loved her from my boyhood; she to me
> Was as a fairy city of the heart,
> Rising like water-columns from the sea,
> Of joy the sojourn, and of wealth the mart;
> And Otway, Radcliffe, Schiller, Shakespeare's art,
> Had stamp'd her image in me . . . (4.18).

Both texts date the author's love of Italy to childhood and name Radcliffe and Shakespeare as literary mediators.[5] In forgetting the *Childe Harold* subtext or simply choosing not to acknowledge it, Dostoevsky was disassociating himself from Russian Byronism as he had characterized it in an article of January 1861. An early formulation of the messianic nationalism Dostoevsky would express in his famous Pushkin speech of 1880, the 1861 article portrayed Russia's emulators of Byron as sneering, idle men too lazy to cope with the frustrations of Russian life. They respond to everything with "demonic laughter" and retreat into "fine dining" and card-sharping (18:58).

Although Dostoevsky was loath to recognize Byron as a soul-mate in any respect, his letter to Polonsky conformed to the Byronic aesthetic cosmopolitanism that had long made Italy a Russian tourist priority. Turgenev's representation of Venice in *On the Eve* (1860) illustrated the Romantic tradition's continuing productivity. A long paragraph of authorial speech begins on a worldly, in-group note: "Anybody who has never

4 Commentary in Dostoevskii, *PSS*, 28.1:406–7n6.

5 Along with works by Radcliffe and Shakespeare, Byron's annotations to his poem cite Thomas Otway's (1652–85) verse-drama *Venice Preserv'd* and Schiller's Gothic tale *The Ghost-Seer*. Dostoevsky's obscure allusions may refer to characters in Matthew Lewis's *The Monk* and Charles Maturin's *Melmoth the Wanderer*: see Robin Feuer Miller, "Dostoevskii and the Tale of Terror," in *Dostoevskii and Britain*, ed. W. J. Leatherbarrow (Oxford and Providence, RI: Berg, 1995), 140.

seen Venice in April can scarcely know all the ineffable charm of that magical city. The gentleness and tenderness of spring suit Venice just as bright summer sun suits magnificent Genoa and autumn's gold and purple suit Rome, the great elder."[6] Turgenev further shored up the aesthetic cult of Italy in "A Trip to Albano and Frascati" (1861), a memoir of meeting the Russian artist Alexander Ivanov during an 1857 Roman holiday with Botkin. In addition to visualizing the Italian countryside through the art of Claude Lorrain and Poussin, Turgenev's essay features the travelers admiring paintings with Ivanov in the Vatican museum.[7] Vyazemsky also contributed to the Byronic aesthetics of travel in 1862, when he dusted off for publication lyrics written during his 1853–54 sojourn in Italy. The cycle includes "Venice," a tourist poem highlighting the San Marco basilica (marred by English graffiti), an animated international crowd of visitors, and the author's delight in seeing the city's most celebrated painters—Titian, Tintoretto, and Veronese.[8] To judge by Dostoevsky's letter to Polonsky, his yearning to visit Italy remained unabated and consonant with aesthetic cosmopolitanism. We may recall as well that Dostoevsky at this time was on very good terms with Turgenev from whom he received advice about traveling before going abroad.[9]

But after visiting Italy, Dostoevsky excluded it from his *Winter Notes on Summer Impressions*, the travel narrative he serialized in *Time* in February and March 1863. For the ostensible purpose of seeking medical treatment, he set off on his journey in June 1862 and toured Western Europe for two and a half months. The opening paragraph of *Winter Notes* states that the author "was in Berlin, Dresden, Wiesbaden, Baden-Baden, Cologne, Paris, London, Lucerne, Geneva, Genoa, Florence, Milan, Venice, and Vienna—twice in some places" (5:46).[10] And yet, except for a few remarks on Berlin and Cologne and a passing mention of Dresden,

6 Turgenev, *SS*, 3:134.

7 Ibid., 11:308–18. On the essay's painterly treatment of landscape, see Molly Brunson, *Russian Realisms: Literature and Painting, 1840–1890* (DeKalb: Northern Illinois University Press, 2016), 77.

8 P. A. Viazemskii, "Venetsiia," in his *V doroge i doma. Sobranie stikhotvorenii* (Moscow: Tipografiia Bakhmeteva, 1862), 177–81 (the Italy cycle includes "Gondola" and other conventional topics). Modern-day editions reprint many of those works without using the collective 1862 title.

9 Frank, *Dostoevsky: The Stir of Liberation*, 178–79.

10 Since Dostoevsky left no travel diary and wrote only one pertinent letter that has survived, some efforts to map his actual trip treat *Winter Notes* as a document and entail much speculation. See M. I. Brusovani and R. G. Gal'perina, "Zagranichnye

Winter Notes ignores every place except Paris and London. The decision to scrap Italy reflected Dostoevsky's nationalist ideology of the Native Soil (*pochvennichestvo*), formulated in *Time* before his trip. Resting on the grassroots implications of "soil" (*pochva*), Dostoevsky's political thinking envisioned a Russian synthesis of European "civilization" and the "native principle" of brotherhood that he considered innate to the *narod* and institutionalized in the Russian village commune.[11] His 1861 preview of the Pushkin speech designated the English and the French "the main Europeans," the leading "representatives of Europe" (18:46). What Dostoevsky perceived as most "representative" was class struggle and a loss of religious faith that he believed made the entire capitalist West "incapable of developing a humane and Christian social order."[12] The Native Soil idea of Western Europe set the anti-tourist parameters of *Winter Notes*. Prioritizing "negative sightseeing," the author gravitates to slums, law courts (represented as showcases of empty eloquence), and cheap entertainments (Paris vaudevilles, a London dance hall) that shore up his animus toward the West.[13]

Dostoevsky's travel narrative has moved from near oblivion to canonization. Perhaps because *Time* had made the Native Soil ideology so familiar to the contemporary readership, *Winter Notes* seems to have aroused no public reaction prior to the 1883 memoir of Nikolai Strakhov (1828–96), a critic, philosopher, and journalist-collaborator of Dostoevsky. Having visited Florence with Dostoevsky in July 1862, Strakhov recollected him as a visitor with a humanitarian eye and also remarked the impact that Herzen's anti-Western Russian socialism exerted on *Winter Notes*.[14] According to

---

puteshestviia F. M. Dostoevskogo," *Dostoevskii. Materialy i issledovaniia* 8 (1988): 273–81; and Frank, *Dostoevsky: The Stir of Liberation*, 180–96.

11 Dostoevskii, "Ob"iavlenie o podpiske na zhurnal 'Vremia' na 1861 god," *PSS*, 18:36–37. A sense of being a rootless person "with no soil beneath his feet" epitomizes self-definition as a "cosmopolitan" on the part of the eponymous *raznochinets* hero of Nikolai Pomyalovsky's (1836-63) novel *Molotov* (1861): see Derek Offord, "Literature and Ideas in Russia after the Crimean War: the 'Plebeian' Writers," in *Ideology in Russian Literature*, ed. Richard Freeborn and Jane Grayson (London: Palgrave, 1990), 64.

12 Linda Ivanits, *Dostoevsky and the Russian People* (Cambridge: Cambridge University Press, 2008), 34.

13 On "negative sightseeing" as confrontation with disgusting aspects of local life, see MacCannell, *The Tourist*, 40–41.

14 N. N. Strakhov, *Biografiia, pis'ma i zametki iz zapisnoi knizhki F. M. Dostoevskogo* (St. Petersburg: Izdanie A. S. Suvorina, 1883), 240–45. During his exile abroad with his wife Anna Grigorievna in 1867–71, Dostoevsky in fact "made no attempt [. . .] to understand sympathetically local inhabitants." He made no European friends, never

Strakhov, Dostoevsky in Florence had focused his attention on local people rather than "nature, historical monuments, or art works, except for the very greatest" (Strakhov specified none of the masterpieces but claimed Dostoevsky had found the Uffizi so daunting that he left after only a few minutes, without having seen the *Venus de' Medici*).[15] Today's critical literature has underscored the humanitarian, anti-capitalist, anti-bourgeois themes of *Winter Notes* and situated the work in the larger history of nationally affirmative Russian travel writing from Karamzin and Fonvizin into the late nineteenth century.[16]

Unnoticed in their time, the innovative dialogic and parodic properties of *Winter Notes* have also inspired much excellent modern criticism. Dostoevsky's creation of a travel persona anticipates the masterpiece *Notes from the Underground* (1864), the prelude to his major novels. Engaging various imagined listeners and conducting mini-dialogues inside his own head, the *Winter Notes* narrator is a direct antecedent of the underground man, complete with a liver complaint.[17] Hyper-conscious of Russia's

---

sought to meet European writers, and according to his wife, he referred to local people abroad as "the foreigners." He spoke French but had relatively little German and no Italian or English: see Malcolm V. Jones, "Dostoevsky and Europe: Travels in the Mind," *Renaissance and Modern Studies* 24 (1980): 51–52.

15 Questioning Strakhov's reliability about the Uffizi, Frank suggested that Dostoevsky might have already been "an interested and assiduous" visitor of art museums during his 1862 trip because his wife's diary said that he was during their life abroad: *Dostoevsky: The Stir of Liberation*, 193–94. See also the claim that "every trip Dostoevskii made to Europe turned into a trawl round the museums," in Konstantin Barsht, "Defining the Face: Observations on Dostoevskii's Creative Process," in *Russian Literature, Modernism and the Visual Arts*, ed. Catriona Kelly and Stephen Lovell (Cambridge: Cambridge University Press, 2000), 33. Barsht provides no evidence for the 1862 journey. Furthermore, when Dostoevsky was in Rome with Suslova in 1863, he apparently visited no museums or cathedrals besides St. Peter's.

16 Il'ia Pomerantsev, "Dostoevskii i literatura puteshestvii," *Russian Literature* 47 (2000): 93–109; Frank, *Dostoevsky: The Stir of Liberation*, 233–248; Offord, *Journeys to a Graveyard*, 197–220; Sarah Hudspith, *Dostoevsky and the Idea of Russianness: A New Perspective on Unity and Brotherhood* (London and New York: Routledge Curzon, 2004), 13, 51–55; Kleespies, *Nomad Nation*, 36–46; Gary Saul Morson, *The Boundaries of Genre: Dostoevsky's Diary of a Writer and the Traditions of Literary Utopia* (Evanston: Northwestern University Press, 1981), 22–26; James P. Scanlan, *Dostoevsky the Thinker: A Philosophical Study* (Ithaca: Cornell University Press, 2002), 60–61; and W. J. Leatherbarrow, "Introduction: Dostoevskii and Britain," in Leatherbarrow, *Dostoevskii and Britain*, 2–12.

17 See especially Pomerantsev, "Dostoevskii i literatura puteshestvii," 94–97; Frank, *Dostoevsky: The Stir of Liberation*, 233–38, 247–48; and Hudspith, *Dostoevsky and the Idea of Russianness*, 177, 182–88.

cultural marginality in Western eyes, Dostoevsky's persona manifests xenophobia but also generates ironic incertitude about the prospects for Native Soil nationhood and the actuality of his ideal of the spiritual, altruistic Russian personality. The exposition of *Winter Notes* is messy and modernistic. It violates the Russian travel genre's norms of linear structure and spatial orderliness, most notably in holding forth on Paris after the author has reached London. In its parody of travel literature and parodies of such parodies, *Winter Notes* becomes a "metaparody" of the genre.[18] These novelistic properties of style and structure make *Winter Notes* a "pseudo-travelogue," "at the boundary between polemical journalism and fiction."[19]

My investigation of *Winter Notes* casts new light on the development of Dostoevsky's anti-cosmopolitanism. Prior to his 1862 journey, his Russian tourists were laughable. Afterwards, starting with *Winter Notes*, he entwined leisure travel with cosmopolitanism perceived as a threat to Russia's self-realization as a nation. This discourse had antecedents in the writings of Pogodin, Belinsky, and Zagoskin. Most pertinent is Zagoskin's conflating the "world citizenry" tradition of Enlightenment philosophy and the tourist craze for the West that entailed disparaging Russian culture. Zagoskin saw these as mutually reinforcing phenomena luring Russians away from family and the homeland. Dostoevsky would make a similar conflation. In the crisis-ridden years following the emancipation, he condemned the socialist radicals' philosophical cosmopolitanism that elevated universality over the nation but came to perceive Russian tourism, too, as a step on the slippery slope of deracination. Dostoevsky's alienation from leisure travel stemmed from an ideal of Christian brotherhood, and in that respect overlapped Tolstoy's *Lucerne*. Nonetheless, the anti-aesthetic and anti-elite inflections of *Winter Notes* chimed with the position of Dostoevsky's political bête noire Chernyshevsky, the socialist atheist who advocated travel for studying foreign peoples and societies, instead of taking pleasure in art, monuments, and scenery, especially in Italy.[20]

18 Morson, *Boundaries of Genre*, 25–26, 61, 142–43.

19 Quotes respectively from David Bethea, "*The Idiot*: Historicism Arrives at the Station," *The Shape of Apocalypse in Russian Fiction* (Princeton: Princeton University Press, 1989), 73n28; and Offord, *Journeys to a Graveyard*, 201–2.

20 On the ideological clash, see Derek Offord, "Dostoevsky and Chernyshevsky," *Slavonic and East European Review* 57, no. 4 (October, 1979): 509–30.

A Confusion of Cosmopolitanisms

When Dostoevsky returned to Petersburg from Siberia, the 1846–47 Maikov-Belinsky clash on cosmopolitanism was back in the public eye. Dostoevsky had closely followed the exchange at the time when he and his friend Maikov were both under the sway of the Christian "socialist cosmopolitanism" of Charles Fourier and like-minded utopian thinkers.[21] Turgenev's *Rudin* paraphrased the debate, and an 1856 article by Chernyshevsky quoted at length the position on national individuality that Belinsky had taken in response to Maikov's essay on Koltsov.[22] Leaving Maikov aside, Chernyshevsky hailed Belinsky as the promoter of literature engaging questions of nationhood and the national good, especially the abolition of serfdom. Dobrolyubov, on the other hand, spoke up for universality in an 1858 evisceration of *Russian Civilization*, a book by the sentimental Slavophile Nicholas Zherebtsov. Without naming Maikov, Dobrolyubov extended his concept of sensible cosmopolitanism and pitted it against "nationalist prejudices" and "arrogant pseudo-patriotism" that he located in Zherebtsov's book as well as in Pushkin's anti-Polish poem "To the Slanderers of Russia."[23] After likening chauvinism to the mentality of a home-bound child, Dobrolyubov argued that education and travel produce a knowledge of foreign peoples that allows a person to "grasp the abstract concept of humanity," so that upon meeting somebody, one "sees first of all a human being rather than a German, Pole, Jew, Russian and so on." One develops a higher, "genuine patriotism" rooted in "a love of humanity," a mentality Byron manifested in joining the Greek struggle for independence, instead of serving an exclusively English cause (263–64, 266). Along with dismissing *Russian Civilization* as a product of "pseudo-patriotism," Dobrolyubov wryly dissected Zherebtsov's idea of Russianness as an "incarnation of eclecticism," attesting to a unique national gift for assimilating influences from other peoples. To Dobrolyubov's ears, this "eclectic" nationalism sounded very similar to the "cosmopolitanism" that Slavophiles professed to abhor (278–79).

21 Frank, *Dostoevsky: The Seeds of Revolt*, 202–15; and quotation from Wayne Dowler, *Dostoevskii, Grigor'ev, and Native Soil Conservatism* (Toronto, Buffalo, London: University of Toronto Press, 1982), 66.

22 Chernyshevskii, *Ocherki gogolevskogo perioda russkoi literatury*, *PSS*, 3:284–91.

23 Dobroliubov, *SS*, 3:257, 263.

Dostoevsky's idea of Russian "panhumanity" (*vsechelovechnost'*) similarly mimicked cosmopolitanism (of an increasingly toxic variety).[24] The 1861 antecedent of his Pushkin speech argued that the "universal ideal" of Christianity had faded in the West, leaving a void in international spiritual leadership that Russia was destined to fill by virtue of its unique amalgamation of Western "civilization" and the native values of the *narod*. In characterizing panhumanity, Dostoevsky asserted that Russians alone have an "instinct" for assimilating foreign cultures. Russians "speak all languages," "get along with everybody," and "sympathize with everything human" (18:55). Although Russians in the aggregate exhibit this "universalizing receptiveness," panhumanity reached its "artistic plenitude" in Pushkin's writings—"the golden fruits" of "the tree of civilization" whose seeds came from the West but grew in Russian "soil." Native "soil" proved paramount in Dostoevsky's scheme. He foresaw the *narod* contributing ninety percent to the realization of panhumanity. The remaining ten percent—the imported "civilization"—would come from the educated classes (68–69).

While Dostoevsky would have balked at the term "aesthetic cosmopolitan," an objective use of the term covers the receptiveness he displayed toward the West European arts that he privileged as emblems of the "civilization" component of Russian panhumanity. In polemics against the utilitarianism of the radicals (specifically Dobrolyubov), Dostoevsky's *Time* essay "Mr. ___bov and the Question of Art" (February 1861) articulated a philosophy of art indifferent to an art work's national provenance. Joseph Frank neatly termed the outlook "an aesthetics of transcendence."[25]

---

24 As Dostoevsky's messianic nationalism intensified, his idea of the "panhuman" appeared "a thinly disguised rationale for Russian imperialism" (see Geraci, "Islam," 217). In response to Dostoevsky's Pushkin speech, Konstantin Leontev's (1831–91) essay "On Universal Love" attacked "panhumanity" as a dissolutive design that would erase ethnocultural Russianness from the globe. An Orthodox monk by that time, Leontev upheld the supremacy of the "cosmopolitanism of Orthodoxy," taking as its prime source of inspiration the "living Person of the crucified Christ": see K. N. Leont'ev, "O vsemirnoi liubvi. Rech' F. M. Dostoevskogo na Pushkinskom prazdnike" in his *Vostok, Rossiia i slavianstvo. Filosofskaia i politicheskaia publitsistika. Dukhovnaia proza, 1872–1891* (Moscow: Izdatel'stvo Respublika, 1996), 315, 322.

25 Frank, *Dostoevsky: The Stir of Liberation*, 76–93; Robert Louis Jackson, *Dostoevsky's Quest for Form: A Study of His Philosophy of Art* (New Haven: Yale University Press, 1966), esp. 126–57; and G. M. Fridlender, "Estetika Dostoevskogo," in *Dostoevskii: khudozhnik i myslitel'. Sbornik statei*, ed. K. N. Lomunov et al. (Moscow: Khudozhestvennaia literatura, 1972), 99–100, 127–39.

Dostoevsky insisted that art's most profound usefulness is not readily measured in a given time and place. Functioning as "spiritual bread," the beauty of "healthy" art fills a universal human need for the "eternal ideals" of harmony and serenity. The *Apollo Belvedere* serves as a major example. Mediated through Pushkin's lyric "The Poet and the Crowd," which presents the *Apollo* as actual divinity immanent in marble, Dostoevsky's thinking intimates a Vatican museum visit occurring but once in a lifetime. "Mr. ___bov and the Question of Art" imagines this "beautiful statue" requiring no more than two minutes to produce in a sensitive young viewer an "inner change," "a galvanizing current that instantly alters something from its former state—transmuting a lump of iron into a magnet." Although the youth may not notice the impact at the time, the "impression of a god" will leave an indelible trace. "And who knows? Twenty or thirty years later, when the young man felt called to serve a great civic cause in which he emerged as a great, progressive public figure [. . .] then perhaps one of the multitude of reasons that prompted him to act that way rather than another might have been the impression unconsciously left upon him by the *Apollo Belvedere* he saw twenty years before" (18: 78). Suggesting an encounter with the original statue, rather than a copy close to home and therefore easily revisited, Dostoevsky's speculations coincided with Melgunov's defense of Russian acculturation tours as undertakings that have the obscure potential to produce citizens eager to contribute to national welfare.

"Mr. ___bov and the Question of Art" situated the *Apollo* within a larger field of Western culture that has enriched and can continue enriching Russian life in unpredictable ways. Dostoevsky asserted that Russia has absorbed "the [Greek] anthologies, the *Iliad*, the Dianas as huntress, the Venuses and Jupiters, the Madonnas and Dante, Shakespeare, Venice, Paris, and London" (100). In referring to Greek divinities and multiple Madonnas, the formulation evoked visual representations as well as literature. The received images of cities might also encompass paintings but most surely point to literary creations such as Shakespeare's Venice, Dickens's London, and Balzac's Paris (Dostoevsky's first publication was his translation of *Eugénie Grandet*). "Mr. ___bov and the Question of Art" singled out two literary personages for special mention: Goethe's Faust and the Marquis Poza of Schiller's *Don Carlos*. Dostoevsky also defended Corneille and Racine against the Russian radicals' charge that they were "old fogies," prized only by French aristocrats in their time and utterly meaningless to modern-day Russian readers (78, 99). In further testimony to Dostoevsky's cosmopolitan receptiveness to art, his *Time* article "Models of Sincerity" (March 1861) extoled the "divine beauty" of the *Venus de' Medici*, a starred treasure in the museum world of the West (19:103).

After equating Western "civilization" primarily to literature, painting, and sculpture, "Mr. ___bov and the Question of Art" stirred up a hornet's nest by accusing the radicals of having forgotten the universalizing significance of Russia's debt to the West. His ventriloquized opponents respond that "universals and historical laws are not what we need right now. We have laundry to do at home, dirty linen to rinse, everything to be made clean. [. . .] This is not the time to write about the Marquis Poza but rather our own affairs" (18:100). Dostoevsky's charge was reckless. As he knew all too well, the radicals had not ceased asking what was common to all humanity. The problem was their scientific, atheistic answer to the question. Faith in science and reason led them to envision a rationally engineered socialist utopia where material well-being would assure individual happiness and social harmony.[26] Dostoevsky considered this version of universal humanity a soul-denying monstrosity. In escalating polemics that began in September with Dobrolyubov's response to "Mr. ___bov and the Question of Art," the *Contemporary* ridiculed Dostoevsky's trope of the expansive Russian soul, his tendency to substitute "soil" metaphors for arguments, and his stress on the Russian people's educational and spiritual needs rather than their material well-being.[27] Dostoevsky's retaliations reflected his growing affinity to the classic Slavophiles Khomyakov and Ivan Kireyevsky, both of whose complete collected works were published in 1861 by Ivan Aksakov, now editor of the journal *Den'* (*Day*).[28] In opposition to the Aksakov camp's myth of Old Muscovy and their demonization

26 Offord, *Journeys to a Graveyard*, 215–19; and Dowler, *Dostoevskii, Grigor'ev, and Native Soil Conservatism*, 76–78.

27 Alluding to Dostoevsky, Dobrolyubov wrote that the so-called "expansiveness of the Russian character" tends to find expression in such behavior as using champagne to make steam in a bathhouse; breaking dishes and mirrors in a tavern; or "devoting one's life to hunting with hounds and in olden days even making it a hunt for people by sewing a low-rank lickspittle into a bear skin and then sicking the dogs on him." See the review of Dostoevsky's writings "Zabitye liudi," *SS*, 7:268.

28 See my bibliography. For special investigation of Dostoevsky's engagement with Slavophilism throughout his career, see Hudspith, *Dostoevsky and the Idea of Russianness*.

of Peter the Great, Dostoevsky's *Time* articles defended westernization as a vital stage of Russia's development and described pre-Petrine society as a realm of "Asiatic, oriental sloth" and bestial treatment of peasant women (18:54–55; 20:12). Dostoevsky nonetheless found in Slavophilism an idealization of the folk, a claim to Russian spiritual superiority to the West, and related ideas to serve his attacks on atheistic socialism.

In February 1862 Dostoevsky branded the radicals advocates of "cosmopolitanism" bent on eradicating national particularities for the sake of a "universal ideal" comprised exclusively of West European attributes. Cosmopolitans, contended Dostoevsky, want to make man uniform throughout the world. Nationality will disappear so that each people will come to resemble a "worn coin" indistinguishable from all the rest.[29] Without quoting the pertinent line from *Don Carlos*, Dostoevsky now refashioned the Marquis Poza as the progenitor of modern-day cosmopolitanism. In declining to pledge loyalty to the Spanish king, the Marquis proclaims himself a citizen of the world (*Bürger dieser Welt*—act 3, scene 20). Dostoevsky thus conflated his immediate target of atheistic socialism with the Enlightenment humanism that Schiller valorized through Poza. This was a point of no return in Dostoevsky's thought. Once and for all, he had defined cosmopolitanism as a sensibility incompatible with nationalism and especially inimical to the besieged project of *pochvennichestvo*.

Prior to Dostoevsky's trip abroad, mounting political crisis in Russia no doubt stoked his worries about the feasibility of Native Soil goals. Upon printing the Emancipation Proclamation in *Time*, he had hailed the liberation of the serfs as a "sublime" step toward forging national solidarity between the elite and the *narod*.[30] In truth, the anticipation of the end of serfdom had provoked the aforementioned flight of Russian nobles abroad. Advantageous to the landowners, the emancipation act sparked peasant unrest and spurred the radical intelligentsia into open revolt against the state. There were student mutinies as well, which led the authorities to close the universities from December 1861 to August 1863. As the case of Mikhailov and the proclamation "To the Young Generation" have indicated, revolutionary leaflets proliferated, agitating against the government and seeking to incite peasant uprisings. The most alarming

29 Dostoevskii, "Dva lageria teoretikov," *PSS*, 20:6–8. For the coin metaphor, see also "Ob"iavlenie o podpiske na zhurnal 'Vremia' na 1862 god," ibid., 19:149.

30 Frank, *Dostoevsky: The Stir of Liberation*, 108.

manifesto was "Young Russia," which called for taking an axe to the tsar, his family, and anybody else who put up resistance. This proclamation went into circulation in Petersburg in mid-May 1862, and Dostoevsky found a copy thrust through the door handle of his apartment.[31] Around the same time "Young Russia" appeared, fires broke out at several sites in Petersburg and burned for two weeks. The origin of the conflagration was never determined, as Dostoevsky made so bold to state in an article the censor prohibited. Rampant rumors, however, blamed students and professors, a testimony to their reputation as promoters of "nihilism." Having witnessed this growing divide between revolutionary forces and the panicked state, Dostoevsky would seek to surmount it in *Winter Notes* with the Native Soil vision of the national cohesion in which he yearned to believe.[32]

Masking the "Mere Tourist"

Dostoevsky's tourist phobia of the post-Siberian years was lighthearted prior to his anti-cosmopolitan turn. In the comic novella *Uncle's Dream* (1859) provincial Russian ladies yearn to see the Alhambra and Malaga, a travel dream that may allude to Botkin's *Letters on Spain*, as well as the fictitious poet Prutkov's amusing lyric "The Desire to Be a Spaniard" (1854).[33] While the silly ladies are armchair travelers, actual tourists make a cameo appearance in *The Insulted and Injured*. A shady ex-school teacher who plays a merciful role in the life of the waif heroine Nellie, Masloboev recalls how he and a friend smashed mirrors in a Paris brothel, as the proprietor Madame Joubert stood by screaming (3:275). Let us also remark the catalogue of modern-day tourists in Dostoevsky's 1861 antecedent of his Pushkin speech. First come "our cavalrymen—jolly, good-natured folk" known at home for parading in "snug buckskin britches." Other tourists who have been "flocking abroad" are "crowds of flighty youths" and "our deep-country landowners with their entire families and their cartons. With a genial, serious air, they have clambered up the towers of Notre Dame, surveyed Paris from there, and chased *grisettes* on the sly,

31 Ibid., 146–47; and Dostoevsky's 1873 recollection of this occurrence in *PSS*, 21:25.

32 On the more general tension in Dostoevsky's writings between "what the writer saw and what he wanted to believe," see Ivanits, *Dostoevsky and Russian People*, 158, 227n54.

33 Commentary in Dostoevskii, *PSS*, 2:517.

away from their wives" (18:47). These are images of rowdies and yokels, meant to help explain why West Europeans have failed to recognize Russians' panhumanity. In one way or another, these literary works and the *Time* article represent tourism as a trivial pursuit rather than a serious national liability.

Dostoevsky's history of ridiculing Russians abroad made him vulnerable to tourist angst, the pangs of which he experienced soon after reaching Paris. In a letter from June 26/July 8 to Strakhov, Dostoevsky reported that he had so far seen "only one wonder of nature—the Rhine River and shore lands, and they really are wonderful." But this was just a postscript and exemplified precisely the kind of aesthetic receptiveness that *Winter Notes* would repress. The main body of the letter concerned Paris and revealed Dostoevsky's self-perception as a "mere tourist" (*prostoi turist*), too pressed for time to produce a distinguished travel account. He had been in the city for ten days, long enough, he believed, to have discovered that the French were a "nauseating people" obsessed with money and that Paris was "extremely boring" despite having "many remarkable things" to see. Dostoevsky had also concluded that "the general level of culture is extremely low." True, the French have "certified scholars," but not many, "and anyway, is scholarship culture in the sense we are accustomed to giving the word?" (28.2:27). Dostoevsky had formed strong opinions. But how was he to put them in writing? How was a self-defined "mere tourist" as belated as he going to stand out in the crowded sphere of postwar Russian travel writing?

Dostoevsky's awareness of the competition produced a dense intertextual fabric. The Bible was the only reading material allowed in the Omsk stockade, and the Russian journals and newspapers that reached Semipalatinsk did so with great delay.[34] But once Dostoevsky returned to Petersburg, he devoured Russian periodicals to catch up on what he had missed. *Winter Notes* bears many traces of those readings. In addition to the much discussed Herzen and Fonvizin connections, Dostoevsky's narrative alludes to post-Crimean War tourist ephemera such as Druzhinin's feuilletons. Most importantly, *Winter Notes* stages a polemic with Mikhailov's *Paris Letters* and *London Notes*. Whereas Mikhailov's revolutionary tourist persona finds much to praise in Paris

34 Frank, *Dostoevsky: The Years of Ordeal*, 241.

and London, Dostoevsky fires back with an anti-tourist, anti-capitalist tale of the two cities.[35]

Dostoevsky's most brilliant response to the dilemma of belatedness was to introduce the *Winter Notes* narrator as a parody of elitist snobbery toward the Russian tourist crowd. Addressing an envisioned audience of "friends" who urged him to go abroad and are now awaiting his travel report, the author begins in a chatty, faux-naïf voice reminiscent of Druzhinin's conventionally parodic introduction to *Our People Abroad*. "What can I say that's new?" asks the *Winter Notes* author. He fears that every Russian reader already knows Western Europe even better than Russia. Dating back to childhood, his own acquaintance with the thickly inscribed territory of "Europe" began with Gothic novels: "Before I could read, I spent long winter evenings listening agog, paralyzed with ecstasy and terror as my parents read me the novels of Radcliffe at bedtime."[36] As the narrator later remarks, however, *reading* is not *knowing*: "You will never learn from books what you can see with your own eyes."[37] And so the dam of pent-up desire to *see* finally broke when he turned forty. He went abroad, to the "land of holy wonders," with the intention of seeing "everything, absolutely everything" in two and a half months. Moreover, he reports, he did in fact see "virtually everything," except Rome and St. Paul's. "It's true. I did not see St. Paul's cathedral," he adds, an omission he calls "rather improper for a traveler" (46–47, 49, 58). This is the first dig at Mikhailov who had asserted that every foreigner visiting London "should, of course, begin his sightseeing tour with St. Paul's."[38] Presenting himself as an indiscriminate gawker,

35 Specific reminiscences of Mikhailov include Dostoevsky's combination of travel "impressions" (*vpechatleniia*) and "notes" (*zametki*—literally "remarks," rather than the more common *zapiski*) and the image of English couples mechanically dancing in a London dance hall: see Mikhailov, *Parizhskie pis'ma* (October 1858), 468 and *Londonskie zametki* (September 1859), 147. In ironically titling the meaty ch. 3 of *Winter Notes* "A Completely Superfluous One" ("I sovershenno lishniaia"), Dostoevsky might be alluding to Mikhailov's refusal to perform the "entirely superfluous task" (*sovershenno izlishniaia zabota*) of describing sights already well covered in Russian travel literature in his "Iz Berlina" ("From Berlin"), *Sovremennik* 87, no. 5 (May 1861), section 2: 200.

36 I have consulted but regularly emended the following translation of *Winter Notes* which contains errors including the mistranslation of *puteshestvennik* (traveler) as "tourist": Dostoevsky, *Winter Notes on Summer Impressions*, trans. David Patterson (Evanston: Northwestern University Press, 1997).

37 *Seeing*, of course, does not necessarily mean *knowing*, either: on this issue, see Siegel and Wulff, "Travel as Spectacle," 111–12.

38 Mikhailov, *Londonskie zametki* (August 1859), 224.

Dostoevsky's parodic persona enters thumbing his nose at the "proper" norms of tourism.

In addition to the feuilleton chattiness, the cosmopolitan cluster of the Cologne episode in Druzhinin's *Our People Abroad* resonates in the opening chapter of *Winter Notes*. Druzhinin's persona visits the city's celebrated cathedral, buys eau de cologne (the commonplace souvenir), he likes speaking German with Germans at dinner, and enjoys listening to the German music. *Winter Notes* stages a xenophobic inversion with a preliminary jab at Panayeva's likening train travel to recovery from an illness. Dostoevsky's storyteller says the train made him sick. His liver is acting up, and the bile will keep churning throughout his sojourn abroad. He cuts short his stays in Berlin and Dresden (his first two stopping points) because he cannot stand the sight of Germans en masse. Still testy in Cologne, he skips the cathedral, which he likens to a "haberdasher's knickknack," a gigantic "paperweight," an image ironically suggesting his tourist interest in souvenirs on sale. In short order, he buys a bottle of eau de cologne, but just to get rid of the relentless German hawker. Equally insufferable is the German toll collector at a new bridge. The German's eyes declare: "You see our bridge, miserable Russian; well you are a worm before our bridge and before every German because you do not have such a bridge" (48). The narrator's retaliatory thoughts deem the bridge overrated, a reaction that asserts his national dignity, while taking a sideswipe at Chernyshevsky's and Mikhailov's enthusiasm for the West's suspension bridges. So begins the *Winter Notes* dispute with Russian tourist cosmopolitanism, especially of the gentlemanly sort that Druzhinin enacted.

Quoting from Karamzin's *Letters*, Dostoevsky's persona states that when he saw the Cologne cathedral a second time under more favorable circumstances, he "felt like 'begging its forgiveness on my knees' for not perceiving its beauty the first time, for exactly the same reason Karamzin had knelt before Rhine Falls" (48). Is *Winter Notes* underwriting or double-voicing Karamzin's quoted sentiment?[39] Is the envisioned genuflection to the cathedral not an ironic, self-parodic act on the part of an author about to deplore Russians' "servile worship" of Western civilization (61)?

39 Sincere emulation of Karamzin is favored in Jackson, *Quest for Form*, 229; and A. V. Arkhipova, "Dostoevskii i Karamzin," *Dostoevskii. Materialy i issledovaniia* 5 (1983): 106. Arkhipova observes, however, Dostoevsky's tendency to ironize about Karamzin's sentimental images of the Russian folk, 101–3.

Political Cosmopolitanism

After the opening parody of the "mere tourist," the rest of *Winter Notes* thoughtfully constitutes the history and current dilemma of Russia's quasi-colonial relationship to the West (the envisioned audience now broadens beyond "friends" to encompass unknown members of the public, concerned with the homeland's problems and quick to talk back when the narrator's opinions offend them). Titled "On the Train," chapter two metaphorically broaches the sociopolitical cosmopolitan tendency to uphold Western Europe as the model for Russia's future. For the first time in Dostoevsky's writing, the railroad appears as an engine of modernization threatening to destroy Russia's spiritual and moral strengths (running from Petersburg through Poland to the Prussian border town Eydtkuhnen, the featured network had opened in June 1862, just in time for Dostoevsky's journey). Riding the train symbolizes the crippling effect that the Russian elite's enthrallment with the West has had on national development, and the self-assertion required to chart a distinct Russian path becomes analogous to escaping a wreck. The passivity of sitting on the train becomes so anguishing that Dostoevsky's persona wants to jump off and sprint alongside the locomotive: "Even if the worse should happen, even if I flag, not being in the habit, even if I get lost—no matter! Because I'm moving on my own two feet, I've found myself a cause, and I'm pursuing it, and because if the cars collide and turn over on the fly, I won't be idly sitting locked inside, paying with my hide for somebody else's mistake" (52). *Winter Notes* thus makes the authorial body analogous to the body politic, grappling with questions of Russia's future. Get off that train, instead of submitting to the modernizing dictates of the capitalist West!

Specifically targeting the Russian radicals' universalistic orientation, chapter three uses a different metaphor to elaborate the symbolic import of train travel. The chapter's main theme is how "Europe came to visit its civilization upon us" in Fonvizin's time. Even then, thinks the narrator, some educated Russians were probably "already getting sick of having nothing to do and being in a foreign halter" for toddlers. But passivity set it, the Europeanizing "phantasmagoria" triumphed, to produce a "masquerade" of "French caftans, cuffs, wigs, little swords, and ungainly fat legs wormed into silk stockings." The halter became "sacred," and the upshot is now an educated class so "civilized, so European" that the *narod* are "ready to vomit just looking at us." "We," in turn, "now despise our people and native principles so profoundly that we even treat them with a new, unprecedented squeamishness." Not entirely sparing the author, the "we" extends responsibility to all educated Russians, historically enlisted in a "civilizing mission" vis-à-vis the folk. But the prime culprits are "our ultra-progressive party," the radicals with their allegiance to universalistic humanism at the expense of Russian national particularity. On the "progressive" view, "there is no native soil, no native people; nationality is merely a system of taxation, the soul is a tabula rasa, a little piece of wax from which one can immediately mold a real man—a global, universal man, a homunculus. One need only apply the fruits of European civilization and read two or three books." Distanced from the narrator through ironic ventriloquism, the "sergeant majors of civilization" proclaim contempt for the *narod* and by extension the Native Soil ideology: "What do you have to teach us, numbskull peasant, when all nationality, all national character, is essentially reactionary, a means of tax apportionment and nothing more!" (55, 59–60).

Sacrificing Art

While the protest against political cosmopolitanism sets *Winter Notes* against the radical camp, the presentation of tourism as an equally insidious pastime overlaps Chernyshevsky's democratic assault on the art-oriented travel regime. Dostoevsky's specimen tourist is a Russian "European," a member of "our privileged upper class" under the "powerful, magical" spell of the West's "holy wonders." *Winter Notes* grants that Russia owes "almost all its development" to those wonders: "the sciences, arts, civic consciousness, and humanity" (*chelovechnost'*), the latter designating not merely humaneness (*gumannost'*) but rather the evolution from a sub-human state of existence. While "privileged" Russians go abroad, the "ordinary" people—"fifty million" strong—"have remained in Russia" (51). As Panayeva's *Russians in Italy*, Druzhinin's "The Majestic Tourist," and other post-Crimean War writings had shown, the *narod* were not as immobile as *Winter Notes* maintains. Shortly after *Winter Notes* appeared, Nikolai Leskov's reportage on the Russian "servant-tourist" even represented sojourns in Paris deracinating former serfs taken there to serve their traveling Russian employers. Having heard stories at home about the hardships of working people in the "rotting" West, the Russian housemaids and lackeys are surprised and envious to discover the high standard of living, the pension plans, and the respect from employers that their Parisian counterparts enjoy.[40] By ignoring the problematic mobility of the *narod*, *Winter Notes* maximizes the socioeconomic and cultural chasm between them and the educated. Instead of lending a hand to the reform effort at home, the pleasure-seekers run off to genuflect to the "holy wonders." These oppositions (upper classes versus the *narod*, hedonism versus work for the Russian commonweal) create the same values as Chernyshevsky had done in critiquing Yakovlev's *Italy*, Botkin's *Letters from Spain*, and Turgenev's *Asya*.

Post-emancipation Russian emigration amplifies *Winter Notes* worries about tourism. The narrator's Russian fellow travelers on the train are a deracinated businessman and an elderly couple of provincial landowners "hurrying off to the exhibition in London," presumably the 1862 International Exposition much discussed in the Russian press at the time and later portrayed in *Winter Notes*.[41] Having just completed a two-week work trip to Petersburg, the businessman is returning to London, his place of residence and professional activity for a decade. He seems to have "lost the very idea of homesickness" (52). For all practical purposes, he is an émigré, and his juxtaposition to tourists blurs the boundary between them. The two groups would remain linked in Dostoevsky's imagination, a conjunction that may reflect the September 1862 publication of statistics showing that the number of Russians either traveling or living in Western Europe had grown from to 23,704 in 1830 to 275,582 in 1860. The dismayed reporter perceived all those people draining wealth away from home, no matter what their purpose in being abroad.[42] Unlike the deracinated businessman on the *Winter Notes* train, the elderly couple plan to stay in London "just a few days," but they "have left their family" behind, a detail contributing

40 In the first version of the article, as serialized in the *Library for Reading*, the sojourn in Paris radicalizes and corrupts the Russian housemaids Sasha and Lena, who both desert their Russian employers. Back in Petersburg, Sasha becomes a nihilist, living with a man to satisfy purely sexual needs. Lena marries a Parisian *restaurateur*. Together in Petersburg, they set up an "ambiguous" restaurant, perhaps doubling as a brothel: see N. S. Leskov, "Russkoe obshchestvo v Parizhe," *Polnoe sobranie sochinenii*, 30 vols. (Moscow: Terra, 1996–2016), 3:206–19, 650–51.

41 On Russian discussion of the 1862 International Exhibition, see Dianina, *When Art Makes News*, 46, 63–68.

42 U., "Zametki o khoziaistvennom polozhenii Rossii," *Russkii vestnik* (September 1862): 239–40. The author was very likely the conservative nationalist historian N. G. Ustrialov.

to Dostoevsky's theme of post-emancipation dissolution of Russia's traditional social and moral order.

While still on the train, the *Winter Notes* narrator mentally zooms ahead to the West, to represent the ineptitude of Russian tourists he *will* have seen on the museum circuit there by the time he returns home and writes up his journey. Dostoevsky now differentiates the writing self from "our tourists" in the manner of post-Crimean War cosmopolitan observers such as Turgenev. In Dostoevsky's description, the modern-day Russian tourist crowd is enslaved to guidebooks and completely ignorant of art:

> They all go about with guidebooks and avidly rush to see the sights in every city, and they really do it exactly as if it were an obligation, a continuation of service to the fatherland. [...] They gawk at a side of beef by Rubens and believe it is the *Three Graces* because that is what the guidebook has ordered. They rush to the *Sistine Madonna* and stand before it in blank expectation: something will happen any second, somebody will crawl out from beneath the floor and dispel their vague yearning and fatigue. Then they walk away astonished that nothing happened. (63)

Desperate as the Russian quest for culture may be, it differs from the "mechanical curiosity" of "English men and women tourists," so absorbed in guidebooks that they hardly glance at the sights. "No," proclaims the narrator, "our curiosity is savage, nervous, and mightily ravenous." Whether Russians are just passing through or have come to the West to "settle down for good" (the émigrés), "this alone can be said of the whole lot: we no sooner pass Eydtkuhnen than we become strikingly similar to unhappy little dogs who run about after having lost their master" (63). The shifting pronouns dissolve the boundary between the traveling self and the others. Moving from "they" to "we," the text entangles the narrator in Russian tourist confusion in the museum world of the West.

In opposition to the adoration of West European painting and sculpture that Dostoevsky's *Time* articles expressed prior to his trip, the *Winter Notes* narrator never honors art, and this differentiates him from the many predecessors who promoted connoisseurship as an aim of travel—Zhukovsky, Myatlev, Annenkov, Zhukova, Turgenev, Herzen, Fet, Botkin, Druzhinin, and Melgunov. While recognizing that ill-prepared tourists reinforce Russia's provincial image in Western eyes, *Winter Notes* treats the foreign museum world as a bastion of discriminatory, colonialist values. Museums consecrate art objects as marvels one *must* see. But why should Russians have to observe those cultural rituals? Why should they have to pass art tests of the sort Myatlev inflicted on Kurdyukova, the yokel to whom Dostoevsky's account probably alludes in mentioning Tambov province as a nest of avid women shoppers (77)? Whatever interest Dostoevsky might have taken in museums on his first trip abroad, *Winter Notes* declines to endorse the art pilgrimage. The author finds the faces of Berlin pedestrians so disgusting that he flees the city "without having even made an attempt on Kaulbach's frescoes" (47), an attraction Fet's *From Abroad* had signposted for Russian tourists. Likewise, when in Dresden, the *Winter Notes* narrator never even thinks about going to see the *Sistine Madonna.* Bypassing art becomes a badge of nationalist resistance to the "worship" of foreign culture.

The Lust of the Eyes versus Humanitarianism

Writing retrospectively, instead of pretending to produce a spontaneous travel diary, the *Winter Notes* author seems to resist art and other sources of aesthetic pleasure because he has witnessed the inhumanity of the capitalist West, especially the London slums. Well before the slum tour, chapter three adumbrates the clash between aestheticism and humanitarianism. There the narrator characterizes "professors of aesthetics" as people unmoved by the Russian lord's practice of flogging his peasants: the aesthete views the atrocity as a mere detail in the big picture of "life's tragedy" (54). This startling formulation reflects the *Winter Notes* assumption that the aesthetic enjoyment of tourism compromises sensitivity to human suffering. Most implicated are Karamzin's *Letters* and Mikhailov's *London Notes.* Although Karamzin's persona observes working class dives and prostitutes, he finds London "beautiful" and enjoys art collections, parks, St. Paul's, and other monuments. Mikhailov's account also features squalid aspects of London but gives more attention to pleasurable sightseeing. Despite the poverty, pollution, and nightmarish scenes of pedestrian crowds and screeching transport, *London Notes* deems the English capital "majestic." *Winter Notes* reverses the aesthetic and humanitarian dynamics. Dostoevsky's London boasts "magnificent public parks and gardens," but they appear anomalous alongside screeching machines and trains, the filthy Thames, the air full of coal dust, and the slums' "half-naked, savage, hungry" inhabitants (69).

The amoral tourist lust of the eyes is conjoined to the Russian radicals' sociopolitical cosmopolitanism in the *Winter Notes* representation of the International Exhibition in London and the associated attraction of the Crystal Palace in Sydenham. Dostoevsky may not have visited either site.[43] Whatever he actually saw, his fictionalized travel account merges the two venues in its famous framing of the exhibition as the realization of a "prophecy from the Apocalypse" (70). Having lost the capacity to establish a Christian social order, the bourgeoisies of Paris and London have revived the worship of ancient Syria's Baal, characterized as a god of material satisfactions. Since the realm of Baal must have *some* social contract in order to prevent mass cannibalism, the idolaters have settled for an "anthill" mode of coexistence. Elaborated in *Notes from the Underground*, the anthill symbol counters the utopian universality that Chernyshevsky and Mikhailov projected onto the Crystal Palace. Chernyshevsky's unsigned article of 1854 stressed the building's beauty.[44] Mikhailov, on the other hand, accentuated social utility. His *Paris Letters* presented the Palace as a prototype for "new cities" that he imagined springing up to serve the democratizing purpose of housing people in a "no-frills" but "comfortable, convenient, and uniform manner."[45]

Tourist receptiveness to the neo-pagan spectacle of the exhibition operates in synergy with the radicals' politics of universality. Dostoevsky's anthill idea is mirrored in the milling, international crowd of visitors. "Millions" of tourists from "all over the globe" have congregated in this "colossal palace" of material and visual consumption. The intensity of the tug to follow the "herd" terrifies the *Winter Notes* narrator: "You feel it would require a great deal of eternal spiritual resistance and denial not to succumb," not to join in idolizing Baal, which means accepting "what *is* as your ideal." Although the author spots no Russians in the crowd, he foretold their presence in featuring the elderly train passengers bound for London. No less than Russian radical enthusiasts of the Crystal Palace, Russian tourists drawn to the exhibition emerge in *Winter Notes* as honorary citizens of the land of Baal, the epicenter of global materialism and "anthill" universality that would erase national particularities.

---

43 Dianina, *When Art Makes News*, 54–55; and Young, "The Crystal Palace," 180.

44 Young, "The Crystal Palace," 179.

45 Mikhailov, *Parizhskie pis'ma* (December 1858), 617.

In the slum tour that follows the exhibition scene, the narrator's instinctive altruism toward a stranger represents the Christian alternative to Baal. Not knowing a word of English, Dostoevsky's persona visits the Haymarket quarter at night and silently gazes upon the prostitutes (including minors pandered by their mothers), drunks, and forlorn children. He gives a coin to a raggedy, bruised little girl who takes to her heels in disbelief. *Winter Notes* presents this act of charity as evidence of Russian spiritual superiority to the English and enlists Dickens to help make the point. Echoing the satire of Mrs. Jellabee in *Bleak House*, Dostoevsky's persona faults comfortable Londoners for neglecting their own children and the poor at home while making great efforts to aid people in Africa. The Anglican "religion of the rich" and a Catholic evangelist distributing leaflets on the street appear equally unconcerned with alleviating the misery of the local slums (71–73). The Russian narrator alone incarnates authentic Christian brotherly love and pits it against London's inhumanity and pseudo-Christian religions.

Right after the slum tour, however, comes a dark comedy of shopping that implicates Russians in materialism. The shopping episode is a wicked parody of Mikhailov, who described French sales personnel ever courteous and gracious, no matter how little the customer spends. In double-voicing this discourse, Dostoevsky provides a foretaste of the underground man's hilarious expressions of self-disgust. The *Winter Notes* narrator fashions himself as a "modest traveler" new to Paris shopping: "Enter a store to buy something, and the lowliest salesclerk will crush you, simply crush you with his ineffable nobility." Given "your unenviable appearance" and the "disgusting ten francs" you meant to spend, "you begin to despise yourself to the utmost," as the clerk proposes a series of expensive items. Filled with remorse, you toss him the entire hundred francs you have in your pocket, "your eyes begging his pardon." The parody is ambiguous. On the one hand, *Winter Notes* characterizes Parisian shopkeepers as money-grubbers and more generally maintains that money alone guarantees respect and self-respect among the French bourgeoisie. But on the other hand, Dostoevsky's "modest" persona plays by the materialist rules in depleting his funds to save face before the salesman. Furthermore, he asserts that "Russians are generally keen to show they have immense sums of money" whenever they shop. Blurring the boundary between Russians and the French, that putative Russian tendency conforms to the motifs of materialism, ostentatiousness, and shopping mania in the writings of Druzhinin, Melgunov, and Panayeva. Likewise, in imagining how "Russian ladies" would flock to the "huge haberdasheries" of Paris, *Winter Notes* extends the stereotype of Russian women tourists as the worst shopping addicts (76–77).

An extended authorial exposition of the Native Soil paradigm of Russian spirituality then reasserts difference between Russia and the West. Dostoevsky now characterizes the "western personality" as too ruthlessly individualistic to build a society based on brotherhood. To Russians, on the other hand, he attributes a "personality at the very highest level of development"—a personality dedicated to brotherly love and ready to sacrifice its "whole self for the sake of all [. . .] without the slightest thought for individual gain" (79–80). The principle of brotherhood has been inbred in Russians over the centuries, whereas the godless West has been attempting to *invent* brotherhood, a project *Winter Notes* likens to an effort to make a rabbit stew without a rabbit. At this point, all intimations of Russian materialism and receptiveness to Baal's anthill palace have disappeared.

But in a final turn of the screw, a cruel farce of authorial xenophobia in Paris compromises the supposition of the spiritualized Russian personality. Dostoevsky's persona first sneers at French parents contentedly watching their children play by a fountain in the Palais Royal garden (a jab, perhaps, at Pogodin's positive description of such a scene). *Winter Notes* then recounts the narrator's atypical visit to a tourist attraction, the Paris Pantheon, where an old guide delivers by rote a high-flown speech about Rousseau, Voltaire, and Marshal Lannes. Dostoevsky's narrator finds the empty eloquence so absurd that he repeatedly interrupts, to complete the guide's sentences or to inject disrespectful remarks about the illustrious dead men of France. Sounding "slightly hurt" when the badgering begins, the guide's voice eventually grows so "mournful" that the author feels "ashamed" of himself (90). Like the underground man who enjoys inflicting pain on others, if only as a civil servant gnashing his teeth at petitioners, the *Winter Notes* storyteller cannot resist tormenting an aged French halfwit. If this is Russian brotherly love in action, foreigners would surely prefer Russians keep it at home.

In short, *Winter Notes* muddles educated Russians' relationship to the "Russian personality," and this in turn casts doubt on the viability of the Native Soil vision of nationhood. Dostoevsky's narrative sustains throughout the ideal of the *narod* imagined as perpetual homebodies, the uncontaminated guardians of "authentic" national traditions. But since the *narod* cannot achieve national cohesion alone, *Winter Notes* ultimately confesses deep pessimism about Russia's future. This occurs in the midst of the final chapter devoted to satirizing the French bourgeois marriage and marital infidelity. Digressing from the subject, the author asserts that Frenchmen and other foreigners appear more naïve than Russians: "We are more cynical, we value less what is ours, we do not even like it, or at least do not duly respect it, failing to understand what it is. We get involved in European interests and the universal interests of humanity and belong to no nation" (93). This digression reflects the drastic extent of post-emancipation disunity in Russia. *Winter Notes* recognizes that reality but has sought to overcome it through belief in *pochvennichestvo*. Lest the reader linger on the existential problem of Russian nationhood, the text immediately resumes vilifying the French bourgeoisie. To the vice of materialism, the author adds hypocrisy, licentiousness, and vulgarity epitomized in popular vaudevilles and melodramas. He also ridicules the importance Parisians attach to taking a holiday "to roll in the grass" or go to a sea resort—a reminiscence perhaps of Mikhailov's "estimable bourgeois" who drowns in Trouville. *Winter Notes* thus trails off, instead of summing up its central issues of national identity and Russian relation to the West.

While slipped in as a digression, the jarring characterization of Russians as a nationless people intimates an unstoppable tide of cosmopolitan values and concerns threatening to scupper the Native Soil project. The perceived threat comes from two directions: the atheist global design of the radicals, on the one hand, and on the other—Russian cultural tourism cast in a colonial mold of worshipping European civilization. As in Zagoskin's anti-cosmopolitan writings, the politics and culture of westernization operate in synergy in *Winter Notes* to attenuate attachment to the Russian homeland, instill contempt for what is "ours," and pose the ultimate risk of deracination. Dostoevsky's moral and political anxieties about the divisiveness of Russian tourism were so acute that they prompted the *Winter Notes* suspension of his supranational aesthetics of transcendence. Far from extolling the *Apollo Belvedere*, Raphael's Madonnas, or any other art work as "spiritual bread" of universal value, Dostoevsky's travel narrative strategically sacrificed the museum world of the West to the exigencies of *pochvennichestvo*, and that politicizing of culture coincided with Chernyshevsky's contention that Russian art "worshippers" value objects of aesthetic interest more than human beings, whether at home or abroad.

As Dostoevsky subsequently resumed his battle against the radicals' utilitarian aesthetics, he reasserted his religious engagement with Western art. In problematizing beauty's status in the modern world, especially the disjunctions between metaphysical and physical beauty, his novel *The Idiot* (1868–69) and other writings would describe or allude to paintings he contemplated on repeated museum visits during his self-imposed exile abroad with his wife Anna Grigorevna.[46] When Dostoevsky returned to Russia in 1871, he loathed the capitalist West more than ever but had acquired lasting reverence for Hans Holbein's *The Body of the Dead Christ in the Tomb*, Raphael's *Sistine Madonna* and *Madonna della seggiola*, the Ghiberti doors of the Florence baptistery, Carracci's *Head of Christ*, and other works. The Italian masterpieces in particular sublimated his youthful reverie of Italy's "magic," as mediated by Radcliffe, Shakespeare, and Byron. Art acquired for Dostoevsky the votive significance of an icon.[47] The corollary, however, was his antagonism toward the disinterested aestheticism and quest for connoisseurship that motivated cosmopolitan Russian cultural tourism. Dostoevsky could not accept "the museum attitude" toward art.[48] In conformity to that outlook, he never again underwrote the tourist acculturation regime as "Mr. ___bov and the Question of Art" appeared to do in discussing the *Apollo Belvedere*. His novels countenance Russians displaying reproductions of foreign religious art at home (Holbein's: *The Body of the Dead Christ in the Tomb* in *The Idiot*, the *Sistine Madonna* and a photograph of the Ghiberti doors in *The Adolescent*), and Dostoevsky himself acquired a reproduction of the central, icon-like detail of the Virgin and child of the *Sistine Madonna*.[49] During this phase of his career, he expressed

---

46 A. G. Dostoevskaia, *Vospominaniia*, ed. L. P. Grossman (Moscow and Leningrad: Gosudarstvennoe izdatel'stvo, 1925), 102, 126, 137. Representative studies are Jackson, *Quest for Form*, 213–17; Tatiana Goerner, "The Theme of Art and Aesthetics in Dostoevsky's *The Idiot*," *Ulbandus Review* 2, no. 2 (Fall 1982): 79–95; Robert L. Belknap, "On Dostoevsky and the Visual Arts," in *Depictions: Slavic Studies in the Narrative and Visual Arts in Honor of William E. Harkins*, ed. Douglas Greenfield (Dana Point, CA: Ardis, 2000), 58–60; and Brunson, *Russian Realisms*, 162–96. Brunson's illuminating analysis of Dostoevsky's treatment of art in *The Idiot* exaggerates in describing the novel as "a virtual Grand Tour through Europe's greatest museums" (168). *The Idiot*, after all, features no paintings belonging to museums in Italy, Paris, Vienna, Madrid, or London.

47 On the "iconographic power" West European paintings had for Dostoevsky, see Richard Peace, "Dostoevsky and the 'Golden Age,'" *Dostoevsky Studies* 3 (1982): 68–74.

48 Fridlender, "Estetika Dostoevskogo," 136.

49 See *The Idiot*, pt. 2, ch. 4 and pt. 3, ch. 6; and *The Adolescent*, pt. 1, ch. 6.

a receptiveness to West European religious art in other ways as well. In 1873, for instance, he declared the Christ of Titian's *Tribute Money* aesthetically and spiritually superior to what he considered the mundane guise of the Savior in the Russian painter Ge's *The Last Supper*, recently exhibited in Petersburg.[50] Nevertheless, to Dostoevsky's way of thinking, the secularized Russian art tour of the West would always smack of pernicious cosmopolitan disengagement from the nation.

50 Dostoevskii, "Po povodu vystavki," *PSS*, 21:76–77.

Part Three
Embourgeoisement and Its Enemies

9

The Rising Tourist Tide: Foreign Travel from *Winter Notes* to *Anna Karenina*

Dostoevsky returned to Western Europe in August 1863, just a few months after *Winter Notes* appeared. He told his dying wife he was going for medical treatment but had in fact arranged to meet his lover Suslova in Paris. Before they left France for Germany and Italy, Dostoevsky sent his younger brother Nikolai a letter that confirmed the strategic nature of negative sightseeing in *Winter Notes*. Dostoevsky now reported a new impression:

> This time I like the look of Paris, the architecture. The Louvre is a tremendous thing, and the entire stretch of the quay to Notre Dame is amazing. It is too bad, Kolya, that you, in preparing to be an architect, have never gone abroad. An architect ought not to miss going abroad. No drawing can convey the impression of reality.[1]

The remark was off the record, but *Winter Notes* had made public this belief that truly knowing a foreign place requires seeing it with "your own eyes." Doing more than armchair traveling soon became difficult for Russians. The value of the ruble declined, the seven-week Austro-Prussian War (1866) erupted, and then came the Franco-Prussian War (1870–71) when the working-class radicals of the Paris Commune burned the Tuileries palace and the French army converted Paris hotels into hospitals.[2]

1 Dostoevskii: *PSS*, 28.2:38.

2 On the economic factor, see Joseph Frank, *Dostoevsky: The Miraculous Years, 1865–1871* (London: Robson Books, 1995), 54.

Russian tourism picked up again after the war. Tchaikovsky went abroad four times between 1874 and 1879. He loved Paris where he stayed in a pension popular among Russians. In 1875 the poet and budding religious philosopher Vladimir Solovyov encountered many other Russian visitors in London on his way to Egypt. As Solovyov reported to his parents, he was staying near the British Museum in a cheap, pleasant hotel that provided a Russian-speaking guide for tours of the city. Contrary to an unnamed "idiot" Russian journalist who had recently claimed that dirt, dust, and disease made London unlivable, Solovyov found the air remarkably clean, especially in the parks he adored. He was practicing his English with shoe-shine boys and patronizing English, French, German, and Italian taverns.[3] During his museum tour of Italy in 1876, the founder of the Wanderer school of realist Russian painting, Ivan Kramskoy (1837–87), saw Russian tourists in Naples but shied away from them, apparently regarding them as idlers incapable of appreciating his serious, professional reason for traveling.[4] The familiar letters show that Russians remained active on the Continental tourist circuit and that tourist services were expanding to accommodate them. The fashion for publishing travel "notes" had passed, however.

Writings about Russian tourism abroad thinned out during the years between *Winter Notes* and *Anna Karenina*. In addition to Tolstoy's novel of adultery, with its episodes in Italy and Germany, the two other relevant classics were the 1867 spa stories of Dostoevsky and Turgenev, *The Gambler* and *Smoke*. *The Gambler* takes place in Roulettenburg, a coinage the Russian translator of *The Kickleburys on the Rhine* had invented to approximate Thackeray's Rougetnoirebourg [sic]. Dostoevsky's supporting cast of Russians fritter away time at the spa, while the central protagonist and narrator Alexei Ivanovich indulges his addiction to roulette.[5] Pitted against stuffy Germans and the staid English gentleman Mr. Astley, Alexei performs the authorial ideal of the devil-may-care Russian national character, jeering at Western bourgeois materialism and gentility. *Smoke*, on the other

3 V. S. Solov'ev, letters from June 30/July 12, July 1/13, July 17/29, August 14/26, 1875, in his *Pis'ma*, 4 vols. (St. Petersburg: Tipografiia tovarishchestva "Obshchestvennaia pol'za," 1908–1923), 2:3–6, 9.

4 I. N. Kramskoi, letter to S. N. Kramskaia, May 7, 1876, *Pis'ma, stat'i*, 2 vols. (Moscow: Iskusstvo, 1965–66), 1:328. In a letter to Pavel Tretyakov sent the same day, Kramskoy wrote: "If I were a cosmopolitan, I would choose to live in Naples," 1:330.

5 For a professional assessment of the pathology, see Richard J. Rosenthal, "Gambling," in Martinsen and Maiorova, *Dostoevsky in Context*, 148–56.

hand, delivers a blistering critique of Russia as a civilizational wasteland at a political impasse and draws a Byronic boundary between the author and compatriot tourists. Turgenev's Russian crowd in Baden-Baden includes a frenzied gambler (a dig at Dostoevsky), a fat, overdressed Tambov landowner, military mediocrities, and clothes-horse women. Blathering Russian intellectuals are salient, too, and the smoke of locomotives carrying them back and forth from home to the West symbolizes their "confused state of mind and the obscurity" of Russia's future.[6] For all their artistic and political interest, *The Gambler* and *Smoke* treat tourism conventionally. As for classics of Russian travel writing about the West, no successor to *Winter Notes* would come along until the satirical novelist and journalist Mikhail Saltykov-Shchedrin (1826–89) published *Across the Border* (1880–81).[7] This work recounted his recent visit of Germany, Switzerland, and France but was short on touristic detail, with the exception of the author's ecstatic response to Paris and his little fits of tourist phobia. His Russian women tourists, for instance, love spas because they get to change their clothes five times a day, while the men (lawyers, civil servants, merchants) radiate a pathetic sense of purpose as they "feverishly" follow Baedecker's sightseeing programs.[8]

The paucity of masterpieces has made the years between *Winter Notes* and *Anna Karenina* an under-studied part of the literary-cultural history of Russian tourism abroad. A closer look, though, reveals that this was a pivotal period when Russian gusto for cultural tourism in the West won a markedly middle-class public stamp of approval in the face of fierce politico-moral contestation and the ascendance of cultural nationalism in Russian literature and painting. As in the past, the extremes of Russia's political spectrum met in opposition to pleasure trips and tended to fixate on the putative tourist personality. At the conservative nationalist pole, a Slavophile protest against foreign travel erupted in mid-1863 in connection with the Polish rebellion against Russian rule. Suppression of the uprising produced a wave of Russophobia in Western Europe, a development that Russian patriots interpreted as an imperative to stay home. In the 1870s, when Dostoevsky perceived an ominous, accelerating "fragmentation" of

6 Billington, *The Icon and the Axe*, 383.
7 See detailed analysis of the work in Offord, *Journeys to a Graveyard*, 221–48.
8 M. E. Saltykov-Shchedrin, *Sobranie sochinenii*, 20 vols. (Moscow: Khudozhestvennaia literatura, 1965–77), 14:7–8, 42–44.

Russian society, he too would take up a nationalist cudgel against tourists, as he had done in the past.[9]

Along with nationalists, populist revolutionaries continued to malign tourism in the 1860s and 1870s. Broadly speaking, "populism" refers to the preoccupation with the *narod* that began with Herzen's anti-bourgeois, anti-capitalist ideology of Russian socialism.[10] Chernyshevsky and Dobrolyubov radicalized the socialist program after developing frictions with Herzen, and this entailed Chernyshevsky's articulation of an anti-tourist commitment to the *narod.* Lesser radicals of the sixties contributed by using travelogues as "weapons for attacking Russia's social and political order," a strategy Radishchev's *Journey from St. Petersburg to Moscow* had pioneered. The *Contemporary* collaborator Vasily Sleptsov's *Vladimirka and Klyazma* (1861) reported on factories along a stretch of the convict route to Siberia. A later contribution to the genre was the economist Vasily Bervi-Flerovsky's *The Condition of the Working Class in Russia* (1869).[11] Exposing the misery of Russian peasants, factory workers, and prisoners, such accounts put tourism to shame. Chernyshevsky's novel *What Is to Be Done?* did its part by sculpting a revolutionary-minded Russian traveler abroad. Written in prison and serialized in the *Contemporary* in 1863 (overlapping *Winter Notes* in *Time* in February), the novel galvanized generations of Russian radicals with its representation of "new people" out to build the harmonious socialist future of cooperative labor, equality between the sexes, a utilitarian ethics of "rational egoism," and shining cities of crystal and steel. To further this project, the central masculine protagonist, Rakhmetov, takes a trip to learn about workers' conditions outside Russia. He visits the Slavic countries, Romania, Hungary, Germany, Austria, Switzerland, France, and England. He thinks about going to Spain, Italy, and the United States as well. In conformity to Chernyshevsky's prior treatments of Italy as a magnet for corrupt Russian aesthetes, Rakhmetov does not consider Italy a "necessary" destination.[12] It can be skipped, along with Spain, the land that Botkin staked out for Russian tourism.

---

9 Dostoevskii, "Obosoblenie" (1876), *PSS*, 22:80–3.

10 See the classic Franco Venturi, *Roots of Revolution: A History of the Populist and Socialist Movement in Nineteenth-Century Russia*, trans. Francis Haskell with intro. by Isaiah Berlin (New York: Grosset and Dunlap, 1960) .

11 Offord, *Journeys to a Graveyard*, 15–16. See also Ely, *This Meager Nature*, 152.

12 Nikolai Chernyshevsky, *What Is to Be Done?*, trans. Michael R. Katz, annotated by William G. Wagner (Ithaca and London: Cornell University Press, 1989), 291.

Moral fervor at odds with vacationing intensified among the self-described Russian populists of the 1870s, the *narodniki*. Dobrolyubov died in 1861, Chernyshevsky was sent to Siberia for eighteen years in 1862, Pisarev was imprisoned from 1864 to 1866 and drowned shortly after his release, either by accident or suicide. Pyotr Lavrov (1823–1900) and Nikolai Mikhailovsky (1842–1904) were the major spokesmen of the moderate populists who succeeded the "men of the sixties." An army colonel and professor of mathematics at the artillery academy in Petersburg, Lavrov formulated in his *Historical Letters* (1868–69) a philosophy of self-sacrificing social service based on the conviction that Russia's privileged classes should repent and atone for all they had attained from peasant labor. Following his arrest for revolutionary activity, Lavrov escaped abroad in 1870. Mikhailovsky, a nobleman trained in sociology, held the fort at home as a journalist propagating the populist credo in *Annals of the Fatherland*, the leading organ of radical opinion after the authorities closed the *Contemporary* in 1866. Although Lavrov and Mikhailovsky were positivists, they "claimed Christ on occasion as the source of their moral ideas" and respected the religious faith of the folk.[13] The idealistic spirit of service prompted the famous "going to the people" episode in the summer of 1874, when large numbers of young men and women descended on villages to preach socialism to the peasantry. After this project came to naught and even led to the arrest of some activists, the populist movement split between moderates and terrorists, who began in 1878 to pursue their single-minded goal of assassinating government officials. Despite the disagreements about violence, both branches of populism claimed the mantle of a "moral crusade." As one activist recollected, the more he and his comrades learned about the suffering of the *narod*, the more ashamed they felt of their own "miserable bourgeois plans for a happy personal life!"[14] In expressing at their trial a like-minded, self-abnegating ambition to eradicate social justice and inaugurate an age of brotherhood, the terrorists who assassinated Alexander II in April 1881 would arouse much sympathy in the Russian public.

Nationalist and revolutionary disaffection to tourism provoked a backlash sympathetic to *embourgeoisement* as of the late 1860s, primarily in the periodical press. Unashamed of desiring a happy personal life, travel lovers started talking back. They reaffirmed the Byronic practice of tourism as a tonic and a ticket to acculturation, not only for the rich but for the

13 Billington, *The Icon and the Axe*, 435; and Joseph Frank, *Dostoevsky: The Mantle of the Prophet, 1871–1881* (Princeton: Princeton University Press, 2002), 67–68.

14 Ibid., 69–70.

expanding middle orders of Russian society. This chapter maps that new field of conflict between tourists and their detractors. The investigation starts with Ivan Aksakov's and Saltykov-Shchedrin's disagreement about leisure travel in the wake of the Polish revolt. In that context, Aksakov branded Russian tourism a cosmopolitan threat verging on treason. Unlike Aksakov, Saltykov-Shchedrin had never been abroad but had a gift for corrosive satire that would find a rich vein in tourism. In addition to the journalism of patriotic Aksakov and Juvenalian Saltykov-Shchedrin, writings by Herzen, Dostoevsky, and Lavrov enlarged the anti-tourist gallery of Russians abroad in the late 1860s and the 1870s.

But the very intensity of the contempt for Russians disporting themselves in the West bespoke the strength of a rising tide of narratives and pictures that popularized the *embourgeoisement* agenda of "leveling upward" through tourism. My main representative of the new trend is Evgeny Markov, a novelist, journalist, and Russia's preeminent travel writer of the seventies. While *Sketches of the Crimea* was his major contribution to the travel genre, he also recounted his trips to Switzerland, Italy, Austria, Germany, the Caucasus, Central Asia, the Far East, Serbia, and Montenegro. His debut as a travel writer, to be examined here, was "From a Foreign Clime" (1869), a narrative that celebrates tourism in Switzerland as an opportunity for middle-class Russians such as himself to live as kings and queens on vacation in the bourgeois West. From a modest family of nobles in the Shchigrov district in the Kharkhov province, Markov pursued a socially middle sort of career in education between 1857 and 1870, first as a teacher in Tula and then as director of schools in Simferopol and the Shchigrov district. After a flirtation with populism, he became the editor and prolific contributor to *Russkaia rech'* (*Russian Speech*), a journal that propagated moderate, mainly English-style liberalism. His vast, diverse output as a journalist included three articles from 1879 that shored up the pro-bourgeois vacation mentality announced in "From a Foreign Clime." "Bourgeois Ideals" attacked the Russian intelligentsia's anti-bourgeois prejudices, "Literary Depression" saw negativity and outright disdain for life poisoning Russian literature ever since Pushkin, and "The Psychiatrist Novelist" blamed Dostoevsky most of all for creating a killjoy world of neurotics, psychopaths, and religious fools such as Prince Myshkin, the hero of *The Idiot*.[15]

15 Markov, "Burzhuaznye idealy;" "Literaturnaia khandra," *Russkaia rech'* (February 1879): 233–60; and "Romanist-psikhiatr," ibid. (May 1879): 243–75, (June 1897): 151–206.

Other highbrow publications of the seventies shared Markov's vacation mentality. Most pertinent to the issue of *embourgeoisement*, however, was the hitherto unremarked promotion of scenic, curative, and cultural tourism in *Grainfield*, the phenomenally popular illustrated weekly that Soviet commentators justly labeled a "bourgeois magazine."[16] As Jeffrey Brooks stressed, *Grainfield* was the most widely read Russian periodical of its kind. Enjoying success of an order previously unknown in Russia, the magazine's circulation went from 9,000 copies in 1870 to 120,000 in 1890, a figure "double that of its closest competitors, including newspapers." *Grainfield* promoted Russian cultural nationalism. It showcased the life and works of Russian writers including Pushkin, Lermontov, and Tolstoy.[17] The magazine also publicized Ilya Repin, Kramskoy, Ivan Shishkin, and other Russian artists.[18] But jostling against such features were articles commending the scenery, spas, cathedrals, and museum world of the West. The present chapter concentrates on the scenic and curative options, while reserving most of the art-related material for discussion of *Anna Karenina*. The common thread, though, is *Grainfield*'s opening cosmopolitan vistas of travel in an idiom aimed at Russia's middle social strata, eager to emulate the lifestyle of the aristocracy.

Near Treason or Mere Vulgarity?

The Polish rebellion began in January 1863 when insurgents killed Russian soldiers sleeping in their barracks in Warsaw. The violence ignited a wave of imperialist Russian nationalism, the most influential spokesman of which was Mikhail Katkov, future editor of *Russkii vestnik* (*Russian Herald*, 1856–1906). Katkov interpreted the crisis as a challenge that would prove Russians' superior national strength and consequent right to rule the empire.[19] And yet as the revolt proceeded, some Russians expressed envy of the mobilizing force of Polish nationalism. The most arresting articulation of this sentiment was Strakhov's article "A Fateful Question" published in April in Dostoevsky's journal *Time*. Strakhov perceived Poland as a vibrant and dynamic nation with a highly advanced culture, as opposed to "a Russian

16 Commentary in Tolstoi, *PSS*, 70:14n2.

17 Brooks, *When Russia Learned to Read*, 111–14, quote 113.

18 Ely, *This Meager Nature*, 197–98; and Dianina, *When Art Makes News*, 234, 276.

19 Andreas Renner, "Defining a Russian Nation: Mikhail Katkov and the 'Invention' of National Politics," *Slavonic and East European Review* 81, no. 4 (October 2003): 669–72.

nation that remained passive and insufficiently developed."[20] In Strakhov's view, Poland's sense of cultural superiority to Russia was justified: having been "part of Europe and progressed along with it for centuries," the Poles regarded Russians as "barbarians" and hence found their imperialist domination a moral affront. Russians had created a huge state, continued Strakhov, but savages can do no less. "A Fateful Question" concluded that the insurrection had demonstrated the urgent need for Russians to develop their own national "soul," a "national life" to underwrite their position as the empire's predominant people. Once again, the "national life" of ethnic Russians entailed ruling the empire.

The nationally demoralizing shock of the Polish uprising resonates in the negative modeling of Russian tourists on the part of both Aksakov and Saltykov-Shchedrin. Writing under the pseudonym Kasyanov, Aksakov sent the travel letters "From Paris" to his journal the *Day* in the spring of 1863. The March installment presents Russian tourist behavior as a symptom of a national inferiority complex, and Aksakov sadly finds the complex well-founded. After all, he asks, what have we ever contributed to European civilization, except for the now modish entertainment of Russian ballet?[21] But if seeing justification for Russia's "barbarian" image in the West, Aksakov blames today's tourists for exacerbating the problem. Especially in Paris, the best educated, most linguistically accomplished Russians dress, talk, and act so as to appear French, English, Dutch, Swiss, Italian, "even German"—anything but Russian. Even Russian priests, claims Aksakov, discard their nationally distinct vestments abroad, whereas the clergymen of all other denominations, including the Armenian Orthodox, "freely

---

20 Maiorova, *From the Shadow of Empire*, 100. This paragraph also draws on Andrzej Walicki, "The Slavophile Thinkers and the Polish Question in 1863," in *Polish Encounters, Russian Identity*, ed. David L. Ransel and Bozena Shallcross (Bloomington: Indiana University Press, 2005), 90–91; and Edyta M. Bojanowska, "Empire by Consent: Strakhov, Dostoevskii, and the Polish Uprising of 1863," *Slavic Review* 71, no. 1 (Spring 2012): 1–24. After "A Fateful Question" prompted the authorities to close *Time*, Dostoevsky and Strakhov maintained that the article ironically mimicked Poles' pretensions to superiority over Russia. That contention is accepted at face value in Frank, *Dostoevsky: The Stir of Liberation*, 211–12, but demolished in Bojanowska, "Empire by Consent," 6–12. On the complicated treatment of Poles in *Memoirs from the House of the Dead*, see Nina Perlina, "Dostoevsky and His Fellow Prisoners from the House of the Dead," in Ransel and Shallcross, *Polish Encounters, Russian Identity*, 100–109. Dostoevsky's later writings lampooned Poles, as in *The Gambler* where they figure as spongers, sidling up to winners in the casino.

21 Aksakov, "Iz Parizha," in *SS*, 2:98–99.

stroll around Paris" in their churchly garments, unafraid of "provoking snickers." Among all the nationally diverse visitors of Paris, only Russians "in the presence of foreigners feel ashamed of their clothing, their customs, their nationality, themselves!" This shame-driven and shameful masquerade deepens European contempt for Russia (97–98).

The upshot is that in Western eyes, Russians abroad collectively incarnate "Molière's *bourgeois gentilhomme*, bourgeois philistines amidst the nobility" (*meshchane vo dvorianstve*). Such has been the case, claims Aksakov, ever since Russians "flooded" Europe right after the Crimean War. But nowadays, the bourgeois air of Russians cropping up abroad has purportedly acquired more geographical specificity: "What place in boundless Russia has not sent its representatives over here? Urzhum, Belebey, Sterlitamak, all districts of the Orenburg, Kazan, and Vyatka provinces now extend to Paris, to Dresden, to Baden!" Displaying patrician Muscovite snobbery, Aksakov conjures an ill-bred specimen in the shape of "some Mamadysh lady who has spent her life pickling mushrooms and lording it over the muzhiks" but now *must* see Paris. Repeating the figure published in 1862, Aksakov states that 275,582 Russians are now roaming the West. "An awful number!" he shrieks. Be they Petersburg dandies or gauche arrivistes, these Russian travelers strike Aksakov as a new breed of *guliashchie liudi*, a seventeenth- and eighteenth-century social category of footloose people such as freed serfs or escapees from captivity. Feeling beholden to no authority, modern-day Russians in the West are exerting an "economically, morally, and socially" pernicious impact on their homeland (98–101, 103).[22]

Aksakov's on-the-scene report from April accuses the whole lot of cosmopolitanism bordering on treason in light of the Polish uprising. Expressing an outlook close to Dostoevsky's (but disapproving the notion of "panhumanity"), Aksakov contends that the "Russian cosmopolitans" abjectly idolize European civilization as a source of so-called "universal" values that exclude Russian culture. These people, he argues, have changed the *homo sum* adage to: "I am human, and so everything Russian is alien to me; give me the panhuman!" Europeans, however, now treat the Russian as "an unwelcome guest," guilty of persecuting Poles (101–2). Weaving a fabric of rumors, the article represents Polonophilia as "a serious danger" for Russians abroad. Aksakov recounts that customers at the Bal Mabille

22 Ibid., 98–101; and April installment, 102–3.

recently roughed up and snatched off the hat of a Pole mistaken for a Russian. When the Pole asserted his nationality, the crowd accosted him for not staying home to fight the Russians. Paris cancan dancers insult Russian tourists, and Polish sympathizers in Dresden incite local boys to throw mud at Russian children traveling with their parents. Having reported the hearsay as fact, Aksakov sees only two choices for Russians abroad: come home or else "renounce your nationality and your solidarity with your people and your fatherland." The ultimatum applies to "plain travelers" and "emigrants" (103–4, 106). No less than the expatriates, Aksakov's tourists appear as actual or potential turncoats.

Aksakov's patriotic hysteria made him an irresistible target for Saltykov-Shchedrin. From an impoverished noble family in Tver province, Saltykov-Shchedrin started building his literary reputation with *Provincial Sketches* (1856–57), a cycle targeting the bureaucratic absurdities he observed as a civil servant. He upheld the utilitarian view of art but was a political maverick who quarreled with Pisarev as well as Dostoevsky. After Chernyshevsky's arrest, Saltykov-Shchedrin became the major contributor to the *Contemporary* until it was suppressed, when he moved to *Annals of the Fatherland.* Throughout his career, he "saw little worth preserving in Russian society and criticized it mercilessly."[23] True to that rule, he portrayed vulgar Russian tourists but rooted them in sociopolitical malaises at home.

Saltykov-Shchedrin's riposte to Aksakov's letters from Paris rings a change on *guliashchie liudi*, to give it the connotation of "gallivanting people," imagined as a strictly male arriviste collective. Paris is the specimen Russian tourist's "alma mater," "his second homeland, if not the first." Patriots need not worry, however, because the gallivanting abroad poses no political threat: all it amounts to is "grub and sex cosmopolitanism" (*kosmopolitizm zheludochno-polovoi*).[24] In preparation for the trip, the

23 Charles A. Moser, "Mikhail Evgrafovich Saltykov-Shchedrin," in Terras, *Handbook of Russian Literature*, 381–83.

24 Saltykov-Shchedrin, "Podvigi russkikh guliashchikh liudei za granitsei" ["Feats of Gallivanting Russians Abroad"], *SS*, 6:101–2. Russian tourist gluttony also features in *Our People in Paris* (1863), an anonymous play serialized in *Iskra* (*Spark*), the illustrated satirical supplement to the *Contemporary*. These "steppe" Russians breakfast on oysters, radishes, and anchovies, *eau de vie chartreuse*, partridge pâté, Swiss, Roquefort, and Parmesan cheese, Madeira wine, and champagne. The men are the biggest eaters, while their womenfolk make fools of themselves overdressing, going about unescorted, and otherwise experimenting with a social freedom they have never known at home: see *Nashi v Parizhe*, *Iskra*, no. 22 (June 14, 1863): 308–11.

Russian tourist learns to "slurp down oysters without swallowing the shell" and to eat "artichokes without pricking his mouth." Having mastered those arts, he "drifts through Europe with his belly, arousing astonishment at his carnivorousness and the jolly lewdness of his morals" (100, 106). While positing gluttony as a generic tourist trait, Saltykov-Shchedrin maps the sex quest along the generational divide that Turgenev's *Fathers and Children* established. The sex tourists are men of the forties frustrated by their loss of *le droit du seigneur* over serf women. Deprived of serf harems, the tourists frequent the Bal Mabille and similar establishments where money buys the "comfort" they used to get at home for free. The men of the sixties, on the other hand, refrain from sex tourism. They are basically talkers, "racking their brains about how to raise the enigmatic curtains of the future" (107–8). However, this image of Russian youths, lost in empty philosophizing, has a satiric thrust against the authorities who had not yet reopened Petersburg University, closed in April 1861.

The theme of idleness that bridges the generational divide ramifies to cover domestic malaises that Russian tourists both reflect and flee. With feigned incomprehension, Saltykov-Shchedrin wonders why his compatriots abroad continue to embarrass their homeland now that the Great Reforms have effected so many improvements. We have glasnost, he writes, "total glasnost, at that, with no constraints except censorship." Judicial reforms and local self-government are in the offing, and we already have "the right to spend the whole day in a dressing gown!" The allusion to the Oblomovian *khalat* epitomizes the enforced passivity of Russian citizens still deprived of a meaningful role in political decision-making. The psychosis about Poland disfavored progress in that direction and was orienting Russian nationalist myth-making toward the glorification of war as the "only arena for mass participation" in the political process.[25] Camouflaging the political issue as a question of entertainment, Saltykov-Shchedrin suggests that in order to coax gallivanting people to stay home, we need to "come up with some extraordinary food" or perhaps "turn all Russia into a dance class" resembling the Bal Mabille (106–7). Tourist escapism is gross and trivial but reflects the frustrations of life in a rigid autocracy, losing reformist élan and maintaining an iron grip on power in the face of nationalist and revolutionary discontent.

25 For an analysis pitting the imperialist vilification of Poland against the ethnocultural Russian nationalism of Tolstoy's *War and Peace* (1865–69), see Maiorova, *From the Shadow of Empire*, 94–127, quote 127.

Nikolai Solovyov's essays on "Work and Leisure" (1866) shed an equally harsh light on the escapist factor. A member of the nationalist cohort at Dostoevsky's journal *Epokha* (*Epoch*, 1864–65), Solovyov interpreted travel as a reaction to "an awful emptiness" in Russian life. He perceived a commonplace absence of family happiness and a scarcity of refined leisure activities, especially but not only in the provinces: "Public entertainments have taken on the character of a show-booth," and crowds flock to Izler's (a Petersburg amusement park mentioned in Dostoevsky's *Crime and Punishment*).[26] The emptiness of daily life and the consequent need to fill time have produced "an urge to escape; people rush from the provinces to the capital, and from the capital—abroad." No less than Russian émigrés, escapist Russian tourists display cold and egotistical "cosmopolitan tendencies" incompatible with "love for the homeland."[27] Solovyov's indictment resembles Aksakov's, but the cross-cultural dynamics have changed. Aksakov saw Russians running *to* Paris and other Western tourist centers; Solovyov sees them simply running *away* from home. They have no agenda beyond being elsewhere.

Herzen's Cosmopolitan Snobbery

Accused of near treason, gluttony, sexual opportunism, and mindless desertion of hearth and home, Russian tourists came in for further abuse from Herzen. His *Ends and Beginnings* (1861–62) expressed a bilious vision of "conglomerated mediocrity" in every sphere of bourgeois existence in the capitalist West. He borrowed the term from John Stuart Mill, kept it in English in his text, and projected a mob scene of Cook-style travel as one of its manifestations: crowds cramming into carriages and charging onto trains "like a beast."[28] Russian holidaymakers fare especially badly in Herzen's estimation. *Ends and Beginnings* stereotypes them as boors, yelling in hotel lobbies, guffawing at the top of their lungs, and objecting

26 Named Mineral'nye vody (evoking the Caucasian spas), Ivan Izler's park included artificial springs and various entertainments regularly advertised in newspapers. It was extremely popular in the 1860s: commentary in Dostoevskii, *PSS*, 7:377.

27 N. I. Solov'ev, "Trud i naslazhdenie. Stat'ia tret'ia i posledniaia," *Otechestvennye zapiski* 167 (July 1866), section 1: 138–40 and 148. On the author, see Charles A. Moser, *Esthetics as Nightmare: Russian Literary Theory, 1855–1870* (Princeton: Princeton University Press, 1989), 111.

28 Herzen, *SS*, 16:137–39, 141.

to no-smoking rules in dining rooms (169). First aired in the *Poliarnaia Zvezda* (*Polar Star*) and then incorporated into Herzen's memoirs *My Past and Thoughts*, his "Russian tourist zoology" of the late sixties catalogues other specimens. An anecdote set in the Alps highlights Russians' reverence for rank. Herzen writes that if two English women tourists meet at a mountain resort, one will say to the other, "There's the Jungfrau!" But if the tourists are Russian, the observation will be, "There's Reutern." Referring to the Russian minister of finance, the remark pinpoints the social-climbing vacationer's joy to be in the same place as high Russian officials. Another vignette mercilessly depicts the abject behavior of a low-class Russian in a European train-station cafe. A young Russian woman with her belongings in bundles cowers at a table in a corner. She orders no food, and then as other customers prepare to rush back to the train, she wolfs down soup that children unknown to her have left behind. "Did you see that?" the author asks a waiter. "How couldn't I?" shrugs the man, "They're Russians."[29]

Herzen's snobbery entails nostalgia for a remembered past of cultural authenticity abroad. Of special note is "La colonie russe" (1867), an essay commissioned in France for a collection of foreign writers' responses to Paris and made immediately available to the Russian readership via an *Annals of the Fatherland* translation entitled "Herzen on Paris." Just as *Letters from France and Italy* represented the railroad as the annihilator of Europe, so too does "La colonie russe" assert that train-assisted mass tourism has destroyed the "real" Paris. Herzen's "real" French capital is actually two time-space entities: the genuinely "foreign city" that the Russian army and its allies encountered in 1814, and the still proudly foreign, monolingual city that the author and his friends visited in the 1840s. Truly "foreign" Paris has now disappeared, ravaged by the "big army of tourists" that supplanted Napoleon's *grande armeé*. Russian visitors are much implicated in the loss of authenticity. Herzen writes that the discerning eye can spot the arrivistes of the sixties. However well-dressed they may be, they "radiate an atmosphere of Russian leather, Turkish tobacco, Siberia, and Tatary." They are "Turanians" (figurative members of the Uralo-Altaic ethnic group that includes the Samoyed). Yet, these tourists confidently stroll about Paris "as if they were in Kazan or Ryazan." And well they might, because Paris has adapted to *them*. The Russian guest now finds signs in Cyrillic, caviar, and a semblance of the samovar. "There was nothing of the sort in my time,"

29 Ibid., 11:441–42.

laments the author.[30] *My Past and Thoughts* reiterates the theme of vanished authenticity. Present-day Paris, asserts Herzen, "is not the *real* Paris, but a new one. Having become a hub of the globe, Paris has ceased to be a preeminently French city [. . .]. It has blurred into being a hotel for the world, a caravanserai, and it has lost its original individuality." Instead of authentic foreignness, today's "nomadic tourist" finds an "old Babylon decked out, rouged, and powdered, not for her own sake but for his."[31] Paris has sunk to the level of an aged prostitute, marketing herself indiscriminately.

Herzen's nostalgic snobbery is a striking antecedent of the views of Paul Fussell and Daniel Boorstin, two frequently cited American commentators who recalled the pre-World War II period as a golden age of travel—*their* time when the "going was good," the time before mass tourism hit Europe.[32] But in parallel to the changes Fussell and Boorstin observed, Russian travelers young enough to be Herzen's children were not about to shun Western Europe just because it was not like it used to be. To the contrary, those signs in Cyrillic and the availability of samovars beckoned to the home-plus type as gestures of hospitality, easing and enhancing the visit. As noted in connection with the Caucasus, tourists tend to seek the authenticity effect but define "foreignness" subjectively. The point is to have the personal experience of being elsewhere, and the fact that many other tourists have been to the place and written about it is not necessarily off-putting. Instead of being a warning to stay away, the destination's popularity may function as accreditation.

Markov's Swiss Arcadia

Markov represented that kind of tourist, placing a big premium on amenities, comfort, and scenery made famous through literature, pictures, and guidebooks. After serializing several of his Crimea sketches in *Annals of the Fatherland*, he turned to Switzerland in a way suggesting he already had Dostoevsky in his cross hairs. Published in Russia while the Dostoevskys were living abroad, *The Idiot* challenged the conventional tourist image of Switzerland as the invigorating, spiritually uplifting homeland of

30 Ibid., 19:303.

31 Ibid., 10:325.

32 Fussell, *Abroad*, 42, 47–50; Daniel J. Boorstin, "From Traveler to Tourist: The Lost Art of Travel," in *The Image or What Happened to the American Dream* (New York: Antheneum, 1962), 78–79, 117.

uncorrupted shepherds. Karamzin had projected that bucolic vision, and its landscapes persisted in Tolstoy's *Lucerne.*[33] Far from having a salubrious effect, the Switzerland of *The Idiot* makes Prince Myshkin feel like a stranger in a strange land, where his epilepsy and psychological abnormalities have landed him in the care of a doctor. The mountains and the perpetually blue sky make Myshkin uneasy, and the village is a nest of vipers. When the young woman Marie returns home after being seduced and abandoned, the villagers abuse her. Even her mother mistreats her, and local children pelt her with stones and filth. Myshkin pities her, befriends her, and teaches the children Christian charity. The children then prompt their elders to love and comfort Marie and to ask her forgiveness for their cruelty before she dies. The Russian "idiot" thus acts as a curative agent in a Switzerland of brutal villagers and an alien, anxiety-producing natural environment.

Markov's "From a Foreign Clime" takes a stand against the Switzerland of *The Idiot*, as well as the anti-bourgeois, anti-capitalist critique of Swiss society and the Swiss tourist industry in *Lucerne.* "From a Foreign Clime" appeared in *Annals of the Fatherland*; and as we will see, Markov threw a sop to Russian populism's pro-worker ideology. In the main, however, "From a Foreign Clime" spotlights the enjoyment of tourism and defies the tendency of Russian aristocrats and the intelligentsia, right and left, to turn up their noses at the bourgeoisie. "From a Foreign Clime" begins with the author's train ride from France to Neuchâtel, a charming Swiss "bourgeois town" (*meshchanskii gorod*), where watch-making and lace-making have long been the dominant economic activities. The traveler perceives Switzerland as "a land of freedom," where "everything seems to become better and purer." The crystalline lakes, the blue sky, "picturesque villages," and the "chain of the Alps from Mt. Blanc to Saint Gothard" create the impression of "a fairy-tale Arcadia of eternal peace, where everybody loves one another, and there is no illness, sadness, or lamentation." Here the visitor begins to "believe in happiness, brotherhood, and Nature's maternal love."[34] This pastoral vision disputes Tolstoy's and Dostoevsky's shared scenario of a Russian visitor encountering local victims of Swiss inhumanity: the homeless busker of *Lucerne* and Marie of *The Idiot.*

---

33 On *The Idiot* as a response to *Lucerne* and other works of Tolstoy, see Orwin, "The Return to Nature," 87–102.

34 E. L. Markov, "S chuzhoi storony (Pis'ma v Shchigrovskii uezd)," *Otechestvennye zapiski* 184, no. 6 (June 1869), section 1: 225, 228–29.

Whereas *Lucerne* denounced the Swiss republic as a fraud, "From a Foreign Clime" represents Neuchâtel as a paradigm of democracy, multi-confessional religious harmony, and bourgeois communality. After three weeks of "the insane whirlwind of life in Paris," the tranquillity and simplicity of Switzerland revitalize the author, as though he has escaped "a factory's din" (234). True, he grants, life in Neuchâtel may strike many outsiders as boring. The town has "no world-class museums such as the Louvre, no Champs-Elysées, no Bois de Boulogne, no Versailles" or other famous tourist attractions. Neuchâtel is off the tourist circuit but provides its hard-working residents with everything that comprises "the true essence of civilization." There are excellent schools, a natural history museum that trains scientists, hospitals, an orphanage, a well-stocked public library, banks that readily extend credit, an art gallery showcasing local artists, a transport network of buses, trains, and steamships; a telegraph, swimming pools, fountains, and serviceable shops that do not try to lure people into buying things they do not need, as happens in consumerist Paris. Finally, the town's hotels are a bargain. They do not have the "stupendous lobbies" and "astounding doormen" one finds in Paris but are comfortable, cheaper, and cleaner. They have polite staffs and good food and afford "picturesque views" of the scenery (235).

Hiking, Byron, and a Luxury Hotel

As the author hikes into the Bernese Alps, he finds a population exploited and corrupted by the tourist industry. He recommends trekking as the only way to get to know Switzerland, the only way to make an "alpine pilgrimage" that will give you a store of memories you will cherish for the rest of your life. You must work to attain this: "The greater the labors and dangers of your route, the happier shall be the recollection of what you experienced." The author finds the "grandiose picturesqueness" of the Alps exhilarating until he notices the capitalist enslavement of the modern-day natives of the land of liberty-loving William Tell.[35] An emblematic figure is a Swiss shepherd boy trudging for miles across mountains to deliver "a heavy jug of milk or a tub of butter" to a hotel for tourists' breakfast (7). In high season the whole region resembles a huge "residence for lackeys" working as "porters and food-servers for European guests." Operating "in the wings"

35 Ibid., 185, no. 7 (July 1869), section 1: 1–2 (subsequent citations appear parenthetically).

of the "stage" of Swiss tourism, those poorly paid menial laborers alone know the cost of the "enticing decor," the "elegance and convenience" of the production (9–10).

The author's compassion for local people wears thin, however, as he encounters more and more of them pestering tourists to hire them as guides, to buy souvenirs, or to give them tips for little services such as opening a gate. No sooner does a hiker reach a lookout point and prepare to gaze at the scenery, when "up pops some middle-aged Swiss, or a girl or boy," proposing to sell you a picture of the panorama before your eyes, or else a sandwich, seltzer water, a wooden paperweight in the form of a deer, or a paper knife bearing the locality's name. Even old crones try to sell you plates. "In Switzerland," concludes our hiker, "this commerce is really prostituting the beauties of nature and the originality of national life." One can only be grateful that the whole county has not been so contaminated (10–13). Having paid his dues to populist sympathy for beleaguered working people, the author stops noticing the exploitative and corrupting aspects of tourism. He has identified a problem but never implicates himself in it.

The narrative now becomes a paean to hiking, with an intermediate tribute to Byron, and a deluxe hotel at the end of the day. Markov and his brother set off at dawn with their trusty Swiss guide Ulrich to scale a glacier near the Matterhorn. Naturally, when they reach their goal, they find a well-equipped Englishman already installed. Per stereotype, he ignores the view until he has eaten (at his habitual time) the breakfast he brought along in his *nécessaire* (17). Markov then digresses on the prominence of English men and women in Europe's increasingly fashionable mountain-climbing clubs, a form of active leisure he likens to "conquering the oceans and discovering America." Mentioning his hikes in the Crimea as a preparation, he places himself among Switzerland's most energetic tourist-explorers, working up a sweat, conversing with companions each evening about "the dangers, labors, and beauties of the day's route," and then sleeping a sweet sleep, "enveloped in a fresh pillow." Markov is equally pleased with the relatively easy hiking that attracts "entire caravans of tourists" around the Kleine Scheidegg mountain pass and the highly rated Hôtel de la Jungfrau. The tourist crowd is not a form of pollution but rather a source of "animation, vivacity, colorfulness, and conversation" that makes every climb seem effortless (21–25). The author's self-fashioning as a happy member of the crowd clashes with his admiration of Byron. The poet's shrine is the hotel where he stayed while writing *Manfred*, a work reflecting the awesome alpine scenery (26). “From a Foreign Clime” thus incorporates an element of a Byron pilgrimage, but like Madame Kurdyukova, seeking Byron’s Venice while unaware of his detestation of the tourist flocks she represents, Markov sees no contradiction between adoring Byron and enjoying the crowd experience of tourism.

The cosmopolitan pleasure in mingling with upmarket tourists of various nationalities reaches a peak at the end of the day at a deluxe hotel. Despite Ulrich’s advice to stop for the night in Byron country, the hikers press on to accommodations near the Grosse Scheidegg. When the exhausted tourists finally glimpse the hotel lights twinkling in the distance, they feel like weary Walter Scott horsemen finding nocturnal shelter from the elements. The Swiss hotel, however, is no humble Scottish tavern but rather a monument to “the civilization of our time,” an outpost of “pampered big-city life.” From the quiet, night-time world of rugged nature, the hikers step into a brightly lit hall filled with “conversation, bustling, fashionable outfits, smiling faces, steaming food, and clinking glasses.” Markov now imagines mountain peaks and the stars gazing with “Olympian contempt” at this “audaciously swarming anthill” of merry tourists (29–30). The remark probably takes a sideswipe at Dostoevsky’s anthill symbolism. The polemical thrust at Tolstoy’s *Lucerne* is beyond doubt. The English dominate Markov’s luxury tourist scene, and he enjoys this as much as *Lucerne* hated it. Far from perceiving English tourism as cultural imperialism, “From a Foreign Clime” marvels at the success with which the English have transported “their island” to Switzerland. The English language and English civilization—the books, newspapers, churches, port wine, banks, and so forth—have thoroughly penetrated Switzerland, to reach into every “good-sized hamlet” (31).

In Markov’s view, the creation of little Englands on the Continent demonstrates the democratizing appeal of tourism. No matter who you are, he writes, as the client of an excellent European hotel you command a level of service and comfort available only to the highest ranks of royalty in the past. Without revealing that he refers to his native realm, the author envisions a Russian reader in the provinces:

> At home in the Shchigrov district you think about going to Switzerland, to the Rigi, to the Faulhorn, and once you are there, in Switzerland, they have divined what you want, they have understood your tastes, and they have been vying with one

> another to have ready for your arrival above the clouds everything you like to eat, drink, and look at. You tumble from your journey right into a bed made for you and are put in the hands of servants who have been waiting solely for your arrival. (31)

In olden days, adds Markov, many princes and kings could only envy such a level of service and "refined amenities." Saying nothing about the cost of such a vacation, "From a Foreign Clime" leaves one wondering how many Russians could afford it. What matters from the standpoint of cultural history, however, is Markov's status as the first Russian to articulate the Cook philosophy of tourism as an opportunity for the middle classes to feel the "gloss" of the aristocrat lifestyle descend upon their shoulders and boost their hopes of translating that experience into improved social standing at home.

Proclaiming democracy "the engulfing wave" of modern times, the concluding remarks of "From a Foreign Clime" applaud the very processes of *embourgeoisement* that Herzen perceived as conglomerated mediocrity eradicating authenticity. Markov states that the Continental tourist industry has made available to the masses a range of scenic and cultural marvels surpassing what any single French king ever had at Versailles. Democracy "has opened for itself all these parks, castles, galleries, and museums. It has transformed entire towns into its hotels, shops, and casinos. It has transformed entire countries into art museums, pleasure gardens, and factories for its clothing." Markov likewise approves photography and modern technologies of print and reproduction as tools making available for just "a few francs" a pocket-sized "picture gallery" or a book of knowledge formerly reserved only for the "elect, doomed to scholarly monasticism." True, he grants, these aids and the mass tourism they motivate are shallow. "But what breadth!" (32). On that note ends Markov's celebration of tourism in the West as a way for Russians of the diversified middle stratum to learn a little bit about a lot of things, for the purpose of breaking down cultural and social barriers at home.

The Wider Affirmation of Tourism

Markov elaborated his travel values in the preface to the 1872 edition of *Sketches of the Crimea*. Alluding to unnamed killjoys of every sort, he objected to a Russian "monastic mentality" that considers pleasure incompatible with civic responsibility. Tourism, he insisted, is not empty, self-indulgent recreation. To the contrary, it performs "social work" (*obshchestvennoe delo*) in revitalizing the individual:

> The greater the sorrows of personal life, the greater the imperfections of our society, the more vital the need for consolation and recreation. Impressions that can smooth away the wrinkles of care and shake off the despondent somnolence of apathy, if only for a few hours, should not be deemed inimical to essential needs of society. They perform social work just as useful as any other direct source of influence on man's knowledge and beliefs.[36]

"From a Foreign Clime" had made clear that the "social work" Markov had in mind was *embourgeoisement*, a goal that made him anathema to both Dostoevsky and the populists.

Markov's zest for vacationing, his treatment of Switzerland in particular, and his endorsement of Byron ripple more widely through Russian public discourse of the seventies. Revivals of the Byron cult feature in journals and books aimed at the most educated readers. Those contributions included the composer Mikhail Glinka's memoirs of visiting Western Europe in the 1840s, a time when he was in the process of divorcing his unfaithful Russian wife. Glinka focuses on his development as an artist (Berlioz, for instance, was among the many composers he met). But Glinka also frankly recounts how he went about establishing relationships with nice, respectable local girls to take care of his sexual needs, first in Paris and then in Spain. Reminiscent of Byron's public persona, the memoirs' twin motifs of aesthetic and erotic self-expansion most tightly intertwine in Spain, where the Prado and other art museums, the Alhambra, the scenery, and an Andalusian woman singer make Glinka's sojourn pure pleasure.[37] Byron motivates literary tourism in the 1874 account of a Russian traveling to study educational institutions in England. The author raves about the "idyllic pictures of rural nature" observed on a smooth express-train trip from London to Harrow, the elite school nestled amidst "emerald waves of hills, gardens, and woods." In a flight of cultural ecstasy, the Russian visitor describes the school's cemetery as "a national shrine" because Byron, as a

36 Markov, "Neskol'ko slov dlia vstupleniia k pervomu izdaniiu" (1872), reprinted in his *Ocherki Kryma* (1884), iii–iv.

37 "Zapiski Mikhaila Ivanovicha Glinki. Chast' chetvertaia," *Russkaia starina*, no. 2 (1870): 421–22, 426–32.

student, wrote poetry there while seated on a tombstone now protected by a metal grid to limit the tourist practice of hacking off pieces of that most "*fortunate* grave site."[38] As for Byron's literary influence, a surprising example is Strakhov's account "From a Trip to Italy" (1876), to be examined more closely in connection with *Anna Karenina*.[39] Strakhov's narrative deviates from the contemporary nationalist tendency for literature and painting to endow central Russia's "meager" terrain with spiritual value, making it more precious than spectacular scenery abroad, particularly of the Mediterranean variety.[40] Bucking that trend, "From a Trip to Italy" imitates the preference for the foreign that *Childe Harold's Pilgrimage* and *Beppo* performed. As Markov had done by designating a Byron shrine in the Alps, so too did Glinka's memoirs, the semi-anonymous pilgrimage to Harrow, and Strakhov's *récit de voyage* reassert Byron's authority for Russian tourists.

Customarily at odds with pro-tourist tendencies, Dostoevsky dismantled the Byron cult in *Demons* (1873), his novel inspired by the revolutionary Sergei Nechaev's instigation of the murder of a member of his conspiratorial cell. Modeled as a brooding Byron surrogate devastatingly attractive to women, the central protagonist Stavrogin confesses to sexually abusing a young girl and is associated with the Marquis de Sade, vampirism, political murder, plain murder, and other crimes.[41] He is also a traveler. Advised to see Italy, Stavrogin roams "all over Europe" and visits Egypt, Jerusalem, and Iceland. Prior to the trip, which lasts about three years, his mother instructs him to keep a journal recording his "impressions of nature, local people, cities, customs, art, and industries"—a project in keeping with Byron, the writing wanderer.[42]

---

38 K. N. M. [sic], "Obshchestvennoe vospitanie v Anglii. Iz dorozhnoi zapisnoi knigi," *Russkii vestnik* (January 1874): 245, 248–49.

39 N. N. Strakhov, "Iz poezdki v Italiiu," in *Bratskaia pomoshch' postradavshim semeistvam Bosnii i Gertsegoviny* (St. Petersburg: Izdanie peterburgskogo otdela slavianskogo komiteta v tipografii A. A. Kraevskogo, 1876), 75–92.

40 Ely, *This Meager Nature*, 160–91, 223–29.

41 Frank, *Dostoevsky: The Miraculous Years*, 465–71; Miller, "Dostoevskii and the Tale of Terror," 146, 149–50; W. J. Leatherbarrow, "*The Devils* in the Context of Dostoevsky's Life and Works," in *Dostoevsky's* The Devils. *A Critical Companion*, ed. W. J. Leatherbarrow (Evanston: Northwestern University Press, 1999), 45–49; and Lynn Ellen Patyk, "The Byronic Terrorist: Boris Savinkov's Literary Self-Mythologization," in *Just Assassins? The Culture of Terrorism in Russia*, ed. Anthony Anemone (Evanston: Northwestern University Press, 2010), 169–73.

42 Dostoevskii, *PSS*, 10:42, 45, 56; *Demons*, pt. 1, ch. 3, 4, 6.

5. Against the backdrop of Mt. Pilatus and picturesque architecture, swanky tourists congregate by a dock for boat rides on the lake in Lucerne.

If not immediately contributing to the Byron cult, *Grainfield* supported Markov's broader program of giving Russians the preparation and confidence to enter the glamorous world of foreign travel. Swiss sites, including Lucerne, garnered most of the magazine's recommendations for scenic tourism abroad in this decade (fig. 5). In 1870, its inaugural year, the magazine ran a piece beginning: "Tourists who want to visit the summit of the Rigi-Kulm in the Swiss Alps usually spend the night in a small hotel half-way up the mountain and then undertake the rest of the climb with one or two local Swiss guides." As the hikers pass through "lush, green alpine pastures" and "shady pine groves," they see "the deep blue sky," smell "aromatic herbs," and hear "the rustling of the forest" and "melodic little bells of herds." Every sense is awakened, to produce "a feeling of blissful pleasure." One encounters other tourists "every step of the way," especially "the sons of Albion" but also German, French, Dutch, and Swedish hikers.[43] In conformity to Markov's "From a Foreign Clime," the *Grainfield* reference to the density of visitors is not a phobic warning but rather a celebration of collective pleasure. The site is being touted as a fashionable one for the armchair traveler who has enough money to contemplate a trip abroad but needs guidance about where to go, where to stay, what to see. An 1873 number of the magazine similarly treats the tourist development of Vevey as a positive phenomenon: "It is hard to imagine anything more inviting and elegant that this little town," located in "an astonishingly beautiful site" on Lake Geneva and conveniently close to the railroad station. More and more hotels and pensions are springing up to accommodate the "flood of visitors" that make Vevey "an animated" vacation spot.[44]

Locomotive technology features as a boon to tourism in Switzerland in an illustrated article from 1874 on the Rigi-Kulm cogwheel railway from Vitznau to Rigi Staffelhöhe, a service that opened to the public in May 1871. Impressing upon readers the engineering ingenuity of the West Europeans who had conceived and executed this project, *Grainfield* made the train an emblem of unqualified progress rather than a diminution of nature's authenticity. A joy ride had supplanted the toil of "clambering to the summits of the Alps" with "donkeys, mules, and alpine walking sticks." The train passengers behold a "delightful," unfolding panorama of mountains, roaring streams, deep forests, pastures, Lake Lucerne, and Mt. Pilatus against the "azure sky." At the end of the line, "tourists relax a while in a Swiss hut" and then either hike to the summit or ride back down on the train. The passengers include women and children, as shown in an illustration of sightseers on the viewing platform of the train's rearmost car, moving through spectacular terrain (fig. 6).[45] Russian families as well as Markov's sort of athletic male hiker are being invited here to consider taking a Swiss vacation. Addressed to "the Russian tourist," a subsequent article charted

43 *Niva*, no. 18 (1870): 284. On Bad Pfeffers, Lucerne, and Firvaldshtet lake, see also ibid., no. 28 (1873): 445–46; ibid., no. 24 (1876): 411, 417, and ibid., no. 51 (1876): 859, 866.

44 Ibid., no. 49 (1873): 769–70.

45 Ibid., no. 5 (1874): 74, 76.

6. The Rigi-Kulm railway. Opting for comfort instead of hiking, tourists with a child ride on the train's platform for scenic viewing.

railroad service to the Rigi town by town, both from St. Petersburg and Moscow, through Germany, France, and Switzerland.[46]

A strategy of polite initiation into cultural as well as scenic tourism guides the *Grainfield* marking of cathedrals. An article of 1875 begins: "Who does not know St. Peter's in Rome, that most magnificent of all Christian cathedrals? Who has not seen St. Peter's square in pictures at least, if not in reality?" This is inclusive discourse, careful not to talk down to the many socially middle readers who had surely not visited Rome. After establishing this semblance of parity, the author describes the cathedral and the festivities of Holy Week, when "60,000 to 70,000 foreigners, mainly English and American" pour into Rome for church services and choral performances in the Sistine Chapel. The accompanying illustration shows St. Peter's square jammed with fine carriages and tastefully dressed people on Easter Sunday.[47] The image and the text expand Russian readers' knowledge of the cathedral as a tourist destination, while pretending they already "know" it. Other *Grainfield* numbers of the seventies offer excellent engravings of the cathedrals of Cologne, Milan, Reims, Strasbourg, and other European cities.[48] If not directly pitched toward travel, those images, too, add to the

46 Ibid., no. 14 (1878): 241, 247.

47 Ibid., no. 27 (1875): 418–19.

48 Ibid., no. 44 (1871): 697, 700, 701; and ibid., no. 16 (1873): 250–51, 253.

magazine's archive of starred monuments for culturally and socially ambitious Russians to think about going to see.

Dostoevsky and the Ems Dilemma

The magazine's coverage of West European spas gave special attention to Ems, the fashionable cure center where Dostoevsky took treatments for emphysema in the summers of 1874, 1875, 1876, and 1879. Just prior to his first trip there, the magazine ran an article describing Ems as "one of Europe's most important spas" and the "most famous" in its region of Germany. Accentuating scenery, the accompanying illustration shows a bird's-eye view of the site with the River Lahn dominating the center.[49] Might Dostoevsky have noticed the *Grainfield* publicity?[50] Be that as it may, his repeated patronage of the Ems spa would complicate his anti-cosmopolitan public persona. He had returned to Russia in 1871, the tenth-year anniversary of the abolition of serfdom, the evaluation of which preoccupied commentators of every political stripe. Nationalists, populists, and Western-style liberals all had their organs of opinion, but for one reason or another, virtually everybody considered the emancipation a failure. Dostoevsky assumed a big role in this journalist fray, most importantly as author of the *Diary of a Writer* (1873, 1876–77, 1880, 1881), the irregular publication that featured some of his best short literary works, such as "The Dream of a Ridiculous Man" (1877), and reams of his commentaries on current events at home and abroad.

If relatively marginal in Dostoevsky's late writings, tourist trips abroad are implicated in several central concerns of the *Diary*: capitalism's spawning a materialistic middle class (worshippers of "the moneybag"), the socioeconomic and cultural chasm between the elite and the *narod*, and the spread of pro-Western liberalism hostile to *pochvennichestvo* and other Slavophile orientations.[51] A *Diary* installment from August 1873,

49 Ibid., no. 27 (1874): 417–18.

50 A full-page ad in Dostoevsky's name, calling for subscriptions to the ultra-conservative journal *Grazhdanin* (*Citizen*) under his editorship, appeared in ibid., no. 41 (1873): 656 and ibid., no. 42: 672.

51 On the "moneybag" image in the *Diary* and antecedents in the treatment of capitalism in *The Idiot*, see Vadim Shneyder, "Myshkin's Million: Merchants, Capitalists, and the Economic Imaginary in *The Idiot*," *Russian Review* 77, no. 2 (March 2018): 244–45, 253–54. For surveys of Dostoevsky's journalism in the 1870s, see Frank, *Dostoevsky:*

for example, typifies the Russian tourist as a liar, seeking status at home by pretending to have seen things abroad that he never saw. Panayev's *Russians in Italy* had featured such tourists, but Dostoevsky gave the topic new energy in his unending struggle to convince the readership that their "native soil," their national "soul" was superior to Western civilization. Similarly, the *Diary* represents Russian tourists so ashamed of their nationality that they try to pass as French or English abroad. This theme, too, was old but had new urgency in the crisis-wrought, fissiparous period of the seventies. The same text also quotes a passage from Herzen's memoirs protraying Russians' lack of savoir faire in the West: "They talk loudly when they should keep quiet; and when they should speak, they cannot utter a word properly and naturally" (21:118–20). Whether deliberately or not, this deference to Herzen as an authority on Russian tourism mollifies Dostoevsky's January 1873 attack on him as "a born émigré." That article granted Herzen's brilliance as a writer and thinker but epitomized him as a deracinated atheist: a "gentilhomme russe et citoyen du monde" fond of "comfort and domestic tranquillity." A "product of the old serf system," he professed love for the Russian people but in fact "despised them and severed himself from their ideals," the native soil (21:8–9).[52] Embroidering upon this *pochvennichestvo* scenario, the *Diary* from 1876–77 conjures a fifth column within the nation in the triadic form of Russian tourists traipsing abroad, Russian émigrés, and Russian domestic consumers of cayenne pepper and other foreign luxuries.[53]

Dostoevsky confronts his own participation in curative tourism in the *Diary* for July 1876, just two months after a *Grainfield* article about the growing prestige of Ems. The magazine stated that the town's "favorable location in the picturesque valley of the rapid little River Lahn, its delightful, cool forest surroundings, and its proximity to the Rhine—that

The Mantle of the Prophet, 87–119, 254–81, 322–37; and Kate Holland, "Dostoevsky's Journalism in the 1870s," in Martinsen and Maiorova, *Dostoevsky in Context*, 288–94.

52 *Demons* allusively entangles Herzen in its web of political psychopathology. Kirillov, the suicide-philosopher of god-manhood, figures as *un citoyen du monde*, while Stavrogin acquires citizenship in the Swiss canton of Uri. Mikhailovsky took Dostoevsky to task for maligning Herzen and for failing to recognize that the biggest threat to the welfare of the *narod* came from Russia's railroad magnates, factory owners, and bankers rather than a "handful of madmen" as imagined in *Demons*: N. K. Mikhailovskii, "Iz literaturnykh i zhurnal'nykh zametok" (1873), *Sochineniia*, 6 vols. (St. Petersburg: Tipografiia V. M. Vol'fa, 1896–97) 1:867–72.

53 On the latter see his *PSS*, 25:20–21.

highway for tourists and high state officials on private visits—have made Ems a deluxe spa, drawing some 10,000 to 11,000 ill people every year." In addition to meeting medical needs, the spa has "a multitude of good hotels" and "rich sources of entertainment"—a reading room, concerts, and theater in German and French. Along with a panoramic view of the town, illustrations show spa facilities and a turreted chateau where the Russian emperor stays.[54] During his first visit in 1874, Dostoevsky initially considered the scenery of the Ems region "the most enchanting in the world."[55] He was also glad to find a refined clientele, including "remnants of the public in the entourages of the emperors" (Alexander II and Kaiser Wilhelm I). But after royalty and their courtiers departed, "God knows who" flooded the spa—"despicable Germans" and "even worse Russians," leaving Dostoevsky determined "not to meet anybody."[56] Among the Russian celebrity-seekers who pursued him, he found most insufferable a garrulous "cosmopolitan," a liberated, atheist Russian woman who was the director of an institute in Novocherkassk.[57] In addition to the Russian crowd, the regimented spa routine and the price gouging in shops and accommodations soon made Dostoevsky detest Ems.

The tourist hater would establish a superior plane of existence for himself in the *Diary*. Entitled "Heading Abroad. Something about Russians on Trains," the dialogic installment begins defensively. The author is en route to Ems "but not for a holiday, oh no!" (23:54). Apparently following doctor's orders, he writes that if he had had his "own way," he would have gone to a Russian spa in the South. Such a journey had in fact become easier in 1875, when railroad service opened between the Russian capitals and Sevastopol and some of the Caucasian spas.[58] The trip south was feasible, if not as convenient as taking a train to Germany. But instead of explaining why he opted for the ritzy German watering hole, Dostoevsky's persona goes off on a tangent that leaves no doubt about his imperialist ethnic-Russian

54 *Niva*, no. 20 (1876): 355–57.

55 Dostoevskii, letter to his wife A. G. Dostoevskaia, June 12/24, 1874, *PSS*, 29.1:327.

56 Letter to A. G. Dostoevskaia, July 8/20–9/21, ibid., 348. For Dostoevsky's prior description of the Russian spa crowd as self-satisfied, idle mediocrities, see the letter to his wife from June 28/July 10, 339. Alexander II had spent roughly a month in Ems shortly before Dostoevsky arrived, and Kaiser Wilhelm was there between June 15 and July 9: see ibid., 535.

57 Letter to A. G. Dostoevskaia, June 16/28, ibid., 332.

58 On the rail links, see Mal'gin, *Russkaia Riv'era*, 142; and Phillips, "Health and Sociability," 68–69.

allegiance: he argues that since many Crimean Tatars have emigrated to Turkey, the Russian state should colonize the Crimea with Russian farmers, lest the fertile territory fall prey to "the Yids" (55).

In resuming his account of the train trip, he explains that he equipped himself with "two brochures and some newspapers, just in case. Precisely 'in case,' because I always dread being stranded in a crowd of Russian strangers of our intelligentsia class, and no matter where, on a train, a steamship, or at some sort of meeting" (55). The dreaded fellow traveler to shut out by reading does not belong to the intelligentsia understood as thinking people alienated from the state. The object of tourist phobia appears rather as an educated Russian with pretensions to sophistication and perhaps even a misplaced sense of membership in what Elizabeth Valkenier called the emergent concept of the intelligentsia as "that glamorous elite of the mind and spirit which guided the rest of the nation" and which was beginning to include the generally low-born Wanderer painters hitherto regarded as an "artistic proletariat."[59]

"Heading Abroad" expresses tourist phobia in the ironic spirit of Apollon Maikov's anti-cosmopolitan poem *Two Fates* but diagnoses the compatriot other in a manner recalling *Notes from the Underground.* "What joy! You are ready to embrace as a friend / All Holy Russia in a stranger!" said Maikov's Vladimir. "How pleasant it is to encounter compatriots in a foreign land," dialogically echoes Dostoevsky's narrator. Vladimir finds Russian compatriots in Italy uncongenial but has the space to avoid them. In "Heading Abroad," on the other hand, the narrator is confined to a train compartment with an intrusive, competitive neurotic who exhibits an underground-man sort of inferiority-superiority complex. Touchy and suspicious, he acts out "a hidden lack of self-esteem combined, of course, with boundless self-importance and vanity." The chatty stranger seems friendly at first but soon starts playing power games. He will establish, for instance, that *he* but not *you* met or at least heard about an illustrious Russian at a European spa. Conversation breaks off in mutual enmity, and the initiator of contact now ignores you, to talk instead to "some German baker" on the train. And then at night when the arrogant Russian stretches out to sleep, his feet will touch you, "perhaps deliberately" (55–57). He disembarks in the morning without even giving

59 Elizabeth Valkenier, *Russian Realist Art: The State and Society: The Peredvizhniki and Their Tradition* (Ann Arbor: Ardis, 1977), 11, 17.

you a nod. You have ceased to exist because you refused to compete in tourist social climbing.

The competitive syndrome makes the author partial to traveling in the company of Russian generals because they speak to nobody on trains, for fear of exchanging a word with a person beneath them in the Table of Ranks. The philistine general and hyper-consciousness of rank were clichés of Russian travel writing, but Dostoevsky adds the piquant touch of masquerading in mufti. Dressed in civilian clothing purchased from "Petersburg's finest tailor," the Russian general plays the dandy at foreign spas which attract "many pretty ladies from all over Europe." As evidence of his participation in fashionable tourism in the West, he comes home with photographs of himself in his Petersburg suit, made into visiting cards to distribute to friends and subordinates. While the generals' aloofness on trains is a point in their favor, their conformity to banal tourist norms makes them as alien to the author as the talkative neurotics. The *Diary* concludes that no matter who the others are, "a book or newspaper helps tremendously on a trip, precisely against Russians: 'I'm reading, you can say, so leave me in peace'" (57).

"Heading Abroad" thus resolves the Ems dilemma through Byron's role-distancing strategy but with existential inflections of the Dostoevskian "underground." In parallel to Childe Harold, Dostoevsky's persona is alone in the crowd. He feels obliged to participate in curative tourism in Ems but will not do so in the ordinary way. Beginning on the train, the crowd cannot claim him as one of their own: he sits "among them but not of them," "in a shroud of thoughts" which are not theirs (*Childe Harold's Pilgrimage*, 3.113). Dostoevsky's persona invents the others' voices and dialogically inserts the reader into the theater of his imagination. This performance of anti-tourism has much in common with the psycho-philosophical dynamics of Jean-Paul Sartre's *Huis clos*: hell is other people, and a train compartment can become a torture chamber where your best defense may be a flimsy newspaper.

Lavrov's Tourist Other

Reluctant as Dostoevsky's patronage of Ems may have been, it blunted his critique of Russian vacationing abroad. Lavrov faced no such dilemma and seems to have been the first author anywhere to have articulated the idea of "the tourist" as a surface-skimmer in any endeavor.[60] This occurs in "A Russian Tourist of the 1840s," printed under the initials P. U.[61] The article is a hatchet job on Annenkov's book *Memoirs and Critical Essays* (1877), a collection of five of his previous publications. The two outstanding pieces concern Annenkov's 1841 sojourn in Rome with Gogol and Annenkov's presence in Paris during the 1848 revolution.

Lavrov begins with the assertion that "one can divide writers into three different groups: the stay-at-homes, the travelers with a plan, and the tourists." Running the gamut from reactionaries to ultra-radicals, the stay-at-homes are dogmatists chained to their "homestead," the convictions they never alter.[62] The travelers with a plan are intrepid explorers of "the realm of thought." They "leave home not for entertainment but to educate themselves" for the purpose of realizing their sociopolitical goals. Indifferent to "their own personal comfort," they are history's "tragic" servitors. They tread a hard road, bypassing "merry, calm places where they might find respite from their cares" (25–27). The "traveler" category clearly refers to dedicated populist revolutionaries. Finally come "the tourists," the contemptible ilk that Annenkov epitomizes in Lavrov's estimation. Incapable of any serious engagement, the tourist dabbles in art, thought, and politics. Whatever the topic, he glides along the surface "merely skimming off the cream" (28). When Lavrov parses Annenkov's essays, the charge of superficiality is usually on target. In the anecdotal Paris memoir, for example, Annenkov reports snatches of conversation overheard in Tortoni's cafe and recollects watching demonstrators clashing with the police in the streets.[63] Lavrov has grounds, then, for judging the essay politically "pathetic" (39).

The generic portrait of "the tourist" is most brutal in trashing Annenkov's bonhomie. "The tourist," asserts Lavrov, fancies his motto *homo sum: humani nil a me alienum puto*. In fact, however, he is just indiscriminate and morally flabby. He extends a friendly hand to "stockbrokers, proletarians, scholarly thinkers, habitués of fashionable cafes on Paris boulevards, a secretary of Napoleon III, and comrades-in-arms

60 A modern-day example is Samuel Beckett's likening ordinary consciousness to "the position of the tourist" by contrast to the "enchantment" of Marcel Proust's art: see Buzard, *The Beaten Track*, 2–3.

61 P. L. Lavrov [P. U.], "Russkii turist 40-vykh godov," *Delo*, no. 8 (1877): 24–40 (cited in Offord, *Portraits of Early Russian Liberals*, 244n2).

62 Lavrov, "Russkii turist," 24 (subsequent citations appear parenthetically).

63 P. A. Annenkov, *Vospominaniia i kriticheskie ocherki. Sobranie statei i zametok* (St. Petersburg: Tipografiia M. Stasiulevicha, 1877), 242, 245–51.

of Garibaldi." From his global roaming of culture and thought, the tourist emerges "content, jolly, and rosy, without the slightest wrinkle on his face." With "an ironic smile" at "stubborn, narrow-minded homebodies" and "the tormented, wounded, and maimed travelers-with-a-plan whose brows are all wrinkled," the tourist goes on his merry way, "whistling arias from Offenbach's latest opera" (28–29). Lavrov designates the 1840s and 1850s the decades when the number of Russian "tourist" types exploded, in print and reality (31–32). However, the reference to Offenbach expands the time framework into the present era of populist struggle. This notably accords with Mikhailovsky's 1871 essay pairing Offenbach with Darwin as the authorities who unleashed an atavistic process that has destroyed all moral boundaries and made Paris the capital of licentious hedonism.[64]

Lavrov more directly evokes the Russia of the seventies when he relates his metaphor of tourism to actual sociocultural phenomena:

> [The tourist] is a product of a new economic period of the mobility of capital and a craving for comfort; a product of a new stratum of people with no traditions or a serious, definite purpose in life; a product of the spread of their interests in the realm of knowledge, art, and politics. However, they could not care less about deepening knowledge, making art more productive and inspired, or having political interests prompt more self-sacrifice. To the contrary, all these "higher" spheres are important to them only as a fleeting adornment of life and an augmentation of their personal comfort. (27–28)

The references to capitalism, "a new stratum" of leisured people, their rage for comfort, and their incapacity for self-sacrifice all address Russian circumstances of the seventies and the revolutionary populist agenda. Relevant, too, is Lavrov's jab at Markov's idea of tourism erasing creases of worry and performing useful "social work." In stern rebuke, Lavrov makes the wrinkled brow a proud attribute of self-sacrificing political actors, construed as society's sole performers of truly "constructive work" (31).

Complementing Lavrov's position, Mikhailovsky found alarmingly materialistic the *embourgeoisement* that Markov advocated.[65] Markov's

---

64 Mikhailovskii, "Darvinizm i operetki Offenbakha," *Sochineniia*, 1:391–422. On the popularity of Offenbach in Russia as of 1869, see McReynolds, *Russia at Play*, 231–38.

65 Mikhailovskii, "Literaturnye zametki" (1879), *Sochineniia*, 4:801–27.

"Bourgeois Ideals" stepped most directly on the populists' toes by arguing that the best way to combat proletarian and peasant misery in Russia was to spread healthy creature comforts such as good food and a soft, clean bed. On this view, anti-bourgeois ideologies reeked of hypocrisy in degrading the material well-being that all civilized people desire. More generally, "Bourgeois Ideals" protested against the widely spread Russian assumption that "a railroad tycoon" or anybody else with disposable income must be an exploiter rather than "a genius of sorts, with a head for business." In sum, Markov insisted that instead of nurturing their blind prejudice against the French bourgeoisie, the German *Philister*, and English shopkeepers, Russians should wake up to the fact that if their country is to prosper, it needs a hard-working middle class dedicated to trade, commerce, and business, those spheres of endeavor that provide the necessary, material foundation for the development of every civil society.[66]

This chapter has shown continuity and change in the public image of Russian leisure travelers abroad from the mid-1860s into the late-1870s. Writings of the period by Aksakov, Herzen, and Dostoevsky added new mutations of familiar species to the "zoological" taxonomy of Russian tourists: the dangerous cosmopolitan often attempting to disguise Russianness, the gauche provincial, the social climber, the slave to the Table of Ranks, the vainglorious general, and the self-aggrandizing liar. In developing "the tourist" into a metaphor of superficiality, Lavrov struck an original note of international historic significance. Saltykov-Shchedrin's "grub and sex cosmopolitanism" was also an innovation. As Pope illustrated in connection with Karamzin, eighteenth-century English literature already featured sex as a tourist priority. In Russia, however, censorship kept a cork on the topic. Saltykov-Shchedrin broke the taboo, to open the way for more narratives integrating sex into the tourist experience. An early example was Viktor Burenin's travesty of Turgenev's *Spring Torrents* (1872), a tale of travel, romance, and lust that Philip Roth in a *New York Times* interview (November 17, 2012) called one of "the greatest short stories ever written." Engaged to marry Gemma, a charming Italian whose family has a pastry shop in Germany, Turgenev's idealistic Russian traveler-narrator, Sanin, is seduced by the Russian wife of a Russian acquaintance. In self-disgust he breaks off his engagement and

66 Markov, "Burzhuaznye idealy," 202–13.

endures years of erotic enslavement to his seducer. As an aging bachelor, he still feels the agony of having lost Gemma. Burenin's travesty portrays an aristocratic Russian traveler of the forties, enthralled with German Romanticism and Byronism, and blithely pursuing his pleasures abroad while his serfs break their backs at home. The tourist is engaged to marry a lovely Italian but loses her and ruins his life when she catches him copulating naked on the banks of the Rhine with an equally naked, unknown woman who just happened to come along.[67] Crude as it was, the travesty of Turgenev may qualify as a harbinger of late nineteenth-century Russian stories about vacation flings in the Caucasus and the Crimea, a mini-genre that included Burenin's own *A Love Affair in Kislovodsk* (1886).[68]

The more momentous change that occurred by the mid-seventies and horrified Lavrov and Mikhailovsky was not about sex in particular but rather the bigger agenda of Russian writers counting tourism as a human right to the pursuit of happiness, not only for aristocrats but for the middle social orders. In writing about his first trip to the West for a spa cure in 1875, Saltykov-Shchedrin said that his Russian fellow travelers on the train included a lawyer, who made him think about the proliferation of public prosecutors, railroad agents, and bank directors in Russia.[69] This marginal but socially significant detail reflected the ongoing expansion of the legal profession and entrepreneurialism in Russia, two indices of a growing middle class. It was Markov, however, who emerged as the chief promoter and performer of tourism as an energizer of the *embourgeoisement* that he—a teacher and educational administrator—desired for Russia. His Switzerland narrative and then his *Sketches of the Crimea* represented leisure travel as a tonic and a respite from the "burning questions" of Russia's future. This entailed emulation of Byron, the titan of romantic cosmopolitanism who haunted other accounts of the time.[70] In the same period, upwardly mobile Russian visitors of Western Europe acquired in *Grainfield* a journal that

67 Viktor Burenin [Opasnyi sopernik Turgeneva, pseud.], "Pirozhnitsa beregov Reina, ili Russkii dvorianin za granitsei. Turgenevskaia povest'" ["A Pastry Shop on the Rhine, or a Russian Nobleman Abroad"], *Iskra*, nos. 7–9 (February 13, March 5, 1877): 104, 137–38.

68 Compare Soviet experience: as of the 1920s, the domestic health spa became a "symbol of the casual extramarital affair": see Koenker, *Club Red*, 36–37, 202–3.

69 Saltykov-Shchedrin, "Kul'turnye liudi" (1876), *SS*, 12:321.

70 For Byron's persistent hold on Russian imagination with respect to Venice, see Evgenii Markov, "Tsaritsa Adriatiki. Iz puteshestviia po evropeiskomu iugu," *Vestnik Evropy* (January 1903): 231. Markov writes of Byron that "in speaking of Venice, in contemplating Venice, it is impossible not to recall her brilliant celebrator."

spoke to their needs and desires. The aficionados of tourism were at daggers drawn with the stay-at-homes, concerned with national welfare and domestic nest-building. In inimitably novelizing the work-play division, *Anna Karenina* would interrogate anew old questions of tourism's capacity to acculturate Russians and deepen their self-knowledge.

10

Anna Karenina and the Tourist Passion for Italy

The rising tourist tide of the seventies sets a new context for *Anna Karenina*, the anti-tourist Russian masterpiece of the decade. Serialized in the *Russian Herald* from January 1875 through April 1877, Tolstoy's novel displays anachronistic indifference to the contemporary buzz about leisure travel as an engine of *embourgeoisement. Anna Karenina* features Russian lawyers, bankers, entrepreneurs, and railroad agents but depicts leisure travel as an overwhelmingly high-society phenomenon. In giving aristocrats center stage, Tolstoy's novel deviates from the proliferation of arriviste tourists in post-emancipation writings: Aksakov's blanket category of bourgeois philistines and his hordes from Siberian boondocks, Saltykov-Shchedrin's lewd gluttons, Herzen's "Turanians" desecrating Paris, Dostoevsky's neurotic social climbers, and Lavrov's "new class" of Russians with disposable income but no serious interests. The aristocratic slant of *Anna Karenina* reflected Tolstoy's limited experience. He vacationed abroad in the late 1850s, when he spent time in Soden with his brother Nikolai who was taking a cure for tuberculosis, the disease that killed him in 1860. By 1873, when Tolstoy started *Anna Karenina*, eleven years had passed since his second journey abroad, when he mainly investigated schools in Germany and England but also spent roughly five weeks in Italy. In 1876 Tolstoy remarked that he had not even been in Petersburg for sixteen years and hoped to keep it that way.[1] The portrayals of tourism in *Anna Karenina* are thus based on the reclusive author's memory, imagination, reading, and impressions gleaned from acquaintances.

1 Tolstoi, letter to Ia. P. Polonskii, November 17–18, 1876, *PSS*, 62:292.

Along with the sociology, the aesthetics of anti-tourism in *Anna Karenina* have roots in the early reform era, when Tolstoy went abroad and when Chernyshevsky and Pisarev inaugurated the populist interpretation of vacationing as self-indulgence of the morally corrupt, idle rich. The radical killjoys condoned nothing but knowledge-building journeys, and they expressed special scorn for cultural tourism in Italy. The anti-tourism of *Anna Karenina* extensively overlaps the populist discourse. Whereas *Lucerne* lacked an anchoring ideal of home to outshine tourism in the West, the parallel-plot structure of Tolstoy's adultery novel produces precisely such a mutually illuminating clash. Tolstoy's aforementioned 1860 letter to Chicherin upheld work at home in the country as the only morally defensible life for an educated Russian and set that choice against "roaming" abroad for aesthetic self-expansion, especially in Italy. *Anna Karenina* builds on those convictions. The novel's most pertinent representation of divergent life-paths appears in the first thirteen chapters of part five. In counterpoint to the "Italian fantasy romance" that Anna and Vronsky chase in a vain effort to escape the stresses of adultery, Levin and Kitty marry in an extensively described church ceremony.[2] They skip the standard honeymoon abroad and begin a life in the country that will center on family, estate management, Levin's relations with peasants, his writing a book on farming, and his search for religious truth. If particularly pointed in part five, Tolstoy's disapproval of tourism permeates the novel through his "labyrinth of linkages"—the multiplication of words, images, juxtapositions, and "situation rhymes" that create big patterns of meaning.[3] Representing vacationing as a dead end, Tolstoy seeks to demonstrate that one discovers and nurtures the authentic self only by leading a responsible, work-centered life at home in the manner of Levin and Kitty.[4] The *Anna Karenina* "narrative of family life" overwhelms "the narrative of passion."[5] "Moral work" triumphs over high-society hedonism.[6] Such is the crux of the anti-tourist value system that Tolstoy creates.

2 Quote from Gustafson, *Resident and Stranger*, 253.

3 Letter to N. N. Strakhov, April 23 and 26, 1876, *PSS*, 62:269.

4 Gary Saul Morson, "Work and the Authentic Life in Tolstoy," *Tolstoy Studies Journal* 9 (1997): 36–48. Morson's concept of authentic work also covers Dolly's parenting and Mikhailov's painting.

5 Julie A. Buckler, "Reading Anna: Opera, Tragedy, Melodrama, Farce," in *Approaches to Teaching Tolstoy's* Anna Karenina, ed. Liza Knapp and Amy Mandelker (New York: Modern Language Association of America, 2003), 132.

6 Gina Kovarsky, "Moral Education," ibid., 170.

The valorization of the Levin-Kitty world of farming, family, and moral work made *Anna Karenina* a heavy weight on the nativist, populist side of a cultural controversy raging in the seventies. First published as a book in 1871, Nikolai Danilevsky's *Russia and Europe* codified the nationalist position on culture, while underwriting ethnic Russian domination of the empire and Slavic peoples everywhere.[7] Danilevsky blamed Peter the Great for Russia's cultural lag behind the West: debasing the *narod* and the entire native realm, the tsar's reforms infected the elite with the "disease" of "aping Europe" (*evropeinichanie*) and thereby arrested the development of authentic national culture. Danilevsky, however, now saw all the Russian arts coming into their own: Pushkin, Gogol, and Tolstoy's *War and Peace*, Glinka's music, Ivanov's painting, and the sculptor N. S. Pimenov's *Transfiguration* in St. Isaac's cathedral.[8] Danilevsky's ideas, especially on literature and science, favorably impressed Tolstoy at the time.[9] Dostoevsky, for his part, declared *Russia and Europe* nothing less than a "future handbook for all Russians."[10] Independently, Dostoevsky's journalism continued championing the *narod* as "diamonds in the mud," morally and spiritually superior to educated Russians, the self-appointed "cultured people." Calling the elite "depraved" beneath their veneer of refinement, Dostoevsky made the very term "culture" an invective.[11] Compatible rants against engagement with West European culture as a pernicious "cosmopolitan" practice punctuated Stasov's articles promoting Russian art as a builder of civic consciousness and national pride. The same tendency reigned at the populist journal *Delo* (*Cause*, 1866–84), where Lavrov and others urged Russian artists "to do their share" in exposing social injustice.[12] Few artists were as civic-minded as the agitators would have liked. Most notably, an ideal of art as an agent

7 On Danilevsky's imperialist, Pan-Slavic dimension, consult Maiorova, *In the Shadow of Empire*, 184–90.

8 N. Ia. Danilevskii, *Rossiia i Evropa* (Moscow: Drevnee i sovremennoe, 2002), 503–508.

9 Tolstoy, letter to N. N. Strakhov, March 22 and 25, 1872, *PSS*, 61:277. Danilevsky was a friend and associate of Strakhov, who prepared a posthumous edition of *Russia and Europe*: see ibid., 65:161.

10 Maiorova, *In the Shadow of Empire*, 184.

11 Dostoevskii, "O liubvi k narodu" and "Kul'turnye tipiki," *PSS*, 22:42–45, 105–19. This article was a response to V. G. Avseenko [signed A.], "Opiat' o narodnosti i o kul'turnykh tipakh," *Russkii vestnik* (March 1876): 365–87 (see remarks on this critic below). On his clash with Dostoevsky, consult Frank, *Dostoevsky: The Mantle of the Prophet*, 259–61; and Dianina, *When Art Makes News*, 31–32.

12 Elizabeth Kridl Valkenier, "The Intelligentsia and Art," in Stavrou, *Art and Culture in Nineteenth-Century Russia*, 163.

of moral regeneration dominated Kramskoy's aesthetics.[13] But whatever the artist's individual outlook, Ilya Repin's iconic *Haulers on the Volga* (1870–73) and other Wanderer paintings of the *narod*, village life, and Russia's natural environment were readily harnessed to populist and nationalist agendas.[14]

Coexisting with the nationalist and populist discourses that *Anna Karenina* buttressed was the moderate liberal conviction that the industrialized, democratized West remained the best model for Russian development. From the liberal standpoint, the *narod* appeared mired in a brutish life of passivity and ignorance, a life awaiting transformation through cultural transfer from the West. The westernizing outlook focused on "culture" understood as a whole way of life, a modern civilization, including the railroad. As for "culture" understood as the arts, liberals simply assumed their uplifting impact on well-bred people. Equally significant, their cultural bias was Western. A prominent representative was Vasily Avseenko (1842–1913), the chief literary critic of the *Russian Herald*. Avseenko propagated an aesthetics of beauty and gentility. He abhorred the politicizing of art, dismissed much of Nekrasov's writing as "counterfeit" poetry suiting "flea-market tastes," and used his own novel *The Milky Way* to bolster bourgeois Russian attraction to the luxury, refinement, and pleasures of aristocracy.[15] Avseenko's elitism and fastidiousness made him odious to both Tolstoy and Dostoevsky.[16] On the other hand, once the critique of the westernized elite became unmistakable in *Anna Karenina*, the *Russian Herald* editor Katkov chose to serialize *The Milky Way* between October 1875 and July 1876, concurrently and abrasively with Tolstoy's novel.

Russian vacationing abroad was part of the cultural kaleidoscope, and the passion for Italy in particular was surging in the *Anna Karenina* years. Despite the rise of Russian cultural nationalism in the early reform era, the *paese felice* cult had persisted, as illustrated by Yakovlev's *Italy*, the Venice

---

13 Valkenier, *Russian Realist Art*, 116–17.

14 On the work's stature as "the icon of populism," see Billington, *The Icon and the Axe*, 406.

15 V. G. Avseenko, "Literaturnaia zametka," *Russkii vestnik* (February 1876): 885–86. On this critic, see Olga Maiorova, "Avseenko, Vasilii Grigor'evich," in *Russkie pisateli, 1800–1917. Biograficheskii slovar'* (Moscow: Sovetskaia entsiklopediia, 1989), 20–22; and Kenneth Lantz, *The Dostoevsky Encyclopedia* (Westport: Greenwood Publishing Group, 2004), 25–26.

16 Tolstoi, letters to Strakhov, July 6, 1872 and November 17–18, 1876, *PSS*, 61:292; and 62:292–93.

episode of Turgenev's *On the Eve*, his "A Trip to Albano and Frascati," and Vyazemsky's "Venice." A continuation of the tradition was Herzen's memoir "Venezia la bella," first published in London in 1869 and then incorporated into *My Past and Thoughts*. Herzen's title referred to a song popular in Venice at the time but may have also alluded to Apollon Grigorev's "Venezia la bella: A Roving Romantic's Diary" (1858), a poem dramatizing the author's intoxicated discovery of the city (he traveled as a tutor in the pay of wealthy Russians).[17] Herzen's memoir concerns his sojourn in Italy in 1867, and it opens with an evocation of Venetian architecture and the paintings of Titian, Tintoretto, and Veronese. "Venezia la bella" then homes in on Garibaldi and his comrades hailing Herzen as *il poeto russo* on the Piazza San Marco at carnival time. In the crush of visitors, the author meets glamorous Polish exiles—a count and a ravishing woman ("a blood-stained flower torn from Lithuanian fields by a hurricane").[18] Herzen the revolutionary tourist is in his element, feted in Venice after his support for the Polish insurrection damaged his reputation in Russia and decimated Russian subscriptions to his journal *Kolokol (Bell)*.[19] In 1873 Markov published a conventional Venetophile narrative, adding one of his characteristic appreciations of tourist amenities: "I really like Italian hotels. The beds are veritable altars to the god of sleep—ceremonial and gala as an altar, and comfortable as the featherbeds of Gogol's Korobochka. The cuisine is all but better than the French—abundant, subtle, and varied, with wonderful desserts," served in "richly sculpted, gilded halls" of old palaces.[20]

Zestful Russian accounts of visiting Italy steadily appeared during the publication of *Anna Karenina*. Strakhov's "From a Trip to Italy" stands out for the dubious distinction of having irked Tolstoy. Now Tolstoy's closest friend and confident, Strakhov sent him a copy of his *récit de voyage* in March 1876. Having once mentioned to Strakhov that he personally had "liked Rome" in 1861, Tolstoy initially told him that he found "From a Trip

17 A. A. Grigor'ev, "Venezia la bella. Dnevnik stranstvuiushchego romantika (Otryvok iz knigi 'Odisseia o poslednem romantike')," *Izbrannye proizvedeniia* (Leningrad: Sovetskii pisatel', 1959), 349–69.

18 Herzen, *SS*, 11:468–71, 471.

19 Isaiah Berlin, "Introduction" to *My Past and Thoughts: The Memoirs of Alexander Herzen*, trans. Constance Garnett, revised by Humphrey Higgens, abridged with preface and notes by Dwight Macdonald (New York: Vintage Books, 1974), xli.

20 E. L. Markov, "Vena, Miunkhen, Venetsiia: Khudozhestvennye zametki," *Vestnik Evropy* (August 1873): 453.

to Italy" an "interesting" article.[21] Soon afterwards, though, Tolstoy declared it nauseating. He explained to an acquaintance enjoying Venice at the time:

> You won't believe that I would rather live in Mamadysh than in Venice, Rome or Naples; those towns and the life in them have such a conventional and invariably identical grandeur and elegance for everyone else, but such vulgarity for me, that it makes me sick to think about them, and it's unbearable to read about them (Strakhov recently sent me his article about Italy and art).[22]

However sickening Tolstoy found Strakhov's *récit de voyage*, it seems to have made a big impact. The first three words of "From a Trip to Italy" are "What is art?"—the title of the aesthetic treatise Tolstoy published in 1898.[23] In Strakhov's view, anybody seriously interested in exploring the question had to visit Italy. Does the notoriously self-contradictory Tolstoy disagree in *Anna Karenina*? Or at least suspend judgment in depicting the Russian artist Mikhailov who paints a portrait of Anna in Italy?

Among the Russian journals, the *Russian Herald* warrants special attention since its publication of happy-tourist narratives about Italy directly jostled against *Anna Karenina.* As remarks on Avseenko have already hinted, there was much friction between Tolstoy's novel and the orientation of the *Russian Herald* under Katkov's editorship. Among other tensions, the journal advocated Russia's emulation of England and accordingly promoted railroads and the expansion of industry alongside agriculture.[24] Most perversely, Katkov's support of the recruitment of Russian military volunteers to join Serbia's 1876 war against the Ottoman Empire prompted him to suppress the final part of *Anna Karenina*, where Tolstoy belittled Pan-Slavism and the volunteers. Cultural tourism in Italy was yet another bone of contention between Tolstoy and his editor. Moreover, some of the journal's pertinent *récits de voyage* featured trains as ideal vacation transport, a motif that jarred with the soul-killing symbolism of the railroad in

---

21 Tolstoi, letters to Strakhov, March 30–31, 1875 and February 14–15, 1876, *PSS*, 62:164, 246.

22 Tolstoi, Letter to P. D. Golokhvastov, March 17–20 (?), 1876, *PSS*, 62:259 (*Letters*, 1:293). Mentioned in Aksakov's first letter "From Paris," Mamadysh had become a Tolstoy code word for the Russian boondocks.

23 Strakhov, "Iz poezdki v Italiiu," 75.

24 For detailed analysis including the Slavic question, consult Susanne Fusso, *Editing Turgenev, Dostoevsky, and Tolstoy: Mikhail Katkov and the Great Russian Novel* (DeKalb: Northern Illinois University Press, 2017), 164, 170–78, 192–99.

Anna Karenina. If sparring with Tolstoy's novel distinguished the *Russian Herald*, the magazine *Grainfield* was noteworthy for the huge readership it reached in underwriting vacationing in Italy as part of its larger promotion of foreign travel as a leveling-upward acculturation project for the bourgeoisie. Anti-tourist *Anna Karenina* resisted all the pro-Italy trends, although not as belligerently as Tolstoy's Mamadysh remark might suggest.

Tourism and the Russian Way of Life

In the *Anna Karenina* typology of tourism, the curative type alone proves beneficial but not in the mandated manner. The designated purpose of spas is to cure or palliate physical ailments and disease. Pinning hopes on that potential, Levin early in the novel advises his sick brother Nikolai to seek treatment at a spa abroad (2.12).[25] But the author's low opinion of curative tourism soon begins surfacing when the reader learns that Anna's husband Karenin goes to a West European spa every spring in order "to restore his health" after months of "strenuous" work as a big wig in a government ministry (2.26). Later comes the Soden episode, where Kitty learns a lesson about personal authenticity. While her mother agonizes about dressing so as to look "European," Kitty falls into imitating Mlle. Varenka, an orphaned lady's companion who effectively ministers to the sick. Kitty's playing nurse causes marital strife when the artist Petrov becomes infatuated with her. This embarrassing outcome makes her realize that she has been falsifying her personality, and she becomes impatient to return to Russia, the place where her path to true self-realization lies.[26] Like her father, who considers "everything abroad vile" (and yet endures it with charm and generosity), Kitty yearns to go home for "fresh air" (2.34–35). Her prescribed cure served no purpose since she was not sick in the first place but rather heartbroken by Vronsky's desertion. Nor does any other Soden client benefit from taking the waters. Curative tourism most pointedly fails to help Nikolai, doomed to die of tuberculosis in "Death," the novel's sole titled chapter (5.20).

25 Parenthetical references cite the novel's parts and chapters. Quotations follow *Anna Karenina*, trans. Richard Pevear and Larissa Volokhonsky (London: Penguin Books, 2001).

26 For representative commentary, see Vladimir E. Alexandrov, *Limits to Interpretation: The Meanings of* Anna Karenina (Madison: University of Wisconsin Press, 2004), 226–28.

Whereas Kitty was too immature to decline the fashionable spa trip that the family doctor prescribed, Levin actively rejects tourism and prior to his marriage undertakes a knowledge-building journey germane to his work. Drafting his peasant-centered book on agriculture makes him wonder about Russia's entire labor force, and so he decides to go observe economic conditions in the West. On the day of his departure, he reveals his obsessive thoughts of death to Kitty's cousin Shcherbatsky, encountered by chance at the train station. Shcherbatsky jovially proposes a remedy: "Come with me to Paris instead of some Mulhouse" (3.32). Levin, of course, is not in the least tempted to go frolicking in Paris. He wants to see factories, way off the tourist circuit. He returns convinced he has benefited from the trip, as he reports to Anna's brother Stiva Oblonsky: "I was in Germany, in Prussia, in France, in England, not in the capitals but in the manufacturing towns, and saw many new things. I'm glad I went" (4.7). Levin has made an anti-tourist journey of the sort Chernyshevsky and Pisarev recommended. Planned for the purpose of advancing his chosen line of work, Levin's travel program closely resembles Rakhmetov's in *What Is to Be Done?* The externals are much the same, despite the crucial difference between socialist materialism and Levin's concern with the inner being of the Russian peasants whose lives he wants to improve. Contributing to this pattern of anti-tourist meaning is the married Levin's resentment of the "holiday attitude" he observes in city-dwelling visitors of his estate. The guests value the country for rest and recreation, while the host remains absorbed in the business of farming. Holiday-making is the idler's luxury, set against the farm work without which "life is impossible" (6.6, 6.8).

Narrated in the last chapter of part three, Levin's embarking on his knowledge-building trip is juxtaposed to vulgar vacationing in the first chapter of part four.[27] A few months after Vronsky has become Anna's lover, he is put in charge of escorting a "foreign prince" in Petersburg, an episode that satirizes cultural tourism in the anthropological sense of traveling to discover a foreign way of life. Vronsky takes the prince sightseeing and caters to his demands to sample Russia's "national pleasures." The tourist has already cozied up to a mandolin-playing Spanish woman in Spain, shot a chamois in Switzerland, hunted pheasants in England, visited a harem in Turkey, and ridden an elephant in India. The Russian specialties selected for him are trotting races, Russian pancakes, bear hunts, troikas, gypsies, and Russians smashing

27 Serialization in the *Russian Herald* made a break between those two chapters.

crockery. Parodying the tourist quest for authenticity, Tolstoy describes the prince doing as the natives do: smashing "whole trays of crockery" and perching gypsy girls on his lap, while wondering all the while if he has truly penetrated "the Russian spirit." In the end, his favorite "Russian pleasures" are French actresses, a ballet dancer, and the "white label" brand of champagne. This episode sardonically captures the tourist tendency to take the part for the whole, to interpret a selected range of staged spectacles as the essence of a foreign culture. In Vronsky's eyes, the prince comes to appear "a dangerous lunatic" with an offensively instrumental attitude to Russian women, and this perception triggers Vronsky's self-examination about his own stag behavior prior to falling in love with Anna.

Although the visiting "lunatic" has no exact Russian counterpart, his tourist hedonism chimes with sixty-year-old Prince Pyotr Oblonsky, a relative of Anna and her adulterous brother Stiva. In Moscow a few weeks before Anna's suicide, Stiva encounters the prince recently returned from Baden-Baden. When vacationing abroad, Pyotr wines, dines, and feels frisky with young women; but after a couple of weeks back home in the country with his wife, he dons a *khalat* and settles into old-age vegetation. Nothing left for me now, he thinks, except for "saving my soul." After a recent relapse of that sort, he went to Paris where he "rallied again." Still in his thirties, Stiva shares this outlook and envies other Petersburg acquaintances who squander money, live on credit, and commit adultery with nary a thought for soul-saving. Stiva believes that is how "a decent human being" lives (7.20) —one of the novel's most ironic authorial formulations. Linked here to Baden-Baden, a *haut lieu* of spa tourism, the scorn for soul-saving is not an isolated element but rather a leitmotif of Tolstoy's caricatural presentation of Russian liberalism. Already in the novel's first chapter Stiva appears as a habitual reader of "a liberal newspaper" whose "majority" views on everything he adopts, just as he follows the latest fashions in clothing.[28] In the newspaper articles he sees justification for his immorality: the "liberal party" considers marriage "an obsolete institution" and regards religion as "just a bridle for the barbarous part of the population" (1.3). Levin sarcastically paraphrases that latter view in the context of Russia's agricultural problems: even among his well-educated neighbors, trying to

28 Oblonsky may read *Golos* (*Voice*), "known as 'the barometer of public opinion'" and favored by moderately liberal functionaries, or alternatively, as Nabokov proposed, *Russkie vedomosti* (*Russian Gazette*): see Pevear and Volokhonsky, trans., *Anna Karenina*, 819n3.

make a living at farming, Levin encounters the conviction that the "stinking" Russian muzhik occupies a "transitional step of development between ape and man" (3.26, 3.27).

The concern with soul-saving alien to Stiva and old Prince Oblonsky flourishes in the country milieu that *Anna Karenina* portrays as the authentic national way of life, the "real" Russia that tourists never see. The core of this world is Levin's estate Pokrovskoe. Derived from *pokrov* (protection), the name has religious and folk associations central to the farming life Levin has chosen. Pokrovskoe evokes "the idea of protection granted by the Mother of God to the Russian people" but also brings to mind the Russian Orthodox feast of the Pokrov "when cows are mated."[29] In this framework Tolstoy represents the peasant not as the embodiment of nature but rather as "the repository of moral truth."[30] Long before Levin's religious crisis grows acute, his peasant housekeeper Agafya Mikhailovna articulates her simple credo of being mindful of God. Unlike the peasants that Kitty's sister Dolly finds packing the church when she takes her children to communion near Pokrovskoe, Levin does not regularly attend church, has no icons in his house prior to his marriage, and considers peasant belief naïve. Interaction with the *narod* nonetheless reinforces his spiritual longings and supplies the ultimate epiphany when the peasant Fyodor (equivalent to Theodore, a name meaning in Greek the gift of God) tells him that a good man lives "for the soul" rather than the belly (8.11).[31]

Myriad scenes of quotidian life build up Russian authenticity at both the upper-class and the peasant cultural levels. Both before and after his marriage, Levin registers favorable impressions of traditional folkways of communal labor and the preservation of the intergenerational peasant household. He loves looking at his fields of ripening grain and the "honeyed grasses" in meadows at harvest time, but he relates nature to agricultural concerns and dislikes city folks' habit of aestheticizing the land (3.2). Rural people travel by carriages and carts instead of trains, the central *Anna Karenina* symbol of spiritually deadening material progress. The representation of Levin and Kitty's wedding in traditionalist Moscow incorporates distinct Russian Orthodox rituals and customs that translators

29 Knapp, "The Setting," in Knapp and Mandelker, *Approaches to Teaching Tolstoy's* Anna Karenina, 28; see also there remarks on Levin and his prize cow Pava.

30 Orwin, *Tolstoy's Art and Thought*, 144–47.

31 Inessa Medzhibovskaya, *Tolstoy and the Religious Culture of his Time: A Biography of a Long Conversion, 1845–1887* (Lanham, MD: Lexington Books, 2008), 177.

typically explain in ethnographic-type footnotes. Russian infants are swaddled, an intimate cultural detail illustrated when the midwife turns Levin's and Kitty's newborn son into "a stiff little doll" (7.16). As a married couple, Levin and Kitty host their extended families and friends at Pokrovskoe, where mushroom picking, bee-keeping, and the making of jam occur. The lengthy description of the jam-making has special significance as an object lesson about refined Russians in the country producing their own food from "local resources" instead of consuming expensive imports.[32]

The consumption of food contributes to the Russian cultural profile in a manner that situates Pokrovskoe in an imperial framework. During their lunch early in the novel, Oblonsky fancies oysters, French cuisine, Parmesan cheese, champagne, and a "classic Chablis" (1.10), while Levin would just as soon eat cabbage soup and kasha, the staples of the Russian peasant diet.[33] This principled preference for simple Russian fare reaches an extreme on Levin's first day of mowing with the peasants. Instead of returning home for lunch as planned, he heartily shares a peasant's "mash"—salted bread mixed with water (3.5). But when Oblonsky visits him in the country, the bachelor Levin proudly provides a Russian feast, courtesy of Agafya Mikhailovna. The "hors d'oeuvres" are *polotok* (split smoked or dried chicken), pickled mushrooms, bread and butter; then come nettle soup and chicken with white sauce. Levin serves homemade herb liqueur and "white Crimean wine" (2.14) with the meal. Set against the high-society preference for French drink, the Crimean wine is contextualized as Russian, a product of "our" country construed as the empire. Levin's unexamined satisfaction with being a beneficiary of empire finds expression as well in the distinctly Russian relation to the land that he attributes to the peasantry: to maximize agricultural productivity, he argues, one must not impose European methods but rather let the peasants work in "the one way natural to them," a way rooted in their calling "to inhabit and cultivate vast unoccupied spaces"

32 Catriona Kelly, *Refining Russia, Advice Literature, Polite Culture, and Gender from Catherine to Yeltsin* (Oxford: Oxford University Press, 2001), 135.

33 On this episode within the broader literary context, consult Lynn Visson, "Kasha vs. Cachet Blanc: The Gastronomic Dialectics of Russian Literature," in *Russianness: Studies on a Nation's Identity: In Honor of Rufus Mathewson, 1918–1978*, ed. Robert L. Belknap (Ann Arbor: Ardis, 1990), 60–66. On gastronomy's entanglements with sex, see Helena Goscilo-Kostin, "Tolstoyan Fare: Credo à la Carte," *Slavonic and East European Review* 62, no. 4 (October 1984): 481–95; and Ronald LeBlanc, "Unpalatable Pleasures: Tolstoy, Food, and Sex," *Tolstoy Studies Journal* 6 (1993): 1–3, 9–19.

of the East (3.29, 7.3). This "calling" is of course one that native peoples of those regions might dispute.

The National Irrelevance of Art

The edifice of imperial Russian civilization that *Anna Karenina* erects largely dispenses with "culture" understood as the arts. Tolstoy was an accomplished pianist, and domestic music-making was a regular part of life at Yasnaya Polyana.[34] Not at Pokrovskoe. Levin enjoys the happenstance experience of hearing peasants singing in the meadows of his estate at hay-making time (3.12). Thinking that Kitty is about to play the piano at the Oblonsky dinner party where they agree to marry, Levin says to her: "That's what I lack in the country: music" (4.13). But shortly after their marriage, when Kitty suggests they "play four hands" (the first indication that Levin has a piano), he broods in silence, wishing he could write his book in peace, and mentally reproaches *her* for sinking into domesticity at the expense of music "at which she's quite good" (5.15). Not long before Kitty gives birth, Levin thinks longingly of Bach after suffering through a Moscow performance of *King Lear on the Heath*, a fictive title conjuring the new Russian fashion for tone poems (7.5). In short, Levin likes some music, but not enough to integrate it into his life.

Visual art is even more expendable than music at Pokrovskoe. Levin expresses no interest in the visual arts until he goes to the concert in Moscow in part seven of the novel.[35] During the intermission, he criticizes the tone poem for having attempted to create a picture: the composer has strayed from the proper domain of music into the sphere of painting. Likewise, adds Levin, a sculptor should not try to carve "poetic images" in marble (7.5), an apparent allusion to Mark Antokolsky's (1842–1902) model for a Pushkin monument, exhibited at the art Academy in 1875.[36] One may

34 Caryl Emerson, "Tolstoy and Music," in *Anniversary Essays on Tolstoy*, ed. Donna Tussing Orwin (Cambridge: Cambridge University Press, 2010), 10–11; and Ian Saylor, "*Anna Karenina* and *Don Giovanni*: The Vengeance Motif in Oblonsky's Dream," *Tolstoy Studies Journal* 8 (1995–1996): 113.

35 Levin most notably shows no awareness of the Wanderer school, much discussed in the press at the time for their pictures of peasants, village life, and Russian scenery: national subject matter that one might have expected him to find interesting. But in his isolation at Yasnaya Polyana, Tolstoy himself, of course, saw none of the Wanderers' exhibitions.

36 Pevear-Volokhonsky, trans., *Anna Karenina*, 835n10.

constitute metaliterary parallels between Levin's segregating the arts and Tolstoy's aesthetics and discursive practices in *Anna Karenina*.[37] At the mundane level of characterization, however, Levin observations surprise because he has hitherto shown no awareness of painting or sculpture. The second time he does so comes shortly afterwards, when he meets Anna and discusses with her the history of French pictorial art (7.10). If meant to give Levin's character a new facet, those remarks seem too little, too late. Prior to this penultimate part of the novel, visual art never crosses Levin's mind, and this contributes to elevating his moral standing above Anna's. An avid hunter, Levin has stag antlers displayed on the wall of his Pokrovskoe study. On the other hand, "an oval portrait of Anna, beautifully executed by a famous painter" (1.26, 3.14), hangs in Karenin's study in Petersburg (and makes him shudder, after she has strayed). Mikhailov's stunning full-length portrait of Anna adorns the Moscow apartment she shares with Vronsky; and when it mesmerizes Levin, after his evening of dining and drinking with men following the zemstvo elections, Tolstoy structures and characterizes the episode as a visit of a brothel.[38]

Art lacks national relevance not only at Pokrovskoe but everywhere else in the Russia of *Anna Karenina*. The status-conscious leisured class are the principal art consumers, and their preferences are decidedly Western, *not* "cosmopolitan" in the sense of cross-fertilizing consumption of national and foreign art but rather in the distorted Soviet interpretation of "cosmopolitanism" as disparagement of Russian culture. Tolstoy establishes the snobbish, nationally alienated sensibility in introducing Karenin when he meets Anna at the train station upon her return from Moscow. The author states that Karenin likes to talk about "Shakespeare, Raphael, and Beethoven" (1.33), not from genuine interest but merely to project a cultivated image in society. Exemplifying the upper-class norm, other references include Johann Strauss's *Die Fledermaus*, Offenbach's *La Belle Hélène*, the painter Kaulbach, and the pre-Raphaelites (1.11, 2.6, 3.13, 3.27, 4.4, 7.5). Produced on Anna's and Vronsky's Italian "honeymoon" (5.28), the Mikhailov portrait is the hub of a pattern that links adultery to cultural preference for the West. The philandering Stiva enters the novel

37 See for example Amy Mandelker, *Framing* Anna Karenina: *Tolstoy, the Woman Question and the Victorian Novel* (Columbus: Ohio State University Press, 1993), 104–108; and Alexandrov, *Limits to Interpretation*, 89–90.

38 Ronald LeBlanc, "Levin Visits Anna: The Iconography of Harlotry," *Tolstoy Studies Journal* 3 (1990): 2, 6–14.

having just had a dream that features "Il mio tesoro" from *Don Giovanni*. Near the end of Anna's life, Tolstoy associates her (6.32, 7.10) with Emile Zola, Alphonse Daudet, the Bible illustrations of Gustave Doré, and the writings of Hippolyte Taine (1828–93), a critic, art and aesthetics professor, a historian, and travel writer whose books included *Voyage en Italie* (1866, 1874). High-society corruption compromises Pushkin, too, the only identified Russian art-producer that *Anna Karenina* situates in his native land. Stiva is the character fondest of quoting Pushkin, the poet Tolstoy uses to set a paradigm of problematic relationships between sexual and artistic energy. As David Herman argued, the subtext is *Egyptian Nights* (1837), Pushkin's unfinished story about a brilliant Italian in Petersburg improvising verses about Cleopatra and her lovers.[39] In combination with the reference to *Don Giovanni* and Italy's beckoning to Anna and Vronsky, the *Egyptian Nights* allusions interrelate Italy, sensuality, adultery, and tourism—a cluster anticipated in Tolstoy's *Family Happiness* (1859), in which the Russian heroine Masha nearly succumbs to an Italian marquis while at a European spa without her husband.

Lofty Art, Unworthy Tourists

Anna and Vronsky carry into Italy high society's morally tainted, recreational engagement with art. When the reader first sees them in a "little Italian town," they have been touring Europe for three months, most recently in Venice, Rome, and Naples. At this point, they believe that have seen everything they need to see, none of which is mentioned. Moreover, in a reminiscence of *Lucerne*, Tolstoy dismisses sightseeing as something the English alone are shallow enough to enjoy: being a "Russian and an intelligent man," Vronsky cannot grant sightseeing the "inexplicable importance" it has for Englishmen (5.8). And besides, Vronsky has already visited Italy, as he remarks when he first meets Levin (1.14). Desperate to fill time, Anna and Vronsky seize upon art. A draft of the novel underscores the role-playing. Beginning in Rome, states the author, Anna conformed to Vronsky's tacit desire to give the impression of being "well-to-do tourists" and "patrons of the arts."[40] Tourism's invitation to experiment with social

39 David Herman, "Stricken by Infection: Art and Adultery in *Anna Karenina* and *Kreutzer Sonata*," *Slavic Review* 56, no. 1 (Spring 1997): 16–19, 24–25.

40 Tolstoy, *PSS*, 20:396.

identity ostensibly captivates Vronsky more than Anna (whose flair for theatricality is vividly dramatized in other episodes of the novel).[41] Enjoying the vacation freedom to discard his uniform, Vronsky first dresses in ordinary civilian clothes.[42] After taking up painting and becoming fascinated by Italian life of the Middle Ages, he sports a medieval-style cape and floppy hat. The adulterers' rented palazzo serves as a stage set, where Vronsky harbors the "agreeable illusion" that he has become a different, more interesting person than he was at home.[43] Living amidst the frescoes, mosaic floors, stucco moldings, urns, a guidebook-marked Tintoretto, and other paintings, Vronsky no longer thinks of himself as "a Russian landowner" or "chief equerry without a post" but rather as "an enlightened amateur," a "patron of the arts," and "a modest artist himself," having "renounced the world, connections, ambition for the woman he loved" (5.9).

Set against the villa of the idle tourist-dauber, Mikhailov's studio is a chronotopic site of artistic authenticity emblematic of times past, when Russian painters generally revered Italy. Prior to taking Anna and Vronsky to see Mikhailov's paintings, the windbag art historian Golenishchev describes him as a representative of "the same old Ivanov-Strauss-Renan attitude towards Christ and religious painting" (5.9), a reference to Russian controversy about historicized interpretations of Jesus (discussed further below).[44] Moreover, both Golenishchev and Anna nurture Vronsky's illu-

41 On Anna's theatricality, see David A. Sloane, "Anna Reading and Women Reading in Russian Literature," in Knapp and Mandelker, *Approaches to Teaching Tolstoy's* Anna Karenina, 125–26; and Buckler, "Reading Anna," 131–36.

42 Vacation costuming in Italy already features in Tolstoy's comic drama *An Infected Family* (1863–64), *PSS*, 7:212–13. Contemplating a deluxe vacation in Italy, a Russian imagines how he and his wife will be *seen* there in stylist outfits: for her—"silk and velvet" on daytime outings and at the Florence opera in the evening—a décolleté gown, with him beside her in a "black English frockcoat."

43 My observations on Vronsky's dilettantism conform to the critical consensus exemplified in Rimvydas Šilbajoris, *Tolstoy's Aesthetics and His Art* (Columbus: Slavica Publishers, 1990), 153–54; and Alexandrov, *Limits to Interpretation*, 85–87, 222–24. An original but implausible interpretation appears in John Burt Foster, Jr., *Transnational Tolstoy: Between the West and the World* (New York and London: Bloomsbury Academic, 2013), 25–26: on this view, Vronsky's medieval trappings imitate "what he must have seen in Italian paintings from a period when some of this art was indebted to Byzantine mosaics," and thus intimates a "glimmering awareness" of Russian icons' and Western painting's "shared source in Eastern Christianity." This presumptive speculation ignores Tolstoy's irony.

44 David Friedrich Strauss's (1808–74) book *Das Leben Jesu* (1835–36) acquired notoriety in Russia as of the 1830s when Ivan Kireyevsky made it a cautionary example of the "fallacious" Western principle of "formal reason:" see Nicholas V. Riasanovsky,

sion that he is already producing paintings as good, if not better than what Mikhailov has labored for years to achieve. Tolstoy, however, makes the professional painter the conduit of his own conceptions of artistic creativity and the complexities of interpreting a work of art.[45] As underscored in the Italian episode, painting operates in *Anna Karenina* as literature's "supporting sister art," creating metaliterary parallels between Mikhailov's portrait of Anna and Tolstoy's novel as a verbal "picture" of the heroine.[46] In professional fellowship with the author, Mikhailov mentally formulates the damning metaphor of Vronsky as a repulsive dilettante, caressing and kissing "a big wax doll" before the eyes of a man with his beloved (5.13).

Critics have speculated about Mikhailov's prototype without interrogating his unexplained choice to work in Italy. The consensus holds that Mikhailov was "probably modelled on Kramskoy," who painted Tolstoy's portrait at Yasnaya Polyana in 1873.[47] As was generally true of the Wanderers, Kramskoy was a semi-educated, socially insecure provincial.[48] Mikhailov's plebeian appearance, limited education, and defensive aloofness toward the traveling aristocrats conform to that profile. Kramskoy's aesthetics are equally relevant. At the time he did Tolstoy's portrait, he had recently exhibited his *Christ in the Wilderness* (1872), an image of an intensely contemplative Jesus sitting alone on a boulder in a bleak landscape. Although an agnostic, Kramskoy attributed a quasi-religious mission to art and said

---

Russia and the West in the Teaching of the Slavophiles: A Study of Romantic Ideology (Cambridge, MA: Harvard University Press, 1952), 96. Ernest Renan (1823–92) was a Hebrew scholar, philologist, critic, and historian of religion whose *La vie de Jésus* (1839–40) denied the divinity of Christ. The book was banned in Russia, but as librarian at the Russian national library, Strakhov obtained a copy for Tolstoy in 1878: see Christian, *Tolstoy's Letters*, 1:309n5, 322n1.

45 E. N. Kupreianova, "Vyrazhenie esteticheskikh vozzrenii i nravstvennykh iskanii L. Tolstogo v romane *Anna Karenina*," *Russkaia literatura*, no. 3 (1960): 126; and Svetlana Evdokimova, "The Drawing and the Grease Spot," *Tolstoy Studies Journal* 8 (1995–1996): 34–41.

46 Caryl Emerson, "*What Is Art?* and the Anxiety of Music," *Russian Literature* 40, no. 4 (November 1996): 443. The major study of ekphrasis is Mandelker, *Framing Anna Karenina*.

47 John Bayley, *Tolstoy and the Novel* (New York: Viking Press, 1966), 234. See also Justin Weir, "Tolstoy Sees the Truth but Waits: The Consequences of Aesthetic Vision in *Anna Karenina*," in Knapp and Mandelker, *Approaches to Teaching Tolstoy's* Anna Karenina,176; and Jefferson J. A. Gatrall, "Tolstoy, Ge, and Two Pilates: A Tale of the Interarts," in *From Realism to the Silver Age: New Studies in Russian Artistic Culture: Essays in Honor of Elizabeth Kridl Valkenier*, ed. Rosalind P. Blakesley and Margaret Samu (DeKalb: Northern Illinois University Press, 2014), 83.

48 On the Wanderers' social insecurities, see Valkenier, *Russian Realist Art*, 13–17.

that he wanted this painting to inspire people "not to recede an inch before evil," when confronting moral dilemmas.[49] If he articulated those views to Tolstoy, he may well have influenced the representation of Mikhailov and his painting *Pilate's Admonition to Christ*. Furthermore, although it is not clear when Tolstoy first saw *Christ in the Wilderness*, in 1894 he would designate it "the best [painting of] Christ that I know." [50]

But on the other hand, Ivanov rather than Kramskoy corresponds to Mikhailov's going to work in Italy and having no nationalist axe to grind against Raphael. Tolstoy's painter believes that *Pilate's Admonition* is a truly original, authentic work sprung from the depths of his soul but that it is not "better than any of Raphael's" images of Christ (5.10). This supranational, cosmopolitan spirit of artistic fellowship rather than competition points to Ivanov. As a gold medalist with a stipend from the Academy, Ivanov went abroad in 1830, a time of strong Russian "artistic allegiance to Italy." Ivanov sought out Titian in Venice, Giotto and Masaccio in Florence, and spent twenty-eight years in Rome producing *The Appearance of Christ to the People*, a painting whose prototypes "were entirely Western, ranging from the Raphael/Romano *Transfiguration* at the Vatican to various works by Poussin and the Nazarenes."[51] First exhibited in Petersburg in 1858, the work stirred hot debate. The Slavophile Khomyakov hailed it as the achievement of "a sacred artist" who had captured the "transcendent" Russian Orthodox Christ. For their part, Chernyshevsky and Herzen contended that by the end of his life Ivanov had lost religious faith, adopted Strauss's view of Christ, and aspired to make painting advance the cause of socialism.[52]

While resisting such tendentious political agendas, the Wanderers felt social and moral obligations that they believed required working at home, and they had a parochial, nationalistic grudge against Italy in particular. Their movement originated in revolt against the official tutelage of the Academy and its tradition of privileging Italy as a training ground and source of subject matter. Kramskoy declared in 1864 that the Russian artist "should finally stand on his own feet in art. It's time to throw away

49 Ibid., 116–17.

50 Letter to P. M. Tret'iakov, November 14, 1894, *PSS*, 67:175.

51 Bowlt, "Russian Painting," 117, 126–28.

52 Laura Engelstein, *Slavophile Empire: Imperial Russia's Illiberal Path* (Ithaca and London: Cornell University Press, 2009), 158–71. See also Valkenier, *Russian Realist Art*, 15, 20; and Pamela Davidson, "Aleksandr Ivanov and Nikolai Gogol': The Image and the Word in the Russian Tradition of Art and Prophecy," *Slavonic and East European Review* 91, no. 2 (April 2013): 167–68, 201–2.

the foreign diapers—thank God, we've already grown a beard but are still walking in an Italian toddler's harness. It's time to think about the creation of our own Russian school, our own national art!"[53] On his museum tour of Italy in 1876, Kramskoy "passed through Venice at night," saw nothing, and never went back. He "did not like Rome," where he turned an indifferent eye on everything except "some antiquities," the *Apollo Belvedere*, and "ancient busts" in the Vatican.[54] In the opinion of the painter Grigory Myasoedov, Russian artists could skip Italy entirely because "Italian art, with all its 'Raphaels,' was 'dead'"; and anyway, he added, the Hermitage has the best samples of foreign art.[55] The landscape painter Shishkin chipped in, to declare Italian scenery "too sweet."[56] As of the early 1860s, Stasov stoked the flames with a barrage of articles that consigned "all preceding art—whether it was Raphael or Bryullov—to the dustbin of history."[57] That campaign was going strong in 1875. Shortly before the Italy episode of *Anna Karenina* appeared, Stasov published excerpts from letters that Repin had sent him from Rome. Raphael figured there as a nauseatingly "infantile" painter, and Michelangelo's *Moses* was proclaimed the only art worth seeing in Italy.[58] Carrying Stasov's seal of approval, those opinions outraged Russian art-lovers, including Dostoevsky who jotted in a notebook: "The Repins are idiots, Stasov—even worse."[59] Another reaction was Turgenev's *Virgin Soil* (1877) caricature of Stasov as the art critic Skoropikhin, "hissing on the boil."[60]

Anna Karenina, too, opposes the militant cultural nationalism of Stasov, no less than the anti-aesthetic tendencies of Russian radicalism. In the 1850s, when Chernyshevsky began maligning upper-class tourists and their passion for Italy, Tolstoy's writings upheld Romanticism's Byronic construct of Italy as "mother of the arts," especially music (the Tyrolean busker of *Lucerne*, the violinist of *Albert*).[61] Tolstoy was not ready to renounce that

53 Dianina, *When Art Makes News*, 164. For Kramskoy's later concessions to Raphael, see Valkenier, *Russian Realist Art*, 16.

54 Kramskoi, letter to P. M. Tret'iakov, April 23, 1876, *Pis'ma, stat'i*, 1:324–25.

55 Dianina, *When Art Makes News*, 158.

56 Ely, *This Meager Nature*, 175.

57 Valkenier, *Russian Realist Art*, 19.

58 V. V. Stasov, *Izbrannye sochineniia*, 3 vols. (Moscow: Iskusstvo, 1952), 1:266–67; and the half-hearted apologia "Priskorbnye estetiki" (1877), ibid., 288–89.

59 Dostoevskii, *PSS*, 25:390–91.

60 Turgenev, *SS*, 4:183.

61 *Albert* and drafts for the story cast romantic light on *Lucia de Lammermoor*, *La Sonnambula*, *La Traviata*, "divine" Rossini's *William Tell*, Paganini's "Venetian Carnival," and Mozart's *Don Giovanni*—material that reflected Tolstoy frequenting the

vision in *Anna Karenina* and perhaps never fully did so (into old age he had hanging at Yasnaya Polyana a set of five icon-like reproductions of the figures in the *Sistine Madonna*: the upper half of the Virgin holding the child; St. Sixtus, St. Barbara, and each of the two cherubs).[62] Lofty visual art, necessary to somebody's life, exists in *Anna Karenina* in Italy alone.[63] This, however, is a matter of what Bakhtin termed "surplus" knowledge that an author keeps to himself.[64] Anna and Vronsky have not gained the insight that Tolstoy conveys to the reader. The adulterers' trip has in fact taught them nothing. They buy Mikhailov's cute-sounding painting of two boys fishing. They take home narcissistic souvenirs (the portrait of Anna, a photo of Vronsky in Rome), and Anna cherishes the memory of watching horse races on the Corso (5.31). But they have achieved no moral self-awareness, an outcome that mocks the penitential pilgrimage that Anna had imagined in the delirium of puerperal fever: she shall "go to Rome" in emulation of the sinful woman who became a "holy martyr" seeking redemption in the "deserts" (4.17), an allusion to the fifth-century St. Mary of Egypt, a prostitute who converted to Christianity and was venerated in the Orthodox Church.[65] Tolstoy's unedified adulterers establish a touristic life at home. At first they live in one of Petersburg's "best hotels," where they feel isolated as in a "foreign city" (5.28); and when Dolly visits them at Vronsky's country estate, she finds an alienating array of French and English imports and an overall luxury befitting the "best hotels abroad" (6.18).[66] The estate resembles a "vacation resort," where Anna first appears astride a horse, an

opera in Petersburg at the time: see Tolstoi, *PSS*, 5: 39–40, 157–58. Note also the draft of *Childhood* in which Nikolenka's grandmother tells him that Italy is *la patrie des poètes* ("the homeland of poets"): Tolstoy, *PSS*, 1:192. On the importance of music in the *paese felice* cult, see Deotto, "Materialy dlia izucheniia ital'ianskogo teksta," 198, 208–9.

62 Illustration "Au travail," in *Tolstoj, 1910: Échos, résonances, interprétations*, special number of *Revue des études slaves* 81, no. 1 (2010), between 32 and 33.

63 Cf. the stimulating but undeveloped assertion that *Anna Karenina* anticipates *What Is Art?* in expressing Tolstoy's "distrust for art that is not an involuntary and necessary part of the life of the individual and society," in Bayley, *Tolstoy and the Novel*, 234.

64 Gary Saul Morson and Caryl Emerson, *Mikhail Bakhtin: Creation of a Prosaics* (Stanford: Stanford University Press, 1990), 53–54, 241–42.

65 Pevear-Volokhonsky translation of *Anna Karenina*, 829–30n17.

66 On such details contributing to Tolstoy's representation of "alien residences," see Suzanne Osbourne, "When a House is not a Home: The Alien Residences of *Effi Briest* and *Anna Karenina*," *Tolstoy Studies Journal* 5 (1992): 67–69, 71–72.

image conforming to Victorian literature's trope of the "fast" woman, the "provocative equestrienne."[67]

Irresistible Italy

Tolstoy's denial of Italy's capacity to enhance Russian tourists' lives met resistance in a variety of publications encouraging acculturation trips to Venice, Florence, Rome, and other Italian cultural centers. In May 1875, roughly a year before the Italy episode of *Anna Karenina* appeared, the *Russian Herald* featured Henryk Struve's (1840–1912) "Art and Positivism," an essay arguing that art operates as religion to elevate the soul. A Polish professor of aesthetics, art history, and ethics at Warsaw University, Struve maintained that to obtain the spiritual uplift, one must learn to discriminate between *La Belle Hélène* and Beethoven's *Eroica*, for instance. But how is sophisticated taste acquired? Glossing over the question, Struve recommended learning to appreciate his world-class canon. Nothing Russian made the cut: Homer, Dante, Shakespeare, Milton, and Goethe; Bach, Gluck, Haydn, Mozart, and Beethoven; Leonardo, Michelangelo, Raphael, Titian, Correggio, and Kaulbach for painting; the *Venus de Milo*, the *Apollo Belvedere*, Michelangelo's *Moses*, and statues by the Danish sculptor Bertel Thorvaldsen; and finally, the Parthenon, Gothic cathedrals, and the Alhambra.[68] Whereas *Anna Karenina* linked the elite's foreign cultural bias to immorality, "Art and Positivism" gave it a spiritual aura, and dominating Struve's list for the visual arts were guidebook-starred attractions on the Italian tourist circuit.

Contemporaneous Russian *récits de voyage* concurred with Struve's conventional assumptions about Italy as the top destination for uplifting cultural tourism. "From a Trip to Italy" was especially intriguing in light of Strakhov's Slavophilism. An anti-Western orientation found expression in his contributions to Dostoevsky's native-soil program. In 1868 Strakhov had also argued that the battle of Borodino, as depicted in Tolstoy's *War and Peace*, demonstrated how Russia's "concept of the nation" triumphed over the Napoleonic "cosmopolitan idea" that rested on "general principles" used to legitimate "violence and the murder of peoples." [69] Furthermore, in

67 Edwina Cruise, "Tracking the English Novel in *Anna Karenina*: Who Wrote the English Novel that Anna Reads?" in Orwin, *Anniversary Essays on Tolstoy*, 171–74, 177.

68 G. Struve, "Iskusstvo i pozitivizm," *Russkii vestnik* (May 1875): 360–65, 370, 374–79.

69 N. N. Strakhov, *Kriticheskie stat'i ob I. S. Turgeneve i L. N. Tolstom (1862–1885)* (St. Petersburg: Tipografiia brat. Panteleevykh, 1885), 268. So also the intensely

sympathy with Pan-Slavism, Strakhov contributed "From a Trip to Italy" to a literary collection marketed to aid civilian Serbian victims of the ongoing war against the Turks in Bosnia and Herzegovina. His travel account, however, left politics at home.

Strakhov's tourist persona sets out to explain "how Italy conquered me." During the month and a half he spent roaming Venice, Naples, Rome, and Florence with a Baedeker, spells of bad weather prevented his seeing much of that "translucent air" that "we denizens of the North know only from postcards." But satisfactions of the trip far outweighed that disappointment. At first, writes Strakhov, a traveler feels attached to home and seeks to stay in touch through letters (the Karamzin precedent). Already on the train, however, the journey begins liberating "one's mind, one's self" from quotidian existence; and after two or three weeks, "the tie has completely snapped. You wake up free of Petersburg in thought and feeling; your soul is utterly pure, utterly free. And all around you—in Rome, let's say—2,500 years of history, the most thunderous in the world; all around you—an assemblage of art works unique in their abundance and beauty. Come and look," adds the author in guidebook parlance, "and if you still do not understand what art is and wherein lies its essence, then it is unlikely you will ever have a better time to do so" (75–76, 78). Highlights of Strakhov's art tour are the San Martino monastery in Naples, Fra Angelico in the San Marco monastery in Florence, and the Uffizi, which he explores with the aid of Taine (an authority only subsequently mentioned in *Anna Karenina*). Quoting a long extract from *Voyage en Italie*, Strakhov seconds Taine's representation of the art museum as an enclosure "suspended in time and space, suspended from real life" (89–92). The self becomes a purely aesthetic receptor, pleased to consume beautiful objects from diverse times and places and to forget existence outside the museum. In sum, Strakhov's account performs ecstatic aesthetic escape from morally and politically messy reality, joyful release from home and the home-bound self, and satisfaction with the railroad as an enabler of those experiences.

As if that were not enough to make Tolstoy "sick," Strakhov falls into Byron's habit of degrading his homeland and compatriots. Taking St. Isaac's cathedral as the exemplary case, Strakhov's tourist persona prefers Italian architecture to what he knows in Petersburg. In Italy, he asserts, cathedrals

---

anti-cosmopolitan views in Strakhov, *Bor'ba s zapadom v nashei literature: Istoricheskie i kriticheskie ocherki*, 3 vols. (St. Petersburg: Tipografiia S. Dobrodeeva, 1882, 1883, 1896), 2: 39, 3: 293–94.

and other buildings are often plain on the outside (a dubious generalization) but always resplendent within. In Russia, he writes, it is just the opposite (Strakhov had apparently never visited the Hermitage). "From a Trip to Italy" etched an equally skewed North-South dichotomy of national character. In Naples he proclaims that "politeness is an immutable characteristic of Italians." The gracious Neapolitans appear happy, serene, and unlike Russians, they still look young in their fifties. The blessed Italians' immediate foils are a pair of miserable-looking, aged tourists who turn out to be Russian. But only after the traveler goes home, does the contrast fully sink in. Back in Petersburg, he lands amidst a sour-faced, bustling herd of pedestrians. Furthermore, after the bay of Naples and the magnificent architecture throughout Italy, Petersburg seems drably Germanic. "And in addition, all the streets are under repair, on every one of them construction and painting are underway, scaffolding and wheelbarrows are everywhere, the canals are full of barges loaded with bricks and limestone, and not a single person is out walking." Goodbye vacation. Having left behind the "elegance, grace, luxuriance, and majesty" that is Italy, the traveler has come home to the dispiriting workaday world (77, 81, 82, 88).

A variant of Strakhov's metaphor of Italy's conquering the tourist figures in the ecclesiastic Dmitry Kasitsyn's (1838–1902) "Travel Observations: From Berlin to Rome," a two-part account published in the *Russian Herald* in April 1876 (alongside the Italy episode of *Anna Karenina*) and in February 1877. While officially abroad to study religion during his two-and-a-half month stay in Rome, the author comes across as a museum hound. His purview is supranational, truly cosmopolitan, valorizing the foreign without debasing the national. He quotes Goethe's *West-östlicher Divan* on cross-pollination between the Orient and Occident, and he inserts Russia into that framework: "Venice and Petersburg, San Marco and Vasily the Blessed, Moscow and Genoa, Rome and Kiev all meet and unanimously profess '*Artem non odit nisi ignarus*' [only the ignorant despise art]," the motto emblazoned in gold letters over the entrance of Berlin's Neues Museum.[70] Art knows no national boundaries. Italy, all the same, is *primus inter pares*. Quoting Schiller's *Maria Stuart*, Kasitsyn asserts that only after stepping on "Italian soil" does one feel the full "force of art." Great art wields conquering

70 D. Kasitsyn, "Iz nabliudenii v Rime," *Russkii vestnik* (February 1877): 509–10. On this author's advocacy of a Russian national museum, see Katia Dianina, "The Return of History: Museum, Heritage, and National Identity in Imperial Russia," *Journal of Eurasian Studies* 1, no. 2 (2010): 117n37.

power, and nowhere as mightily as in Italy. Art fails to overpower Anna and Vronsky, but Kasitsyn invites Russian tourists to test the "force" for themselves. He also names guidebooks that might help them: Baedeker and the obscure Meyers's *Italy in 50 Days*.[71] Finally, "From Berlin to Rome" resists Tolstoy's symbolism of the railroad. For Kasitsyn, no less than for Strakhov, the train is simply the most convenient, rapid way to reach Italy.

The *Russian Herald* showcased the allure of Venice in particular in Victor Tissot's (1844–1917) "Travel Impressions: A Journey through Lombardy and Austria," translated from French. Identified for Russian readers as a "famous tourist," Tissot was a Swiss journalist, professor, and travel writer whose *Voyage dans le pays des milliards* (1875) had become a best-seller in Western Europe.[72] The "Venice" section of his travelogue tawdrily imitates Byronic tropes of the city as a woman possessed by the man traveler whose heart she has won.[73] "Travel Impressions" begins with the author's arrival by boat at 7:00 am, when the sunrise makes Venice glow with the blush of a young woman "caught naked in the bath."[74] Just passing through, the author devotes himself entirely to *flânerie*. Street life is a series of "picturesque pictures" (*zhivopisnye kartiny*), local women are prototypes of Veronese and Titian beauties, Jews in the ghetto are Rembrandts, and all Venice appears a "stage-set for a Shakespearean tragedy" of love. In her "veil of azure air," the "queen of the seas" promises erotic adventure at every turn (768–73). The aestheticizing tourist eye multiples bromides fast and furiously. And yet in sponsoring Tissot, Katkov's journal was contributing to revival of the *Venezia la bella* tradition that the Kovalevsky-Pisarev duo had sought to overturn with descriptions of the city as a "slattern" and "cesspit."

To conclude this survey of selected *Russian Herald* material, let us consider "Russian Diplomat-Tourists in Italy in the 17th Century." The author was Professor Alexander Brikner (1834–96), a Baltic German historian specializing in the history of the westernization of Russia. The tone and style of Brikner's scholarly essay could not have differed more

---

71 D. Kasitsyn, "Iz dorozhnykh nabliudenii ot Berlina do Rima," *Russkii vestnik* (April 1876): 717, 724.

72 Editorial note on Victor Tissot, "Putevye vpechatleniia: Poezdka po Lombardii i Avstrii," *Russkii vestnik* (August 1877): 767. Tissot's other books included *La Russie et les russes: indiscrétions de voyage* (1882) and *De Paris à Berlin: Mes vacances en Allemagne* (1886). As editor in chief of the *Le Figaro* of Paris during 1888–93, he inaugurated the newspaper's literary supplement.

73 On this trope and its extension in Henry James, see Buzard, *The Beaten Track*, 132–38.

74 Tissot, "Putevye vpechatleniia," 767.

from Tissot's *récit de voyage* but nonetheless had a heartening message for Russian tourists worried about properly responding to the "queen of the seas" and other marvels of the West. Brikner was reporting his research into travel diaries of Russian contemporaries of Pyotr Tolstoy. The sample included Prince Boris Sheremetev but also an anonymous official in close contact with Francis Lefort during Peter I's Grand Embassy. After finishing his assignment in Holland, this traveler followed "his own desire" to Italy, where he spent nearly four months in Venice and visited Bologna, Florence, Rome, Livorno, Genoa, and Milan.[75] Although Brikner considered that unofficial trip the most purely touristic, his basic concept of the "diplomat-tourists" dignified tourism in general, and this happened within months of Lavrov's stereotyping "the tourist" as a traveling idiot, morally and politically clueless. Speaking of his entire sample, Brikner stated that "our travelers [...] were insufficiently prepared," they had not mastered foreign languages, for example, and so their recorded impressions were "superficial and scanty." Nonetheless, "they learned a great deal." In Italy and elsewhere, they saw "works of art, scientific collections," monuments, and luxury items unknown in Russia at the time; they observed festivals, customs, and religious practices.[76] In etching a program of pleasurable, passive learning, Brikner revived the pre-emancipation assumption that Russians could osmotically accrue benefits from visiting Italy. As Melgunov had put it, encountering the art and scenery of Italy was bound to have a good impact on Russians and could even inspire them to further the Great Reforms. Dostoevsky expressed a compatible idea about the *Apollo Belvedere*. Tolstoy, on the other hand, never made such concessions.

Grainfield Contributions

After Katkov ended publication of *Anna Karenina* with the heroine's suicide, Tolstoy considered *Grainfield* as one of the alternative venues for part eight, the novel's final nineteen chapters.[77] Just as Dostoevsky had advertised the *Citizen* in *Grainfield*, so too was Tolstoy evidently drawn to the magazine's large circulation, despite its bourgeois-friendly editorial program which encouraged Russian tourism, nurtured the attraction to Italy in

---

75 A. G. Brikner, "Russkie diplomaty-turisty v Italii v XVII stoletii," *Russkii vestnik* (March 1877): 41–42.

76 Ibid. (July 1877): 6–7.

77 Tolstoi, letter to N. N. Strakhov, May 21–22, 1877, *PSS*, 62:326.

particular, and recommended trains as facilitators of vacationing. An 1877 article on Naples relied heavily on clichés befitting an envisioned audience of *nouveaux riches*, culturally lagging but eager to acquire the sheen of aristocracy. The author explained that Italy's status as a "terrestrial paradise" has made it the object of a "cult" among foreigners with a passion for art and beautiful scenery. And thanks to the railroad, Italy is now more accessible than ever. Its hotels have improved, bandits are no longer a threat, and the Naples *lazzaroni* are well dressed and ready to serve as guides for tourists. Go to Naples to see an Edenic panorama of scenery and villas worthy of the brush of Salvator Rosa! Vedi Napoli e poi muori!—translated into Russian for the sake of any stragglers.[78] Another *Grainfield* article publicized the railroad route from Naples to Pompeii, mentioned convenient hotels, and recommended the Naples national museum for its collection of artefacts excavated from the destroyed ancient town.[79] An illustrated piece from 1880 hailed a "railroad" to the top of Vesuvius. The enterprise of a banker with a concession from the Italian government, the mechanized transport actually consisted of cable-drawn wagons, but the point was capitalist progress serving tourist comfort: the "railroad" has eliminated all the former "difficulties and inconveniences" of the ascent to the crater. Sightseers no longer need to spend time trudging (or being carried) along steep, narrow paths, where every step sends up a cloud of volcanic ash, and sharp slag destroys footwear.[80]

Grainfield designated many other scenic sites, art, and monuments to see in Italy. The aforementioned treatment of St. Peter's as the "most magnificent of all Christian cathedrals" ushered the reader inside to gaze upon Michelangelo's *Pietà* and other art works. Additional tourist attractions featured in the magazine were Pisa for its "magnificent" cathedral and other architecture; the Roman Forum, Genoa, Tivoli, and Florence, ranked as highly as Rome for the "grandeur of its historical and artistic monuments. [. . .] Everywhere you look, you will see marble palaces, bronze statues, columns, frescoes."[81] Illustrations invited men and women readers to project

78 *Niva*, no. 43 (1877): 690–91, 694.

79 Ibid., no. 42 (1879): 831.

80 Ibid., no. 29 (1880): 588. The service was apparently short-lived. See Chekhov's account of his slog up and down Vesuvius (the ascent alone took three and a half hours): letter to his family, April 7/19, 1891, *PPSiP*, *Pis'ma*, 4:212–23.

81 *Niva*, no. 49 (1876): 831; ibid., no. 10 (1878): 170–71; ibid., no. 6 (1879): 117–18; ibid,. no. 7: 130, 133.

7. *Evening in the Environs of Florence.* What Russian travel-lover of means would not have wanted to visit Florence after seeing this panoramic, two-page spread in the mass-market magazine *Grainfield*?

themselves into the position of sightseers. A lady of fashion and two men with walking sticks gaze at a cascade at Tivoli.[82] A similar touristic situation was pictured in *Evening in the Environs of Florence*, an engraving of a work by the Viennese artist Albert Rieger (1834–1905). The image shows a man and an elegantly dressed woman, both on horseback, in the "lush greenery" of a garden on a hill above the city in moonshine (fig. 7). Looking toward his companion, the man gestures toward the panorama of monuments, including Brunelleschi's cathedral dome and the tower of the Palazzo Vecchio.[83]

The *Grainfield* coverage of painting in the seventies placed Russian national achievements in the shadow of world-class Italian art. In 1873 the magazine covered the Wanderers' much publicized maiden exhibit in Petersburg. The critic praised the portraits of Russian writers (including Vasily Perov's of Dostoevsky) and a Shishkin landscape but "saw nothing remarkable" in Kramskoy's "morbid" *Christ in the Wilderness*, a work the author admitted was widely considered the exhibit's "pearl."[84] On the other hand, another 1873 issue of the magazine raved about the "extraordinary

82 Ibid., no. 26 (1879): 512, 515.
83 Ibid., no. 51 (1879): 1020–21, 1030.
84 Ibid., no. 5 (1873): 80.

richness of the psychological motifs" of Leonardo's *Last Supper*, provided a full-page reproduction of the painting, and located several of his other works in galleries in Florence and Rome.[85] *Grainfield* also recommended Raphael's *Madonna della seggiola* in a manner even more redolent of a guidebook for neophytes: "If you cross the Arno from the Uffizi—that world-famous museum built by Vasari in 1574 and the finest ornament of Florence, then right before you will be the magnificent building that served as the Medici residence from the time of Cosimo I: this is the *palazzo* Pitti, the Pitti palace, one of the treasure-troves of Florentine art and Italian art in general."[86] An illustrated feature from 1876 described Michelangelo's *Last Judgment* as a masterpiece that along with the works of Raphael draws "masses of admirers" to Rome and arouses "the justified amazement of the whole world."[87] Using the universalistic idiom of aesthetic cosmopolitanism, the commentary insisted on the global preeminence of Italian art. The tendency culminated in an 1880 article with a two-page reproduction of the *Sistine Madonna*. Screaming italics informed readers that "*This is the greatest work by the greatest artist*." The author quoted Zhukovsky's 1823 advice on how to look at the work and added that Raphael's paintings are scattered among "all the European museums" (not a word here about Raphael in the Hermitage). In conclusion, *Grainfield* observed that in three years' time, "the entire cultivated world [*obrazovannyi mir*] will be celebrating the four-hundredth anniversary of Raphael Sanzio's birth."[88] The remark targeted Russians of means, receptive to the acculturation opportunity (or merely the snob appeal) of making the jubilee the occasion for a Raphael tour abroad.

In pitting the Levin-Kitty way of life against modernizing liberalism and tourist hedonism, Tolstoy attained his full stature as the national literature's chief debunker of Hazlitt's Byronic ideal of traveling "to be free of all impediments" and "inconveniences," traveling "to leave ourselves behind," and to reinvent ourselves as a "more enviable individual, all the time we are out of our own country." *Anna Karenina* unmasks this Romantic credo as self-delusion in counterpoint to the process of authentic self-realization that comes from the day-to-day striving to be a good Christian at home in

85 Ibid., no. 13 (1873): 197, 199–200.
86 Ibid., no. 52 (1873): 828.
87 Ibid., no. 48 (1876): 805, 810–11.
88 Ibid., no. 11 (1880): 212–13, 219.

Russia. Despite Vronsky's prominence as the tourist engaged in self-reinvention, Tolstoy's women most frequently provide the cautionary examples: Masha in *Family Happiness*, naïve Kitty at the German spa, and above all—Anna in Italy, free of the impediments that curbed full release of adultery's "criminal joy" at home (2.9). In the *Anna Karenina* ranking of travel, Levin's knowledge-building factory tour comes out on top, and curative tourism occupies the middle ground: it may meet genuine medical needs but may merely serve as a pretext for a holiday. Beyond redemption is the idle, Anna-Vronsky kind of travel undertaken in hopes of escaping personal problems. The adulterers have chosen tourism's reputedly most cultivated agenda, but they appear as dupes of fashion bearing the full brunt of Tolstoy's moral condemnation.

Anna Karenina nonetheless did not kill Italy's tourist allure. Just as much as the "little town" where Vronsky and Anna rent the palazzo, "Italy" in Tolstoy's novel is a word with no pictorial content. The abstractness left a blank for the reader's imagination to fill, and Russian ephemera of the time assisted that process through a proliferation of verbal "pictures" and actual images of tourist attractions such as *Evening in the Environs of Florence*. Those inducements to leisure travel constituted a counter-narrative to Tolstoy's censorious recounting of the adulterers' Italian "honeymoon." Moreover, *Anna Karenina* itself contributed to the aesthetic inducement by situating Mikhailov in Italy. Within the geographical confines of Russia in Tolstoy's novel, the consumption of art is mainly a bad, high-society ritual of worshipping the foreign to the near exclusion of the national, thus anticipating the Soviet idea of "cosmopolitanism" as the disparagement of Russian culture. Mikhailov in Italy, however, conveys a genuinely cosmopolitan consciousness of the universality of art. Anna and Vronsky prove incapable of learning from Mikhailov, and yet their story sustains Italy's age-old reputation as a land of exceptional aesthetic significance, where more sensitive Russian tourists might experience a profound cultural expansion.

Italy's hold on the imperial Russian tourist imagination continued to grow. Beginning with the formulaic pretense that readers already "know" the place about to be recommended, the 1881 *Grainfield* article "In Venice" conformed to the magazine's self-appointed mission to prepare the bourgeoisie for aristocratic touring: "Who has not heard about, who does not know the Queen of the Seas? [. . .] Who has not imagined nocturnal rendezvous on silent canals in the moonlight?" Although Venice has lost its political power, it "remains for the entire cultivated world a wondrous granary of art and eternal beauty," attracting masses of tourists every year.[89] Other publications of note were the Grand Duke Konstantin Romanov's "Venice" cycle of lyric poems (lots of dreamy gondola rides), and an abridged Russian translation of Henry James's "Venice" (first published in 1882).[90] Accompanied by twelve fine illustrations of architecture, street scenes, canals, and the doges' ceremonial barque, this was a classic essay on the exquisite pleasure of lingering long in Venice to savor the art in museums and churches. During his three-day stay in Venice in 1891, when Dmitry Merezhkovsky and his wife Zinaida Gippius were also there for the first time, Chekhov raved in his letters:

> For sheer enchantment, brilliance, and joie de vivre, I have never in all my life seen a more wonderful city than Venice. [. . .] Merezhkovsky, whom I met here, has gone mad with ecstasy. It's all too easy for a poor, humble Russian to go mad in this world of beauty, riches, and freedom. You want to stay here for ever, and when you stand in a church, listening to the organ, you want to convert to Catholicism.[91]

In Rome several days later, after visiting Florence and Bologna, Chekhov generalized his impressions: "Italy is the sole country where you become convinced that art does indeed reign supreme, and that conviction lifts your spirits."[92] Those letters summed up the aesthetic exaltation that would keep the Russian cultural elite flocking to Italy and Venice in particular throughout the rest of the imperial era.

89 Ibid., no. 41 (1881): 908, 910.

90 Grand Duke Konstantin Romanov [K. R., pseud.], "V Venetsii," *Vestnik Evropy* (September 1882): 287–90; and translation of Henry James, "Venetsiia (Stat'ia Genri Dzhemsa)," *Istoricheskii vestnik* 14 (December 1883): 576, 588–92.

91 Chekhov, letter to I. P. Chekhov, March 24/April 5, 1891, in *Pis'ma*, 4:201–202.

92 Chekhov, letter to his family, April 1/13, ibid., 4:210.

11

Tatars and the Tourist Boom in the Crimea: Markov's *Sketches of the Crimea* and Other Writings

Chekhov's exhilaration in Venice led him to scoff at the Russian cliché of the Crimea as a substitute for Italy. Having found accommodations and food in Venice "ten times cheaper than in the Crimea," he added that, in any case, "the Crimea is to Venice as a cuttlefish to a whale."[1] Many Russians might have agreed from the cultural point of view and yet defended the Crimea's reputation as the empire's Italianate garden by the sea. That old vision acquired new appeal to pastoral imagination as Russia embarked on industrialization under Alexander II. We see this in "A Husband's Confession" (1867), a story by Konstantin Leontev (1831–91), a fascinating author, critic, and religious thinker who served as a military surgeon during the Crimean War, pursued a diplomatic career (1863–73), and eventually became an Orthodox monk. The narrative is written as the diary of the Russian hero long resident in Ay-Burun on the Crimea's southern coast. Living in ear-shot of the bombardment of Sevastopol in 1854–55, he wants Russia to win the war, not only to uphold its military honor but also to block europeanization and industrialization of the Crimea. He assumes that if Russia loses, that will mean "goodbye, mountain paradise!"

---

1 Chekhov, letter to his family, March 25/April 6, 1891, *Pis'ma*, 4:204.

> The French and English will build railroads and factories everywhere here, there will be no end to the steamboats, they will trample everything Russian into the mud, they will deck out Tatars in frockcoats and ladies' jackets, and the stench of local newspapers will spread: "*Courrier de la Tauride!*" Goodbye, then, to the Crimea's wild, forgotten poetry![2]

By the 1870s Russians themselves were building the railroads. After the first train pulled into Sevastopol in 1875, there began a tourist boom that ultimately transformed the southern Crimean coast into Russia's answer to the French Riviera.[3] Among the anti-tourist reactions was that of the Symbolist poet Valery Bryusov (1873–1924). In memoirs written between 1911 and 1913, he recalled early childhood trips to the Crimea with his parents and then summer vacations with his wife in Alupka in 1898 and 1899: "Those were the last years before the Crimea became a bustling bazaar, a European-style resort crammed from Balaklava to Kerch with a swanky, tiresome crowd of people from Petersburg and Moscow. Alupka was still a genuine village, where there were many more Tatars than visitors (now, people say, it is just the opposite)." Long open to tourists, Vorontsov's estate at Alupka could still feel "secluded," and "big hotels" had not yet defiled the beaches.[4] Leontev and Bryusov were lamenting not only the despoliation of nature but also the vanishing of a pre-capitalist Crimean way of life, an outcome imagined in "A Husband's Confession" and recollected as fact in the poet's memoir.

Those authors' dismay about the vanishing of "genuine" Tatar ways of life deviated from the well-known Russian tendency to regard Tatars as backward Muslim objects of a civilizing mission. Successors of the Golden Horde of Genghis-Khan, the Tatar khans ruled the Crimea from 1427 into the late fifteenth century (independently at first, and then as vassals of the Ottoman Empire), and they regularly raided Slavic domains to the north for booty, slaves, and women for harems. Following four Russian invasions of the Crimea between 1771 and 1782, Catherine II annexed the territory in a relatively peaceful manner by pressuring the Tatar khan to abdicate.[5]

---

2 K. N. Leont'ev, "Ispoved' muzha," *Polnoe sobranie sochinenii i pisem*, 12 vols. (St. Petersburg: Izdatel'stvo "Vladimir Dal'," 2000–14), 2:356.

3 Mal'gin, *Russkaia Riv'era*, 142.

4 V. Ia. Briusov, *Iz moei zhizni* (Moscow: Terra, 1994), 171.

5 For detailed military history, consult Alan Washburn Fisher, *The Russian Annexation of the Crimea, 1772–1783* (Cambridge: Cambridge University Press, 1970). See also Fisher, *The Crimean Tatars* (Stanford: Stanford University Press, 1978), 54–69; and

During her triumphal tour of the territory in 1783, she was delighted with the scenery, the Greek antiquities, and her accommodations in the khans' old Bakhchisarai palace, redecorated in "the current Western European style."[6] But on the other hand, Catherine and her retinue of Russian and foreign dignitaries considered "indolence, prurience, despotism, and fanaticism" the dominant traits of Tatars, past and present.[7] Similarly, in his *A Journey across Tauris in 1820* (published in 1823), the classical scholar and diplomat Ivan Muravyov-Apostol (1762–1851) represented the Bakhchisarai palace as "an empty, enchanted castle" transporting the visitor to the *1001 Nights.* The Orient was benighted, however: thanks only to "victorious Russian arms" did the "dawn of enlightenment" begin in the Crimea.[8]

The Russian state's project to "civilize" the Crimea worked against the different-from-home kind of tourist's quest for scenic and cultural exotica. That tension, rather than the crowd phobia that Bryusov articulated, is this chapter's central concern. The investigation carries us into the reign of Alexander III (1881–94), a period of counter-reform and "nationalist imperialism"—an intensification of the hegemonic Russian stance toward the empire's minority peoples.[9] The ground was well prepared for that development. Patriotic rhetoric during the Crimean War enflamed imperial patriotism, underwriting ethnic Russian possession of the Crimea. The completion of the conquest of the Caucasus (1817–64), expansion into the Amur River region (1858–60), the suppression of the Polish insurrection, the piecemeal conquest of Central Asia (1865–95), the rise of Pan-Slavism and its outlet in the Russo-Turkish war (1877–78) strengthened that ideological orientation. And all the while, Russian nationalists were clamoring for official recognition of Russians as the dominant people of the empire.[10] Alexander III fulfilled their heart's desire. The Russian government first affirmed the "principle of ethnic supremacy" in relation to Jews, following

---

Edward J. Lazzerini, "The Crimea under Russian Rule: 1783 to the Great Reforms," in *Russian Colonial Expansion to 1917*, ed. Michael Rywkin (London and New York: Mansell Publishing, 1988), 123–25.

6 Simon Karlinsky, "Two Pushkin Studies," *California Slavic Studies* 2 (1963): 108.

7 Sara Dickinson, "Russia's First 'Orient'. Characterizing the Crimea in 1787," in David-Fox et al., *Orientalism and Empire in Russia*, 91–92.

8 Schönle, *Authenticity and Fiction*, 148, 152.

9 Wortman, *Scenarios of Power*, 2:217.

10 Maiorova, *From the Shadow of Empire*, 162–73; and Nathaniel Knight, "Imperiia napokaz: Vserossiskaia etnograficheskaia vystavka 1867 goda," *Novoe literaturnoe obozrenie* 51, no. 5 (2001): 111–31.

the 1881 pogroms in Ukraine.[11] More generally, however, Alexander III's reign was a period of ethnic and religious persecution, Russification in the Baltics and other borderlands, anti-Semitism in the highest echelons of state, and a "growing eurocentric feeling of superiority of the Russians towards the Asiatics" of the empire.[12] On the Russian home front, the revolutionary movement was quashed, fifteen major journals including *Annals of the Fatherland* were suppressed, the universities lost autonomy, court reforms were reversed, and the scope of the zemstvos reduced.

With respect to culture and tourism, the reign of Alexander III presents a mixed picture. On the one hand, the prosperous bourgeoisie now had an emperor whose preferences in culture and leisure matched their own. Elizabeth Valkenier even made a good case for calling Alexander III "the bourgeois tsar." He most notably co-opted the Wanderer movement. In their early years as rebels whose paintings so often exposed dark sides of the life of the common people, the Wanderers had made it "a point of honor [. . .] to turn down orders from the Court." During the eighties, however, that independent spirit gave way to reliance on the "bountiful, prestigious and jingoistic patronage" of Alexander III. Wanderer painters fell into a "comfortable groove of plain naturalism" and "bland depictions of the Russian scene" that shored up nationalism and found ready takers among the bourgeoisie. Architecture was mobilized as well. Churches, civic buildings, and railroad stations now imitated the allegedly "authentic" national spirit of old Muscovy, an aesthetic that Moscow businessmen also favored for their private residences.[13] Entertainment was another area in which Alexander III's tastes matched those of the bourgeoisie. Festivities for the people were a traditional part of coronation ceremonies. For that august occasion in 1884, Alexander III appointed as director of the entertainment Mikhail Lentovsky, the highly successful owner of a nightclub in a Petersburg pleasure garden that catered to the well-heeled middle class.[14] Lastly and importantly, Alexander III's public image had a bourgeois-friendly touristic dimension. In harmony with the paterfamilias aspect of his scenario

---

11 Wortman, *Scenarios of Power*, 2:237.

12 Andreas Kappeler, *The Russian Empire: A Multiethnic History*, trans. Alfred Clayton (Harlow, England: Pearson Education Limited, 2001), 207; and Wortman, *Scenarios of Power*, 2:217, 283.

13 Valkenier, *Russian Realist Art*, 124–27. See also discussion of the big market in arts and crafts meant to symbolize Russianness, such as the faux-folk nesting *matryoshka* doll, invented in 1891, in Dianina, *When Art Makes News*, 235.

14 McReynolds, *Russia at Play*, 215.

of power, the tsar usually spent one to several months with his wife and children in Livadia, the royal estate in the Crimea, a vacation destination increasingly dominated by the middle class.[15]

But while the prosperous bourgeoisie had much to celebrate under Alexander III, his reactionary politics created among liberal-minded Russians the despondency evoked by the customary characterization of the period as a time of "small deeds," lost hopes, and diminished expectations by comparison to the utopianism of revolutionary populism.[16] The stories and plays of Chekhov most famously captured the mood of "humdrum boredom and impotent aspirations" widely spread in the 1880s and 1890s in the lesser strata of the middle class.[17] Another symptom of dissatisfaction with Russian reality was an appetite for escapist reading, including historical novels and exotic travel literature.[18] Examples were Leskov's fictions of ancient Egypt and Byzantium, the popular historical novelist Daniil Mordovtsev's (1830–1905) travel account *On Spain* (1884), and Helena Blavatsky's travelogue *From the Caves and Jungles of Hindostan*, serialized in a Russian newspaper and a journal between 1879 and 1886.[19] Most significant for the history of tourism was the populist Gleb Uspensky's (1843–1902) 1883 account of his recent vacation in Georgia, to be examined here. Presuming to speak for all his disheartened educated compatriots, Uspensky asserted that "a worm of anxious, unrelenting ennui is devouring the Russian soul."[20] People were itching to get away.

The second edition of Markov's *Sketches of the Crimea* (1884) was the era's outstanding performance of vacationing to escape quotidian existence in central Russia. Dedicated to "my Tatar friends," the book reflected the years the author lived in Simferopol as a teacher and education administrator. Markov's tourist persona is exceptionally complicated, however, due to his decision not to do any retrospective editing. We thus observe him both before and after his valorizations of *embourgeoisement* in "From a

15 Wortman, *Scenarios of Power*, 2:281.
16 Billington, *The Icon and the Axe*, 435–36.
17 Mirsky, *A History of Russian Literature*, 347.
18 Billington, *The Icon and the Axe*, 439.
19 D. L. Mordovtsev, "Po Ispanii," *Vestnik Evropy* (January 1884): 73–118. On Leskov and Blavatsky, see respectively Dan Ungurianu, *Plotting History: The Russian Historical Novel in the Imperial Age* (Madison: University of Wisconsin Press, 2007), 153, 171; and Hokanson, "Russian Women Travelers in Central Asia," 2.
20 G. I. Uspenskii, "Melochnye putevye vospominaniia" ["Trifling Travel Recollections"], *Polnoe sobranie sochinenii*, 14 vols. (Moscow: Akademiia nauk, 1940–54), 8:197.

Foreign Clime" and "Bourgeois Ideals." The first book edition of the *Sketches* appeared in 1872, after most of the material was serialized in Russian journals as of 1867 under the title *Crimean Impressions: A Travel Diary*. Prepared to commemorate the centenary of the annexation of the Crimea, the second, 1884 edition contained the new concluding chapter "From the Southern Coast (Travel Letters)" but was otherwise unrevised. Reprinting in 1902, 1906, and 1911 left the text unchanged, even though Markov's political outlook had evolved. His novel *Black-Soil Fields* (1876–78) concerned a westernized Russian's dedication to farming, his rapport with peasants, and his disapproval of his neighbors' "cosmopolitan" practice of bringing French and English governesses to their country estates (unlike Levin in *Anna Karenina*, however, Markov's hero enjoys tourism in the West before his marriage and spends a joyous honeymoon in Switzerland).[21] *Black-Soil Fields* idealized central Russia's rural topography, whereas *Sketches* left unaltered what Markov retrospectively described as the "naive," unbounded passion for the Crimea's natural environment that he felt in his youth.[22]

More comprehensively than any other pre-Soviet writer, Markov packaged the Crimea for tourist consumption by the dominant Russians. The book's often ornate literariness made a big impression on the readership. Accentuating aesthetic engagement, Markov's prefaces presented the *Sketches* as an "artistic creation," a "travel album" recording "poetic impressions," rather than a guidebook dispensing practical information.[23] In addition to its literary appeal, the narrative also offered wide-ranging content that closely overlaps modern-day guidebooks. Highlights of the *Sketches* are the spectacle of scenery, battle sites of the Crimean War, Bakhchisarai, Chufut Kale and other cave towns, Chatyr-Dag, the Alupka palace and gardens, Yalta, Russian monasteries, and the remains of ancient Christian monuments (on the other hand, Markov's account says little about the Crimea's much-touted Greek antiquities).[24] Both content and style made the *Sketches* a good read to take along on vacation or merely savor as an

---

21 Markov, *Chernozemskie polia*, accessed November 2020, az.lib.ru/markow_e_l/indexvote.shtml.

22 Markov, *Ocherki Kryma*, 3rd ed. (1902), preface, xi.

23 Prefaces to the first and second editions in ibid., 2nd ed. (1884), i, vi.

24 On the elite Russian fashion for a "pilgrimage" to the Crimea's ancient ruins as of Pushkin's time, see Mara Kozelsky, *Christianizing the Crimea: Shaping Sacred Space in the Russian Empire and Beyond* (DeKalb: Northern Illinois University Press, 2010), 46.

armchair traveler. By the late eighties, general enthusiasm for the book had won Markov the reputation of the "Amerigo Vespucci" for the Crimea.[25]

The poetically rendered, eventful vacation of the *Sketches* achieves a romantic perfection of time and space "apart" from the commonplace. With the exception of the 1884 addendum, the book recounts a one-way trip. It begins with the author's journey from Moscow to the Crimea and ends with his return to his Crimean seaside dacha after his seven-day horseback "Odyssey" to the cave towns in the company of a Tatar guide. Elsewhere in the book, women, too, participate in active leisure. The concluding scene, though, separates masculinized "adventure space" (Homeric, no less) from home represented as the woman's realm.[26] Awaiting the author on the dacha balcony is a "feminine figure" in summery white silhouetted against cypresses.[27] The image of this apparently happy couple epitomizes the enjoyment of vacationing in the South, and no return to central Russia is foreseen.

Markov's unrevised tourist persona holds greatest political interest for the ambivalence that he articulates about the Russian state's drive to modernize the Crimea. He mounts a strong critique of ruthless imperialism but adheres to liberal imperialism, a position that precludes entertaining even the thought of Russia giving up the Crimea. While that is a foregone conclusion, Markov's conflicted persona interestingly wrestles with big questions of scenic and cultural authenticity. With its railroads, steamships, and proliferation of Russian hotels, villas, and dachas, is the advance of capitalist tourism enriching or impoverishing the vacation experience, especially for the middle-class tourist? And is the increased contact with westernized Russians improving or contaminating Tatars and their "authentic" culture?

Tatars from Pushkin to 1870

Khan Mengli-Girei of Pushkin's *The Fountain of Bakhchisarai* is not merely Russian literature's most famous Tatar but probably the only one most

25 Anna Moskvich, *Prakticheskii putevoditel' po Krymu s prilozheniem*, 2nd ed. (Yalta: Tipografiia N. P. Petrova, 1889), 9. See also the glowing review by Professor Pypin in *Vestnik Evropy* (January 1884): 424–25.

26 On the general issue of gendered spatial division, see Helena Goscilo, "Women's Space and Women's Place in Contemporary Russian Fiction," in *Gender and Russian Literature: New Perspectives*, ed. Rosalind J. Marsh (Cambridge: Cambridge University Press, 1996), 341–43.

27 Markov, *Ocherki Kryma*, 556–57 (subsequent citations appear parenthetically).

readers could name.[28] He is a conventional oriental potentate whose story mixes Romantic melodrama with eighteenth-century decorum on the order of Mozart's *Abduction from the Seraglio.* Girei first appears smoking an amber pipe as he broods about his newly abducted Polish princess Maria. The situation parallels that of Mozart's Pasha vis-à-vis Constanza: both Asian despots desire the voluntary submission of their Christian captives. Maria intimidates Girei, who allows her to keep her crucifix and gives her a private bedroom. Before Maria's falling captive in a recent raid on Poland, Girei had been living in an improbably monogamous state of "endless rapture" with Zarema, the harem queen abducted as a child from Georgia. Zarema begs Maria to dispel the "criminal thoughts" that have caused Girei's "betrayal." In the fragmentary denouement that curtains the violence, Zarema kills Maria and meets the death of a murderer, thrown into the sea. Girei then resumes rampaging in "peaceful Russian villages" and the Caucasus. He is "bloodthirsty." But the poem's evocations of Tatar raids place no emphasis on the abduction of women and rely mainly on metaphor rather than pictorial detail (an engulfing "river," raging "fire," "orphaned" villages). Moreover, melodrama reaches farcical proportions after Maria's death. Girei betrays his grief in the heat of battle. Having hacked down foes, he stands stock-still, turns pale, whispers something, and may even let flow "a river of hot tears."[29] In recalling how one of his men friends ridiculed this passage, Pushkin publicly attributed it to artistic immaturity.[30]

The khan erects at the palace the funeral monument to Maria, "the fountain of tears" that provides the transition to the poem's splendid touristic finale in present time. The poet declares the Crimea an "enchanted realm." He roams the dilapidated Bakhchisarai palace, the gardens, and the old cemetery. On further rambles, he breathes in the fragrance of roses and fixes his "greedy gaze" on the "merry banks" of the Salgir River, the mountain Ayu-Dag, and the ocean "waves of Tauris."[31] Taking in scenery and

28 In the Russophobic novel *Kudeyar* (1875) by the eminent historian Nikolai Kostomarov (1817–85), a Muscovite nobleman flees the tyranny of Ivan the Terrible to go serve "the noble Crimean khan." Swearing allegiance to the khan, the defector vows to "accept Islam and become one of your Tatars." The novel's anti-Slavophile undertones won it a place in the prestigious liberal journal the *European Herald*: see Ungurianu, *Plotting History*, 132–33.

29 Pushkin, *SS*, 3:148, 153–56.

30 Pushkin, "Oproverzhenie na kritiki i zamechaniia na sobstvennye sochineniia," ibid., 6:343–44.

31 Ibid., 3:158.

the Islamic monuments, this coda was a powerful inducement to Russian leisure travel. The discourse-as-power logic has crowned Pushkin's persona the "conqueror" of the Crimea. As the "sovereign of all he surveys," he allegedly demonstrates that the Crimean landscape "now exists only for the purposes of Russian poetry."[32] However meaningful that argument may be, the poem's more socially significant achievement was surely to represent the Crimea as a place that real Russian state power has made secure for Russian tourism.[33]

"Bloodthirsty" Girei overshadowed Pushkin's Crimean travel romance in the mind of Belinsky, a critic of standing who probably influenced some Russian readers on this point. In 1844 Belinsky labeled the khan "an Asian barbarian" who does not molest Maria because he recognizes her civilizational superiority. Girei shows restraint, but this "still does not make him a human being, even though the animal in him has died, and he has ceased to be a Tatar *comme il faut*"—a "proper" Tatar.[34] Belinsky continued to exclude Tatars from humankind in 1846 while in the Crimea taking a cure for tuberculosis. Writing to Herzen, he reported that his trip had acquainted him with "three new nations: Crimean sheep, Crimean camels, and Crimean Tatars. I think they are different members of the same species, different generations of the same clan, so similar are their physiognomies."[35]

Russian publications of the 1850s contain both negative and positive images of Tatars.[36] In recounting her 1853 journey to the Crimea, Tur raved about the scenery but pronounced the first Tatar she saw "ugly," found the food nauseating when a dinner guest in a Tatar home, and belittled the Tatar host for speaking poor Russian.[37] On the other hand, Tatars were showered

---

32 Katya Hokanson, "Pushkin's Captive Crimea: Imperialism in *The Fountain of Bakhchisarai*," in *Russian Subjects: Empire, Nation, and the Culture of the Golden Age*, ed. Monika Greenleaf and Stephan Moeller-Sally (Evanston: Northwestern University Press, 1998), 147–48.

33 For an example of Pushkin's writings as a lens in a guidebook, see F. V. Livanov, *Putevoditel' po Krymu s istoricheskim opisaniem dostoprimechatel'nostei* (Moscow: Tipografiia T. Ris, 1875), 1–13, 20–37.

34 Belinskii, *PSS*, 7:379–80.

35 Ibid., 12:316.

36 For additional Russian and Western travelers' perceptions of Tatars prior to the Crimean War, see Brian Glyn Williams, *The Crimean Tatars: From Soviet Genocide to Putin's Conquest* (London: Hurst and Co., 2015), 1–4.

37 Evgeniia Tur, "Krymskie pis'ma," letters 1 and 2, accessed November 2020, https://lavanda.life/o-kryme/444-krymskie-pisma-evgenii-tur.html. Originally published in 1853–54 in *Sankt-Peterburgskie vedomosti* (*Saint Petersburg Gazette*).

with compliments in Prince Anatole Demidov's *Travels in Southern Russia and the Crimea, through Hungary, Wallachia, and Moldavia during the year 1837*, a work first published in 1853 in Russian and English. Married to a niece of Nicholas I, Demidov lived in Paris and undertook his journey primarily for mineralogical research. The combined narratives of the author and his fellow travelers present Crimean Tatars as "gracefully built men" who radiate intelligence, have excellent irrigation skills, and produce many fine commodities: felts, camel-hair cloaks, woolens, belts, saddles, slippers, and other "articles of supple, brightly dyed leather." Tatar hosts in Koutlak served Demidov and his companions a "skillfully prepared and good-naturedly offered" supper ("patriarchal in its simplicity") in an "extremely clean apartment" hung with tapestries. And needless to say, all these interactions with Tatars occurred amidst "truly enchanting" scenery that "Italy itself would not disavow!"[38] A similar appreciation of Tatar hospitality figured in the memoir that the Russian military artist Nikolai Berg (1823–84) published in the *Contemporary* in 1856. In recounting the horseback tour of the Crimea that he took after the fall of Sevastopol in the company of a French officer and a North African interpreter in French service, Berg spun an atmosphere of multiethnic conviviality around the hospitality the travelers enjoyed in a Tatar home full of local people.[39]

Appearing the same year as Berg's memoir were Russian lyric poems that marginalized Tatars to reclaim the Crimea for vacationing in a Pushkinian mode. A tourist-friendly notion of the Crimea as a Russian place was forged in the crucible of war-time rhetoric. Government proclamations, sermons, journalism, and imaginative literature established an "imperial patriotism" that "symbolically [anointed] the Crimean penisula with Russianness." Originally published in 1855 in the official military organ *Russkii invalid* (*Russian Veteran*), Tolstoy's "Sevastopol in December" played a major role with its rousing portrayal of heroism on a Homeric scale. The narrative evoked the Crimea's exotic landscapes but made the fearlessness of Sevastopol's inhabitants a sign of the town's "thoroughly Russian character."[40]

---

38 Anatole de Demidov [Demidoff], *Travels in Southern Russia and the Crimea, through Hungary, Wallachia, and Moldavia during the year 1837*, 2 vols. (London: J. Mitchell, 1853), 1:294; 2:4, 168–70, 258, 266.

39 N. V. Berg, "Iz krymskikh zametok," *Sovremennik* 58, no. 4 (April 1856), section 1: 143–45.

40 Maiorova, *From the Shadow of Empire*, 28–37 (quotes 36–37).

The postwar poetry of tourism enacted the pleasure of rediscovering scenic Crimea made newly interchangeable with the Russian nation. Resembling vacation snapshots, the lyrics comprising Vladimir Benediktov's (1807–73) cycle "Travel Notes and Impressions (The Crimea)" chart the poet's solitary visits of tourist attractions. Those include the mountain "Chatyr-Dag," "A House amidst Flowers: Alupka," "Oreanda" (soon to become a retreat of the Russian royal family), and "Bakhchisarai." In bidding the "enchanted realm" farewell, the poet underscores his vision of the Crimea as a Pushkin place, minus Tatars.[41] Alexei Tolstoy's cycle "Crimean Sketches" (1856) likewise depicts crowd-free sightseeing but features Tatars as tourist helpers. Reflecting the poet's trip that spring with his future wife and some friends, the lyrics represent an enamored couple's vacation. Their visits include the Baidar valley in moonlight, Greek antiquities, a settlement of Karaite Jews, and Chatyr-Dag. During their wanderings, the poet spots a Tatar with animals in a pasture. Other Tatars, however, are working for the tourists. After the Chatyr-Dag excursion, the poet orders the silent natives Hussein and Ali to help him take off his boots, fetch a samovar, build a fire, and prepare a picnic by the sea. As though speaking to serfs at home, the Russian vacationer addresses the Tatars in the familiar second-person singular, a detail contributing to the impression that the Crimea has a submissive native population ready to assure tourist pleasure.[42]

Mild-Mannered Tatars and Tourist Solitude

By the time those poems appeared, the Crimean Tatars had begun emigrating en masse to Turkey. During the Crimean War no less than the Russo-Turkish Wars of the eighteenth century, Russian authorities regarded Tatars as a potential fifth column.[43] Convinced that Tatars had committed treason

41 V. G. Benediktov, *Stikhotvoreniia*, 2nd ed. (Leningrad: Sovetskii pisatel', 1983), 185–203. See also Ia. P. Polonskii, "Noch' v Krymu" ("Night in the Crimea", 1857), *Stikhotvoreniia* (Leningrad: Sovetskii pisatel', 1954), 195.

42 A. K. Tolstoi, *Sobranie sochinenii*, 4 vols. (Moscow: Khudozhestvennaia literatura, 1963–64), 1:102–5, 107–9.

43 Brian Glyn Williams, *The Crimean Tatars: The Diaspora Experience and the Forging of a Nation* (Leiden: Brill, 2001), 141. On the complex religious motivations of the migrations, see Williams, "'*Hijra*' and Forced Migration from Nineteenth-Century Russia to the Ottoman Empire: A Critical Analysis of the Great Tatar Emigration of 1860–1861," *Cahiers du monde russe* 41, no. 1 (2000): 79–108. For cautioning against simplistic views of colonial victimhood, see investigation of the Crimean intellectual

during the war, Alexander II welcomed their departure. As the officer and military engineer Edward Totleben stated in his 1860 investigation, some Tatars had no doubt collaborated with the enemy but to a degree and in numbers disproportionate to the suffering that Russian military and civilian authorities inflicted on the general population.[44] While Totleben's report was not made public until 1893, a preliminary version appeared in 1861 under the title "The Persecution of the Tatars" in Herzen's London journal the *Bell*. A "damning exposé," the report detailed a prolonged process of Russian persecution that began with Cossacks given free rein to rob, rape, and murder Tatars at the outset of the war.[45] If more restrained, a chapter of Markov's *Sketches* serialized in 1867 also denounced Russia's treatment of Tatars. The author explained that before visiting the Crimea, he shared the common Russian view of Tatars as traitors. After talking to local people, however, he changed his mind. Summing up what he had learned, he asserted that the Tatars were the Crimea's "most honorable" ethnic group, overwhelmingly "loyal" to Russia during the war. They had been persecuted by Cossacks, Russian soldiers, officers, and "civil-service thieves" who cheated them out of land, property, and valuables. The Tatars were compelled "to abandon their homeland," and in their absence, "the Crimea has perished." It has become a land resembling "a house after a fire," "a desert" along "the shores of the Dead Sea" (111–16). By 1867 when Markov's essay appeared, over 190,000 Tatars had emigrated, leaving more than 780 villages empty and hundreds of mosques abandoned.[46]

In 1870 *Grainfield* evoked the mass emigration as a tourist-friendly event. One of the periodical's first issues retold a familiar tale of Tatars' transformation from "dreadful Mongols" of the thirteenth century to fortunate subjects of "Russia's wise ruler" Catherine. The magazine asserted that Catherine's enlightened policies had assured the preservation of Tatar monuments in Bakhchisarai: "Even today, if the ancients khans were to return to their native hearth and home, they would find little changed."[47] The article

---

Ismail Bey Gasprinskii (1851–1914) in Edward J. Lazzerini, "Local Accommodation and Resistance to Colonialism," in Brower and Lazzerini, *Russia's Orient*, 171–85.

44 E. I. Totleben, "O vyselenii tatar iz Kryma v 1860 godu," *Russkaia starina* 78 (June 1893): 532–46.

45 Mara Kozelsky, *Crimea in War and Transformation* (New York: Oxford University Press, 2018), 200–201, 252n8.

46 Fisher, *The Crimean Tatars*, 88–89; and Kozelsky, *Christianizing the Crimea*, 153.

47 *Niva*, no. 6 (1870): 86. Besides being redecorated for Catherine II, the palace had been renovated after a fire in the 1820s and after the Crimean War, when it served

eventually acknowledged the big demographic difference from "ancient" times but interpreted it as an invitation to Russian tourism: "Ever since the majority of Tatars emigrated to Turkey after the Crimean campaign, gardens are deserted and have gone to seed, and yet there is a certain charm in the solitude of this southern vegetation which is difficult to imagine without having seen it." Come see the "charm" of wild bushes and flowers growing unchecked in the absence of Tatar gardeners! And "over it all, like an overturned cup, is etched the low, round cupola of the [Bakhchisarai] mosque and the Gothic minaret from which the muezzin made his prolonged call to prayer." Clouds envelop "lilac" mountains, forests appear "dark blue," and "golden specks of light" dance in the "pure, translucent air" (87). Furthermore, welcoming hosts await visitors:

> The few remaining Tatars are a mild-mannered people [*smirnyi narod*], engaging in small-scale trade at markets, mainly in fruit. They also maintain saddle horses for people desiring to take excursions into the mountains. They haul wood from forested mountain slopes, work on tobacco plantations and in shops selling the tobacco, which we smoke for the most part under the name "Constantinople." Finally, they make fermented mare's milk and so forth. (87)

Providing psychological preparation for the tourist invasion that the railroad would enable, this article constituted an enticing vacation venue: a scenic paradise that one has to see to believe, exotic ancient monuments allegedly preserved in all their authenticity, and "mild-mannered" natives. These Tatars' cultural profile significantly differs from the one Russians drew for North Caucasian peoples. The latter made daggers, swords, silver knickknacks, and other things that Russians prized as tourist booty. On the other hand, the *Grainfield* Tatars are agriculturists and manual laborers, producing nothing for the souvenir trade. That profiling denied the Tatar creativity that Demidov observed (articles of "brightly dyed, supple leather," and so on). Some readers might have also remembered the ornamentalized Tatars that Gustav-Feodor Pauli (1817–67) contributed to the *Narody Rossii* (*Peoples of Russia*, 1862) album published as part of the

---

as a Russian military hospital. On its sorry condition after the fall of Sevastopol, see Ekaterina Bakunina, "Vospominaniia sestry miloserdiia Krestovozdvizhenskoi obshchiny (1854–1860 gg.)," *Vestnik Evropy* (May 1898): 75.

8. With their pale skin, gorgeous clothing, and elaborate jewelry, these Tatars coincide with two prominent aspects of Markov's widely read *Sketches of the Crimea*: the author's craving to see pre-capitalist Crimean exotica and his presentation of Tatars as an ethnic type more Caucasian than Mongol.

lavish celebration of the millennium of the Russian state.[48] While pictured next to a simple dwelling, Pauli's Tatar family and a mullah are all beautifully dressed, good-looking, and dignified. Looking straight at the viewer,

48 N. M. Shtukaturova, *Kostium narodov Rossii v grafike 18–20 vekov* (Moscow: VRIB Soiuzreklamkul'tura, 1990), 46. As Nathaniel Knight observed, the well-attended all-Russian ethnographic exhibition organized in Moscow in 1867 represented nearly sixty groups of mannequins dressed "in authentic national costume and surrounded by objects typifying their ways of life." The colorfulness of the empire's minority peoples appears to have contributed to the exhibition's failure to give ethnic Russians the aura of being the multinational state's leading people: Knight, "Imperiia napokaz," 111.

the man of the house in particular has an aura of calm self-assurance and self-esteem (fig. 8).

Tatars were apparently so "mild-mannered" that a tourist might ignore their existence, the option chosen in Strakhov's "Crimean Impressions" (1870). In parallel to his "From a Trip to Italy" published six years later, the Crimean account resists the contemporary nationalistic tendency to valorize central Russia's natural environment by contrast to spectacular Mediterranean or Swiss sites. Strakhov explains that only by visiting the Crimea did he learn the joy of living in "southern climes" blessed with "an abundance of light and pure air." Petersburg, on the other hand, seemed dismal and deprived when he returned there in late October. Not even realizing that they "lack the majestic pleasure of full daylight," Strakhov's Petersburg residents are "unfortunates," doomed to spend much of the year in the "unbelievable monotony" of fog, "grey air," and "twilight at noon." Having come to know the Crimea through direct, sensory experience, Strakhov assumes a proprietary attitude: he has now seen with his own eyes "our southern coast."[49] He projects the Crimea as Russian homeland, the meridional edge of the empire affording escape from Petersburg and by extension all the rest of the hibernal, ethnic Russian "core" of the empire. During his vacation Strakhov notices "little houses and huts of Tatars," but the inhabitants never make an appearance. That absence contributes to the depiction of the Crimea as a Russian possession for Russian leisure.

In conformity to the tendency already observed in Russian Crimean-vacation poetry of the 1850s, Strakhov represents himself alone with the scenery. Actual circumstances minimized the risk of encountering other tourists. One factor was the nature of accommodations. In the North Caucasus, Russian tourists clustered at spa hotels and rented rooms, a situation that provoked the crowd phobia that Druzhinin's stories articulated. In the Crimea, on the other hand, accommodations were often private since many Russians acquired property there beginning in Catherine's reign and especially during the big land-grab of the 1860s. The extensive Russian ownership of property meant that tourists might be staying in a home of their own or a friend's. Such was Strakhov's situation. "Crimean Impressions" refers to a kindly "host," without specifying that he was Danilevsky, who had an estate on the coast at Mshatka. Danilevsky would also host Leo Tolstoy there in 1885; and in 1901–02, on doctor's orders, Tolstoy lived with

49 N. N. Strakhov, "Krymskie vpechatleniia," *Niva*, no. 2 (1870):23–25.

his family for nearly a year in Countess Sofya Panina's luxurious seaside villa at Gaspra.[50]

In addition to the housing situation, the dispersal of marked sites also minimized the crowd factor in the Crimea. Caucasian spas brought tourists together in a constricted area, whereas the Crimea's tourist attractions were more widely dispersed. It was apparently easier to avoid other Russian sightseers. Early on, however, the Yalta area became the big sticking point, as illustrated in Nekrasov's *Russian Women* (1872–73), a narrative poem about two wives of Decembrists who followed their husbands into Siberian exile. One of the heroines, Princess Maria Volkonskaya, visited Gurzuf in her girlhood when the exiled Pushkin was there. In imagining her enraptured discovery of the Crimean landscape dotted with "Tatar huts," Nekrasov's poem situates Pushkin there, too, reading Byron, roaming alone by the sea at night, and hugging a cypress to express of his love of the region (Nekrasov's annotation about the tree refers to a well-known bit of biographical lore contained in one of Pushkin's published letters). *Russian Women* appropriates the cypress as a Pushkin shrine suffering desecration: ever since Pushkin's death, "tourists" have been tracking down the tree and tearing off "fragrant branches" as mementoes.[51] In designating Gurzuf a Pushkin haunt, Nekrasov's poem expresses conventional tourist phobia: the author and other sensitive souls can properly appreciate Pushkin's cypress, but the vandal crowd despoils it.

Getting Away

The marginality of tourist phobia in Russian writings about the Crimea intimated a national cohesion consonant with the official nationalist imperialism of Alexander III. Then in 1883 along came the tortured populist Uspensky with a travel narrative that disputed all possible imaginings of Russia as a happy nation at peace with itself. In Uspensky's estimation, the Russian desire to be elsewhere was now pandemic. He knew well the urge to get away. Acclaimed for his nonfiction and fictional treatments of rural life, especially *The Power of the Soil* (1882), he gradually broke with the populist

50 Tolstoi, letter to Strakhov, March 31, 1885, *PSS*, 63:220–21; and commentary in Christian, *Tolstoy's Letters*, 2:513.

51 N. A. Nekrasov, *Polnoe sobranie sochinenii i pisem*, 15 vols. (Leningrad: Nauka, 1981–2000), 4:168.

dogma of the Russian peasant as "an ideal communist."[52] His struggle oscillated between disillusionment and the will to believe. The strain contributed to his mental breakdown, a prolonged disintegration brought on by syphilis.[53] As Uspensky descended into madness, he became a compulsive traveler. Between 1880 and 1893, he took 349 trips, mainly within central Russia but also to Kazan, Odessa, the North Caucasus, and Constantinople in addition to Georgia.[54]

Uspensky's assessment of Russian wanderlust as a malaise of the time starts with a parody of sentimental patriotism, alluding mainly to Karamzin's *Letters* and Pogodin's *A Year in Foreign Lands*. Uspensky writes that the old-time travel memoir was written in tears rather than ink and began with the author's eviscerating departure:

> The homeland, the fatherland, the native shores were so dear to him; his whole heart, his very entrails had grown so inseparably attached to them that the boat carrying him from his native shores (a boat bearing some dreamy name such as *Evrianta or Aglaida*) seemed to him a heartless, cruel creature ripping the traveler by the flesh from everything dear to which he was "spliced" in body and soul.[55]

Uspensky likens the departure to the sight of Caspian Sea fishermen tearing out the innards of sturgeon to prepare isinglass. The imagined traveler stares at Kronstadt until it disappears, and the thought of Russia will never leave him. He produces a multivolume account of his impressions of cities, antiquities, pictures, cathedrals, and "national dances," but becomes unbearably homesick as the sojourn abroad winds down. On the return trip, his "greedy gaze seeks signs of the homeland," and the first glimpse of Kronstadt or the cupola of St. Isaac's makes his "heart pound like a hammer." And then comes the reunion with friends, the tears of joy, the "unending ecstasy" of homecoming! Finally, the spiritually healing return to rural

52 Mirsky, *A History of Russian Literature*, 299–300.

53 Henrietta Mondry, *Pure, Strong and Sexless: The Peasant Woman's Body and Gleb Uspensky* (Amsterdam and New York: Rodopi, 2006), 9.

54 "Opyt itinerariia Uspenskogo," in *Gleb Uspenskii*, ed. A. S. Glinka-Volzhskii and G. A. Leman (Moscow: Gosudarstvennyi literaturnyi muzei, 1939), 228; and Richard Wortman, *The Crisis of Russian Populism* (Cambridge: Cambridge University Press, 1967), 97.

55 Uspenskii, "Melochnye putevye vospominaniia," 193 (subsequent references appear parenthetically).

Russia: "the blessed, quiet crop lands, the weeping birches, the willows, grainfields, thatched roofs, the plowman, the dear homestead, a samovar on the riverbank, a fishing pole at hand, the still river, and a broad-brimmed straw hat" (194–96). Shifting weepiness from willows to birches and evoking samovar coziness even outdoors, Uspensky heightens the irony with the untranslatable jingle of *ivy/nivy*: willows juxtaposed to grainfields.

The returning traveler of the 1880s knows none of the old-time joy. As Russia draws near, a growing anguish assails him. His journey had given him the sense of "being released from an infirmary or gotten up from bed after a prolonged illness." Claiming that he felt well away from home, he now dreads a relapse:

> How dire it is to think that one will again have to put on a hospital gown, sleep on a hard cot, and become material for all sorts of medical experiments by a drunk paramedic, the kind about whom the *narod* tell incredible stories: about "accidentally" amputating the leg of a man in for bleeding and putting the leech on the man who was supposed to receive sodium carbonate, or in the end, quite simply but also "accidentally," just up and bury a sleeping man, taken for dead. (196)

Uspensky grants the hyperbole but insists that his scenario conveys a truth that fully hits a Russian after being away. His elaboration of the issue resorts to another metaphor: "some hulking, heavy thing has come to rest on the collective Russian soul," the same soul that the "worm" of anxiety and ennui is chewing (197).

It turns out, though, that much of the author's agony bespeaks his discovery that vacations, too, can be depressing. He recalls leaving the "grey, leaden, dead" North with high expectations of resuscitation in Georgia with its "delicate air," new "colors, lines, flowers, and faces." At first glimpse, Tiflis promises satisfying "novelty and originality." Yet, as soon as the visitor begins strolling through town, he concludes that "the 'civilized' people of Tiflis and the rest of the Caucasus do not like the uncivilized, Asiatic Tiflis and Caucasus," the exoticism the author has come to see. The catalogue of disillusionments begins with a Georgian encountered on Golovin Prospect. The man is wearing a Chancellery clerk cap and carrying a briefcase. "It would have been better," declares the tourist, "if he were just a plain Georgian, rather than a Georgian with a briefcase." The civil servant's inauthenticity is being gauged, of course, against Russian literary and visual images of Georgians as a merry, gregarious, hospitable people, fond of wine, music, dancing, and lolling in the sun. Uspensky's persona finds equally distressing a "handsome, distinctive, typical Shahseven lad" on the house-keeping staff of a French hotel: "rabbit-like timidity shines in his beautiful eyes," as he moves a lamp under the "drill-sergeant" supervision of a self-important "French café assistant." The last straw is a "Mengrelian beauty" in "picturesque national dress" working the counter at the train station café. She has become a "lackey," making cheese sandwiches for boorish tourists from Moscow (218–20).

Uspensky's account represents the local working people not as free agents exercising choice but rather as dupes in the coils of capitalism. Symbolized by the railroad, modernization with its promise of material bounty has seduced the Caucasians into cooperating in the eradication of "all national features and particularities from the face of the earth." Economic development disfigures the Georgian landscape, too, and not only in the form of railroads. Amidst the "ruins of ancient castles and towers" near Mleta, the author finds a Russian merchant's shop with a sign announcing vegetables, tea, and sugar for sale (220–21). As perceived by the populist tourist, the encroachments of capitalism are drastically, rapidly eroding Georgia's scenic and cultural authenticity. In flight from central Russia imagined as a clinic run by alcoholic paramedics, the author attempted but failed to ameliorate his existence by taking a vacation in the sunny South.

Markov's Crimea as a National Unifier

Markov's *Sketches*, on the other hand, maintains that tourism in the Crimea *does* revitalize the individual and deliver civic benefits, to boot. As noted earlier, Markov's 1872 preface declared that leisure travel performs "social work," in addition to erasing "wrinkles of care." Especially through the performance of battlefield tourism, the *Sketches* suggests that the most important "social work" in this case is the promotion of Russian national solidarity through remembrance of the Crimean War. Markov dramatizes the conviction that Russia bought Crimea with Russian blood. The trope had a long history. Lermontov's lyric "Homeland" ("Rodina," 1841) rejected patriotism based on "glory purchased with blood."[56] In 1844, however, Belinsky

56 Lermontov, *SS*, 1:70.

wrote of the North Caucasus being "purchased with the precious blood of [Russia's] sons."[57] Similar rhetoric appeared in the 1860 travel account of a Russian army doctor conducting home to Tiflis a group of Georgian cadets whose health had deteriorated during training in Petersburg. The doctor described the Georgian Military Road as a route that "Russian bayonets built for us"—a product of bloody work for the sake of Russians.[58] During the Polish insurrection Katkov intensified the nationalist inflections of the metaphor of blood as currency. He contended then that any territory conquered by "streams of Russian blood" was a sacrosanct possession of the "nation." This was a pivotal formulation that solidified the "nationalization of traditional imperial patriotism."[59]

Not even in the North Caucasus did blood-and-soil Russian nationalism intertwine so tightly with tourism as in the Crimea. A visitor's account issued soon after the Crimean War asserted that the "fresh blood" of Sevastopol's defenders was still to be seen on the city walls, an image sanctifying Russian sacrifice for the territory.[60] The depiction of battlefield tourism in Markov's *Sketches* extends the traditional symbolism. Three chapters (serialized in 1867) are devoted to the author's pilgrimages to Sevastopol, Inkerman, and Balaklava.[61] Sevastopol features as "A Dead City," a city of ruins.[62] Markov's narrator expresses the frustrations of seeking storied sites in the imperial era. Among other disappointments, he finds hardly a trace of the tower on Malakhov hill (overrun by the French in the final storming) or the ultra-dangerous fourth bastion prominently featured in Tolstoy's *Sevastopol Stories*. Markov's persona nonetheless finds sufficient prompts to patriotism.

57 Belinskii, "Stat'i o Pushkine. Stat'ia shestaia," *PSS*, 7:373.

58 Degaev, *Rukovodstvo dlia edushchikh v Piatigorsk i Tiflis*, 33–34.

59 Renner, "Katkov and Russian National Politics": 671.

60 D. M. Afanas'ev, "Sevastopol' i ego okrestnosti v nastoiashchee vremia," *Sovremennik* 62, no. 3 (March 1857), section 1: 134.

61 On battlefield tourism's tensions between vacationing and patriotic pilgrimage, see David W. Lloyd, *Battlefield Tourism: Pilgrimage and the Commemoration of the Great War in Britain, Australia and Canada, 1919–1939* (London: Berg, 1998); and my "Russian Military Tourism," 58–62.

62 Markov, *Ocherki Kryma*, 62–78. For similar descriptions of devastated Sevastopol, see D. N. Sokolov, *Progulka po Krymu s tsel'iu oznakomit'sia s nim* (Odessa: Tipografiia L. Nitche, 1869), 60; Livanov, *Putevoditel' po Krymu s istoricheskim opisaniem dostoprimechatel'nostei*, 110–11; and Fet's account of his visit in 1880, in his *Vospominaniia*, 2:375.

As he wanders through ruins and rubble, he raises Russian military sacrifice to the level of the Crucifixion. In Sevastopol he feels as "a medieval pilgrim in Golgotha," and Russia's fallen soldiers assume the guise of holy martyrs: "The bones and blood of zealots sanctified for me these earthen embankments, and I was barely capable of carrying away a handful of the holy dust as a relic" (94). A later chapter enfolds the 1783 annexation into the symbolic pattern of blood and soil. Having detailed his wonderment at the sight of the Baidar valley, the tourist author declares that this "paradisiac region," this "pearl in the Russian crown" was "obtained for [Catherine] by the blood of her northern sons" (373). The statement telescopes one era into another, implying that streams of Russian blood have been sanctifying possession of the Crimea for decades. The Crucifixion-linked symbolism of sanctification through blood reinforces the Christian pedigree that Markov's persona constitutes for the Crimea in recounting his visits of Russian monasteries and the ruins of ancient Christian churches. As already advanced through architecture, poetry, and official utterances during Catherine's reign, the *Sketches* motifs of Christianity affiliate the Russian Empire with the West, to underwrite the ideology of a civilizing mission in Asia. Inspired by the Crimea's Greek antiquities, the Hellenic accents of eighteenth-century discourse served the same affiliation function.[63] The latter persists in the *Sketches* through references to Iphigenia, Odysseus, the Cyclopes, and Achilles, even though the author does not visit the Greek antiquities (147–49, 170).

Markov's patriotism clashes with his portrayal of Tatars as victims of Russian imperialism, a motif irreconcilable with the 1884 edition's presentation of the book as an "expression of Russian society's fraternal relations with the Crimea" (vii). The critique of empire-building is concentrated in the ironically titled chapter 15, "Christian Civilization's Boon to the Infidel Tatar," first published in the 1872 book edition. Deviating from the dominant, artistic "album" texture of the travel narrative, this chapter reads as polemical journalism, contending that Russian rule has done Tatars no good. With his professional eye for schooling, Markov argues in particular that the Tatar population's general level of enlightenment was higher before the annexation: "During the past eighty years, we Russians have not spread among the Tatars the least bit of learning, not the least bit of technical knowledge" (357). "We *could not* give anything, because in the first place we ourselves

63 Wortman, *Scenarios of Power*, 1:137–39; and Kozelsky, *Christianizing the Crimea*, 45.

were poorer than the Tatar in many respects" (the second alleged reason is that a corrupt "bureaucratic machine" hijacked the colonial administration: 358). The critique does not spare the infrastructure policy that facilitates the author's own rambles: "We have made paved highways for tourist outings to dachas on the southern coast, we have erected Orthodox cathedrals and maintained Orthodox parish schools but left the Tatars' mountain roads as they were under Khan Mengli-Girei" (357). The centenary of the annexation occasioned public debate about the extent to which Russia had succeeded in bringing enlightenment to the Crimea.[64] "Christian Civilization's Boon to the Infidel Tatar" was a provocative liberal contribution to that public conversation, but neither here nor anywhere else did the *Sketches* come close to envisioning Russian power sharing with Tatars.

Moreover, the very next chapter, "The Southern Coast," is the book's apogee of exulting in Russian possession of the territory. The chapter epitomizes the author's tendency to glorify the Crimea's topography and climate at the expense of central Russia's, and the big point is that paradisiac Crimea is "ours," an integral part of the Russian homeland (*rodina*): "Reader of our quotidian, prosaic age! If you would like to plunge for a few weeks into the living source of incomparable beauty and authentic poetry, you need not seek Italy or Andalusia. You will find all you crave in your homeland, on the Crimea's southern shore." Leave behind "the North's low grey sky," snow and cold, the monotonous steppes, the "continuous wall" of grainfields, and the "dry, starchy" grain itself. Come south to your seashore, bright sunshine, a "deep blue Italian sky," "Swiss precipices and Swiss mountain peaks," rich green foliage, flowers, fruit, and "fragrant oils" (374–76).

Relishing the Foreign

The Crimea is "ours," purchased with Russian blood, but it thrills the author most when providing previously "unknown" experiences and sights outside the "routine of our Europeanized Russian life." The *Sketches* author

64 For a critical assessment, see G. P. Levitskii, "Pereselenie Tatar iz Kryma v Turtsiiu," *Vestnik Evropy* (October 1882): 604–39. For patriotic discourse and official celebration of the annexation, see Lazzerini, "Local Accommodation and Resistance to Colonialism," 172–74; and Kozelsky, *Christianizing Crimea*, 171. Evocation of Russia's overthrow of the Tatar yoke figured in the festivities organized for Alexander III's coronation in 1884: see Wortman, *Scenarios of Power*, 2:225–26, 231.

is a self-described "inveterate tourist," ready to try weird Tatar food, even though he may spit it out (250, 254). Active leisure is a huge part of his program (horseback riding, cave explorations, swimming), and he takes pride in roughing it. Even in 1884, he warns, visitors of the Crimea should have "Spartan tastes" and be well-schooled in "stoicism," unless they choose to stay in overcrowded Yalta (567). The best illustration of roughing it comes during the author's two-day horseback excursion to Chatyr-Dag with his companions, a solidly middle-class group of "men and women professors." At nightfall the tourists are caught in a thunderstorm with their Russian-speaking Tatar guide Osman, who negotiates shelter in the earthen bunker of shepherds. Sitting around a cauldron with the tourists, the shepherds serve them shashlik scrumptious as "candy," and the author considers this supper scene worthy of "the brush of Salvator Rosa." The Russians bed down on the ground and doze off to the howling of wolves drawn to the animals. As the other tourists sleep, the author alone has the special pleasure of consuming ultra-authentic foreign culture in ultra-foreign surroundings: he hears shepherds singing a "well-sung Tatar song" to the accompaniment of pipes and bellows (249, 250, 253). In the morning the happy campers eat their remaining provisions for breakfast, while sitting with their legs dangling over a cliff.

Markov's adventure-tourist persona never remarks that in publicizing his exciting discoveries of the new and different, he endangers their novelty. High in the mountains about four miles from Yalta, the Uchan-Su cascade ("flying water" in Tatar) might still have qualified as *terra incognita* when the *Sketches* first appeared.[65] Quoting Markov's long description of seeing the falls, an 1876 issue of *Grainfield* conjured the risks of creeping along a cliff edge in the wilderness to obtain the best view. Tourists up to the challenge would have an unforgettable experience uniquely available in the Crimea: "So many picturesque sites in Switzerland have been photographed thousands of times and are known to all and sundry through those pictures or drawings." Hotels and viewing platforms spring up to accommodate the "thousands of tourists" flocking to see the cascades. Uchan-Su, on the other hand, remains relatively pristine. But alas, the site will "no doubt" soon "lose its wild, isolated character and be equipped with assorted amenities to suit the fussiest tourist." Development is in fact well underway: an "excellent road" from Livadia to Uchan-Su is being constructed, a gazebo

65 See Markov, *Ocherki Kryma*, 395–401.

for the empress is in place at the site, and a recently cleared path affords "an entrancing view" of the falls.[66] The journalist regretted the shrinkage of adventure space. But he failed to recognize that Markov's constitution of Uchan-Su as a sight encouraged tourist development and that *Grainfield* was accelerating the process by putting the *Sketches* episode into mass circulation and by advertising the Russian royal family's involvement, a part of the story that gave the waterfall an aristocratic cachet, consistent with the magazine's general ethos of middle-class sociocultural advancement at home through tourism.

The shepherds' shelter and the Uchan-Su cascade circa 1872 produce the authenticity effects that Markov's tourist persona seeks in the Crimea, with respect to local culture as well as scenery. His images of Tatars are contradictory, stressing both difference and similarity between "us" and "them." His first impressions of Tatars are filtered through the spectatorial prism of romantic orientalism. He takes "great pleasure" in looking at "unfamiliar sort" of people: "a *hadji* in a white turban" and chador-clad women gathered around "simple but noble" fountains, a "theatrical scene" that "transports the tourist's fancy to Syria and Asia Minor, the true realm of oriental life." Thrilling cultural difference coexists with a motif of somatic proximity. The author reports that the "Caucasian rather than the Mongol element" is more pronounced in Crimean Tatar physiognomy (26–29). They have "white" skin, their "faces are entirely European [. . .]. Their children in particular resemble ours. There is not a drop of Mongol blood in them." In short, they look as Caucasian "as we do." Moreover, there are cultural affinities between us and them: a sedentary way of life; a liberal form of Islam, permitting women considerable freedom; and the assimilation of some Christian festivals (214–15). Elsewhere, however, the *Sketches* favors a Mongol connection. Remarking that "nothing authentic" remains of the Bakhchisarai palace, the author denies the monument the world-class stature of the Alhambra or Cordoba's Mezquita. He nonetheless grants that "the ancient khans built themselves an excellent capital." But on the other hand, Bakhchisarai's present-day residents "cannot even construct decent huts" for themselves. "And to think we were enslaved by these people for nearly two and a half centuries!"(54). The formulation collapses present time into the past, to the double disadvantage of the living: "these people" before the author's eyes are descendants of Genghis-Khan, but they lack

66 A. Nevel'skoi, "Vodopad Uchan-Su v Krymu," *Niva*, no. 23 (1876): 395.

their mighty ancestors' building skills. This assertion is anomalous, for as a general rule, the *Sketches* depicts Tatar homes as "remarkably clean," airy dwellings, infinitely superior to the stuffy, cramped "Russian muzhik hut," infested with cockroaches and bedbugs (31, 215).

The *Sketches*'s configuring of Crimean cultural authenticity makes "patriarchal" an antonym of "civilized." Prominent among the exemplars of the allegedly traditional way of life are *Naturmensch*: herdsmen and shepherds as thoroughly at home in the "virginal bosom of nature" as wild deer and goats, domesticated horses and cattle. Spotted from the author's hotel balcony in modernizing Alushta, "an authentic Tatar bazaar" with ox-carts and "unpretentious little stalls" also represents "patriarchal," pre-capitalist culture (223–26, 568). A clash between "civilization" and putative Tatar tradition is dramatized when the guide Osman arranges for the author to visit the home of his acquaintance Selyamet, a well-to-do Tatar in Korbekly. First impressions satisfy the author's ambition to see places normally inaccessible to tourists. He drinks in the material culture: the "clean, multicolored carpets," "oriental sofas" with pillows along the walls, "brightly polished pewter tableware," and other fine possessions tastefully displayed in keeping with local custom. The arrival of Selyamet's brother destroys the authenticity effect. The youth is barefoot and cannot speak Russian but has a modern-day cloak of a fancy fabric draped "city style" over his Tatar jacket. The author is "utterly scandalized" by this young Tatar "fop" aping the fashions of "Simferopol civilization and the urban boulevards" (216–18). What the author perceives as a corrupted attitude to money further taints "civilization." Speaking in Russian, which none of the other Tatars present understand, Osman prompts the author to offer Selyamet some money for the visit. The tourist feels certain his host will refuse. The Tatar gladly accepts, however, provoking an authorial aside to the reader: "So much for Tatar hospitality and the patriarchal way of life!" (219)

Symbolic Architecture

Selyamet's disillusioning attitude to money belongs to the book's larger theme of capitalist development as a menace to authentic Tatar culture as well as the landscape. Tourist pleasure is the paramount concern on both counts. The author fears that the "patriarchal charm of life" will entirely vanish along the seashore, as it already has in Yalta and some other densely visited places. The entire coast, "Russia's pearl," risks turning into "one solid dacha for the Russian capitals." Thirsting for "the simplicity and truth of nature," dissipated high-society people (especially women) will overrun the territory in search of the rejuvenating powers of sea, sun, and "the salubrious juice of Crimean grapes" (378). The triumph of capitalism will mean "paradise lost":

> The restless spirit of commercial exploitation will hum in luxuriant, balmy valleys whose main charm lies in the stillness of their semi-wild seclusion and the primordial simplicity of the local way of life. Forests will be cleared, wildlife will flee, babbling mountain streams will fall silent, and the Tatar in his Asiatic attire will be seen only in circuses. Instead of the ox-drawn *mazhar* [Tatar cart], there will be locomotive traffic; instead of clay Tatar huts with their brushwood smoke-stacks, comfortable little European houses will appear everywhere, wilderness will be transformed into towns, and the silent forest become a bazaar. But who, *who* will gain from that, reader? Will it make this delightful corner of the Crimea more delightful? Will outfitting the Asian in German clothing make you happier? (379)

The apprehension about capitalist development comes to rest on the contentment of Russian spectators. The feelings, needs, and opinions of Tatars are never addressed. The next paragraph makes this negligence all the more glaring. Instead of probing the rhetorical question about Russian happiness, the author launches into yet another of his hymns to the scenic glory of the southern coast: sheltered by mountains from "the polar blizzards and scorching droughts of the steppes," the Crimean coast is a "titanic greenhouse," a "blessed little corner [. . .] of our fatherland," "a patch of Italy," where cypresses, olive trees, laurels, magnolias, and oleanders thrive (379–80). Markov's opposing geographical semiospheres replicate a pattern of medieval Russian literature: with its climate assuring agricultural abundance, the Crimea is paradise, while central Russia with its climate of "ice and fire" corresponds to hell.[67]

The passage quoted above evokes a circus act for Russians as the ultimate degradation of Tatar authenticity. *Sketches* itself, however, presents underprivileged Tatar villages as exhibits for tourists. The author writes that from Foros to Alupka, the towns on the southern coast are "all equally

67 On the medieval precedent, see Lotman, *Universe of the Mind,* 173.

picturesque, equally pristine," with their "elegant villas" nestled in wild mountain flora. The Russian buildings' proximity to Tatar villages affords a piquant contrast for the tourist: "Now you ride along the dirty, winding street of a Tatar village amidst flat roofs, dirty shops, and the wide-spreading branches of gigantic walnut trees, and then you are passing the intricately decorated grilles, the gates, and gazebos of aristocratic gardens" (385, 388). Juxtaposed to the "dirty" villages, the "elegant" Russian properties proclaim the implantation of civilization in a form the author admires. He expresses no concern here with improving the villages and even seems to assume their usefulness as a sort of Asian theme part for European visitors. The "dirty" villages provide the spectacle of a "static native world" alongside the dachas, villas, gardens, and gazebos that demonstrate the dynamism of Russia as the agent of European enlightenment in Asia.[68]

Markov's treatment of the Alupka palace further suggests that Russians alone have the vision and know-how needed to improve Tatars' lives. Designed by the English architect Edward Blore (1787–1879), the palace was built between 1828 and 1848 in a style blending Scottish baronial and Moorish elements. *Sketches* depicts Alupka as an integral part of the landscape mirroring the mountain Ay-Petri that looms behind it (fig. 9). Ay-Petri itself is "an enchanted castle in the clouds. With its ruined towers, steep precipice walls, and merlons gnawed by time, it is a castle worthy of Byron's Manfred." Poetically enhanced by the Byron connection, the figurative description normalizes the presence of "Prince Vorontsov's real castle" in the Crimea: it is as though it were not a building but a creation of nature. Alupka is thereby declared an organic part of the landscape, but the recommended view creates the opposite impression of colonial penetration. Seen from the seashore against "all the magic of Ay-Petri," the palace grounds include an imposing "Orthodox cathedral in the noble style of the Parthenon" (409–10).

Close description of Vorontsov's palace amplifies the architectural symbolism. The building incorporates "all the marvels of civilization" in a synthesis of Europe and the Orient (411). Reflecting "the oriental particularity of Islamic Crimea," the palace is "a fragment of the Alhambra of the Arab caliphates" (an honorific comparison *Sketches* withheld from

68 I borrow here from the analysis of old versus new Tashkent under Russian rule, in Robert D. Crews, "Civilization in the City: Architecture, Urbanism and the Colonization of Tashkent," in Cracraft and Rowland, *Architectures of Russian Identity*, 117–29 (quote 129).

9. *Alupka: Prince Mikhail Vorontsov's Estate in the Crimea.* First published in Russia in 1857 in one of the almanacs regularly prepared for court and state officials, this image stupendously symbolized the enduring solidity of Russian possession of the Crimea in the immediate aftermath of the humiliating Crimean War.

the Bakhchisarai palace). In conducting the reader ("you") on a tour of connecting parts of the palace, the author calls special attention to a long, narrow passage-way that he likens to "a street in oriental cities. Only instead of dirt and stench–all is clean and fragrant here, as in a garden." Carrara marble abounds in the palace, and the dining room chairs have upholstery from Lyon. In sum, the European designers and decorators of Alupka have assimilated but enhanced the Crimea's Asian idioms to produce a cross-cultural monument that outshines all other palaces the author has seen: Versailles, Sanssouci, Schönbrunn, and Peterhof. The supremely grand Alupka befits its visionary proprietor Vorontsov, "one of the few genuine statesmen of our fatherland, a man of enlightened mind," a man who liked to retreat to the palace library to pour over his extensive collection of books and documents now on view for tourists to see (415–19). European enlightenment has overmastered static Asia, and the metaphor of the passage-way as a cleaned-up, perfumed oriental street attributes exclusively to the colonizers the skills and energy required to improve the prototypal "dirty" Tatar village.

Unwelcome Native Agency

On the other hand, when Markov's Tatars make unsupervised efforts to better their lives, they become corrupt, especially in Yalta. While touring in the 1870s, the author perceives Yalta from afar as "a miniature Naples." Up close, though, the town is already not so pretty:

> In the summer season Yalta is chock-full, prices are high, and amenities minimal. Music, the boulevard scene, and dancing in the evening have transformed it into a fashionable "spa" kind of town, where there is no use seeking the authentic rural life that brings health and tranquillity. The streets are dusty, courtyards stink, and houses are damp and dirty inside. In Yalta the simple-hearted Tatar is already a civilized exploiter and cheat. (395)

More adamantly than ever, "civilized" signifies betrayal of "patriarchal" values, with the implication that the native is overstepping his proper place as a dogsbody in the colonial economy.

Before specifying how Tatars cheat, Markov establishes an ultimately positive, distinctly bourgeois perspective on capitalist development in his 1884 addendum "From the Southern Shore." After a ten-year "intermission" since his last sojourn in the Crimea, the tourist author finds the "stage-set" (*dekoratsiia*) enormously changed. The transformation is already palpable in the Novorossiya steppe, where "romanticism" has succumbed to capitalist greed, symbolized by the railroad with its "gleaming stations, French cuisine, big crowds, and hustle-bustle." On the railroad line, Simferopol has become "a genuine European town" with lots of shops. "Once an isolated, sparsely populated hamlet," Alushta has turned into "a sort of *station balnéaire* of the kind for which Switzerland, Italy, and France are famous." By the time the author reaches Yalta, he sounds not merely reconciled but seduced by the transformations. He writes that Yalta has become "a quite large, pretty city reminiscent of fashionable European tourist centers—Menton or Baden" (560–66, 576). *Sketches* expresses a special, bourgeois receptiveness to the Rossiya, "Russia's first luxury hotel."[69] After mentioning the hotel's gas lights, running water, and other amenities, the author describes a classy social atmosphere:

69 Quote from the caption of a picture of the Rossiya in McReynolds, *Russia at Play*, 175.

> A colorful, swanky crowd always fills the Rossiya's numerous balconies and galleries. When the tsar's court is at Livadia, there is a large assembly of generals here. The formerly poor, naive Yalta is beyond recognition. Cavalcades of cavaliers and horsewomen, and carriages of the fashionable set push past at every step. Music of orchestras resounds in three different places nearly every night. (577)

Having presented himself as a professor on vacation, the author inserts himself into a ritzy, aristocratic scene, most resplendent when the royals are in town. In parallel to Markov's travel writings about Switzerland and Venice, the Yalta episode of the *Sketches* reveals an authorial soft-spot for a luxury hotel fit for the tsar.

Markov's idea of Tatar "corruption" now turns out to mean their profiting from tourism by working as guides and tradesmen. In the Yalta streets, teams of Tatars with carriages and horses for hire await "profitable orders" as guides. Among them is a certain "cunning lad" who has made a specialty of private, horseback excursions for Russian women visitors "looking for adventures that leave a powerful impression. [. . .] Our Russian ladies are evidently seeking more than grapes and sea-bathing on the southern coast." Frustrating readers who would like to hear more about "our ladies" lusting after Tatar guides, *Sketches* immediately switches to providing information about the availability of dachas in "a wide variety of architectural styles, mainly Italian and Moorish," and the costs of renting carriages and hiring servants (577–79). When the narrative returns to the issue of corruption, it focuses on the fashion for westernized Russian clothing:

> The Yalta Tatars are fully civilized, that is, fully corrupted. "Progress" in the vulgar sense has made an astonishing conquest in ten years. It is now no longer rare to encounter even a simple Tatar in a European coat and starched shirt with cufflinks; youths have ceased shaving their heads and look like gypsies, and even older men are letting their black locks grow. In general, the conviction that there is no need to distinguish themselves from Russian customs is penetrating Tatar thinking more and more deeply. Probably as a consequence of this principle, the pure-blood Yalta Tatar has become almost as much of a swindler as a Muscovite at the Sukharev Tower. (584)

Referring to a Moscow market, the remark warns would-be tourists to practice their haggling skills at home to prepare for dealing with Tatars liable to cheat them.[70] Navigating a Crimean market sounds all the more treacherous since the "swindler" may display the "typical chic of the true Tatar," wearing one of the traditional "gold-braided jackets" (585).

Tatar Lessons for Russia

After framing Tatar work as the "exploitation" of tourists, *Sketches* does another ethnographic volte-face, to pronounce Tatars an admirable people whose way of life holds lessons for Russians. Markov first summarizes his conversations with local Russian coachmen and brick-layers, representatives of the *narod*, who are "not quick to praise, especially 'heathens.'" According to Russian folk consensus: "The Tatar is a reliable man, he'll never deceive. [. . .] Among their kind, you don't find drunkards or foul-mouths like in Russia, they don't insult people either, they live together like friends, just like brothers." The Russian laborers also "speak enviously of the Tatar's prosperity, the fact that each has his own land, horses, and oxen" (585). Reliability, sobriety, civility, brotherhood, and prosperity trickling down to the working class: a utopian distortion perhaps, but one bespeaking personal characteristics and social conditions that the *Sketches* implies the Russia of Alexander III can only envy.

The book then adds cleanliness to the catalogue of Crimean virtues in a manner dramatizing authorial respect and even affection for an individual Tatar human being. While riding through Yalta one day, the author hears his name called. The man hailing him is a Tatar who, as now mentioned for the first time, worked for him years ago during his residence in the Crimea. For fifteen rubles a month, the man served as a factotum, often running through mountains four miles a day. They embrace, overjoyed to see one another. The Tatar has come up in the world. He has a lumber business and has acquired local celebrity as the unnamed guide of the *Sketches* who conducted the author to the cave towns when those places were way off the beaten path. In visiting the

70 Erected under Peter I, the tower initially housed government facilities and became a marketplace in 1812. The Soviet authorities destroyed the monument in 1944. See description of this tower of "gigantic dimensions" in Lermontov, "Panorama Moskvy," *SS*, 4:255.

man's home, Markov's persona is struck by the "astounding cleanliness," even in the kitchen:

> This Tatar cleanliness is very instructive for us Russians. The Tatar's tidiness, meticulous decoration of rooms, the display of taste, and even a touch of luxury in the homeowner's belongings bespeak respect for one's human worth, a belief in one's right to a life on an equal footing with others. And indeed it would be difficult to find another people whose social intercourse shows greater simplicity and more natural freedom of relations between the upper and lower strata. (589)

The *Sketches* image of the "Russian muzhik hut" full of cockroaches and bedbugs resonates here, to imply that people aspiring to raise the Russian lower classes' standard of living might profit from the Tatar example.

Markov also expresses respect and empathy for Tatars in a section that anticipates modern-day tourism planners' concern with managing natural resources so as to strike a balance between tourists' enjoyment and the welfare of local people. The tourist author reports that Tatars resent the Vorontsov family's possession of vast properties, stretching from Alupka to Massandra and encompassing the head waters of the Alma River. "The prince has all our water!" they cry. The *Sketches* accuses Tatars themselves of wantonly felling trees in the "most picturesque and valuable forests." As concerns water, however, the book sides with the locals, to argue that too much water, vital for Tatar agriculture, is used to maintain the lawns, flowers, trees, bushes, and "poetically gurgling fountains and cascades" of parks and gardens that are undeniably "beautiful" but serve only tourists (590, 592). The liberal author of the *Sketches* is calling here for Russian authorities to factor Tatar needs into the management of natural resources, even though no participatory role in that process is explicitly foreseen for the local people.

Rather than attempting to formulate a unifying political perspective, the *Sketches* ends instead on a tourist note of exhilarating active leisure. Markov's persona goes galloping up Ay-Petri with his band of "merry, young companions." Stimulated by the exercise and the panoramic vista from the mountain top, he closes with carpe-diem advice to his readers: "Whoever breathes in the Crimea's atmosphere, breathes in joie de vivre, poetry, longevity. Hurry now to the Crimea, whoever can, whoever still has the time" (593). Time is running out for every mortal, and vacations are revitalizing, but paradisiac Crimea's own days are numbered because of the tourist development that the *Sketches* depicts with deep ambivalence. On the one hand, the march of tourism threatens the natural environment and makes the world poorer in cultural diversity. But on the other hand, the "aristocratic" villas, gardens, and dachas, the Alupka palace and park, the Rossiya hotel with its "fashionable" and even royal clientele exercise snob appeal to the middle-class author. Symbolic architecture also contributes to the book's warring stereotypes of Tatars. The *Sketches* situates them both in spic-and-span homes and in "dirty villages" that function as theme parks underwriting the ideology of the civilizing mission. Contesting ruthless imperialism, the author expresses considerable empathy for Tatars. He represents them an exemplars of brotherhood, reliability, sobriety, and civility. He further suggests that even though Russian rule has allegedly worsened Tatars' lives, they have managed to forge a society more humane and democratic than the colonial overlord's. But on the other hand, tourism has made Tatars swindlers and cheats.

The warring stereotypes allowed readers to extract two conflicting scenarios, and the bad Tatar story tragically prevailed. The very year that the second edition of Markov's *Sketches* appeared, *Grainfield* ran a lurid picture inspired by Pushkin's *The Fountain of Bakhchisarai*. Sensationally visualizing what Pushkin left to imagination, the image shows the Polish princess Maria being captured by Mongol fiends with heads on pikes (fig. 10). Her face turned to heaven, she stands stripped to the thighs with her skirt slipping off. A fallen kinswoman beside her tries to protect an infant, and local warriors—stabbed in the back—lie dead in the foreground. A companion 1884 image depicted Khan Girei as a satanic, slouching figure smoking a hookah in a high-vaulted throne room with Mongolian minions in attendance. As a contribution to the centenary celebration of Pushkin's birth, a special 1899 issue of *Grainfield* republished those pictures, along with one of semi-clothed harem beauties observing the arrival of a new captive in the garden (figs. 11, 12). During those same decades, Russian chapbook literature presented Tatars "as objects of scorn or low-dealing cheats," "the most dangerous and terrible of peoples within the empire." In the words of an 1898 tale: "Out of the Asiatic steppe there surged into Russia the Tatar, the remnants of whom you have probably seen."[71] Guidebooks contributed to the denigration of

71 Brooks, *When Russia Learned to Read*, 228–29.

Tatars. An 1889 guide to spa cure facilities near Yalta declared that "Alupka Tatars are extremely lazy people, spoilt and at the same time cunning and gluttonous, having become accustomed to living at the expense of visitors." And the 1904 edition of Grigory Moskvich's guide asserted that "the cult of profit, having reached extremes of insolence, has pushed aside the Yalta

10. The abduction of Pushkin's Maria. The monstrous Mongol raiders appear ready to kill the terrorized woman's baby, a reminiscence of West European paintings of the Biblical massacre of the innocents.

11. The harem scene imitates orientalist French painting. In particular, the nude in the right foreground resembles *La Grande Baigneuse* (1808) of Jean Auguste Dominique Ingres.

Tatar's sentiment of humanity."[72]

While the *Sketches* seeks to counterbalance the Tatar "swindler" stereotype with liberal protest against ruthless imperialism, the book is permeated by a tourist relish for exotic spectacle that is arguably akin to the eye-riveting illustrations for *The Fountain of Bakhchisarai* in *Grainfield*. Whatever Tatars are doing in Markov's narrative, they are supposed to provide visual pleasure, and they repeatedly do so in an aestheticized manner: the chic men in their nationally distinct clothing, for instance, the chador-clad woman assembled around a fountain, or the shashlik supper with shepherds framed as a Salvator Rosa painting. The same quest for the picturesque Orient explained Uspensky's dissatisfaction with Georgia in 1883. Both tourist narratives maintained that capitalism in the service of Russian tourism was not improving the borderland and its native peoples but rather eradicating the cultural and scenic differences that provide the stimulating experience of *dépaysement*. Markov's *Sketches* exhorted readers to hurry if they wanted to catch a glimpse of pristine scenery and precapitalist society in "our" Crimea. In Uspensky's telling, it was already too late for Georgia. His proposition was that capitalist mod-

72 Both sources cited in Mal'gin, *Russkaia Riv'era*, 275.

12. Depicted in lavish, orientalizing detail, this scene of Girei holding court with a scribe among his minions has no counterpart in Pushkin's poem.

ernization has altered formerly exotic Georgian people and scenery to such an extent that Russians might as well stay home. Markov, too, perceived a lamentable loss of cultural authenticity in the Crimea but enacted a countervailing, middle-class pleasure in being in Russified Yalta represented as one of the haunts of the tsar. The populist Uspensky, on the other hand, despised the aspirations and values of the Russian bourgeoisie and would portray a member of that social class as a particularly loathsome tourist specimen in the North Caucasus.

12

Tourist Decadence at the Fin de Siècle: Chekhov, Veselitskaya, and Other Writers

In the years between the mid-1880s and the turn of the century, the public image of Russian tourism ran the gamut from wholesome, nation-building cruises on the Volga to cosmopolitan consumerism in the West and the Russian Empire. Russian folk culture as well as the high literary tradition, including sentimental poetry by Karamzin, had long made the Volga a focal point of Russian national identity. Only in the latter third of the nineteenth century, however, did writers and painters render the Volga a scenic tourist destination. In tracing that development, Christopher Ely remarked that the 1872 introduction of American-style paddle boats provided a technological boost but that the enticements to Volga travel came from guidebooks, articles, photographs, and the paintings of Chekhov's friend Isaak Levitan (1860–1900) and other artists. Prior to the 1870s, writers tended to cast Volga trips as educational projects and a "patriotic duty." Parents were urged to take their children. The appeal to national feeling never ceased but gradually meshed with the ameliorative vacation agenda. Guidebooks of the 1890s maintained that a Volga cruise would reaffirm the visitor's connections to "purely Russian, unpretentious but deeply sympathetic landscapes," while simultaneously affording the "benefits of relaxation and escape from daily life." This proved a particularly winning combination for middle-class customers "who found it both financially and culturally daunting to travel abroad."[1] The Volga boats were literally vehicles of cultural nationalism. By 1892 there were cruise ships named *Pushkin*, *Zhukovsky*, *Gogol*, *Lermontov*, *Nekrasov*, *Turgenev*, *Saltykov-Shchedrin*, and *Tolstoy*. In the first-class salon of the *Pushkin* and the *Zhukovsky* (and presumably the others), there was a portrait of the eponymous writer and copies of his works for reading on board.[2] After describing the Volga alone as sufficient reason for a trip, a contemporaneous series of *Grainfield* articles added that many towns along the river have historical and cultural sites meriting the tourist's full attention. Furthermore, the last ten years of alleged progress meant that Russia's "provincial towns are not as awful as they used to be and may be even more attractive than the capitals."[3]

At the opposite pole of this nationally affirmative vacation option in central Russia, tourism in the West and in southern borderlands of the empire acquired the decadent aura of soulless consumerism and indulgence in sex, fashion, food, and drink. Decadence could be alluring, as in Paul Bourget's novel *Une idylle tragique: mœurs cosmopolites* (1896). Translated that year for the journal *Russkaia mysl'* (*Russian Thought*, 1880–1918), the novel recounts a romantic intrigue ending in a fatal duel in glamorous settings of Monte Carlo. The story opens with a "cosmopolitan throng" of aristocrats, "famous artists," and stylish bourgeois in a casino.[4] The novel's main characters are from Western Europe and America, but secondary personages include a Russian grand duke with a racing yacht named *Jenny*. Bourget's vacationers on the Riviera belong to the international, sociocultural elite. In Russia's own literature, on the other hand, tourist decadence

1 Ely, "The Origins of Russian Scenery," 669, 671, 673–75.

2 N. K. Mikhailovskii, *Literaturnye vospominaniia i sovremennaia smuta*, 2 vols. (St. Petersburg: Tipografiia V. M. Vol'fa, 1900), 1:405. For a big exception to the rule of ameliorative Volga tourism, see E. P. Letkova, "Otdykh" ("Holiday"), *Russkaia mysl'*, no. 1 (1896), 180–96, discussed in Catriona Kelly, *A History of Russian Women's Writing, 1820–1992* (Oxford: Oxford University Press, 1994), 145–46. An orphaned spinster in her forties, the heroine is a Petersburg telegraph operator whose boss arranges the cruise for her health. She eagerly sets off but has an emotional crisis on the boat. Lacking the confidence to speak to other passengers, she grows depressed after overhearing on deck the amorous talk of a young couple and the cynical words of an adulteress telling a distraught young man their affair is over. Feeling her life is empty and that she shall never know love, the heroine jumps overboard. Nobody notices, and the boat moves on.

3 *Niva*, no. 13 (1896): 298; no. 17: 385. On the persistence of the literary trope of hideous provinciality at the fin de siècle, see Lounsbery, *Life Is Elsewhere*, 235–42.

4 Paul Bourget [Pol' Burzhe], *Tragicheskaia idilliia. Kosmopoliticheskie nravy*, trans. M. R. [sic], *Russkaia mysl'*, no. 1 (1896), section 1: 109–12.

bespoke a broader sense of social decline, symbolized by the ascendant bourgeoisie in the North Caucasus and the Crimea.

The term "decadence" entered Russian print in an 1889 article by the Hungarian-Russian law student Vladimir Grabar, studying in Paris at the time.[5] He used the word to approve a new literature that emphasized subjective experience as opposed to the strict formalism of the French Parnassian poets. Best known today as a Russian literary tendency of the nineties, so-called "decadence" reflected the belated assimilation of Baudelaire. A sign of his growing stature among the Russian avant-garde was the 1885 publication of Merezhkovsky's translation of "L'invitation au voyage" from the *Fleurs du mal*. With dreamy musicality, the lyric persona beckons his beloved to a "distant," "wondrous" Orient by the sea.[6] Baudelaire's more familiar themes entered the writings of the Symbolist poet Konstantin Balmont (1867–1942), Bryusov, Fyodor Sologub (1863–1927), and others: aestheticism, spleen and melancholia, immoral sexuality, sadism, morbidity, suicide, pessimism, cynicism, and retreat into mysticism. Finding the new trends disturbing, Tolstoy took to reading Baudelaire in 1892 "to get an idea of the degree of depravity of the fin de siècle."[7] Tolstoy himself, though, helped make sex a public preoccupation with his exploration of lust and homicidal marital rage in the novel *The Kreutzer Sonata*, published in 1891 after two years of circulating illegally.[8] Coinciding with mounting political turmoil in Russia, problematic sex lives and the wider blossoming of *fleurs du mal* in Russian literature generated an apocalyptic sense of doom in society at large. This intensified when the depraved Rasputin gained influence

---

5 My overview closely follows Kristen M. Harkness, "Mariia Iakunchikova and the Roots of Decadence in Late Nineteenth-Century Russian Modernism," in *Decadence, Degeneration, and the End: Studies in the European Fin de Siècle*, ed. Marja Härmänmaa and Christopher Nissen (New York: Palgrave Macmillan, 2014), 145–47. See also Evelyn Bristol, "Idealism and Decadence in Russian Symbolist Poetry," *Slavic Review* 39, no. 2 (1980): 269–80; Georgette Donchin, *The Influence of French Symbolism on Russian Poetry* (The Hague: Mouton, 1958), 15–16, 93–96; and Olga Matich, *Erotic Utopia: The Decadent Imagination in Russia's Fin de Siècle* (Madison: University of Wisconsin Press, 2005), 5, 11.

6 D. S. Merezhkovskii, "Iz Bodlera," *Stikhotvoreniia i poemy* (St. Petersburg: Akademicheskii proekt, 2000), 370–71.

7 Tolstoy, letter to his wife, cited in Orwin, "Chronology," *The Cambridge Companion to Tolstoy*, 30.

8 Peter Ulf Møller, *Postlude to* The Kreutzer Sonata: *Tolstoj and the Debate on Sexual Morality in Russian Literature in the 1890s*, trans. John Kendal (Leiden: E. J. Brill, 1988); and Laura Engelstein, *The Keys to Happiness: Sex and the Search for Modernity in Fin de Siècle Russia* (Ithaca and London: University of Cornell Press, 1992), 218–21.

over Nicholas II and the tsarina through his apparent capacity to ease their son's hemophilia.

Tolstoy's unrivaled moral authority in Russia set the framework for a wave of anti-tourism that tarred vacationing with the brush of decadence. Beginning with his *Confession*, Tolstoy castigated the Russian upper classes as "parasites" (the term Pisarev had used), living high while the masses work and die. Tolstoy's subsequent writings contain glimpses of tourism as a form of elite feasting during a plague among the have-nots. Based on his participation in the Moscow census, his essay "What Then Must We Do?" (1889) detailed the shocking poverty of the slums.[9] When famine ravaged the countryside in the early nineties, Tolstoy organized relief in the Yasnaya Polyana region. His theological treatise *The Kingdom of God Is within You* (1893) placed Russian vacationing abroad alongside tobacco and alcohol as addictive substances that "numb" the moral consciousness of people refusing to face divergences between their behavior and their professed humanitarian values.[10] Intimating sexual excess, the murderer-narrator of *The Kreutzer Sonata* recalls his honeymoon in Paris as a shameful time. *Resurrection* (1899) recounts the moral crisis of Prince Dmitry Nekhlyudov, the Petersburg aristocrat whose casual seduction of the housemaid Katyusha leaves her pregnant, outcast, and drawn into prostitution and crime, leading to her deportation to a Siberian prison camp. Before accidentally learning her fate as a member of the jury at her trial, Nekhlyudov appears a refined man of fashion, an art lover, and amateur painter who goes to Italy to perfect his craft (a throw-back to Vronsky). And "Italy" is the name of the hotel where the Russian judge has an extramarital tryst that leads him to cut short legal deliberations that might have lightened Katyusha's punishment. Finally but importantly, *What Is Art?* (1899) took a sideswipe at cultural tourism by inveighing against the idle rich patronizing art exhibits, museums, concerts, and theaters for opera, drama, and ballets performed by "half-naked women."[11]

Tolstoy's religious philosophy made him a beacon of moral ideals in the fin de siècle public arena more replete than ever with invitations to enjoy

9 Daniel R. Brower, *The Russian City between Tradition and Modernity, 1850–1900* (Berkeley: University of California Press, 1990), 148; and Richard Wortman, "Tolstoy and the Perception of Poverty: Tolstoy's 'What Then Must We Do?,'" in his *Visual Texts, Ceremonial Texts, Texts of Exploration: Collected Articles on the Representation of Russian Monarchy* (Boston: Academic Studies Press, 2013), 320–32.

10 G. M. Hamburg, "Tolstoy's Spirituality," in Orwin, *Anniversary Essays on Tolstoy*, 144.

11 Tolstoi, *PSS*, 30:27–28, 31.

tourism and to indulge in cosmopolitan cultural and material consumption at home. This chapter begins with a thematic survey of that arena. Next will come analysis of the counter-construction of Russian tourist decadence in Uspensky's Caucasian spa sketch "The *Burzhui*," Veselitskaya's witty society tale *Mimi at the Spa*, Burenin's *A Love Affair in Kislovodsk*, and stories by Chekhov. Previous commentary has interpreted his "The Lady with the Little Dog" as a riposte to the tragic adultery plot of *Anna Karenina*. My analysis argues that Chekhov's primary target was not Tolstoy's novel (published over twenty years earlier) but rather Veselitskaya's story, which contains parodic echoes of *Anna Karenina*. Engaging *The Kreutzer Sonata* as well, *Mimi at the Spa* assesses upper-class Russian "sexual morality from a basically Tolstoian position."[12] Chekhov's "Ariadne" (1895) and his Tatar-guide sex scandal story "A Loose Tongue" (1899 version) will then provide examples of his decadent tourists.[13]

Although the North Caucasus and the Crimea predominate as vacation venues in this group of writings, the authors' politics differ greatly. In their preoccupation with the tourist sex hunt, "The *Burzhui*" and *Mimi at the Spa* ignore or demean colonized people of the Caucasus. Those works thus give the impression that the officially propagated principle of Russian hegemony over the empire's ethnic minorities had become normalized in Uspensky's and Veselitskaya's imagination. In sharp contrast, Burenin's novella and Chekhov's "A Loose Tongue" shine a harsh light on Russian tourists disrespecting local people in the North Caucasus and the Crimea. The authors further suggest that vacationing Russian women, lacking political power at home, may derive special satisfaction from treating colonized men as lackeys and even exotic erotic objects.

The Allure of the Foreign

As of the late 1880s, the Russian cultural scene took a cosmopolitan turn, and a growing array of imported commodities encouraged cosmopolitanism in everyday life. What appears to have been the first exhibit of modern French painting opened in Petersburg in 1888, and "exhibits from Europe became quite frequent in the following decade." In Elizabeth Valkenier's assessment, those events contributed to the Wanderer school's loss of

12 Møller, *Postlude to* The Kreutzer Sonata, 191, 227.

13 Censorship mutilated "A Loose Tongue" ("Dlinnyi iazyk") when first published in 1886: see Chekhov, *PSSiP*, 5:653,

authority among the new generation of artists, who included Mikhail Vrubel and Levitan. By comparison to the Wanderers, the younger painters were better educated, born into more secure socioeconomic positions, less defensive about the West, and receptive to the idea of art for art's sake.[14] Stasov and other nationally engagé traditionalists were at loggerheads with this innovative, modernist sensibility that produced the *World of Art* circle and journal. The long-planned opening of the Russian Museum in Moscow in 1898 precipitated a big showdown between the two camps. A pet project of Alexander III, the museum elicited scathing reactions from Sergei Diaghilev (1872–1929) and a like-minded commentator: the catalogue was "patriotic howling" about trivia, the collection contained "rubbish," and visiting it was tantamount to landing in "the sticks" (*zakholust'e*), the stultifying, stagnant Russian provinces.[15] Along with the cosmopolitan turn in the art world, a supranational approach to literature developed as well. The significant new journal was *Cosmopolis* (1896–98), a multilingual literary review addressed to worldly, metropolitan audiences.[16] Founded in London, *Cosmopolis* had a distribution center in Petersburg as well as New York, Paris, Vienna, and Amsterdam. Among its offerings was Chekhov's story "U znakomykh" ("The Visit", 1898) a relatively lighthearted depiction of provincial Russian banality.[17]

Russian material consumption also integrated the foreign into daily life. Stores in the capitals sold grand pianos from England and Austria, English bicycles and saddles, "Huste-Nicht" German cough drops, French toothpaste, perfumes and cosmetics, and many other Western commodities.[18] By the first half of the nineteenth century, a "fashion press" for women had arisen in Russia. It reached into the provinces, privileged Paris styles, and provided patterns to make clothing at home or else pay a dressmaker to do so.[19] That tradition reached a new level in 1890 when Printemps, the Paris *grand magasin*, ran an illustrated ad in French in the mass-market *Grainfield*. Russian shoppers could order a free

14 Valkenier, *Russian Realist Art*, 128–30, 211n52.

15 Dianina, *When Art Makes News*, 186–88.

16 On the journal's history and target audience, see Agathocleous, *Urban Realism and the Cosmopolitan Imagination*, 53–56, 62–68.

17 Publication information in Chekhov, *PSSiP*, 10:355.

18 Ads in *Niva*, no. 1 (1890): 29; no. 23: 520.

19 Christine Ruane, *The Empire's New Clothes: A History of the Russian Fashion Industry, 1700–1917* (New Haven: Yale University Press, 2009), 132–33; and Ruane, "European Fashion in Russia," in Kivelson and Neuberger, *Picturing Russia*, 119–23.

Printemps catalogue and samples of cloth and have their custom-made clothing shipped anywhere in the world. Printemps also announced it could conduct correspondence *en toutes langues*.[20] Russians desiring to write letters or keep records in a worldly manner had only to acquire "The Cosmopolitan," a "new invention" put on the market in Petersburg in 1892. It was a patented typewriter with interchangeable typefaces in the Cyrillic and Roman alphabets.[21] Moscow's Oriental Department Store (*bol'shoi vostochnyi magazin*, a calque of *grand magasin*) sold "genuine Persian rugs," dressing gowns, ottomans, and other commodities that brought elegant touches of Asia into the everyday lives of well-to-do Russians.[22] Petersburg had Tatar eating houses (of unenviable reputation) and popular Georgian restaurants. Patrons of the latter included the venerable populist Mikhailovsky. French restaurants retained a high-class presence in the capitals (Dostoevsky had dined at Les Rochers de Cancale in the 1870s, and the eponymous haut-bourgeois judge of Tolstoy's *The Death of Ivan Ilyich* [1886] patronizes Donon's).[23] But even Petersburg taverns catering to cab drivers had names to stir reveries of travel: "Venice," "Naples," "Caucasus," "Persia," and others.[24] Perhaps those establishments had themed menus and pictures on the walls?

Lipson's travel agency, opened in Petersburg in 1885, sought to make tourist dreams come true. For clients looking to rough it, the company offered excursions to (undesignated) places with no railroads or communication networks. In such cases, the agency would provide "guides, translators, cooks, domestic servants, pack animals, animals to ride, porters, sedan chairs, tents, and beds for over-nights." Lipson's specialty, though, was the Cook-style tour of foreign cities where customers would stay "in the best, most comfortable hotels." Among the offerings was a forty-day train trip of Italy, departing from Petersburg and priced at 775 rubles. The pace sounded hectic. After speeding through Warsaw and Vienna so as "to leave behind the severe North as fast as possible," the tourists would spend two and half days in Venice, one in Florence (with two more on the

20 *Niva*, no 13 (1890): 367.

21 "Novoe izobretenie (Pishushchaia mashina 'Kosmopolit')," *Niva*, no. 37 (1892): 819.

22 Ibid., no. 1 (1899): 2.

23 Iu. B. Demidenko, *Restorany, traktiry, chainye. Iz istorii obshchestvennogo pitaniia v Peterburge XVIII–nachala XX veka* (Moscow: Tsentrpoligraf, 2011), 73–77, 115, 117; and Tolstoi, *PSS*, 26:70.

24 Julie A. Buckler, *Mapping Saint Petersburg: Imperial Text and Cityshape* (Princeton: Princeton University Press, 2005), 106.

return journey, including a side trip to Pisa), six days each in Rome and Naples, two days each in Genoa, Turin, and Milan, a day in Verona, two days each in Munich and Vienna, and half a day in Warsaw. The brochure requested customers to conform to the norms of "polite society," especially with regard to the "female sex," and to refrain from discussing politics and religion so as to maintain "amical accord" throughout the trip. Lipson's flashiest offer, costing 2,500 rubles, was a 120-day round trip via trains and steamboats from Petersburg to the "Orient," including Constantinople, Bucharest, Beirut, Cairo, a cruise on the Nile, and on the return journey Naples, Rome, Florence, and other Italian cities, Paris, Brussels, and Berlin.[25] While wealthy people might take such a vacation, Chekhov stories of the 1890s feature teachers in rural Russia earning only twenty-one or twenty-five rubles a month.[26]

Narratives of leisure travel contributed to the allure of the West. In 1890 a new translation of *Childe Harold's Pilgrimage* began serialization in the journal *Russian Thought*, reintroducing into the public sphere Byron's glamorization of flight from home.[27] On the same wavelength, Balmont published in 1896 a poem in the spirit of Harold bidding good-riddance to England ("Adieu, adieu! my native shore," 1.28).[28] For the first time since they appeared in the 1840s, Annenkov's *Letters from Abroad* were reprinted posthumously in 1892. Professor Pypin's review heartily recommended the book and quoted Annenkov's passage about the enjoyment of West European spas as "centers of cosmopolitanism." In Pypin's estimation, Annenkov regaled the reader with "stories of a tourist" but "a tourist who knew how to integrate details into a general scheme and evaluate broad phenomena of intellectual and literary life" abroad.[29] With his joie de vivre, emulation of Byron, and praise for trains as transport for cultural tourism, the Annenkov of *Letters from Abroad* was back as an acclaimed model for Russian leisure travel. Madame Kurdyukova, the cosmopolitans' laughing double, was also

25 Lipson, *Pervoe v Rossii predpriiatie dlia obshchestvennykh puteshestvii*, 3–4, 8–13. The number of Russians who actually took Lipson trips appears to remain a mystery.

26 Chekhov, *Dom c mezoninom* (*The House with the Mezzanine*) and "V podvode" ("In the Cart"), in *PSSiP*, 9:176, 341.

27 *Chaild Garol'd* [sic], translation by Pavel Kozlev of *Childe Harold's Pilgrimage*, 1.1–93, in *Russkaia mysl'* (January 1890), section 1: 1–16; and ibid. (February 1890), section 1: 1–16.

28 K. D. Balmont, "V more" ("At Sea"), *Nesobrannoe i zabytoe iz tvorcheskogo naslediia*, 2 vols., ed. A. Iu. Romanova, (St. Petersburg: Izdatel'stvo Rostok, 2016), 1:48.

29 A. N. Pypin, "P. V. Annenkov," *Vestnik Evropy* (March 1892): 302, 308.

back. Reprintings of Myatlev's tourism feuilleton appeared in 1893, 1894, 1898, 1904, and 1907.[30] The nineties also saw the debut of Nikolai Leikin's derivative but extremely popular comic sagas *Where Oranges Ripen* and *Our People Abroad*.[31] In representing Russian merchants bumbling through Berlin, Vienna, Paris, the French Riviera, and Italy, Leikin copied Myatlev's practice of transliterating foreign words into Cyrillic. As in earlier times, the yokels' slipshod speech and general ineptitude invited fin de siècle readers to feel sophisticated, and that satisfaction no doubt contributed to the books' continuing success in the Soviet period. Leikin's journeyman style, however, and his anti-Semitic jokes about Viennese Jews placed his books aesthetically and morally far below the poetry and cosmopolitan spirit of Myatlev's classic invention of an innocent abroad.[32]

Georgia and the Rest

Among the Russian Empire's exotic alternatives to vacationing abroad, Georgia began acquiring four-star status in the 1880s. Tiflis and Borjomi were the most popular destinations, made increasingly accessible by improvements and expansion of infrastructure. A *Grainfield* article of 1885 publicized the ease of travel along the refurbished Georgian Military Road, smoothed and restructured in zig-zags that eliminated the scary steep ascents and descents of the past. With prices starting at fifteen rubles, public carriages and omnibuses now operated from Vladikavkaz to Tiflis, and private coaches could be rented at higher cost. The trip took two days and passed through the "most picturesque spots in daytime." Tourist were further assured that all along the route, they would find good hotel-quality accommodations in posting stations offering "healthy and tasty," if not

30 Kovarskii, "Poeziia Miatleva," 616; and data from library catalogues.

31 N. A. Leikin, *Gde apel'siny zreiut. Iumoristicheskoe opisanie puteshestviia suprugov Nikolaia Ivanovicha i Glafiry Semenovny Ivanovykh po Riv'ere i Italii* (St. Petersburg: S. P. Iakovlev, 1898); and *Nashi zagranitsei. Iumoristicheskoe opisanie poezdki suprugov Nikolaia Ivanovicha i Glafiry Semenovny Ivanovykh v Parizh i obratno*, 16th ed. (St. Petersburg: S. P. Iakovlev, 1899). On Leikin's tourists as "protagonists of modernity," see McReynolds, *Russia at Play*, 155. Both books were reissued many times and recent reprintings are now marketed on the internet. "Where oranges ripen" refers to Seville in Pushkin's "K vel'mozhe" ("To a Grandee"), an 1830 lyric that imagines the scenic and cultural pleasures of the Grand Tour of his addressee, Prince Nikolai Yusupov (1751–1831).

32 For the anti-Semitism, see *Nashi zagranitsei*, esp. 473–78.

"elaborate" fare.[33] In the 1890s, Russia's railroad network extended to all the Caucasian spas, no doubt enlarging the pool of vacationers wanting to go on to Tiflis.[34] At the turn of the century, a Russian youth described the Georgian Military Road as a beaten track much less "picturesque" than the "extremely wild" Ossetian Military Road that he chose to explore.[35] By 1903, the "flood" of Russian tourist traffic to Georgia had become so enormous that a guidebook advised travelers to buy carriage tickets from special agents far in advance or else risk waiting for days in Vladikavkaz to obtain a seat.[36] Serious bikers were also hitting the Road. In August 1903, an organized group of members of the Society of Russian Tourists spent three days cycling and walking from Vladikavkaz to Tiflis. Not lingering to sightsee, they began the return trip the following day.[37] For Russian vacationers choosing to linger, Borjomi as well as Tiflis was a popular spot. Opened in 1872, the Poti-Tiflis railway was publicized in Russia in the eighties as "one of the most beautiful in the world," just waiting for tourists to use it.[38] That railroad facilitated trips to Borjomi, whose late nineteenth-century Russian visitors included Tchaikovsky and Chekhov.

Tiflis was the urban center, and Borjomi was the scenic spa; but both had affiliations to Russian aristocracy, conferring the leveling-upward stature that appealed to middle-class Russians. Furthermore, the absence of an "oriental" quarter made Borjomi the more uniformly European and therefore culturally advanced tourist option in Russian eyes. *Grainfield* interpreted for mass consumption Russia's annexation of Georgia as the start of a long struggle to introduce order, cleanliness, and "houses of purely European architecture" into the "purely oriental" city of Tiflis

33 *Niva*, no. 45 (1885): 1086. On the other hand, John Murray's 1888 guide advised travelers to take provisions because "little beyond black bread is obtainable on the [Georgian Military Road]." Accommodations were deemed merely "tolerable" in generally "dirty" post-houses. The guidebook's next edition (1893), however, spoke of "very fair sleeping accommodations" and "restaurants well supplied with food, wine, and beer." See Murray, *Handbook for Travellers in Russia*, 407–8 and 319, respectively.

34 For an account of such a trip, see S. Zarin, "Gruzinskaia Voennaia Doroga. Putevye vpechatleniia," *Zvezda*, no. 15 (1893): 285–90.

35 A. Mech, "Voenno-ossetinskaia doroga na Kavkaze," *Russkaia mysl'*, no. 27 (1900), section 2: 26. Running between Kutais and Alagir, the road was built between 1854 and 1889.

36 Moskvich, *Illiustrirovannyi prakticheskii putevoditel' po Kavkazu*, 168.

37 The society prepared a detailed round-trip itinerary, indicating the distance between each posting station: see a reproduction of the document in Dolzhenko and Putrik, *Istoriia turizma*, 61.

38 *Niva*, no. 49 (1882): 1166–67.

with its "houses like flat-roofed dug-outs" and "narrow, crooked streets, lanes, and alleys crammed with all sorts of rubbish and filth." Lending the city "some Christian character," the "magnificent cathedrals" alone had "somewhat redeemed the generally unsightly picture" that met the gaze of Russian civilizers in 1801. Vorontsov's urban transformation of Tiflis left a lasting Russian idea of the city as half-European, half-Asian. Symbolizing European progress, one of the new town's tourist attractions was the viceroy's palace on Golovin Prospect. Built under Vorontsov in the 1840s, the palace was "extended and remodeled in Renaissance style" during the period when Alexander II's nephew the Grand Duke Mikhail Nikolaevich was viceroy (1862–82).[39]

The public image of Tiflis as a culturally divided city spatialized time to produce a chronotope of the Orient that functioned as a theme park for tourists. At the fin de siècle the very names of the best-ranked Tiflis accommodations—Hôtel de France, Hôtel d'Europe, and the *Londres* (French for "London")—promised Russian urbanites home-plus. According to Moskvich's guide, tourists staying in the European part of Tiflis would feel as though they were in Petersburg or Moscow.[40] But as *Grainfield* had put it in 1882, modern Europe coexisted with old Asia: "Many traces of old times long past have persisted in the life of the [Georgian] people, but that life is hidden away in remote, secluded nooks." It is concentrated in those quarters where "winding, reeking streets; mosaics, and flat roofs are intact." Artisans work there in the open air, and food vendors grill shashlik and cook pilaf. In guidebook style, *Grainfield* urged Russian tourists to dare to explore and try the food: "Do not be squeamish about those winding, dirty streets. Walk along them and deign to visit the rest of the old town's sights," including the oriental baths and the Armenian bazaar.[41] Unlike the condescending journalist, Tchaikovsky on his stroll through the old quarter of Tiflis in 1886 perceived not Asian dirtiness but rather a charming resemblance to Venice with its narrow lanes lined with little shops.[42] Grigory Moskvich's 1903 guidebook spruced up Tiflis's old town but retained the

39 Ibid., no. 19 (1882): 450.

40 G. Moskvich, *Illiustrirovannyi prakticheskii putevoditel' po Kavkazu*, 8th ed. (Odessa: Tipografiia L. Nitche, 1903), 229.

41 *Niva*, no. 33 (1882): 784.

42 P. I. Chaikovskii, letter to his brother Modest, April 1, 1886, in his *Pis'ma k blizkim. Izbrannoe* (Moscow: Gosudarstvennoe muzykal'noe izdatel'stvo, 1955), 350.

oriental chronotope. In recommending standard sightseeing and culinary adventures, Moskvich singled out Persian restaurants—clean Asian places to eat before returning to the tourist Europe of hotels and other architectural emblems of imperial Russian power.[43]

The architecture of Borjomi blended Western vernaculars with the East, not in its "dirty" guise, however, but rather in exquisite forms befitting the spa's emphatically elite social ambience.[44] In 1871, at the behest of Alexander II, Borjomi became the Grand Duke and Viceroy Mikhail Nikolaevich's personal property in perpetuity. Borjomi was also renowned in Russia as the summer retreat of the "affluent population of Tiflis."[45] Ordinary Russian visitors thus became titular guests of the royal family and rubbed shoulders with well-to-do Georgians. The spa became known as "the Caucasus's Vichy." The finest dachas bore (usually transliterated) names such as "Belle-Vue" and took inspiration from various European architectural traditions, including the Gothic and the Swiss chatlet.[46] Nestled in lush flora by the Kura River, Mikhail Nikolaevich's Borjomi palace became a tourist attraction. Illustrated articles in *Grainfield* in 1884 and 1895 described the royal residence as a "simple but extremely beautiful and elegant" two-story building blending the West and the Orient: its carved ornamentations "in the Caucasian style" are "delicate as Brussels lace," and its balconies, galleries, and terraces bring to mind the Alhambra.[47]

While tourism in Georgia gained a glamorous, elite aura, the spas of the North Caucasus remained subject to often anxious public debate about whether they measured up to West European alternatives. The Pyatigorsk cluster of resorts had been losing aristocratic exclusivity for

43 Moskvich, *Illiustrirovannyi prakticheskii putevoditel' po Kavkazu*, 244.

44 Note the disenchanted reaction to Tiflis in Pavel Zalesov, *Kartinki Kavkaza. Putevye vospominaniia i zametki o kavkazskoi prirode* (Moscow: Tipografiia E. Lissner and Iu. Roman, 1883), 82–83: "For some reason I had created in imagination a picture of one of those marvelous oriental cities that gleam with wonders of art: houses submerged in the greenery of gardens, the plash of fountains, delicate minarets, svelte beauties in captivating chadors—mysteriousness, luxury, and a special oriental poetry. But my God, what did I see?! Just imagine a big pit, hemmed in on all sides by low, muddy-gray hills seared by the sun. Not a bit of foliage, nothing but earth—dry, ash-colored earth."

45 *Niva*, no. 43 (1884): 1035.

46 Dzhanshiev, *Perl Kavkaza*, 79, 82

47 *Niva*, no. 43 (1884): 1020, 1034; and ibid., no. 20 (1895): 476, 480. Much better known today is the Romanov palace in Likani, a townlet of Borjomi. Taken over by the Soviet state, it was used as a vacation retreat by Stalin and other Soviet dignitaries.

a long time.[48] Upper-class Russians exercised their preference for cure centers abroad, which were in fact often cheaper but had snob appeal (and were of course more costly to reach). St. Remo joined the upper crust's list of fashionable spas in the 1870s when Alexander II's wife, Maria Alexandrovna, began frequenting it for treatment of tuberculosis. Lesser Russian aristocrats followed her lead and acquired properties on the Italian coast and the French Riviera.[49] In the same period, a tourist from Petersburg, proud of his familiarity with Europe, published a litany of complaints about the North Caucasus spas. He spoke of dusty air, bedbugs, scarce provisions, lazy Georgian service personnel ("the people here just do not want to work"), and foul but paid-admission privies dotting the Kislovodsk park.[50] The complainer also reported a confrontation with a Pyatigorsk hotel manager who said his establishment catered to "modest, easily pleased" customers, rather than Petersburg people who could and should go abroad.[51]

As in Verderevsky's travelogue of the 1850s, "modest" designated middle-class people from far and wide in central Russia and set them apart from snobs. A *Grainfield* article of 1881 had words to fortify the social underdogs' self-esteem. The magazine described the "district town" (*uezdnyi gorod*) Pyatigorsk as "one of the most remarkable in Europe by virtue of the abundance and variety of its mineral springs."[52] The domestic resort had moved from Asian borderland into Europe.[53] An 1885 guidebook stopped short of this radical realignment of imperial/imperialist Russian imaginative geography: the author did not relocate Pyatigorsk *in* Europe but said it *looked* European.[54] The *look* of Europe derived from civic and religious architecture, improved tourist facilities, and the clustering of westernized Russian visitors. In addition, Pyatigorsk had become Lermontov country. On the fortieth anniversary of his death in a duel in the outskirts of the

48 On a comparable development at British spas in the eighteenth century, see Walton, "British Tourism between Industrialization and Globalization," in Berghoff et al., *The Making of Modern Tourism*, 116–17.

49 Mal'gin, *Russkaia Riv'era*, 311–12, including remarks on the better price-quality ratio abroad.

50 Ivanov [sic], "Poezdka na obnovlennye kavkazskie vody," *Russkii vestnik* (October 1875): 690, 710, 712–13.

51 Phillips, "Health and Sociability," 321n52.

52 *Niva*, no. 11 (1881): 247.

53 Covering Tambov among other towns, *uezdnyi gorod* was typically a lowly category in Russia's "provincial geography," as argued in Lounsbery, *Life Is Elsewhere*, 34, 84, 149.

54 Phillips, "Health and Sociability," 307.

spa, tourists and local residents of the town honored his memory with a requiem in the local cathedral, speeches, the unveiling of a bust, a tour of his house and haunts, a charity bazaar to raise money to add to the 20,000 rubles already gathered for a monument, and, finally, a dinner at a hotel.[55] For happier occasions, tourists by the mid-1880s could enjoy Lermontov-themed meals with pies à la Pechorin and other dishes named for the poet's Demon and Princess Tamara.[56]

By joining the crowd at the North Caucasus spas, Russians of relatively modest socioeconomic standing could perform "European" identity. Middle-class attraction to this practice was demonstrated by the growing numbers of Russian doctors, teachers, lawyers, university professors, and merchants at these (and other Russian resorts) into the late nineteenth century.[57] Noticing lots of doctors vacationing in Yalta, a seasoned visitor declared in the early 1890s that the most common kind of tourist at the "Russian Riviera" was now a *raznochinets*, a designation contextually connoting an arriviste.[58] Soon to move with his sister and mother into his new home in Yalta, Chekhov, too, remarked in 1899 that the southern coast had become the favorite vacation spot for "zemstvo doctors from Moscow province."[59] He made them sound easily satisfied, which gave Chekhov the doctor's observation a supercilious edge. All the same, Chekhov the *meshchanin* had expressed a "conflicted social imagination" much more strongly in 1888 when he first espied Yalta from the deck of an arriving steamboat.[60] He had the impression of "a mixture of something European, reminiscent of aspects of Nice, and something of the Russian petty-bourgeois trade fair" (*s chem-to meshchanski-iarmarochnym*).[61] Thus did the son of a shopkeeper and grandson of a serf initially respond to the town where he would come to live as a celebrated writer, a writer whose story "The Lady with the Little

55 *Niva*, no. 36 (1881): 796.

56 Zalesov, *Kartinki Kavkaza*, 33 (cited in Phillips, "Health and Sociability," 150n51).

57 Phillips, "Health and Sociability," 247–70.

58 A. Petrushevskii, "Russkaia Riv'era," *Vestnik Evropy* (March 1892): 627, 644.

59 Chekhov, letter to Grigory Rossolino, October 11, 1899, in *PSSiP, Pis'ma*, 8:284.

60 For the quoted concept, developed in analysis of Chekhov's plays and theater career, see Edith W. Clowes, "Social Discourse in the Moscow Art Theater," in Clowes et al., *Between Tsar and People*, 271–87 (quote, 285). A similar point about Chekhov is made in discussion of his 1888 Yalta letter (cited below) in Kelly, *Refining Russia*, xxxvi, 205–206; and Phillips, "Health and Sociability," 297n168.

61 Chekhov, letter to M. P. Chekhova, July 14, 1888, in *PSSiP, Pis'ma*, 2:295.

Dog" immortalized the resort as a place of scenic enchantment, minus a vulgar tourist crowd.

The Crimea certainly had its aristocratic affiliations. Tourists could enjoy the parks of the royal estates Livadia and Oreanda, and they even could hope to get a glimpse of the tsar in Yalta. As in the North Caucasus, however, the bourgeois tourists in the Crimea struck snooty observers as people of *mauvais ton* (bad taste). The peripatetic travel writer Mikhail Bernov, for example, deplored the inelegant clothing and demeanor of Russian women tourists in Yalta by comparison to the refinement he had seen in Trouville, Biarritz, and other Western resorts.[62] Russian vacationers in the empire might master *bon ton* and yet take part in the illicit sexual experimentation that resort newspapers, "The Lady with the Little Dog," and lesser imaginative literature depicted as standard behavior in the North Caucasus and the Crimea.[63]

Pig-Man on Vacation

Uspensky's sex-minded *burzhui* in the North Caucasus incarnates Russian domestic tourist vulgarity at its most extreme. In the cataclysmic months leading up to the Bolshevik revolution, Russian socialists used *burzhui/bourgeois* as smear terms not only for a Russian bourgeois strictly speaking but rather for "all property-owners, even those just living comfortably."[64] "The *Burzhui*" begins in Essentuki with the narrator seething about the cure regime. For twenty days straight, he has followed doctor's orders to drink half a glass of water no. 17 and a half glass of warm milk at 6:00 am sharp in the cure house. The regimentation has instilled a terror of being late that he compares to his trepidation as a boy unable to please his "merciless" German tutor who understood no Russian. Having obeyed the rules ever since he arrived, the author is now in revolt. He will *not go* to the cure house this morning: "To hell with them, all those idiotic half glasses!"[65] Like Dostoevsky's underground man, Uspensky's persona has asserted his free

62 M. A. Bernov, *Iz Odessy peshkom po Krymu. Pis'ma russkogo peshekhoda* (St. Petersburg: Glazunov, 1896), 191–92.

63 Phillips, "Health and Sociability," 191–95.

64 Boris I. Kolonitskii, "Antibourgeois Propaganda and Anti-'*Burzhui*' Consciousness in 1917," *Russian Review* 53, no. 2 (April 1994): 190–91. Vladimir Mayakovsky's couplet of 1917 contains the untranslatable rhyme of *burzhui* with *zhui* ("chew"): "Eat your pineapples, chew your grouse, / Your days are numbered, *burzhui*."

65 Uspenskii, "Burzhui," *PSS*, 9:457–58 (subsequent citations appear parenthetically).

will against the system. He nonetheless remains cross without understanding why. Could it be the ghastly "potpourri" of music twice a day (lots of *La Traviata* and army songs)? Or maybe just being amidst many sick people?

The author finally puts his finger on the main source of aggravation: it is the "new-born Russian *burzhui*," emanating the "odor of a cold corpse." There follows a general disquisition on the *burzhui* type. Uspensky's narrator remarks that Turgenev's *Virgin Soil* simply transliterated "bourgeois" into Cyrillic as a description of the character Faleev, a Russian merchant. But a new word is needed, insists the Essentuki author, because contemporary Russia's commercially minded, middle-class men are not comparable to a French bourgeois or German *Bürger*. The difference lies in the existence of a West European *culture* of commerce stretching back for generations. The sketch's emblem of the culture of commerce is a German tongue sausage with pistachios, a product allegedly created to please Friedrich the Great. "The *Burzhui*" suggests there is something absurd about this unnecessarily fancy culinary concoction. But then the author imagines at length the initiative, creativity, and pride in service that went into the conception, production, packaging, advertising, and selling of the sausage. In Russia, on the other hand, the creative commercial ethos of the Western bourgeoisie does not exist. The *burzhui* "horde" have "wads of unearned cash" but no historical roots, no culture whatsoever, no refinement. Accustomed to greasing palms, they live in a "pig-swill" and are driven by "greedy, animal strivings for profit." Their motto is "buy it!" whatever it may be: "somebody else's wife, a ballet dancer," a contract, an elector, a government official (462–70).

The "pig-swill" image carries the author back to the spa and the sketch's eponymous specimen, henceforth designated "*burzhui*-belly." The swinish creature was one of the strangers with whom the narrator shared a second-class compartment of a steamer on a southern Russian river (the others were three Jews whose mumbled praying and failure to clean their crumbs off the common table annoyed the author). A well-dressed, well-coiffed man in his forties, the *burzhui* began showing his voracious appetite on the boat. At the end of the line, while killing time before taking a train to the spa, he had their hotel proprietor bring him a girlish prostitute. In Essentuki he has become a leech, bending the narrator's ear with talk about his sexual and culinary exploits and drinking bouts. For him, vacationing in the Caucasus is indeed a home-plus experience. As in Moscow, he pairs off with floozies, overeats, and gets drunk but does it all on the heights of

Mt. Bermamut, while clouds float past his self-described "mug." If it were not for fear of enraging pig-man, the author would tell him to his face that he is "vile and disgusting, a blood-sucker, an ulcer, and that there is no word in human language to name his existence" (479).

After *burzhui*-belly has left the spa, the narrator recuperates in the peace of a "splendid southern evening" and reaches out for restorative human contact with a Russian man of the folk, an elderly peasant that he has noticed in the courtyard of his accommodations but never spoken to (480). The local man is a mason who has been busy carving a funerary monument, an activity suggesting religious concern. The author's earlier remarks on the inspiring "simple Russians" that he encountered on his boat trip before pig-man invaded his life prepare the ground for the symbolic mason. "Interested in moral questions and trying to solve them," the inspiring people were primarily Mennonites, Baptists, and other sectarians, but also German colonists, and Russian pilgrims going to Kyiv or Jerusalem (471). Unlike the Jewish strangers in the author's boat cabin, the Christian religious seekers speak to his own spiritual yearnings. The memory of those people resonates in his decision to speak at last to the mason. Giving his action the air of a ritual cleansing after *burzhui* contamination, the tourist narrator seeks out the peasant: "I shall go to him now to unburden my soul" (480). The narrative ends with those words, underlining the gulf between "*burzhui*-belly" and the mainly Russian spiritual people, a reminiscence perhaps of the peasant Fyodor's formulation to Levin in *Anna Karenina*.

From Vichy to the Caucasus

Uspensky's gross *burzhui* provides a foil to the classy bourgeois tourists in Veselitskaya's *Mimi at the Spa* (1891), the second of three stories put together as a trilogy after appearing separately in the *European Herald* under the author's pseudonym V. Mikulich.[66] The main characters are all cultivated but fall along a remarkably proto-Soviet fault line between pleasure and purpose. On the one side stand adulterous Mimi and her lover

66 On the trilogy as "a witty picture of the average *jeune fille* of Petersburg bureaucratic society," see Mirsky, *A History of Russian Literature*, 352. "Mimochka—nevesta" ("Mimi the Bride", 1883) related the heroine's life from childhood up the age of twenty, when a financial disaster in her Petersburg family leads to her arranged marriage to an ageing general. The third, longer story *Mimochka otravilas'* (*Mimi has Taken Poison*, 1893) deals with her post-vacation life as an embittered serial adulteress who attempts suicide when her children's French tutor breaks off their affair.

Valerian, both linked to West European fashion and poetry.[67] Mimi takes her cues from Paris, while Valerian represents both the English and French strands of Russia's "culture of the dandy" at the fin de siècle.[68] On the other side of the divide stands Mimi's cousin Vava, an adolescent disaffected to her social class. She is close to her father, but he is too weak to stand up to her disapproving mother and brother, who predicts that if she is not sent to school in France or Switzerland for a couple of years, she will end "on the gallows."[69] Associated with Russian rather than foreign literature, Vava goes to the Caucasus determined to achieve physical, civic, moral, and intellectual self-improvement.

Mimi at the Spa opens with the twenty-six-year-old heroine wasting away after six years of marriage. Her *maman*, her husband General Spiridon Ivanovich, and most of her aunts cannot imagine what ails her. She leads a life of leisure, after all: shopping, socializing, riding about town, and going to exhibits.[70] They assume Mimi must be sick, perhaps suffering aftereffects of childbirth under chloroform. Aunt Mary, however, suggests to other aunts that Mimi's problem is having an "old husband." The omniscient narrator reveals that Mimi is indeed unhappy with marriage and motherhood. Under the constant supervision of the general and *maman*, Mimi feels like "a fly in a spider's web" (51–52). Spiridon Ivanovich is an important military official whose various missions included one to the Caucasus, where he admired Bermamut. *Maman* loves talking to him about his job and politics. Mimi, on the other hand, finds both topics boring. Her screaming infant is also driving her crazy. The boy's name is not mentioned. He is identified either as "baby" transliterated into Cyrillic or else *bebichka*, a diminutive of "baby." Defamiliarizing the child, the strange-looking Cyrillic appellations

67 Valerian's name proclaims him the man to cure what ails Mimi, a habitual user of valerian drops.

68 On the general phenomenon, see Olga Vainshtein, "Russian Dandyism: Constructing a Man of Fashion," in *Russian Masculinities in History and Culture*, ed. Barbara Evans Clements et al. (Basingstoke: Palgrave, 2002), 67.

69 L. I. Veselitskaia [V. Mikulich, pseud.], *Mimochka na vodakh*, in *Mimochka* (Moscow: Eksmo, 2003), 65 (subsequent citations refer to this edition).

70 In the third story of the trilogy, Veselitskaya would specify the heroine's position in the bourgeois hierarchy: Mimi "belongs to the middle sphere of Petersburg society, people with lots of money and no shortage of pretensions, people who come into contact with *la crème de la crème* and the small fry, currying favor with the former and snubbing the latter," 224.

underline Mimi's alienation from motherhood and mock the contemporary Russian bourgeois fad for things English.[71]

Paris rather than London, however, is Mimi's pacesetter for cosmopolitan self-fashioning. Along with the name "Mimi" and the steady usage of *maman*, conversational exchanges and characters' thoughts in French are sprinkled throughout the text to imply the bourgeois family's general preference for that language. In the retrospective representation of Mimi's three-month honeymoon in Vichy and Paris, she experiences France as her true "homeland" (*rodina*). Charmed by all the French people she encounters, she perceives some as authenticators of the French look of her French Petersburg hairdresser and French actors at the Mikhailovsky theater (45–46). Vichy's elite international clientele parade beautiful clothes and go to balls, parties in casinos, and concerts whose offerings include Strauss and Patti.[72] The subsequent stay in Paris was Mimi's "real honeymoon," her peak time of material consumption. The episode epitomizes Veselitskaya's modeling the heroine as Veblen's "ceremonial consumer" type: the married woman whose acquisition and display of expensive clothing, accessories, and "household paraphernalia" publicly proclaim her husband's wealth.[73] After sightseeing every morning with Spiridon Ivanovich, Mimi would "go shopping, buying and buying," while he napped in their luxury hotel. The material consumption includes gourmet food. Apparently having a room of her own, Mimi arises late, dolls herself up, and then joins her husband for breakfast. He kisses her hand, she kisses his balding head, and then they tuck into "cutlets *en papillote*, lobster, and hors d'œuvres" (48–49). Paris, Vichy, French clothing and other commodities for women, French culture, and French cuisine comprise nationally disassociated Mimi's discovery of a "homeland" better than home.

In the eyes of *maman* and Mimi's aunts, her honeymoon experience situates her in Russia's worldly social ranks too lofty to descend to taking a cure in the Caucasus. After Mimi has hysterics at the dinner table one evening, *maman* hustles her off to their doctor, a medical man of

71 On the general fad, see Lovell, *Summerfolk*, 88, 100.

72 Adelina Patti (1843–1919) was an Italian opera singer acclaimed in musical capitals of Europe and America.

73 Veblen, *The Theory of the Leisure Class*, 58; and Mary Louise Roberts, "Gender, Consumption, and Commodity Culture," *American Historical Review* 103, no. 3 (1998): 819–20. On the "extreme view of female acquisitiveness" in *The Kreutzer Sonata*, see also Sally West, *I Shop in Moscow: Advertising and the Creation of Consumer Culture in Late Tsarist Russia* (DeKalb: Northern Illinois University Press, 2011), 140.

high standing in their social circles. Having assumed he would recommend a cure in Western Europe, *maman* is horrified when he proposes the Caucasus. Many people have told her, she says, that those spas are "so primitive, so undeveloped. There are no apartments, no doctors. Just awful quacks, I hear. And there's nothing to eat." Even as the travelers are about to head south on the train, Aunt Sophie says they should be going to Carlsbad instead. Aunt Mary, too, envisages a wasteland: "I know perfectly well what the Crimea means, what the Caucasus means: hunger, boredom, and dirt. It's a waste of money" (58, 68–69). Mimi, however, is secretly pleased because a married woman friend of hers was sexually "emancipated" during a recent summer vacation in Kislovodsk (59). Besides, Mimi now needs a vacation wardrobe. She goes on a shopping spree at Knop's immediately after the medical consultation.[74] Mimi also looks forward to getting away from baby. Since Aunt Julie is exasperated with her rebellious daughter Vava, the family makes a deal: "Vava for baby, baby for Vava" (66). And so, having left the infant with Julie, Mimi goes to the Caucasus with her maid Katya, *maman*, and Vava.

Clashing Plans and the Dogs

No less than Mimi, Vava looks forward to slipping the leash of domestic supervision but for the entirely different reason of having the freedom to pursue self-improvement as she sees fit. At the spa Mimi resists having sex with Valerian until the last third of the story. The long build-up to that consummation elaborates the contrast between Mimi and Vava: decadent hedonism and foreign culture, on the one hand, versus idealism, purposefulness, and Russian culture, on the other. Dogs play important roles as symbolic adjuncts of their owners and as movers of the plot.

When Mimi returns from her honeymoon in France, her relatives perceive her as a new, mature woman. Having shed the demeanor of a "poor bride," she now has the look of the "wife of a general, a division commander." She has become "a lady with a *toilette* fresh from Paris and a position in the beau monde" (49). Mimi the "lady" (*dama*) takes to the Caucasus her "little dog" (*sobachka*) Monichka, a diminutive of Monique. Echoing "Mimochka" (the heroine's nickname), the name "Monichka" lends the pet the air of an

74 Knop's was a Petersburg shop renowned for having the latest styles: see W. [sic] Mikoulitch, *Mimi aux bains de mer*, trans. Charles Simond (Paris: H. Geffroy, 1895), 30.

emblematic totem of her owner. Veselitskaya draws out this implication in the scene of Mimi's relatives gathered at the train station to see her off. The "elegant group" includes Vava's older sister, "Zina the beauty." Wearing "a huge fashionable hat and a wee fashionable blouse," Zina is carrying "two white lapdogs [*bolonki*] who gaze upon God's world just as boldly and disinterestedly as their mistress" (67). The smart set's indifference to religion and moral questions is placed on the level of a dog's.

The spiritual chasm between Mimi and Vava yawns wide on the train trip. Mimi's party travels in a private ladies' sleeping car with velvet sofa-seats, mirrors, blue window-shutters, lilac shades on the night-lights, and tea provided by the conductor. The journey is barely underway when Vava excitedly anticipates seeing the Caucasus (70). She steps into the corridor, gazes up at the sky, and recites Lermontov's Caucasian exile lyric "Tuchki nebesnye, vechnye stranniki" ("Celestial clouds, eternal wanderers"). In the meantime, Mimochka settles into the compartment, takes off her new travel hat, and plays with Monichka. In her berth that night, Mimi finds the train's motion and hubbub lulling, as it speeds away from the Petersburg bed she associates with insomnia, crushing silence, and sheer "torture." She is now alone with her thoughts. But unlike Tolstoy's Anna Karenina, who ponders her life on the train going home from Moscow where she danced with Vronsky, Veselitskaya's consumer heroine thinks about the "new dresses" she bought for the trip (73). She has packed so much that her outfits require planning. Which of her five hats, for instance, will go best with the mousseline frock? She drops off to sleep with such questions in mind.

Vava alone lies awake, full of high aspirations. What a joy it will be to roam through mountains and forests with no French or English governess on her heels! And there in the Caucasus she will find "new people, new acquaintances," great people of the sort she has always wanted to meet: the likes of George Washington, William Tell, Joan of Arc, and above all—Tolstoy, her supreme hero (she once thought of writing him a letter but felt too embarrassed, too much of a "little bug" to "bother such a sun" as he!). Identifying with Pushkin's Aleko of *The Gypsies*, Vava foresees great people in the Caucasus teaching her how to become "a bold, firm, strong person," throwing off the bondage, mendacity, and artifice of high society. She shall rise above the trivial existence of her mother, brother, and Zina the "fashion plate." Thinking ill of her mother and siblings makes Vava feel guilty of "vileness and malice" so grave that she begins "praying fervently to God to forgive her sins" and "grant her strength and health in spirit and body."

With that "prayer on her lips," she falls asleep in a bunk above "placidly snoring *maman* and pale, air-head Mimi" (75–77).

The adultery plot commences the next morning when the travelers have their pre-paid breakfast during a stop to change trains in Rostov. In the "hazy, dusty hall reeking of tobacco," they recoil from their social inferiors: "estate owners from the sticks and dishevelled, haggard provincial women with cigarettes dangling from their mouths" (78, 80–81). At the Pyatigorsk spa, Mimi and *maman* will similarly take pride in the Petersburg "elegance of their *toilettes* and manners" by contrast to the crowd of "all sort of provincials," such as "doctors' wives," no better than "communion-bread bakers" (*prosvirni*, 88–89). The theatrical entry of "a smartly dressed man" with "an enormous black Newfoundland" alters the dismal scene in Rostov. An emblem of hyper-masculinity (versus Mimi's toy dog Monichka), the Newfoundland walks behind his owner, along with a porter carrying "an elegant suitcase with a travel rug attached with brand-new, pretty straps." The dandy's clothing, stylish luggage, and the breed of his dog suggest Anglophilia, an impression confirmed by the pet's name, Rex, printed in Roman letters in the text. The newcomer and Mimi connect right away amidst the provincials. He sits down at their table and gazes into her beautiful dark eyes (the eyes of a "Madonna," remarks the author). *Maman* chats with him as Mimi looks on, silently thrilled by this change from her "monotonous, quotidian Petersburg life" (78–81). Vava pats Rex and feeds him half her chicken, an act anticipating "The Lady with the Little Dog," where Gurov strikes up his acquaintance with Anna by asking permission to give her Pomeranian a bone in the Yalta hotel restaurant. Veselitskaya's dandy and the ladies do not introduce themselves, and even at the spa it takes Mimi a while to learn who he is. Until then, she thinks of him as *l'homme au chien*—the man with the dog (86, 89, 101, 103, 111, 114).

The parodic Tolstoyan motif of the language of the eyes continues after the travelers switch trains for the spa. Mimi juts her head out the window to look in the dandy's direction, only to discover he is doing the same. At every rest stop, the man strolls on the platform, trying to catch her eye. She, though, pretends to be looking at the sky or the station, yet mentally registers his fashionable outfits. He, too, has brought more than one hat and has stylish boots Mimi takes to be French (such details liken him to *The Kreutzer Sonata* violinist Trukhachevsky who wears fashionable clothes attractive to women). On the last night of the journey, Vava and Mimi gaze out at the moon together, but with very different, unspoken thoughts in mind.

Vava wonders if the soul lives on after the body dies.[75] Can the dead see their loved ones on earth? Will people ever solve the mysteries of nature as Goethe's *Faust* sought to do? For Mimi, on the other hand, the moon merely heightens romantic curiosity about the man with the dog. Veselitskaya then imitates the Anna-Vronsky encounter on the railroad platform during a snowstorm. While Mimi's companions are sleeping, she glances out at a station, and there stands the dandy staring right into her window. She returns his penetrating gaze but feels "a little guilty" (*nemnozhko vinovata*) about it (84), just as Tolstoy's Anna felt after the ball (1.28).

Hiking and Sex

Once the tourists are situated in Pyatigorsk, Vava begins a daily routine of hiking. She climbs mountains "like a goat" and stays out late to watch sunsets. Fearing neither snakes nor tarantulas, she lies in tall weeds and nettles in the forest and returns home with tattered boots, scratches on her face and arms, burrs and grass in her hair, and ticks and caterpillars on her clothes. A favorite haunt is a grassy plateau on Mt. Zheleznaya, where she has a panoramic view of the valley, including the shining "gold cross of the church" in Zheleznovodsk. Blissfully removed from the spa's orchestra music and high-society ado, Vava attends to the alternative "choir" and "audience" of nature: grasshoppers play "their waltzes," while "ladybugs, beetles, caterpillars, moths, and bees" listen.[76] As Vava lies in the grass, her soul fills with "peace and joy" she has never known at home. She stares at the sky and the gliding eagles that she imagines carrying "to God her dreams, hopes, and faith" (91–94).

The Zheleznovodsk church is symbolic architecture grounding the spas in the realm of imperial Russian law and Orthodoxy. The symbol is weighty in light of Vava's hiking by herself past sunset. Although *maman* worries about Vava running into "snakes and rabid dogs," vagrant "musicians," or "paupers" known to roam the vicinity (94), the thought of savages lurking in the wilderness never crosses anybody's mind. Decades of imperial Russian rule have made the Caucasus safe for lone women tourists. That situation was already made vivid in an 1882 picture in *Grainfield* (fig. 13). The illustration shows an elegantly dressed woman on horseback descending a

75 Recall Tolstoy's transcendentalist Prince Nekhlyudov gazing at the sky in *Lucerne*.

76 Compare Tolstoy's Olenin listening to insects, as he lies in tall grass in the Caucasus in *The Cossacks*.

13. The original caption of this image translates as "The way of traveling in Abkhazia," a formulation that expressed the perceive normality of a lone lady roaming the Caucasus with native guides and a pack animal for her luggage. She rides sidesaddle in a gracefully draped gown, and her hidden right hand would have been holding a cane or a balance strap to help control the horse.

mountain road in Abkhazia with an escort of tough-looking armed natives that the magazine identified as *gortsy*. Further commentary explained that the pictured tourist was an Italian named Carla Serena. For the readership of *Grainfield*, however, this image would have represented generic European Woman, radiating assurance that imperial Russian law and power are protecting her there in the middle of nowhere with five mountaineers.

In addition to hiking and reveling in nature, Vava joins a gymnastics group at the spa and socializes with thinking visitors. They discuss the "immortality of the soul, the woman question, and the views and ideas of Leo Tolstoy" (125). Vava also makes friends with representatives of the working class—the nannies and *bonnes* (maids) of vacationing families. Finally, she maps out humanitarian work for the future. She plans to stay single and use her dowry to establish a home for abandoned children. A tongue-in-cheek tone creeps into the author's projections of Vava's boundless idealism. And yet under Tolstoy's influence, Veselitskaya clearly sculpts Vava as the story's heroine in opposition to decadent Mimi.

While Vava pursues self-improvement on multiple fronts, Mimi reassumes her Vichy-Paris tourist persona. In spare time she reads Georges Ohnet's (1848–1918) popular but critically panned *La Grande Marnière* (1885), a novel about a man who falls in love at first sight with a woman who turns out to be the daughter of his worst enemy. For the most part, though, Mimi is preoccupied with using clothes to proclaim her husband's wealth and power. Love affairs proliferate at the Caucasian spas, but she vows to remain aloof, to tend to her cure and her appearance. She parades many outfits, dictated by the vestimentary coding of fashion magazines (white for summer, casual for weekends, and so forth).[77] When Mimi lends a hand at a charity bazaar in Zheleznovodsk, for instance, she wears an "exquisite *pêche* [peach-colored] dress." For walking in the Kislovodsk part, she goes for the simpler look of "a white dress and red hat" (113, 136). Clothes make this woman and even convey her assumption that Russians of Petersburg crown the imperial power structure. At the spa *maman* learns that *l'homme au chien* is Valerian Nikolaevich, a thirty-five-year-old lawyer from Kyiv married to the wealthy daughter of big landowner and factory magnate. In his campaign to seduce Mimi, Valerian strategically flirts with a woman of

77 Roland Barthes, *The Fashion System*, trans. Matthew Ward and Richard Howard (London: Hill and Wang, 1985), 15, 249–53. In the first story of the trilogy, Mimi follows what sounds like a newspaper fashion page, *le chronique de l'élégance*, in *Mimochka*, 35.

easy virtue named Lenskaya, the sister of an actress in the Kyiv vaudeville troupe in the Caucasus for the summer. The status-conscious semiotics of clothing counterbalance Mimi's jealousy. With her Parisian-style wardrobe befitting the wife of a Petersburg general, she feels superior in every way to the Kyiv demimondaines in their "gaudy," "vulgar" outfits (90–91).

In parallel to Pechorin in the "Princess Mary" section of *A Hero of Our Time*, Valerian's courtship becomes ever bolder but plays upon Mimi's distinct obsession with clothes.[78] The two fashion plates often go riding together, allowing Mimi to wear the bust-sculpting equestrian outfit whose price outraged her husband. Valerian, too, takes riding as a fashion opportunity: he has brought along his customary vacation assortment of Circassian clothing. In addition to quoting Pechorin, Valerian recites Baudelaire, Byron, Heine, Fet, Alfred Musset, and other French poets with the aim of persuading Mimi that love is exalted even if adulterous (119–21). The *coup de grâce* is his praising Lenskaya as a real woman who has "loved and lived," unlike high-society "dressmakers' mannequins for Parisian *toilettes*" (130). Mimi is crushed, especially since Valerian in his amicable moods has given her much appreciated tips about how to dress.

Mimi adds sex to her vacation regime in Kislovodsk but in a manner that further demonstrates her clothes-horse, consumer character. Valerian persuades her to join him on a night-time ride with the mountaineer guide Osman to see a cliff-side rock formation, the Castle of Treachery and Love.[79] In the canyon-like enclosure of the rocks, the tourists sit on a *burka* and gaze at the site. "*C'est féerique* [it's magical]," Mimi repeatedly murmurs. They embrace and then disappear from the reader's view. The curtain rises again with Mimi's mental question: "How had it happened?" The moonlit scene had seemed so romantic. And now adultery! Yet, soon after "it" is over and they head back to Kislovodsk, Mimi realizes she has lost her new riding crop, a smart accessory. Valerian and Osman hunt down the crop.[80] In Mimi's consciousness, the near loss of a stylish commodity takes precedence over the loss of virtue. Back in her room that night, she assesses the

---

78 On parallels with Lermontov, see Olga Demidova and Mary Zirin, "Lidiia Ivanovna Veselitskaia," in *Dictionary of Russian Women Writers*, ed. Marina Ledkovsky et al. (Westport: Greenwood Press, 1994), 705–6.

79 Valerian explains that the site's name came from a legend about star-crossed local lovers who made a suicide pact to jump off the cliff. The man went first, but the woman chickened out and got married (141–42).

80 As Helena Goscilo reminded me, Veselitskaya's crop motif ties in with *Madame Bovary*.

experience of "sin." It was not at all the horror people pretend it to be. In fact, she feels "happy and calm" (147–49).

After becoming Mimi's lover, Valerian expresses sentiments about his wife echoed in Chekhov's "The Lady with the Little Dog." The day after the fateful horseback excursion, Mimi and Valerian declare that they fell in love at first sight in Rostov. However, the consumerist substitute for authentic intimacy is her access to his closet. They meet regularly in his rented apartment in Kislovodsk, where Mimi discovers the forty neckties and matching socks he has brought on vacation. She is also impressed by his numerous "watch fobs, pins, and rings" (151). There, too are photographs of Valerian's wife and children. Mimi finds the wife distressingly pretty. Valerian tells her, however, that his wife is just "a *samka* [female of the species], *une femelle*," strictly a mother rather than a lover (154). The phrase alludes to *Anna Karenina*. When Oblonsky takes Levin to meet Anna, he defends her as a remarkable woman, "busy with writing," instead of being a "mere *samka, une couveuse*" (7.9). Speaking back to both texts, the concluding paragraph of Chekhov's story elevates the biological terminology in describing Gurev and Anna as a pair of migratory birds, "male and female" (*samets* and *samka*), caught and forced to live in separate cages but determined to make a life together.

Mimi's vacation fling ends in the Caucasus, the outcome that "The Lady with the Little Dog" inverts. During one of his last rendezvous with Mimi, Valerian vows their love will last forever, and he promises to come to Petersburg to take her to concerts and the theater. In actuality, she will never hear from him again. Chekhov's Gurov, on the other hand, finds Anna unforgettable, and he tracks her down in her hometown at a theater. For Mimi and Valerian, materialism and its tokens substitute for love. After hearing her tale of marital misery, he tells her she was prudent to wed a rich man because "money is the key to happiness" (153). Narcissistic consumerism fittingly crowns their affair. They exchange professionally done photographs (he in Circassian dress, she in her riding-habit) and turquoise rings, a commonplace souvenir on sale at the spa.

Veselitskaya's conclusion emphasizes the tawdriness of the anti-heroine's marital infidelity. Back home, Mimi suffers no guilt about adultery but is stuck in an unhappy marriage and has literally nothing to show for her trip (she could lie about the ring but must keep hidden the photo of Valerian). Vava, by contrast, has improved her physical fitness, made new friends with whom she explored intellectual, moral, and civic issues, kept a vacation scrapbook, and brought home a set of stereoscopic images of scenery to share with the family. The vicarious tourism delights Aunt Mary, who finds the Castle of Treachery and Love a particularly marvelous sight. Mimi explains that she saw it with her own eyes: "I took a horse ride there. It's very beautiful. Especially in moonlight . . . *c'est féerique*" (162). Those words end the story, ironically hiding Mimi's shabby secret memory of the place where she first betrayed her husband.

Her vacation report expunges Osman as well as Valerian. The erasure of the native guide, however, is relatively neutral by comparison to Veselitskaya's depictions of Armenians and Georgians, True, the silenced mountaineer has no existence beyond serving tourists; but he is just an incidental figure, and the author is not obliged to make him anything more. Armenians and Georgians are not so lucky. A smarmy Armenian named David is predictably a merchant, a shopkeeper selling souvenirs.[81] He attracts Katya with trinkets, seduces her, and breaks her heart. As Mimi and Valerian settled a tiff during their courtship while sitting on a bench in Zelezhnovodsk, a group of "clumsy Armenian women in muslin veils" passed by and "obtusely goggled poor Mimi with their round black eyes" (132). Veselitskaya's Georgians are mainly ridiculous. Prior to becoming Mimi's lover, Valerian treats her, *maman*, and Katya to shashlik, Kakhetian wine, champagne, and *chikhirtma* (a lemony, egg-thickened soup) at the restaurant of a Pyatigorsk hotel run by the Georgian Chkhikvadze. The author states that Chkhikvadze's rooms all "smell of kitchen fumes and oil for grilling." Furious about the tough chicken he was served, an inebriated Georgian customer threatens Chkhikvadze with a dagger. Musicians in *papakhi* (tall sheepskin hats) pipe out "ear-splitting sounds" that delight the Georgian diners Prince Jomarjidze and Prince Kakushadze, who dances a *lezginka* with Princess Arjevanidze (116).[82] Valerian cracks (unspecified) jokes about the dancers and then starts declaiming Pushkin's lyric "The Caucasus." From the top of a mountain, Pushkin's persona surveys a vista that takes in both the Terek and Aragvi Rivers. Mary Louise Pratt's *Imperial*

81 On Russian stereotypes of Armenians, see Ronald Grigor Suny, "Images of the Armenians in the Russian Empire," in *The Armenian Image in History and Literature*, ed. Richard G. Hovannisian (Malibu, CA: Undena Publications, 1981), 105–37.

82 Zaal Andronikashvili kindly informed me that the names Jomarjidze and Arjevanidze belonged to the lower Georgian nobility rather than the aristocracy and were mentioned in the official 1850 list of the various categories of nobles of the Russian Empire. Kakushadze and Chkhikvadze were not noble but common western Georgian/Imereti names.

Eyes has established a tendency for commentators to politicize each and every representation of a privileged traveler viewing from on high a scene in a foreign country less developed than his/her homeland. In this case, though, I would argue that the Pushkin allusion does operate as an analogue of the imperialist condescension toward Georgians and Armenians that Veselitskaya casually normalizes. Burenin and Chekhov would prove more sensitive to the disparities of power between Russian tourists and native peoples of the Caucasus.

Madonna/Whore

In parallel to "Mimi at the Spa," Chekhov's "Ariadne" stages a clash between decadent Russian tourist preference for the West and Russian national attachment entailing religious concern. The story's position on tourism, however, is closer to the Levin-Anna divide in *Anna Karenina* rather than Veselitskaya's Mimi-Vava binary. The narrative structure of "Ariadne" entraps the eponymous anti-heroine in a prison of "Schopenhauerian misogyny."[83] Ariadne speaks few words, and her thoughts are never revealed. She is knowable to the reader only as constituted by the two men storytellers. They happen to meet on a boat going from Odessa to Sevastopol. The main narrator is Ariadne's disenchanted lover Ivan Shamokhin, a man in his late twenties. He recounts the history of his affair to the frame narrator, a writer who lives in Yalta—an unnamed but well-known writer whom Shamokhin recognizes on sight. In the telling of these men, beautiful Ariadne (a conspicuously odd name for a Russian) reverses the myth of Theseus and the Minotaur: instead of saving a hero, Chekhov's Ariadne assumes the role of the man-eating beast.[84] In recollecting how he succumbed to her erotic appeal despite misgivings about her character, Shamokhin portrays her as a repulsive piece of work: a power-hungry social climber, a greedy spendthrift and glutton, lustful, deceitful, cruel, slothful, and phony to the core. The frame narrator's impressions of Ariadne corroborate Shamokhin's portrait. Between the two of them, the men create an insolvable puzzle.

83 Quote from Donald Rayfield, *Understanding Chekhov: A Critical Study of Chekhov's Prose and Drama* (Madison: University of Wisconsin Press, 1999), 124; see also Virginia Llewellyn Smith, *Anton Chekhov and the Lady with the Dog* (Oxford: Oxford University Press, 1973), 10–35.

84 Thomas Winner, *Chekhov and His Prose* (New York, Chicago and San Francisco: Holt, Rinehart and Winston, 1966), 186–90; and Michael C. Finke, *Seeing Chekhov: Life and Art* (Ithaca: Cornell University Press, 2005), 149.

Ariadne appears to be a psychopath wallowing in sin.[85] But how reliable is the retrospective account of Shamokhin, the main storyteller? Might he not misrepresent Ariadne out of a need to deny his "complicity in making [her] what she is by the time the outer narrator meets her?"[86] Unable to resolve this question of complicated retrospectivity, readers are left to observe the workings of Chekhov's language and its interactions with his cultural environment.

The frame narrator's initial, separate impressions of Ariadne and Shamokhin pit her acquisitive attraction to the West against his high-minded Russianness. Upon returning from abroad, the Yalta writer sees the couple going through customs in Volochisk. They have "a mountain of suitcases and baskets crammed with women's clothing," including "some silk rag."[87] Ariadne makes a scene when this piece of foreign frippery incurs a customs charge. She thus enters the story as a bad-tempered, narcissistic consumer in thrall to Western fashion. Shamokhin, on the other hand, makes a positive first impression on the boat. While Ariadne remains in her cabin, he comes on deck and initiates conversation with the writer, who immediately sees something good in him: "I noticed that the Russian language and Russian surroundings gave him great pleasure. That was probably due to his having missed the homeland during his stay abroad. While praising Russians and attributing to them a rare idealism, he did not malign foreigners, and that spoke in his favor" (108–9). The observation introduces Shamokhin as a worldly, humane man, attached to his native realm but not in the least xenophobic.

In the prelude to his personal history, Shamokhin displays highbrow cultural awareness. He knows Max Nordau's theory of decadence and degeneration but finds it inadequate for explaining a national Russian pattern. The problem in Russia, says Shamokhin, is that we men are idealists: in the rapture of new love, we regard the woman as a "Raphael Madonna." Inevitably, though, intimacy reveals that women are "immeasurably lower

85 For illuminating analysis of the story's conformity to psychiatric literature of the time, see Mark Stanley Swift, "Chekhov's 'Ariadna': A Portrait of Psychopathy and Sin," *Slavonic and East European Review* 86, no. 1 (January 2008): 26–57.

86 Carol Flath, "Writing about Nothing: Chekhov's 'Ariadna' and the Narcissistic Narrator," *Slavonic and East European Review* 77, no. 2 (April 1999): 233. "Complicity" is a fair accusation, but as Swift points out, "Writing about Nothing" sometimes "claims the same intimate knowledge of [Ariadne's] inner life for which the critic faults Shamokhin," in "Chekhov's 'Ariadna'": 40.

87 Chekhov, in *PSSiP*, 9:107 (subsequent citations appear parenthetically).

than we men" (108). Although marginal in Shamokhin's personality, religious consciousness significantly reappears after he becomes Ariadne's lover in Italy. He recalls the "ecstasy" of having her "young, beautiful body" in bed (126). After about a month, though, he believed more strongly than ever that she did not love him. He got "sick" of her and wondered why "God had given her this extraordinary beauty, grace, and intelligence" (128). Shamokhin articulates there a very Russian yearning for union between moral and physical beauty. We saw an expression of the outlook in Karamzin's *Letters*. Dostoevsky's *The Brothers Karamazov* contains the much better known example of Dmitry's discourse on the excessively "broad" human soul, capable of simultaneously accommodating the "ideal of the Madonna" and "the ideal of Sodom."

Shamokhin's recollection of falling in love with Ariadne casts him as a man grounded in a rural Russian way life, as opposed to her modeled as a country girl craving the high life she sampled for a couple of years while living in Moscow with a rich aunt who introduced her into society. After completing university studies, Shamokhin went to live with his father on the family estate in "a northern district" of Moscow province. The estate is the idyllic sort that Turgenev often depicted. The manor house stands on the high bank of a river, and the property includes "a big old garden, cozy flowerbeds, an apiary, a kitchen-garden," "curly osier" along the river, meadows and pasture lands, and on beyond—a vast pine forest inhabited by elk. The estate brings in around 2,000 rubles a year. Being "modest people," the Shamokhins live comfortably on that amount, plus the pension of the father, a retired university professor (109). By contrast to the idyllic, prosperous Shamokhin estate, Ariadne leads a profligate life amidst Western imports with her brother Kotlovich, an eccentric absorbed in spiritualism and homeopathy. On his estate there are pineapples (presumable cultivated in a hot house), lightning rods, and a fountain. This haphazard array of stuff betrays a provincial mind-set hankering after westernized sophistication.[88] Ignoring principles of good estate management, Kotlovich lets Ariadne buy what she wants even if he has to sell his work horses for a pittance so she can have a new riding horse. That purchase is the first stroke in Chekhov's profiling of the anti-heroine as a "fast," horsey woman.[89]

88 On this general syndrome, see Lounsbery, *Life Is Elsewhere*, 92–93.

89 See again commentary on Victorian literature and *Anna Karenina* in Cruise, "Tracking the English Novel," 173–74.

Kotlovich's indulgence of Ariadne's "unbridled desires" (111) has evidently encouraged her to consider her beauty an asset to trade for men's money. The perceived Madonna becomes a powerful whore.[90] *Vlast'*—the power of the state, empires, and dictators—is a signature motif. Even in present time, Shamokhin cannot forget the "powerful impressions" that twenty-two-year-old Ariadne made on him when they first met. She "conquered" him instantly, and months later, as her lover, he remained "in her power" abjectly suffering emotional torment when she was miffed (110, 127). It quickly becomes apparent to readers that what the self-defined "idealist" calls "love" is lust blinding him to Ariadne's need for a cash cow. Shamokhin comes upon the scene when Kotlovich's resources are running dry. Unable to buy new "dresses and hats," Ariadne must "dissemble and dodge in order to hide her poverty." Shamokhin claims that she confessed to him at that time her regret at having refused Prince Makutev's marriage proposal in Moscow: not much of a catch, but at least he was a titled aristocrat. Shamokhin says that her attitude reminded him of a muzhik disgustedly blowing on kvas with cockroaches floating on the surface but drinking it down, all the same. Shamokhin even maintains that he had already perceived in Ariadne a demonic "thirst for power," wealth, and prestige in an imagined sphere of "balls, races, liveries, luxurious drawing rooms, her own salon, and a whole slew of counts, princes, ambassadors, famous artists and actors, all worshipping her and delighting in her beauty and *toilettes*" (111–12). Shamokhin nonetheless remained in her power.

Tolstoyan Anti-Tourism

Tourism looms on the horizon with the arrival of Lobkov, a comrade of Kotlovich's university days. A sponger with the "face of a good-natured bourgeois", Lobkov is perpetually in debt and reneging on financial responsibilities to his wife, their four children, and his widowed mother. His jolly attitude toward everything, including his own fecklessness, alters the atmosphere. Shamokhin likes "calm, idyllic (so to speak) pleasures" of rural life: fishing, evening walks, and mushroom hunting (113–14). Lobkov and

90 A subtext is Gogol's "Nevsky Prospect" (1835) in which a beautiful stranger, perceived on the street as "Perugino's Bianca," turns out to be a prostitute. Gogol was referring to the Madonna in Perugino's *Adoration of the Magi* fresco in the Oratorio di Santa Maria dei Bianchi in Pieve: see his *Sobranie sochinenii*, 7 vols. (Moscow: Khudozhestvennaia literatura, 1984–86), 3:276.

Ariadne, on the other hand, prefer recreation typical of the leisured class: racquet games, fireworks, hunting with hounds, and picnics of oysters and champagne (which Shamokhin dutifully fetches and pays for in Moscow). Vacationing abroad becomes the ultimate extravagance. Ariadne develops an obsessive desire to go Italy after hearing the elder Shamokhin talk about its scenery and museums. The perfumed letters she sends him from abroad imply she is traveling with an aunt when in fact she is with Lobkov. Signed "your forgotten Ariadne," a perfumed letter to young Shamokhin tells him to stop wasting his best years moping in the country and join her in the "paradise" of palms and fragrant orange trees (118). He goes, burning to find out if Kotlovich's suspicions about Ariadne and Lobkov being together are correct.

During the vacation Shamokhin develops a bad case of tourist angst. Expecting a "terrestrial paradise," he finds Abattsia "a dirty, little Slavic town" with a single, smelly street that rain turns into a mud morass requiring galoshes. With its "distant islands, swathed in lilac mist," the bay might have been "picturesque," he thinks, if only "greedy developers had not obstructed the view with the absurd, philistine architecture of hotels and *dependences*." The tiresome park, he continues, matches those in every other European resort: "the dark, immobile, and silent green of the palms, bright-yellow sanded paths, bright-green benches, the glint of blaring military trumpets, and the red trouser-stripes of a general" (118–19). Shamokhin's metonyms for the resort milieu reflect what he *will* learn in touring with Ariadne and Lobkov (the two of them at first, and then with her alone). The Abattsia segment likewise projects retrospectively the tourist angst that Shamokhin is *going* to experience in Italy:

> In roaming involuntarily around those resorts, I became more convinced than ever of how inconvenient and dull are the lives of the sated and rich, how flabby and weak their imagination, how lacking in boldness their tastes and desires. And how much happier than they are the tourists, old and young, who, lacking money for hotels, live where they must, admire the view of the sea while lying in green grass on mountain tops, travel on foot, see up close the forests and villages, observe the country's customs, listen to its songs, and fall in love with its women. (119)

The next paragraph describes Ariadne and Lobkov arriving in the park. Decked out in fashionable new clothes (which Shamokhin assumes were bought in Vienna), they are emulating the tourist crowd that Shamokhin's thoughts have just maligned. Ariadne warmly greets him but then tells him to buy a new hat in order to look *comme il faut* (vacation costuming is important to her, as it was for Vronsky and Anna, who wears a straw hat in Italy). Like Tolstoy's adulterers, Ariadne invents a tourist persona who does not exist at home. She and Lobkov have been socializing with a traveling Russian family; and when she introduces Shamokhin to them, he observes how successfully she has been playing the part of a Russian aristocrat with a profitable estate.

Rain drives the threesome to Italy, where Shamokhin passively conforms to Ariadne's and Lobkov's predilection for the ways of the sated rich. While still in Abattsia, Shamokhin experiences unspoken populist revulsion to upmarket tourism. Just as he felt ashamed at home before the peasants working while he went fishing or picnicking, so too on holiday does he feel ashamed of his idleness in comparison with local service people. His tourist angst gets worse as they wander through Rome, Venice, Bologna, and Florence. Staying everywhere in expensive hotels, they all overeat from morning to night. Even after copious breakfasts, eight-course dinners, and tea at 9:00 pm, Ariadne orders more food at midnight, typically ham and scrambled eggs. The men eat, too, to "keep her company" (as usual, they bend to her will). "In between eating," they drag themselves to museums and churches like "sated boa constrictors," stuffed with so much "greasy" food that they cannot appreciate any of the marvels before their eyes. They find more riveting the "flashy items" in shop windows, and they buy "a pile of unnecessary, worthless things." Ariadne behaves like "a duchess," lavishly spending (121–22). And all the while, Shamokhin is footing the bills and wiring home for money. Once he has ascertained that Ariadne is sleeping with Lobkov, he flees after a distressing early-morning parting with her, seductively disheveled alone in her room in her shift amidst scraps of food.

The mimicry of Tolstoy's Levin then ensues. Jealousy assaults Shamokhin on the train going home from Italy: How could such a "young, remarkably beautiful, intelligent girl, the daughter of a senator" give herself to that "vulgar, uninteresting mediocrity?" (123). Like Levin, wounded by Kitty's rejecting him in favor of Vronsky, Shamokhin recovers his spirits through work in the country. He loves Russian winter and wears a peasant-style "sheepskin jacket and felt boots" as he goes about his tasks. With the coming of spring, he turns his mind to sowing crops and tending new calves. While “working in the fields and listening to larks,” he contemplates marrying “a simple peasant girl” (125). The Tolstoyan daydream instantly evaporates, though, when he receives Ariadne’s letter begging him to come to Rome. Lobkov has flown the coop, leaving her in debt, but Shamokhin again glosses over her mercenary need for him. Her whimpering pleas for love immediately master him.

The inauthentic life of tourism now supplants the Tolstoyan agricultural interlude. Chekhov’s decadent couple spend time in Rome, Naples, and Florence. Ariadne perfects her *grande dame* persona and takes up painting, but unlike Vronsky, she charms her way into artistic circles. Gradually, however, she and Shamokhin lose contact with elite people. Erotic ecstasy disappears leaving behind faked emotion, tourist boredom, and mutual antagonism (the stage that Anna and Vronsky reach only later, at home). Shamokhin discovers Ariadne’s “astounding craftiness,” instinctive as “a cockroach twitching its whiskers” (126–27). He discovers her cruelty, too. “Even in merry moments,” he writes, “she thought nothing of insulting servants or killing an insect” (the implication that he *does* spare insects conforms to Tolstoy’s religious views). Ariadne “liked bull fights and reading about murders and became angry when accused people were pardoned.” But despite Shamokhin’s contempt and loathing for this “tigress” (127–28), he keeps bleeding his father for money, driving him to mortgage their estate, hand over his pension, and exhaust his savings. Seeking to alleviate self-disgust, Shamokhin tries to resist Ariadne’s sinfulness and power. He stops drinking, restricts eating, reads, and most importantly frequents museums and galleries. Cultural tourism, rather than moral work, becomes a strategy for self-improvement, a choice showing that he missed a big point of the parallel plots of *Anna Karenina.*

The conclusion of Shamokhin’s account underscores division between the authentic life of work in Russia and the decadence of trying to make vacation a permanent mode of existence. Taking the bloom off tourism, his dwindling funds resign Ariadne to returning to Russia. He feels compelled to marry her and envisions settling in the country, but she refuses to go back to the rural realm where she knew deprivation and poverty. She now can live only in resorts, and so they have been alternating between Yalta and the Caucasus. “If only you knew how I hate those resorts, how stultifying and shameful I find them.” Shamokin cries to the frame narrator. He histrionically adds that he wants “to work, earn my bread by the sweat of my face, and atone for my mistakes” (130). With five years of effort, he figures, he should be able to repurchase the family estate. He then launches into his concluding harangue against Women. The frame narrator suggests that it is not fair to judge the entire sex by Ariadne, but Shamokhin's shtick is unstoppable. His listener turns a deaf ear and falls asleep in their shared cabin.

The master narrator sculpts the final images of Ariadne as the horsey, deceitful faker in her element in Yalta. He sees her racing on an ambler, followed by "two officers barely keeping up with her" (132). She is beating the men. Wearing an artist's smock and a Phrygian cap (a reminiscence of Vronsky's medieval get-up in Italy), Ariadne attracts an admiring crowd as she paints on the quay. She is introduced to the famous writer. With an enraptured gaze she thanks him for the pleasure his works have given her ("Don't believe her," whispers Shamokhin, "she's never read anything of yours"). Hedging her bets, Ariadne has secretly arranged for her brother to bring Prince Maktuev to Yalta. Laden with hors d'œuvres and fruit, Shamokhin excitedly tells the frame narrator of the men's arrival. He clearly hopes the prince will take Ariadne off his hands. Earlier likened to a cockroach, she now looks set to marry Maktuev, also associated previously with cockroaches. Will Ariadne the insect-killer become the prince's wife? Will she somehow continue to overmaster Shamokhin? Or will he liberate himself at last and start a responsible, productive life in the country with his father? The master narrator cannot say because he leaves Yalta before the denouement. With this open end, Chekhov slyly wields the literary power of making his audience want to reread the story again and again, to reconsider everything as filtered through the cloudy lens of Shamokhin's consciousness.

A Bigoted Jewish Adulteress

By contrast to the narrative intricacies and psychological depth of "Ariade," Burenin's *A Love Affair in Kislovodsk* and Chekhov's "A Loose Tongue" are simply told tales of eroticized vacations, but they pack a political punch. Burenin's novella is largely narrated as the diary of the main protagonist Nikolai Argunin, a Petersburg man who has been living for five years with Elena Nikolaevna, a woman separated from her husband. Their affair has grown stale, and so they agree to take solo summer vacations. They will follow their fancies, keep diaries, exchange them afterwards, and then decide whether or not to resume living together. Elena chooses to rent a dacha in Pavlovsk, a place gaining notoriety as a haunt of "the moneyed bourgeoisie" elbowing aside the titled aristocracy.[91] Argunin opts for the Caucasus because he is tired of Europe's "famous and ultra-famous beauties of nature, the 'holy wonders,' the inelegant table d'hôtes, the prostitutes, and petty-bourgeois ways."[92] A know-it-all kind of snobbery, Argunin's act of disdain for the West will serve Burenin as a tool for satirizing the antithetical position: the unconditional adulation of Europe articulated at the spa by Entrecotov (from the French for "rib steak"), a pompous Russian writer who considers his homeland "barbarous" and insists that the visitors of Vichy, Biarritz, and other foreign resorts are infinitely more refined than the Russian crowd in the North Caucasus.

Soon after arriving in Kislovodsk, Argunin makes the acquaintance of the retired General Pavel Petrovich. After listening to a band of exotically outfitted Cossacks playing arias from *La Sonnambula* and other West European operas, the two men go for a drink on a hotel terrace. The general hails a waiter as though commanding an army recruit:

> "Hey you, come here!" The waiter walked over.
> "What's your name? Akhmet?"
> "No, sir: it's Petro."
> "What?!" Pavel Petrovich asked again with a lightning-bolt glare at the waiter. The waiter repeated his name. "Petro! Why Petro?! Better just say 'Pyotr' instead of Petro, you idiot. What are you, anyway—Circassian?"
> "Ossetian."
> "I could tell right away. Alright, Ossetian Petro. Now bring us some tea and cognac. And they'd better be good." (114)

The waiter brought the drinks, the general "chewed him out again for some oversight" and added that "there's no creature on God's earth any stupider than Ossetian waiters." In an "almost polite" tone, he ended the incident with the words: "Alright, Petro, now get the hell out of here" (115–16).

The general also airs his anti-Semitism. Argunin notices a woman drinking champagne with a big Cossack and a young hussar at the next table. She is a beautiful, "chicly dressed" blonde with dark eyes, a "typically oriental profile," and a "splendidly developed bust" shown to advantage in

91 Lovell, *Summerfolk*, 108.

92 V. P. Burenin, *Mertvaia noga. Roman v Kislovodske* (St. Petersburg: Izdanie A. S. Suvorina, 1886), 95 (subsequent citations to *Roman v Kislovodske* appear parenthetically).

her décolleté blouse (116). Remarking Argunin's interest, the general says, "Pretty little rascal, huh?" He explains that she's been here two weeks, and she's a "Yid." Argunin is skeptical:

> "But she's blonde."
> "No matter! Some of them are blonde. I'm telling you, she's a Yid. Just look at her nose and eyes. Her name's probably Rebecca Zodiakovna."
> "Why Zodiakovna?"
> "Because they're all Zodiakovnas, that's what I call them, anyway. She's got Yid arms, too."
> "Her arms are marvelous."
> "Call them nice as you like! All the same, they're Yid arms."
> "Why?"
> "Just take a good look, there's red fuzz under her bracelets. A sure sign."
> "Really? How did you make that out?"
> "I just did. And anyway, I didn't need to see it. I already found out she's a Yid, a Yid from Taganrog. She's married to a merchant. They're all merchants there. Steal ten thousand or so, and you're a merchant." (118–19)

Argunin has just seen the unhappily married woman from Taganrog (Chekhov's home town) who will become his lover.

An early twentieth-century Russian critic considered *A Love Affair in Kislovodsk* pornographic, an opinion today's reader will find quaint.[93] Opting for the heaving-bosom rather than bodice-ripping style, Burenin takes Lermontov's "Princess Mary" as a template. Like Pechorin, Argunin tells his story in a diary, he obtains vital information through eavesdropping, and he enjoys publicly ridiculing his rival Entrecotov. Argunin, however, conducts his erotic campaign with joie de vivre, much less cynically than Pechorin; and unlike the haughty Princess Mary, Burenin's heroine, Nina Pavlovna, makes the first move. She sits down on a park bench where Argunin is seated. After a few minutes of silence, he pulls out a cigar and asks if she minds if he smokes. She says certainly not, takes out a cigarette of her own, and then makes fun of him for having picked such an unoriginal

93 G. S. Novopolin, *Pornograficheskii element v russkoi literature* (St. Peterburg: Tipografiia M. S. Stasiulevicha, 1909), 55–57.

overture. He inquires why he needs to be original. And so begins the flirtation that will develop over several chapters full of cultivated banter and erotic innuendo. During their first conversation, for instance, Nina asks Argunin if he is taking a cure. He says no: "I'm as healthy as forty thousand brothers, maybe even forty thousand bulls." Then he says he *is* taking a cure, a cure for "hypertrophic loving" (*gipertrofiia liubvi*). Following this suggestion of being oversexed, he explains his decision to take a break from his five-year cohabitation with a married woman (136–38).

He then persuades Nina to recount *her* intimate history, in the course of which he ascertains that she practices birth control. Leaning close to him as they stroll in the park, she explains that she married for love at sixteen and converted from Judaism to Orthodoxy, but her twenty-year-old husband turned out to be a libertine. As subsequently paraphrased in Argunin's diary, the faithless husband's liaisons with tarts such as "provincial operetta singers" (144) deeply humiliated Nina. Her spouse has now run away with a lover. Nina's irregular marital situation prompts Argunin to ask if she is pregnant. She calls him "a crackpot" for posing such "a stupid question." "So, you're not pregnant?" he persists. "Oh, stop it! Of course not, and I don't intend to be" (140). Beneath her sexy exterior, Nina will prove as vulnerable (and as doomed) as Ophelia. After getting to know her better, Argunin will describe her in his diary as "*La Bella* di Tiziano in the Pitti gallery with, of course, a little touch of Russian simplicity and obtuseness."[94]

Nina has a heart condition but enjoys horseback riding (yet another provocative equestrienne) and strenuous walks in the wilderness. During one of those afternoon hikes, she and Argunin have sex outdoors. They conduct their affair discretely, spending much time in his rented apartment at the spa. Passionate bliss prevails, but Nina can be anxious and clingy. She even has fits of hysterics noted in Argunin's diary as "a bad, provincial mannerism" (158). Nina's failing health comes to his rescue. The story ends with her fatal heart attack on Bermamut, where the lovers go with a guide to see the sun rise over twin-peaked Elbrus. Ladies usually take that popular excursion by carriage, but Nina wants to be original. She insists on making the nearly six-hour ascent on horseback, and the strain kills her. Argunin weeps over her body, while the guide goes looking for a village to find a make-shift stretcher and bearers to carry the dead tourist back to the spa.

94 Registering Nina's Jewishness, Argunin uses the adjective for imperial rather than ethnic Russian identity: *rossiiskaia prostota i tupost'* (158).

Put off by malicious gossip, Burenin's sexual adventurer skips the funeral and heads home.

Nina's ethnic prejudice complicates her victimhood. Her favorite guide is Jaiani an extraordinarily handsome Imeretian with blue eyes and light brown hair.[95] In his "chic" native clothing worn in "tasteful, dandy style," he is stunning, especially astride a horse (253–54). Argunin reports rumors that Jaiani has had affairs with Russian women he conducted to the Castle of Treachery and Love and other sites. Through those encounters, he has picked up some French. A Caucasian gentleman, in short. When the trek to Bermamut begins, Nina frets about cloudy weather ruining the sunrise. Jaiani observes that the sky is now clear, so why worry? Her testy response initiates this exchange:

> "Why, why?! What if your Allah wants it?"
> "We don't pray to Allah, madame, we pray to your God."
> "Oh, who knows what god you pray to? You're all Circassians, after all."
> "Why Circassians, madame? We are Imeretians" (255).

Argunin tries to laugh off Nina's insulting remarks. He quips that Georgians, Ossetians, and Imeretians all pray to the same god, the cloud god, so a grand sunrise is assured.

The incident dramatizes Nina's urge to exercise authority over a man whose status as a colonized hired hand puts him lower than she in the imperial Russian power structure. The Taganrog *turistka* has three strikes against her: she is a woman, a Jew by birth, and a provincial, condescendingly perceived as such by Petersburg-proud Argunin. At the Caucasian spas, however, Nina can act like an empress with a page, the Imeretian guide that she finds charming as long as he is doing what she wants and not piping up to say something she does not want to hear.

"You're just a Tatar. . . ."

The married tourist of "A Loose Tongue" combines unabashed bigotry with erotic exploitation of her Tatar guide. Chekhov wrote the tale in 1886, before he visited the Crimea. Having recently been there, Levitan

95 Thanks again to Zaal Andronikashvili for identifying Jaiani as a family name from Svaneti that might have later migrated into Imereti.

is considered a likely source of Chekhov's awareness of Russian women's alleged improprieties with guides.[96] Independently, it seems plausible that Chekhov might have known Markov's treatment of the topic in *Sketches of the Crimea*, which observed that "our ladies" come to the Crimea for something a lot more exciting than sun, sea, and grapes. At any rate, by the fin de siècle when the uncensored version of "A Loose Tongue" appeared, Russian journalists and obscure authors had published several salacious stories about Tatar guides. Timothy Phillips remarked that this collective narrative reversed the conventional gender roles of West European and Russian stories about native men of a colony lusting after "our" women. In a turn-about with vexing implications for Russian male self-esteem and the entire "imperial enterprise," Russian women were reputed to initiate erotic interaction with the Tatar guides they hired.[97] Men seem to have worried more about the state of upper-class sexual relations at home rather than the solidity of the empire. Witness Bernov imagining the woman tourist's position in 1896:

> To some extent, I understand that after haemorrhoidal husbands and puny lovers, the Russian lady accustomed to lots of food and little activity (especially mental activity) would see in such a [Crimean] specimen the kind of man she wants and cannot find in her social circle. Some of those women take up with a foreign, bourgeois, or Jewish secretary, others head for the southern coast.[98]

Bernov's observations bring to mind Uspensky's *The Power of the Soil*, which idealized Russian peasant virility and pitted it against the bourgeois city man's "mouse" sexuality.[99] More importantly, Tolstoy was casting his long shadow here. *The Kreutzer Sonata* posited idleness and a copious, meaty diet of heat-generating food as a combination fueling upper-class male lust.[100] Bernov applied the same logic to upper-class female lust and frustration.

96 Commentary in Chekhov, *PSSiP*, 5:653.

97 Phillips, "Health and Sociability," 200–7 (quote 200). Among other works, Phillips cites Forster's *A Passage to India*. On erotic anxieties about Caucasian mountaineers, see my *Russian Literature and Empire*, 133–55.

98 Bernov, *Iz Odessy peshkom*, 185.

99 Mondry, *Pure, Strong and Sexless*, 35.

100 Tolstoi, *PSS*, 27:23; and commentary in Goscilo-Kostin, "Tolstoyan Fare," 493–94 and LeBlanc, "Unpalatable Pleasures," 19–20.

The moronic anti-heroine of "A Loose Tongue" may not commit adultery but is certainly an imperialistic little flirt who experiences in the Crimea the intoxication of power she has never known at home. Chekhov's three-page story consists mainly of dialogue.[101] Just back from a solo vacation, the young wife, Natalya Mikhailovna, is married to Vasya (the nickname for Vasily), a high-ranking bureaucrat so much older than she that she sometimes calls him "Daddy" (*papochka*).[102] Her report begins with stupendous scenery: mountains "a thousand times higher than a church," mists, huge boulders, and the "parasol pines." Vasya wonders about disturbing newspaper articles he read in her absence. Is there something special about those Tatar guides? Nah, says Natalya, "anyway, I only saw them from a distance" and "paid no attention," having always had "a prejudice against all those Circassians, Greeks, Moors, whatever!" But Tatars are supposed to be "dreadful Don Giovannis," continues Vasya (313). Hemming and hawing, Natalya admits the rumors have some foundation. Her tongue loosens, and she says too much.

First, she recalls witnessing the scandalous behavior of other Russian women. She could not believe their immorality! "And you know, it wasn't plain folks or the middling class but those stuck-up *bon ton* types, the aristocrats!" A big shocker was Natalya's travel companion and hotel neighbor, Julia, a woman with a good husband and two children. Julia, we hear, "always poses as a saint." But just listen to this. One fine day Julia rides off with a Tatar. In the mountains she suddenly shrieks and doubles up with pain and would have fallen off her horse if the guide hadn't put his arm around her waist. Natalya saw it all, she was right there with her own guide Suleyman, and the two of them headed back to town to get medicine for Julia, who stayed put with her guide. "Hold it," interrupts Vasya, "you said before that you saw Tatars only from a distance, and now you're talking about some Suleyman" (314). Caught lying, Natalya flounders and loses control, as illustrated in my abridgement of the ensuing dialogue:

> "There you go again, carping about my every word!"
> "I'm not carping, but why not tell the truth? So, you went riding with Tatars, that's how it was. Why be evasive?"
> "I'd like to see you go riding in those mountains without a guide!

101 Chekhov, "Dlinnyi iazyk," *PSSiP*, 5:313–16.

102 He is a state councilor (*statskii sovetnik*), the fifth rank in the Table of Ranks, just below the generals: see again Reyfman, "Service Ranks," 99.

> I'd like to see it! You can't budge there without a guide."
> "Indeed!"
> "Enough stupid grins, thank you!"

Natalya then shifts to self-presentation as a stern task-master keeping a cowering, backward native in his place:

> "Suleyman never went out of bounds with me. Oh, noooo! Mametkul would sit around with Julia 'til all hours. But as soon as the clock struck 11:00 pm, I'd say: 'Alright, Suleyman. March! Get out!' And my stupid Tatar would leave. I ruled him with an iron hand, Daddy. As soon as he'd start grumbling about money or something, I'd say: 'What's that? Huh? Whaaaat?!' Scared him stiff! Ha, ha, ha!! You know Vasya, he had those black eyes, black like coal and a stupid, funny Tatar mug. So that's how I kept him in line! That's how!"

"I see," mutters Vasya, eliciting further self-incrimination:

> "I know what you're thinking. But I swear, even on excursions he never went out of bounds. For instance, when we'd ride into the mountains or go to the Uchan-Su waterfall, I'd always say: 'Come on now, Suleyman, get behind!' And he always did ride behind me, the poor devil. Even at the most moving sites I'd say: 'All the same, you mustn't forget that you're just a Tatar, and I am the wife of a state councilor!' Ha, ha, ha!!"

The loose tongue wags on, to conclude with Natalya's recollection of inviting Mametkul into her room one evening when Julia was out: "We got to talking. They're very amusing, you know! The time flew by." But then Julia burst in and raised a ruckus . . . and so on and so forth. I see you had "a good time," drily remarks Vasya with a frown. This bring Natalya to a halt: "I know what you're thinking! You've got a dirty mind! I'm not telling you anymore. I'm not!" The story ends in this stalemate, leaving the future to the reader's imagination.

Natalya enacts a lust for power reflecting her belief in Russians' superiority over minority peoples of the empire, not to mention outsiders such as Greeks and Moors. Her whole sense of self and her claim to the imperial privilege of ruling Suleyman "with an iron hand" rest on her husband's position in the Table of Ranks. As an individual at home, she is nobody. Her picking Suleyman as her hired companion and her recollection of "moving" moments with him imply at least a little erotic electricity. But this tourist seems to have gotten her biggest kicks on vacation by playing the dominatrix, whipping "her" Tatar into line. To the ears of today's reader, Natalya's exclamation "you're just a Tatar, and I'm the wife of a state councilor!" says it all. That utterance encapsulates the racism of a Russian tourist compensating in the Crimea for her powerlessness in daily life in her patriarchal

14. This postcard reproduction of Repin's painting of a Russian tourist with a Tatar guide was a popular souvenir.

home society. However, Chekhov's insight into the psychopathology of colonialism was far ahead of his time. Mikhailovsky, for example, writing in 1900, considered "A Loose Tongue" an entertainment eliciting "carefree, jolly, good-humored, superficial, and, if you will, conciliatory laughter."[103] The long string of adjectives protests too much, and the notion of a reconciliation (between whom? to what?) is baffling.

Other efforts to come to terms with Russian women allegedly running wild in the Crimea further illustrate Chekhov's remarkable insights. The early twentieth-century Crimean resort press kept moralizing about the "debauchery" of Russian ladies paying Tatar guides for their "dirty, animal caresses." At the same time, vacation postcards invited people to laugh at the cuckold, a standard butt of jokes. Captioned "Free at Last," one souvenir image showed a Russian woman dressed to the nines and galloping alongside a handsome Tatar equestrian looking toward her with a smile, while a stout white-haired man lags behind on his horse.[104] Repin's beautiful painting *The Crimea: A Guide* elevated his subjects for disinterested contemplation (fig. 14). The guide and the tourist ride side by side on a deserted mountain road. In flattering local attire, the dashing Tatar looks straight ahead, while the lady has eyes only for him. Where have they been? And what is to come? Like the Grecian urn and Keats's ode about it, Repin's painting freezes beauty and desire for all time.

In placing sex and fashion at the heart of tourist decadence, the writings examined here provide new testimony to Tolstoy's stature as the conscience of the nation at the fin de siècle. Among the literary elite, Tolstoy alone witnessed the modernizing expansion of Russian tourism from the 1850s to 1910. Beginning with *Family Happiness*, he construed leisure travel as a pursuit likely to disrupt women's proper development as faithful, subservient wives and devoted mothers. Overcome with tourist angst, the vacationing narrator of *Lucerne* perceived the luxury resort as a playground for people shirking work and family responsibilities. *Anna Karenina* extended this anti-tourism to Russian aristocrats, to suggest that everybody should stay home to live the authentic life of farming, child-rearing, moral self-improvement, and religious searching that furthers the nation-building process of bridging the divide between

103 Cited in Chekhov, *PSSiP*, 5:653.

104 Mal'gin, *Russkaia Riv'era*, 264–65.

the elite and the peasantry. More specifically, Tolstoy's adultery novel cast leisure travel as a source of confusion and misdirection, deflecting Russian women from the hearth-and-home path of true self-realization. Adultery, birth control to assure the enjoyment of pregnancy-free sex, indifference to a baby (Anna toward her daughter Annie), the cult of beauty, fashion consciousness, and consumerism (versus Dolly's patching her old clothes), and finally the symbolism of the "fast" woman Anna on a horse at Vronsky's estate that resembles a resort with a four-star hotel with gourmet cuisine: this cluster of themes resonated in the anti-tourist portrayal of Veselitskaya's Mimi, Burenin's Nina, Chekhov's monstrous Ariadne, his unhappy lady with the little dog, and his power-hungry non-entity Natalya. Uspensky's *burzhui* had no antecedent in the œuvre of Count Tolstoy but nonetheless participated in the soulless consumerism, illicit sex, and fashion-plate habits (nice suit, good haircut) that defined Russian tourist decadence at the fin de siècle.

In going beyond personal morality to challenge the entire political order in Russia, Tolstoy's religious philosophy had further ramifications about the decadence of the state. Modernist art of the sort Russian traditionalists considered decadent jostled against the solemn, spiritual tenor of the celebration of the 1896 coronation of Nicholas II. As Richard Wortman documented, evening festivities in the Bolshoi Theater began with scenes from Glinka's *Life for the Tsar* with its culminating procession to the Kremlin and the Glory chorus. But next came the fairy-tale ballet *The Splendid Pearl*, set in a grotto of pearls on the ocean floor. Reproduced in the coronation album, the illustration of the official program rendered the ballet in the "erotic idiom" of the art nouveau designer Elena Samokish-Sudkovskaya (1863–1924). The artist pictured the pearls as "shapely maidens, wearing diaphanous skirts, their breasts and nipples bared," and in the concluding tableau, "semi-nude sea nymphs and sirens" languorously bathed before "an Adonis-like Triton."[105] Nicholas acquired a different aura of decadence on the day of his coronation when at least 1,350 peasants died in the stampede to the refreshment booths set up for the people in the Khodynka Field. While aggrieved by the catastrophe, the tsar went as planned to a dinner and ball at the French ambassador's residence that evening. Shortly after the event, the Tolstoyan, Aylmer Maude, published this reaction:

105 Wortman, *Scenarios of Power*, 2:358.

"Nero fiddled while Rome was burning, and Nicholas II danced at the French ball on the night of the Khodynskoe massacre."[106]

The royal family's integrating tourism abroad into their public images added yet another nuance of decadence. In 1864, when French officialdom feted Alexander II in Paris, Herzen gleefully remarked in the *Bell* that a French newspaper had relegated the tsar's visit to the society pages: he was being trivialized as a "high-ranking tourist."[107] At the fin de siècle Nicholas II and the heir presumptive, the Grand Duke Mikhail Alexandrovich (1878–1918), accommodated the "tourist" identity that Herzen considered demeaning. With a circulation now approaching 200,000, *Grainfield* in 1896 ran an illustrated article on the heir presumptive's villa in La Turbie on the Riviera. The magazine remarked that La Turbie is "well known to tourists" for its important Roman ruin, the Trophy of Augustus.[108] The article described the villa's "luxuriously appointed" rooms, its huge, terraced garden of exotic plants, and the grand duke's hobnobbing in Nice and Monaco with the yacht-owning set of European aristocrats (234). A few months later, *Grainfield* ran a long feature on the festive week that Nicholas II and the tsarina spent in Paris as guests of the state. The French hosts found a fitting way to entertain the emperor of a multinational realm. As recounted and pictured in the magazine, a military parade on the Place de la Concorde was led by a galloping equestrian contingent of the "elders of Arabic and Tunisian tribes" in their flowing robes and native headgear. But in addition to staging that extraordinary imperial encounter, the French led the tsar along the beaten tourist track, as illustrated by a picture of him staring at Napoleon's tomb.[109] Like the ordinary Russian tourists that Tolstoy maligned as parasites gone "numb" to the suffering of their underprivileged compatriots, Nicholas was fiddling around in Paris, while revolution was brewing at home.

106 Ibid., 362–63.

107 Herzen, *SS*, 18:297, 604.

108 *Niva*, no. 10 (1896):233–34. Built in 6 BCE, the monument commemorates the emperor Augustus's campaigns against tribes of the Alps between 16 and 7 BCE.

109 Ibid., no. 41 (1896):1031, 1034a.

Concluding Observations

This book has investigated how the pursuit of tourism and the controversies it stirred contributed to Russian construction of an imperial-national identity between Europe and Asia. The use of Buzard's *The Beaten Track* as a template has shown Russian conformity to the international pattern of tourist phobia emerging and evolving as a constituent component of tourism. Russian anti-tourism replicated prominent tropes of English writing, especially as of the Romantic era: the traveling cosmopolitan's opposition to the crowd, the valorization of art consumption as the loftiest foreign-travel agenda, and men's disdainful views of women travelers. Those cross-cultural similarities were by no means coincidental. To a great extent, they attested to a process of cultural transfer, the mediating roles that Sterne, Pope, Byron, Thackeray, Dickens, and lesser English writers and journalists played in shaping imperial Russian public images of tourists and tourism as of Karamzin's time. But while conforming to the big pattern of self-versus-other in a tourist setting, Russian writers recurrently specified "provincials" as the opposite of cosmopolitans—a characteristic tendency that both reflected and helped constitute Russia's nationally distinct, broader preoccupation with "the provinces."

The main thread of my investigation of Russian leisure travel abroad has excavated roots of the Soviet idea of cosmopolitanism as disparagement of native traditions and culture. Karamzin's Enlightenment travel model valorized foreign culture and underwrote the cosmopolitan ideals of world literature and conversation among an international elite of learned gentlemen. But his *Letters* also contained a strong element of national pride in the empire, represented as compensation for Russia's cultural lag behind the West. Colliding in unresolved conflict, the supranational and imperial-national sentiments adumbrated the two poles of the subsequent field of controversy about Russian tourism abroad. From its first appearance on the national scene in the 1840s, romantic cosmopolitanism rejoiced in cultural, scenic, and material consumption in Western Europe and eschewed demonstrative patriotism. In Belinsky's phrase, Russians had learned to travel without having passports stamped on their foreheads. But even at home, too many of them seemed content to go on mimicking Europeans, and that bothered him as much as it did conservative nationalists, yearning to see the emergence of a distinct Russian national "personality" strong and confident enough to stand alongside the nations of Europe. Touring the West in this period, the professed "homespun patriot" Pogodin perceived political and moral menace in the Russian "cosmopolitans" that he imagined snubbing the whole nation in his person at the Baths of Caracalla.

After the Crimean War, cultural controversies about vacationing in Western Europe became more complicated. Transmutations of the anxious effort to locate "authentic" Russian culture persisted in such ideologies as Dostoevsky's Native Soil program, but they coexisted with revolutionaries' disaffection to "aristocratic" culture and the visual arts in particular. From the time the Grand Tour became a standard practice of the happy few, seeing art was an indispensable part of the program. In no other country was this aesthetic agenda so subject to political and moral condemnation as it was in Russia. Spurred into action by the growing crevice between the educated classes and the common people, the utilitarian Russian radicals called for journeys aimed at building practical and political knowledge. In setting out to visit factories abroad, Chernyshevsky's Rakhmetov enacted industrial tourism *avant l'heure*, and Tolstoy's Levin did, too, although with a different purpose in mind.[1] But instead of merely ignoring cultural tourism, the revolutionaries attacked it again and again. They denied the universal value of art and the value of aesthetic experience in general. Their proscriptive idiom and animus toward cultural tourism in Italy in particular anticipated Soviet tourist officials' battle against so-called "culture fetishism" (*kul'turnichestvo*), an arch bugaboo signifying "tourism with the politics left out."[2] Imperial Russia's public quarrelling over political versus aesthetic travel

1 In 1918 Lenin instructed authorities to organize excursions to factories, agricultural enterprises, and hydroelectric stations in order to impress upon adults and school children the achievements of Soviet industry and engineering: Dolzhenko and Putrik, *Istoriia turizma*, 82.

2 Diane P. Koenker, "The Proletarian Tourist in the 1930s: Between Mass Excursion and Mass Escape," in Gorsuch and Koenker, *Turizm*, 126 (with *kul'turnichestvo* translated as "culturalism").

regimes never seems to have stopped. Consider the dueling reviews that greeted Alexander Kuprin's (1870–1938) *Lazurnye berega* (*Azure shores*, 1913–14), his collected travel sketches about Marseille, Nice, Monte Carlo, and Venice. The brilliant Modernist poet Vladislav Khodasevich (1886–1939) thought the book could be skipped because it ignored art: Kuprin had apparently feasted his eyes on nothing but "taverns, boxers, pimps, shopkeepers, coachmen, and casino croupiers." But what Khodasevich found boring made commendable reading for the anonymous journalist who praised *Azure Shores* precisely for having focused on "the common people in whom Russian tourists usually take little interest."[3]

My treatment of the rise of the bourgeois tourist has constituted another field of conflicting attitudes and values, a field that anticipated the vein of bourgeois decadence in the Soviet concept of the cosmopolitan. Nationalist and populist discourses (Dostoevsky, Lavrov, Uspensky) reviled the arriviste tourist specimen. On the other hand, Markov promoted tourism both at home and abroad as a democratizing opportunity for the Russian middle class. The fin de siècle brought many stories of tourist decadence that were influenced by Tolstoy but which may have given decadence an illicit allure to the Russian bourgeoisie desiring to make aristocratic pleasures their own (Bourget, Veselitskaya). Such readers very likely would have found more appealing even the grossly spendthrift tourism of Chekhov's Ariadne in Italy rather than the story's alternative of sleeping rough. The amalgamated phenomenon was an antecedent of the fascination that cosmopolitanism acquired in popular Soviet imagination of the 1920s, despite or rather *because* of Soviet officialdom's disapproving attitude. With respect to the United States in particular, the Bolshevik press "reified the views they sought to dispel" in barraging readers with stories about "the bad life abroad."[4]

Russian tourists tended to carry imperial consciousness with them wherever they traveled. Some of their stories even cast Russian-English encounters on the Continent as imperial competition (Karamzin, Myatlev, Tolstoy's *Lucerne*, Melgunov's 1859 essay). But it was in the Crimea, the

3 Commentary in A. I. Kuprin, *Sobranie sochinenii*, 6 vols. (Moscow: Gosudarstvennoe izdatel'stvo khudozhestvennoi literatury, 1957–58), 6:821–22.

4 Brooks, "Official Xenophobia and Popular Cosmopolitanism":1447. See similar reasoning about anti-cosmopolitan propaganda under Stalin failing to control "desires and temptations generated by its own negativity," in Caroline Humphrey, "Cosmopolitanism and Kosmopolitizm in the Political Life of Soviet Citizens," *Focaal* 44 (Winter 2004): 140–41.

North Caucasus, and Georgia that tourists got to test the imperial privilege of being Russian through personal contact with frequently ornamentalized "Asia" and "Asians." However modest the tourists' status at home, vacationing in the South gave them the opportunity to experience themselves as full-fledged members of the European "club" of empire owners, instead of mere mimickers of European civilization. Hegemonic Russian stereotypes of ethnic others, including the lazy native and the native swindler, arose in the context tourism. Markov's *Sketches of the Crimea* made cosmopolitan gestures to Tatars, but Chekhov and Burenin depicted appalling Russian tourist bigotry in action.

Imperialist Russian discourse about local peoples of the southern borderlands resonated in official Soviet efforts to harness socialist tourism to a traditional civilizing mission in the North Caucasus. A Soviet newspaper article of 1930 granted "proletarian" tourists the right to admire "the beauty of the Caucasus landscape" as long as they also investigated "class struggle in the Chechen *aul* [village]."[5] Eva Maurer established that by the late 1930s, Russian mountaineering camps in the North Caucasus had evolved into "cultural outposts—replicating the Stalinist power-relations of the center and periphery." The camps were assigned the task of "civilizing" the "pristine wilderness" and eradicating the "hard and lawless past" of local villages. But after Chechens, Balkars, and other North Caucasian peoples suffered deportation and resettlement during World War II, there were no further efforts to organize contacts between locals and the Russian mountain climbers. In 1949 local men raped and murdered a young Russian hitchhiker en route to her camp, an incident that triggered recall of nineteenth-century Russian mythologies of Caucasian savagery.[6] The natives were restless, and they have remained so with regard to tourists in our time. In 2011 "unidentified assailants" opened fire on a van carrying Moscow skiers to a popular Mt. Elbrus resort in Karbardino-Balkaria. "Three people, including a woman, were killed and two others wounded." A "terrorist" explosion also hit a cable car on the mountain, causing no injuries but "bringing the ski lifts to a halt, after the cables and forty cabins crashed to the ground."[7] Although occurring in very different eras, those

5 Koenker, "The Proletarian Tourist," 119.

6 Eva Maurer, "*Al'pinizm* as Mass Sport and Elite Recreation: Soviet Mountaineering Camps under Stalin," in Gorsuch and Koenker, *Turizm*, 151–52.

7 "Tourism Development to Continue in Russia's North Caucasus: Medvedev"—website now defunct (en.rian.ru/business); contact author for downloaded copy.

outbursts of violence suggest that there is still more to learn about colonized peoples' awareness of demeaning stereotypes and plainer racism that nineteenth-century Russian writing aired in tourist settings ("you're just a Tatar," "you're all Circassians, after all.").[8] Far from smoothing the way for the doctrine of the friendship of peoples formulated under Stalin, imperial Russian tourism's development as a form of imperialism may have conceivably left bitter legacies that endure today.

The expansion of domestic Russian tourism followed the usual pattern of diminishing the territory where a visitor might experience the authenticity effect. The English established "little Englands" on the Continent. Westernized Russians accomplished nothing so grandiose abroad but did establish home-plus in the North Caucasus, Georgia, and the Crimea. In the heart and mind of the tourist, Pyatigorsk becomes Lermontov country and moves into Europe (or at least *looks* European). Modernized quarters of Tiflis feel like Petersburg or Moscow, and symbolic architecture makes visible the advance of Europeanizing Russian tourism in the Crimea. The names of hotels popping up in the borderlands in the early twentieth century keep domesticating Asia for the westernized Russian visitor: the Belle-Vue and the London in Pyatigorsk, the France and the Grand Hotel in Vladikavkaz, the fancy France with its restaurant ranked "the best in Yalta" and nearby on the coast the low-budget Edinburgh furnished rooms.[9] The Georgian Military Road becomes a beaten track, and Veselitskaya's Castle of Treachery and Love morphs into an extensively developed tourist attraction. By 1903, the site had become "the favorite excursion of [Kislovodsk's] high-season clientele," who often chose the spot for "mild flirtation" or the "finale of a sentimental affair." Lanterns lighted the route from the spa to the Castle, and at the base of the cliff there was a small hotel with "an excellent restaurant serving local and European fare," including "fresh trout and traditional shashlik."[10]

The tourist armies marched on, eliminating adventure space. But the Russian Empire remained big enough to offer tourist opportunities to suit every taste. At the turn of the century the Far Eastern provinces and

8 See Manning, *Strangers in a Strange Land*, which, among other achievements, gives voice to Georgians telling their own stories about Russian travelers in Georgia, especially 28–58, 251–85.

9 Moskvich, *Illiustrirovannyi prakticheskii putevoditel' po Kavkazu*, 45,161; and Moskvich, *Prakticheskii putevoditel' po Krymu*, 138–39.

10 Moskvich, *Illiustrirovannyi prakticheskii putevoditel' po Kavkazu*, 132–33.

Central Asia beckoned off the beaten track.[11] And in Tashkent at least, tourist authorities recognized the importance of the authenticity effect. They obeyed Alexander III's order to renovate a fifteenth-century mosque in the old city but otherwise "avoided large-scale repairs that might disrupt a tourist's picturesque view of a timeless Orient."[12] While early twentieth-century Russian adventure tourists slept in yurts in Central Asia, their comfort-loving compatriots flocked to a new luxury hotel at the Imatra resort in Finland. "Poetically named the Cascade" and boasting art nouveau décor, the establishment had made vacationing among "our Finnish neighbors" more and more popular. At holiday times, the well-appointed, direct trains from Petersburg to Imatra were "usually crammed with passengers."[13]

Representations of imperial Russian tourism at home and abroad created the collective phenomenon of aesthetic cosmopolitanism set in mutually illuminating relationship to nationalism and provinciality. But who shall we designate the outstanding individual cosmopolitans? I would propose Turgenev, Herzen, and Chekhov for their shared concern with Russian national culture and national welfare in combination with a love of the art, literature, and other aspects of the civilization of Western Europe. With his mind full of the art he was seeing in Italy, Turgenev in an 1857 letter to Tolstoy contemptuously described Russian painters in Rome at the time. With the exception of Ivanov, he wrote, "our artists are nitwits contaminated by the mess Bryullov stirred up [*briullovshchina*]. They lack talent, or if they have it, they don't know how to use it. They live with sluts, curse Raphael, and that's it."[14] But as an émigré in Paris in the 1870s and 1880s, Turgenev played an active role in organizing exhibits of Russian painting of the period and promoted the French translation of Russian literature. In the eyes of Peter Kropotkin (1842–1921), Lavrov, and other Russian revolutionaries who sought him out in Paris, Turgenev remained one of the monumental contributors to the national literature and a chronicler of

---

11 McReynolds, "The Prerevolutionary Tourist," 40; and Hokanson, "Russian Women Travelers in Central Asia," 1–9.

12 Crews, "Civilization in the City," 129.

13 *Niva*, no. 44 (1903): 882 and illustration 881. For mass-circulation promotion of art nouveau art, architecture, furniture, home décor, and jewelry, see ibid., no. 9 (1905): 171–75. The Imatra resort had been attracting "masses" of Russian tourists ever since 1870 when a railroad connected Helsingford to Petersburg's Vyborg station: ibid., no. 34 (1875): 535–38.

14 Turgenev, *PPSiP: Pis'ma*, 3:171.

sociopolitical problems still to be solved to make Russia a better country.[15] Russian obituaries of Turgenev articulated such views, and his funeral in Petersburg in 1883 drew thousands of mourners.

Herzen loathed the French bourgeoisie but expressed great affinity to Italians (especially his fellow revolutionaries and the common folk), he represented the Vatican museum as a secular temple housing treasures of universal significance, and he filed those proud reports of the concerts of Russian music drawing big audiences in London in 1860. Remarkable as well was Georgi Plekhanov's (1856–1918) commemorative speech at Herzen's grave in Nice in April 1912. Plekhanov said Dostoevsky had been wrong to fling the phrase *citoyen du monde* as an invective against Herzen. Quoting Herzen's public defense of his sympathetic response to the Polish revolt, Plekhanov characterized him as a citizen of the world in the sense of an internationalist:

> "We are not slaves to the love of country," he wrote, "just as we are not slaves to anything else. A free man cannot but recognize independence from his land if it would compel him to participate in a cause against his conscience." So did he speak. Those are truly golden words. We should all remember them now whether the subject at issue be cruel, shameful pogroms against Jews, the abolition of the Finnish constitution, the interdiction against Ukrainian children studying in their own little-Russian language, or, in general, whatever be the persecution against whatever people within the population of our state![16]

By comparison to Turgenev and Herzen, Chekhov was a closet aesthetic cosmopolitan and as such no doubt represented the norm for untold numbers of Russian cultural tourists who left no trace in histories of their country. Chekhov's stories described the misery and brutality of Russian peasant life and the despondency of impoverished school teachers and other Russians stuck in the provinces. It is interesting to keep in mind, though, that he wrote several of these stories of grim existence in the Russian sticks during the eight months of 1897–98 that he spent in Nice for

15 Peter Kropotkin, *Memoirs of a Revolutionist*, ed. James Allen Rogers (Garden City: Anchor Books, 1962), 263–67, 329–30.

16 G. V. Plekhanov, "Rech' na mogile A. I. Gertsena v Nitstse 7 aprelia 1912 g.," in *A. I. Gertsen v russkoi kritike. Sbornik statei*, 2nd ed., intro. and commentary by V. Putnitsev (Moscow: Gosudarstvennoe izdatel'stvo khudozhestvennoi literatury, 1953), 243–45.

treatment of tuberculosis. Chekhov visited Western Europe six times, four of which took him to Italy. Delight overflows in his letters from abroad: the exalted discovery of art and architecture in Italy (as previously mentioned), shopping in Vienna, and frequenting Monte Carlo, just a short hop from Nice. His observations on life in the West are overwhelmingly enthusiastic: "I love the culture," he writes from Nice in November 1897. "And you see culture oozing out of every shop window, every raffia basket; every dog smells of civilization."[17] The "obvious pleasure of [Chekhov's] first encounter with European countries [was] such a source of embarrassment" that the 1963–64 Soviet State Publishing House edition of his correspondence simply omitted the offending letters.[18] To the mind of officialdom, Chekhov's cosmopolitan sentiments smacked of disloyalty, and so the collection was curated to fit his public persona.

Juxtaposed to his home life as a much solicited doctor, an activist in public health, and an organizer of famine relief in the Volga region in 1891, Chekhov's letters suggest how many responsible, nationally concerned but unsung Russian tourists must have brought their cosmopolitan experiences home without losing allegiance to Russia.[19] Abroad in the late 1850s when Russians were flooding the West, Panayeva in *Russians in Italy* admitted that she was encountering many intelligent, cultivated compatriots pursuing tourism as pleasurable learning. But like a novelist stumped by the problem of making virtuous characters interesting, she maintained there was "nothing to say" about those people.[20] Tourist morons, philistines, and villains always made better copy.

17 Chekhov, letter to Anna Suvorina, November 10, 1897, in *PSSiP*, *Pis'ma*, 7:98.

18 Karlinsky, *Chekhov's Life and Thought*, 184–85.

19 The memoirs of Boris Chicherin would reward study from this angle.

20 Panaeva, *Russkie v Italii*, 306.

Bibliography

A__v, Ia. "Zametki na puti iz Moskvy v Zakavkazskii krai." *Otechestvennye zapiski* 41, no. 7 (1845), section 8: 10–35.

Abashev, Vladimir. *Perm' kak tekst. Perm' v russkoi kul'ture i literature XX veka*. Perm': Izdatel'stvo Permskogo universiteta, 2000.

Abrams, M. H. "Art-as-Such: The Sociology of Modern Aesthetics." In *Doing Things with Texts. Essays in Criticism and Critical Theory*, 135–58. New York and London: W. W. Norton, 1989.

Adler, Judith. "Origins of Sightseeing." *Annals of Tourism Research* 16, no. 1 (1989): 7–29.

———. "Travel as Performed Art." *American Journal of Sociology* 94, no. 6 (May 1989): 1366–91.

Afanas'ev, D. M. "Sevastopol' i ego okrestnosti v nastoiashchee vremia." *Sovremennik* 62, no. 3 (March 1857), section 1: 134–46.

Agathocleous, Tanya. *Urban Realism and the Cosmopolitan Imagination: Visible City, Invisible World*. Cambridge: Cambridge University Press, 2011.

Agathocleous, Tanya, and Jason R. Rudy. "Victorian Cosmopolitanisms: Introduction." *Victorian Literature and Culture* 38 (2010): 389–97.

Aksakov, I. S. *Ivan Sergeevich Aksakov v ego pis'makh*. 4 vols. Moscow: Tipografiia M. G. Volchaninova, 1892.

———. *Polnoe sobranie sochinenii*. 7 vols. Moscow: Tipografiia M. G. Volchaninova, 1886–87.

Aksakov, K. S. *Estetika i literaturnaia kritika*. Moscow: Iskusstvo, 1995.

Alatas, Syed Hussein. *The Myth of the Lazy Native: A Study of the Image of the Malays, Filipinos and Javanese from the 16th to the 20th Centuries and Its Functions in the Ideology of Colonial Capitalism*. London: Frank Cass, 1977.

Aldonina, N. B. "Obraz Rima v publitsistike A. V. Druzhinina." In *Obraz Rima v russkoi literature. Mezhdunarodnyi sbornik nauchnykh rabot*, edited by Rita Giuliani and V. I. Nemtsev, 163–78. Rome and Samara: Izdatel'stvo OOO NTTs, 2001.

Alekseev, M. P. "Angliiskii iazyk v Rossii i russkii iazyk v Anglii." *Uchenye zapiski Leningradskogo universiteta. Seriia filologicheskikh nauk* 9, no. 72 (1944): 77–137.

———. "Pushkin i angliiskie puteshestvenniki v Rossii." *Literaturnoe nasledstvo* 91 (1982): 574–656.

Alexandrov, Vladimir E. *Limits to Interpretation: The Meanings of* Anna Karenina. Madison: University of Wisconsin Press, 2004.

Andrew, Joe. "Another Time, Another Place: Gender and the Chronotope in the Society Tale" (1998). In *The Society Tale in Russian Literature from Odoevskii to Tolstoi*, edited by Neil Cornwell, 127–32. Amsterdam and Atlanta: Rodopi, 1998.

———. Introduction to *Asya*, by Ivan Turgenev, edited with notes and vocabulary by F. G. Gregory, 7–24. London: Bristol Classical Press, 1992.

Annenkov, P. V. *Parizhskie pis'ma*. Edited by I. N. Konobeevskaia. Moscow: Nauka, 1983.

———. *Vospominaniia i kriticheskie ocherki. Sobranie statei i zametok, 1849–1868*. St. Petersburg: Tipografiia M. Stasiulevicha, 1877.

Anon. "Angliiskie turisty." *Otechestvennye zapiski* 5, no. 9 (1839), section 8: 99–108.

Anon. "Iz zapisok o Gruzii." *Syn otechestva* 52, no. 4 (1850), section 7: 5–9.

Anon. "Peterburgskaia zhizn'. Zametki novogo poeta." *Sovremennik* 65, no. 9 (September 1857), section 5: 82–103.

Anon. "Pokhozhdeniia provintsiala v stolitse." *Syn otechestva* 37, no. 5 (May 1850), section 7: 49–63.

Anon. "Predislovie k izdaniiu 1904 g." In I. P. Miatlev, *Sensatsii i zamechaniia gospozhi Kurdiukovoi za granitseiu, dan l'etranzhe*, i–xxiii. 2nd ed.. St. Petersburg: Izdanie A. S. Suvorina, 1907.

Anon. "Prodolzhenie letnei emigratsii za granitsu i novyi zakon, oblegchaiushchii prebyvanie za granitseiu." *Otechestvennye zapiski* 113 (July 1857), section 3: 22–23.

Anon. "Znachenie predpolozhennoi seti zheleznykh dorog dlia raznykh kraev Rossii." *Otechestvennye zapiski* 111 (March 1857), section 3: 18–25.

Appiah, Kwame Anthony. *Cosmopolitanism: Ethics in a World of Strangers*. London: Penguin Books, 2006.

———. "Cosmopolitan Patriots." In *For Love of Country: Debating the Limits of Patriotism*, edited by Martha C. Nussbaum and Joshua Cohen, 21–29. Boston: Beacon Press, 1996.

———. *The Ethics of Identity*. Princeton: Princeton University Press, 2005.

Arkhipova, A. V. "Dostoevskii i Karamzin." *Dostoevskii. Materialy i issledovaniia* 5 (1983): 101–12.

Avseenko, V. G. "Opiat' o narodnosti i o kul'turnykh tipakh." *Russkii vestnik* (March 1876): 362–87.

———. "Literaturnaia zametka." *Russkii vestnik* (February 1876): 884–98.

Baehr, Stephen L. "The Troika and the Train: Dialogues between Tradition and Technology in Nineteenth-Century Russian Literature." In *Issues in Russian Literature before 1917: Selected Papers of the Third World Congress for Soviet and East European Studies*, edited by Douglas J. Clayton, 85–106. Columbus: Slavica Publishers, 1989.

Bakhtin, M. M. "From the Prehistory of Novelistic Discourse." In *The Dialogic Imagination. Four Essays*, edited by Michael Holquist, translated by Caryl Emerson and Michael Holquist, 41–83. Austin: University of Texas Press, 1981.

Bakić-Hayden, Milica. "Nesting Orientalisms: The Case of Former Yugoslavia." *Slavic Review* 54, no. 4 (Winter 1995): 917–31.

Bakradze, Dmitri. "Stseny iz gruzinskoi zhizni." *Sovremennik* 25, no. 1 (January 1851), section 6: 1–26; and no. 2 (February), section 6: 65–191.

Bakunina, Ekaterina. "Vospominaniia sestry miloserdiia Krestovozdvizhenskoi obshchiny (1854–1860 gg.)." *Vestnik Evropy* 133 (May 1898): 55–105.

Balmont [Bal'mont], K. D. *Nesobrannoe i zabytoe iz tvorcheskogo naslediia.* 2 vols., ed. A. Iu. Romanova. 2 vols. St. Petersburg: Izdatel'stvo Rostok, 2016.

Baranowski, Shelley, and Ellen Furlough. "Introduction." In *Being Elsewhere: Tourism, Consumer Culture, and Identity in Modern Europe and North America*, edited by Shelley Baranowski and Ellen Furlough, 1–34. Ann Arbor: University of Michigan Press, 2001.

Barrett, Thomas M. "Crossing Boundaries: The Trading Frontiers of the Terek Cossacks." In Brower and Lazzerini, *Russia's Orient*, 227–48.

———. "Southern Living (in Captivity): The Caucasus in Russian Popular Culture." *Journal of Popular Culture* 31 (Spring 1998): 75–93.

Barsht, Konstantin. "Defining the Face: Observations on Dostoevskii's Creative Processes." In *Russian Literature, Modernism and the Visual Arts*, edited by Catriona Kelly and Stephen Lovell, 23–57. Cambridge: Cambridge University Press, 2000.

Barsukov, N. P. *Zhizn' i trudy M. P. Pogodina.* 22 vols. St. Petersburg: Tipografiia M. M. Stasiulevicha, 1888–1910.

Barthes, Roland. *The Fashion System.* Translated by Matthew Ward and Richard Howard. London: Hill and Wang, 1985.

Bassin, Marc. *Imperial Visions: Nationalist Imagination and Geographical Expansion in the Russian Far East, 1840–1865.* Cambridge: Cambridge University Press, 1999.

———. "Russia between Europe and Asia." *Slavic Review* 50, no. 1 (Spring 1991): 1–17.

Bassnett, Susan. "Travel Writing and Gender." In Hulme and Youngs, *The Cambridge Companion to Travel Writing*, 225–41.

Batten, Charles L., Jr. *Pleasurable Instruction: Form and Convention in Eighteenth-Century Travel Literature.* Berkeley: University of California Press, 1978.

Baudin, Rodolphe. "Karamzine critique d'art: la peinture dans les *Lettres d'un voyageur russe*." *Revue des études slaves* 83, nos. 2–3 (2012): 731–58.

Bauer, M. "Cultural Tourism in France." In *Cultural Tourism in Europe*, edited by Greg Richards, 107–21. Wallingford, UK: CAB International, 1996.

Bayley, John. *Tolstoy and the Novel.* New York: Viking Press, 1966.

Bazin, Germain. *The Museum Age.* New York: Universe Books, 1967.

Bechler, Rosemary. "Lord Byron's Grander Tour." In *Transports, Travel, Pleasure, and Imaginative Geography*, edited by Chloe Chard and Helen Langdon, 49–74. New Haven: Yale University Press, 1996.

Becker, Seymour. "Russia and the Concept of Empire." *Ab Imperio* 3–4 (2000): 329–42.

Belinskii, V. G. *Polnoe sobranie sochinenii.* 13 vols. Moscow: Akademiia nauk, 1953–59.

Belknap, Robert L. "On Dostoevsky and the Visual Arts." In *Depictions: Slavic Studies in the Narrative and Visual Arts in Honor of William E. Harkins*, edited by Douglas Greenfield, 58–60. Dana Point, CA: Ardis, 2000.

Benediktov, V. G. *Stikhotvoreniia*. 2nd ed. Leningrad: Sovetskii pisatel', 1983.

Berelowitch, Wladimir. "La France dans le 'grand tour' des nobles russes au cours de la seconde moitié du XVIIIe siècle." *Cahiers du monde russe et soviétique* 34, nos. 1–2 (1993): 193–210.

Berg, N. B. "Iz krymskikh zametok." *Sovremennik* 58, no. 4 (1856), section 1: 131–202.

Berghoff, Hartmut. "From Privilege to Commodity? Modern Tourism and the Rise of the Consumer Society." In Berghoff, Korte, Schneider, and Harvie, *The Making of Modern Tourism*, 159–80.

Berghoff, Hartmut, and Barbara Korte. "Britain and the Making of Modern Tourism: An Interdisciplinary Approach." In Berghoff, Korte, Schneider, and Harvie, *The Making of Modern Tourism*, 1–20.

Berghoff, Hartmut, Barbara Korte, Ralf Schneider, and Christopher Harvie, eds. *The Making of Modern Tourism: The Cultural History of the British Experience, 1600–2000*. New York: Palgrave, 2002.

Berlin, Isaiah. "Introduction." In *My Past and Thoughts. The Memoirs of Alexander Herzen*. Translated by Constance Garnett, revised by Humphrey Higgens, and abridged with a preface and notes by Dwight Macdonald, xix–xliii. New York: Vintage Books, 1974.

———. "Introduction." In Raeff, *Russian Intellectual History*, 3–11.

———. *Russian Thinkers*. Edited by Henry Hardy and Aileen Kelly, with introduction by Aileen Kelly. Harmondsworth: Penguin Books, 1979.

Bernov, M. A. *Iz Odessy peshkom po Krymu. Pis'ma russkogo peshekhoda*. St. Petersburg: Glazunov, 1896.

Bestuzhev-Marlinskii, A. A. *Povesti i rasskazy*. Moscow: Izdatel'stvo "Sovetskaia Rossiia," 1976.

———. *Sochineniia*. 2 vols. Moscow: Gosudarstvennoe izdatel'stvo khudozhestvennoi literatury, 1958.

Bethea, David. "*The Idiot*: Historicism Arrives at the Station." In *The Shape of Apocalypse in Russian Fiction*, 62–104. Princeton: Princeton University Press, 1989.

Billington, James H. *The Icon and the Axe: An Interpretive History of Russian Culture*. New York: Vintage Books, 1970 [1966].

Blakesley, Rosalind P. "Art, Nationhood, and Display: Zinaida Volkonskaia and Russia's Quest for a National Museum of Art." *Slavic Review* 67, no. 4 (Winter 2008): 912–33.

Bojanowska, Edyta M. *A World of Empires. The Russian Voyage of the Frigate Pallada*. Cambridge, MA: Harvard University Press, 2018.

———. "Empire by Consent: Strakhov, Dostoevskii, and the Polish Uprising of 1863." *Slavic Review* 71, no. 1 (Spring 2012): 1–24.

Boorstin, Daniel J. "From Traveler to Tourist: The Lost Art of Travel." In *The Image or What Happened to the American Dream*, 86–125. New York: Atheneum, 1962.

Booth, Alison. "Author Country: Longfellow, the Bröntes, and Anglophone Homes and Haunts." *Romanticism and Victorianism on the Net* 48 (2007): 1–29. https://doi.org/10.7202/017438ar.

Botkin, V. P. "Dve nedeli v Londone 1859 goda." In "*Ia bereg pokidal tumannyi Al'biona . . .*" *Russkie pisateli ob Anglii, 1646–1945*. Edited by O. A. Kaznina and A. N. Nikoliukin, 281–86. Moscow: ROSSPEN, 2001.

———. *Pis'ma ob Ispanii*. Edited by B. F. Egorov and A. Zvigil'skii. Leningrad: Nauka, 1976.

Bourdieu, Pierre, and Alain Darbel, with Dominique Schnapper. *The Love of Art: European Art Museums and Their Public*. Translated by Caroline Beattie and Nick Merriman. Cambridge, UK and Malden, MA: Polity Press, 1997.

Bourget, Paul [Pol' Burzhe]. *Tragicheskaia idilliia. Kosmopoliticheskie nravy*. [Translated by M. R. (sic) from *Une Idylle tragique: mœurs kosmopolites*]. *Russkaia mysl'*, no. 1 (1896), section 1: 109–79.

Bowlt, John E. "Russian Painting in the Nineteenth Century." In Stavrou, *Art and Culture in Nineteenth-Century Russia*, 113–39.

Boyer, Marc. *Histoire générale du tourisme du XVIe au XXIe siècle*. Paris: Harmattan, 2005.

Breckenridge, Carol A., Sheldon Pollock, Homi K. Bhabha, and Dipesh Chakrabarty, eds. *Cosmopolitanism*. Durham, NC and London: Duke University Press, 2002.

Brien, Alan. "Tourist Angst." *The Spectator*, no. 6840 (July 31, 1959): 133.

Brikner, A. G. "Russkie diplomaty-turisty v Italii v XVII stoletii." *Russkii vestnik* (March 1877): 5–44, (April 1877): 561–607, (July 1877): 5–62.

Briggs, A. D. P. *Alexander Pushkin. A Critical Study*. Beckenham, Kent: Croom Helm, 1983.

Bristol, Evelyn. "Idealism and Decadence in Russian Symbolist Poetry." *Slavic Review* 39, no. 2 (1980): 269–80.

Briusov, V. Ia. *Iz moei zhizni*. Moscow: Terra, 1994.

Brooks, James F. "Bondage and Emancipation across Cultural Borderlands: Some Reflections and Extensions." In David-Fox, Holquist, and Martin, *Orientalism and Empire in Russia*, 291–94.

Brooks, Jeffrey. "Official Xenophobia and Popular Cosmopolitanism in Early Soviet Russia." *American Historical Review* 97, no. 5 (December 1992): 1431–48.

———. *When Russia Learned to Read: Literacy and Popular Literature, 1861–1917*. Princeton: Princeton University Press, 1985.

Brower, Daniel R. *The Russian City between Tradition and Modernity, 1850–1900*. Berkeley: University of California Press, 1990.

Brower, Daniel R., and Edward J. Lazzerini, eds. *Russia's Orient: Imperial Borderlands and Peoples, 1700–1917*. Bloomington: Indiana University Press, 1997.

Brower, Daniel R., and Susan Layton. "Liberation through Captivity. Nikolai Shipov's Adventures in the Imperial Borderlands." In David-Fox, Holquist, and Martin, *Orientalism and Empire in Russia*, 270–90.

Brown, Malcolm Hamrick. "Native Song and National Consciousness in Nineteenth-Century Russian Music." In Stavrou, *Art and Culture in Nineteenth-Century Russia*, 57–84.

Bruner, Edward M. *Culture on Tour. Ethnographies of Travel.* Chicago and London: University of Chicago Press, 2005.

Brunson, Molly. *Russian Realisms: Literature and Painting, 1840–1890*. DeKalb: Northern Illinois University Press, 2016.

Brusovani, M. I., and R. G. Gal'perina. "Zagranichnye puteshestviia F. M. Dostoevskogo." *Dostoevskii. Materialy i issledovaniia* 8 (1988): 272–92.

Buckler, Julie A. *Mapping St. Petersburg: Imperial Text and Cityshape*. Princeton and Oxford: Princeton University Press, 2005.

———. "Reading Anna: Opera, Tragedy, Melodrama, Farce." In Knapp and Mandelker, *Approaches to Teaching Tolstoy's* Anna Karenina, 131–36.

Burbank, Jane, Mark von Hagen, and Anatolyi Remnev, eds. *Russian Empire: Space, Power, People, 1700–1930*. Bloomington: Indiana University Press, 2007.

Burenin, V. P. *Mertvaia noga. Roman v Kislovodske*. St. Petersburg: Izdanie A. S. Suvorina, 1886.

Burenin, V. P. [Opasnyi sopernik g. Turgeneva, pseud.]. "Pirozhnitsa beregov Reina, ili Russkii dvorianin za granitsei. Turgenevskaia povest'." *Iskra*, nos. 7–9 (February 13–March 5, 1877): 103–5, 136–38.

Burkart, A. J., and S. Medlik. *Tourism: Past, Present and Future*. 2nd ed. London: Heinemann, 1981.

Buzard, James. *The Beaten Track: European Tourism, Literature, and the Ways to 'Culture', 1800–1918*. Oxford: Oxford University Press, 2001. [1993]

———. "'The Country of the Plague': Anticulture and Autoethnography in Dickens's 1850s." *Victorian Literature and Culture* 38 (2010): 413–19.

———. "The Grand Tour and after (1660–1840)." In Hulme and Youngs, *The Cambridge Companion to Travel Writing*, 37–52.

———. "Race, Imperialism, Colonialism, Postcolonialism, and Cosmopolitanism." In *The Oxford Handbook of Charles Dickens*, edited by John Jordan, Robert L. Patten, and Catherine Waters, 517–36. Oxford: Oxford University Press, 2018.

Byron, Lord George Gordon. *Poetical Works*. Edited by Frederick Page. 3rd ed. Oxford: Oxford University Press, 1979.

———. *Chaild Garol'd* [sic]. Translation by Pavel Kozlev of *Childe Harold's Pilgrimage*, 1.1–93. *Russkaia mysl'* (January 1890), section 1: 1–16 and ibid. (February 1890), section 1: 1–16.

Calhoun, Craig. "The Class Consciousness of Frequent Travellers: Toward a Critique of Actually Existing Cosmopolitanism." In Vertovec and Cohen, *Conceiving Cosmopolitanism*, 86–109.

Cannadine, David. *Ornamentalism. How the British Saw Their Empire*. London: Allen Lane, The Penguin Press, 2001.

Chaadaev, P. I. "Letters on the Philosophy of History. First Letter." Translated by Valentine Snow. In Raeff, *Russian Intellectual History*, 160–173.

Chaikovskii, P. I. *Pis'ma k blizkim. Izbrannoe*. Moscow: Gosudarstvennoe muzykal'noe izdatel'stvo, 1955.

Chard, Chloe. *Pleasure and Guilt on the Grand Tour: Travel Writing and Imaginative Geography, 1600–1830*. Manchester and New York: Manchester University Press, 1999.

Cheah, Pheng, and Bruce Robbins, eds. *Cosmopolitics: Thinking and Feeling beyond the Nation*. Minneapolis and London: University of Minnesota Press, 1998.

Chekhov, A. P. *Polnoe sobranie sochinenii i pisem*. 30 vols. Moscow: Nauka, 1974–83.

Chernyshevskaia, N. M. *Letopis' zhizni i deiatel'nosti N. G. Chernyshevskogo*. Moscow: Gosudarstvennoe izdatel'stvo khudozhestvennoi literatury, 1953.

Chernyshevskii, N. G. *Polnoe sobranie sochinenii*. 16 vols. Moscow: Gosudarstvennoe izdatel'stvo khudozhestvennoi literatury, 1939–53.

———. *What Is to Be Done?* Translated by Michael R. Katz, annotated by William G. Wagner. Ithaca and London: Cornell University Press, 1989.

Chhabra, Deepak. "Staged Authenticity and Heritage Tourism." *Annals of Tourism Research* 30, no. 3 (2003): 702–19.

Clark, Katerina. *Moscow, the Fourth Rome: Stalinism, Cosmopolitanism and the Evolution of Soviet Culture, 1931–1941*. Cambridge, MA: Harvard University Press, 2011.

———. "Political History and Literary Chronotope: Some Soviet Case Studies." In *Literature and History: Theoretical Problems and Russian Case Studies*, edited by Gary Saul Morson, 230–46. Stanford: Stanford University Press, 1986.

Clifford, James. "Mixed Feelings." In Cheah and Robbins, *Cosmopolitics*, 362–70.

———. "Traveling Cultures," in *Routes: Travel and Translation in the Late Twentieth Century*, 17–46. Cambridge, MA and London: Harvard University Press, 1997.

Clowes, Edith W. *Russia on the Edge: Imagined Geographies and Post-Soviet Identity*. Ithaca: Cornell University Press, 2011.

Clowes, Edith W., Samuel D. Kassow, and James L. West, eds. *Between Tsar and People: Educated Society and the Quest for Public Identity in Late Imperial Russia*. Princeton: Princeton University Press, 1991.

Cohen, Erik. "Authenticity and Commoditization in Tourism." *Annals of Tourism Research* 15, no. 3 (1988): 371–86.

Cracraft, James. *The Petrine Revolution in Russian Architecture*. Chicago and London: University of Chicago Press, 1988.

———. *The Petrine Revolution in Russian Imagery*. Chicago and London: University of Chicago Press, 1997.

Cracraft, James, and Daniel Rowland, eds. *Architectures of Russian Identity: 1500 to the Present*. Ithaca and London: Cornell University Press, 2003.

Crawshaw, Carol, and John Urry. "Tourism and the Photographic Eye." In *Touring Cultures: Transformations of Travel and Theory*, edited by Chris Rojek and Urry, 176–95. London: Routledge, 1997.

Crews, Robert D. "Civilization in the City: Architecture, Urbanism, and the Colonization of Tashkent." In Cracraft and Rowland, *Architectures of Russian Identity*, 117–32.

Cross, A. G. *"By the Banks of the Thames": Russians in Eighteenth-Century Britain*. Newtonville, MA: Oriental Research Partners, 1980.

———. *N. M. Karamzin: A Study of His Literary Career, 1783–1803*. Carbondale, IL: Southern Illinois University Press, 1971.

Cruise, Edwina. "Tracking the English Novel in *Anna Karenina*: Who Wrote the English Novel that Anna Reads?" In Orwin, *Anniversary Essays on Tolstoy*, 159–82.

Culler, Jonathon. "The Semiotics of Tourism." In *Framing the Sign: Criticism and Its Institutions*, 153–67. Oxford: Basil Blackwell, 1988.

Dal', V. I. *Tolkovyi slovar' zhivogo velikorusskogo iazyka*. 4 vols. Edited by I. A. Baudouin de Courtenay. 3rd ed. St. Petersburg-Moscow: Tovarishchestvo M. O. Vol'f, 1903–1909; repr. Moscow: Progress/Univers, 1994.

Daly, Kirsten. "Worlds Beyond England: *Don Juan* and the Legacy of Enlightenment Cosmopolitanism." *Romanticism* 4, no. 2 (July 1998): 189–201.

Danilevskii, G. P. *Polnoe sobranie sochinenii*. 24 vols. 8th ed. St. Petersburg: A. F. Marks, 1901.

Danilevskii, N. Ia. *Rossiia i Evropa*. Moscow: Drevnee i sovremennoe, 2002. First published in 1869 in the Slavophile journal *Zaria* (*Dawn*).

David-Fox, Michael. *Showcasing the Great Experiment: Cultural Diplomacy and Western Visitors to the Soviet Union, 1921–1941*. New York: Oxford University Press, 2012.

David-Fox, Michael, Peter Holquist, and Alexander Martin, eds. *Orientalism and Empire in Russia*. Bloomington: Slavica Publishers, 2006.

Davidson, Pamela. "Aleksandr Ivanov and Nikolai Gogol': The Image and the Word in the Russian Tradition of Art and Prophecy." *Slavonic and East European Review* 91, no. 2 (April 2013): 157–209.

Degaev [sic]. *Rukovodstvo dlia edushchikh v Piatigorsk i Tiflis i piatigorskie mineral'nye vody, s prilozheniiem marshruta*. Moscow: Tipografiia Lazarevskogo Instituta vostochnykh iazykov, 1860.

Dekker, George G. *The Fictions of Romantic Tourism: Radcliffe, Scott, and Mary Shelley*, Stanford: Stanford University Press, 2005.

Delanty, Gerard, and David Inglis, eds. *Cosmopolitanism: Critical Concepts in the Social Sciences*. 4 vols. London: Routledge, 2011.

———. "Introduction: An Overview of the Field of Cosmopolitan Studies," in Delanty and Inglis, *Cosmopolitanism*, 1:1-27.

Demidenko, Iu. B. *Restorany, traktiry, chainye. Iz istorii obshchestvennogo pitaniia v Peterburge XVIII–nachala XX veka*. Moscow: Tsentrpoligraf, 2011.

Demidoff [Demidov], Anatole de. *Travels in Southern Russia and the Crimea, through Hungary, Wallachia, and Moldavia during the year 1837*. 2 vols. London: J. Mitchell, 1853.

Demidova, Olga, and Mary Zirin. "Lidiia Ivanovna Veselitskaia." In *Dictionary of Russian Women Writers*, edited by Marina Ledkovsky, Charlotte Rosenthal, and Mary Zirin, 705–6. Westport: Greenwood Press, 1994.

Deotto, Patrizia. "Impressioni, dipinti, visioni: L'Italia nell'immaginario russo." *Europa Orientalis* 15, no. 2 (1996): 51–76.

———. *In viaggio per realizzare un sogno: L'Italia e il testo italiano nella cultura russa*. Trieste: Università degli Studi di Trieste, 2002.

———. "Materialy dlia izucheniia ital'ianskogo teksta v russkoi kul'ture." *Slavica tergestina* 6 (1998): 197–226.

Dianina, Katia. "Passage to Europe: Dostoevskii in the St. Petersburg Arcade." *Slavic Review* 62, no. 2 (Summer 2003): 237–57.

———. "The Return of History: Museum, Heritage, and National Identity in Imperial Russia." *Journal of Eurasian Studies* 1, no. 2 (2010): 111–18.

———. *When Art Makes News: Writing Culture and Identity in Imperial Russia*. DeKalb: Northern Illinois University Press, 2013.

Dickinson, Sara. *Breaking Ground: Travel and National Culture in Russia from Peter I to the Era of Pushkin*. Amsterdam: Rodopi, 2006.

———. "*Otechestvo, otchizna, rodina*: Russian 'Translations' of *Patrie* in the Napoleonic Period." In Offord et al., *French and Russian in Imperial Russia*, 179–96.

———. "Russia's First 'Orient.' Characterizing the Crimea in 1787." In David-Fox, Holquist, and Martin, *Orientalism and Empire in Russia*, 85–106.

Dobroliubov, N. A. *Sobranie sochinenii*. 9 vols. Moscow and Leningrad: Gosudarstvennoe izdatel'stvo khudozhestvennoi literatury, 1961–64.

Dolgobanov, G. I., ed. *Khronika zhizni i tvorchestva A. S. Pushkina*. 3 vols. Moscow: IMLI RAN, "Nasledie," 2001, 2009, 2016.

Dolzhenko, G. P. *Istoriia turizma v dorevoliutsionnoi Rossii i SSSR*. Rostov-on-Don: Izdatel'stvo Rostovskogo universiteta, 1988.

Dolzhenko, G. P. and Iu. S. Putrik. *Istoriia turizma v Rossiiskoi Imperii, Sovetskom Soiuze, i Rossiiskoi Federatsii: 1696 g.-sovremennost'*. Moscow/Rostov-on-Don: Izdatel'skii tsentr "MarT"/Feniks, 2010.

Donchin, Georgette. *The Influence of French Symbolism on Russian Poetry*. The Hague: Mouton, 1958.

Donskov, Andrew, and John Woodsworth, eds. *Lev Tolstoy and the Concept of Brotherhood*. New York, Ottawa, and Toronto: Legas, 1996.

Dostoevskaia, A. G. *Vospominaniia*. Edited by L. P. Grossman. Moscow and Leningrad: Gosudarstvennoe izdatel'stvo, 1925.

Dostoevskii, F. M. *Polnoe sobranie sochinenii*. 30 vols. Leningrad: Nauka, 1972–90.

———. *Winter Notes on Summer Impressions*. Translated by David Patterson. Evanston: Northwestern University Press, 1997.

Dowler, Wayne. *Dostoevskii, Grigor'ev and Native Soil Conservatism*. Toronto, Buffalo, London: University of Toronto Press, 1982.

Draitser, Emil A. *Techniques of Satire: The Case of Saltykov-Shchedrin*. Berlin: Mouton de Gruyter, 1994.

Drozdov, I. I. [doctor]. "Kavkazskie mineral'nye vody." *Sovremennik* 39, no. 5 (May 1853), section 2: 1–60.

Drozdov, I. I. [artillery officer]. "Zapiski kavkaztsa." *Russkii arkhiv* 3, no. 10 (1896): 213–74.

Druzhinin, A. V. *Povesti. Dnevnik*. Edited by B. F. Egorov and V. A. Zhdanov. Moscow: Nauka, 1986.

———. *Sobranie sochinenii*. 8 vols. St. Petersburg. Akademiia nauk, 1865–67.

Ducamp, Emmanuel. "Private Galleries as Public Service." In *Great Private Collections of Imperial Russia*, edited by Oleg Yakovlevich Neverov and Emmanuel Ducamp, with introductions by Mikhail Borisovich Piotrovsky and H. H. Nicholas Romanov, prince of Russia, 10–16. London: Thames and Hudson, 2004.

Dudyshkin, S. Review of Goncharov's "Russkie v Iaponii." *Otechestvennye zapiski* 104, nos. 1–2 (January 1856), section 3: 35–50.

Dumont, Jean-Paul. Review of *The Tourist*, by D. MacCannell. *L'homme—Revue française d'anthropologie* 17, nos. 2–3 (1977): 190–91.

Dzhanshiev, D. G. *Perl Kavkaza: Borzhom. Abastuman. Vpechatleniia i mysli turista*. 4th ed. Moscow: I. N. Kushner, 1900.

Eikhenbaum, Boris. *Lev Tolstoi*. 2 vols. repr. Munich: Wilhelm Fink Verlag, 1968 [1928–31].

Ely, Christopher. "The Origins of Russian Scenery: Volga River Tourism and Russian Landscape Aesthetics." *Slavic Review* 62, no. 4 (Winter 2003): 666–82.

———. "The Picturesque and the Holy: Visions of Touristic Space in Russia, 1820–1850." In Cracraft and Rowland, *Architectures of Russian Identity*, 80–89.

———. *This Meager Nature: Landscape and National Identity in Imperial Russia*. DeKalb: Northern Illinois University Press, 2002.

Emerson, Caryl. "Tolstoy and Music." In Orwin, *Anniversary Essays on Tolstoy*, 8–32.

———. "*What Is Art?* and the Anxiety of Music." *Russian Literature* 40, no. 4 (November 1996): 433–49.

Engelstein, Laura. *The Keys to Happiness: Sex and the Search for Modernity in Fin de Siècle Russia*. Ithaca and London: University of Cornell Press, 1992.

———. *Slavophile Empire: Imperial Russia's Illiberal Path*. Ithaca and London: Cornell University Press, 2009.

Evdokimova, Svetlana. "The Drawing and the Grease Spot: Creativity and Interpretation in *Anna Karenina*." *Tolstoy Studies Journal* 8 (1995–96): 33–45.

Fairweather, Maria. *The Pilgrim Princess: A Life of Princess Zinaida Volkonsky*. London: Robinson, 2000.

Fateev, P. S. "Mikhail Ilarionovich Mikhailov (1830–65)." In M. I. Mikhailov, *Stikhovoreniia*, 5–40. Leningrad: Sovetskii pisatel', 1950.

Fet, A. A. *Iz-za granitsy. Putevye vpechatleniia. Sovremennik* 60, no. 11 (November 1856), section 1: 71–117; ibid. 61; no. 2 (February 1857), section 1: 237–71; and ibid. 64, no. 8 (August 1857), section 1: 81–128.

———. *Stikhotvoreniia i poemy, 1839–1863.* Vol. 1: *Sochineniia i pis'ma.* Kursk and St. Petersburg: Akademicheskii proekt, 2002–

———. *Vospominaniia.* 3 vols. Moscow: Kul'tura, 1992 [1890, 1893].

Figes, Orlando. *Natasha's Dance: A Cultural History of Russia.* London: Penguin Books, 2003.

Finke, Michael C. *Seeing Chekhov: Life and Art.* Ithaca and London: Cornell University Press, 2005.

Fisher, Alan Washburn. *The Crimean Tatars.* Stanford: Stanford University Press, 1978.

———. *The Russian Annexation of the Crimea, 1772–1783.* Cambridge: Cambridge University Press, 1970.

Flath, Carol. "Writing about Nothing: Chekhov's 'Ariadna' and the Narcissistic Narrator." *Slavonic and East European Review* 77, no. 2 (April 1999): 223–39.

Florinsky, Michael T. *Russia: A History and an Interpretation.* 2 vols. New York: Macmillan, 1965.

Foster, John Burt, Jr. *Transnational Tolstoy: Between the West and the World.* New York and London: Bloomsbury Academic, 2013.

Frank, Joseph. *Dostoevsky: The Mantle of the Prophet, 1871–1881.* Princeton: Princeton University Press, 2002.

———. *Dostoevsky: The Miraculous Years, 1865–1871.* London: Robson Books, 1995.

———. *Dostoevsky: The Seeds of Revolt, 1821–1849.* Princeton: Princeton University Press, 1976.

———. *Dostoevsky: The Stir of Liberation, 1860–1865.* Princeton: Princeton University Press, 1986.

———. *Dostoevsky: The Years of Ordeal, 1850–1859.* London: Robson Books, 1983.

Fridlender, G. M. "Estetika Dostoevskogo." In *Dostoevskii khudozhnik i myslitel'. Sbornik statei*, edited by K. N. Lomunov, A. L. Grishunin, U. A. Gural'nik, and V. S. Nechaeva, 97–164. Moscow: Khudozhestvennaia literatura, 1972.

Fussell, Paul. *Abroad: British Literary Traveling between the Wars.* Oxford and New York: Oxford University Press, 1980.

Fusso, Susanne. *Editing Turgenev, Dostoevsky, and Tolstoy: Mikhail Katkov and the Great Russian Novel.* DeKalb: Northern Illinois University Press, 2017.

Galgan, Galina. "The Idea of World Renewal." In Donskov and Woodsworth, *Tolstoy and the Concept of Brotherhood*, 23–37.

Gatrall, Jefferson J. A. "Tolstoy, Ge, and Two Pilates: A Tale of the Interarts." In *From Realism to the Silver Age: New Studies in Russian Artistic Culture*, edited by Rosalind Blakesley and Margaret Samu, 79–93. DeKalb: Northern Illinois University Press, 2014.

Geraci, Robert P. "Islam." In Martinsen and Maiorova, *Dostoevsky in Context*, 209–18.

———. *Window on the East: National and Imperial Identities in Late Imperial Russia*. Ithaca: Cornell University Press, 2001.

Gersevanov, B. N. "Iz putevykh vpechatlenii turista. Pereezd cherez Simplon." *Otechestvennye zapiski* 57 (1848), section 8: 1–21.

Geyer, Dietrich. *Russian Imperialism. The Interaction of Domestic and Foreign Policy 1860–1914*. Translated by Bruce Little. Leamington Spa, Hamburg, New York: Berg, 1987.

Gleason, Abbott. *Young Russia: The Genesis of Russian Radicalism in the 1860s*. New York: Viking Press, 1980.

———. "Westernism." In Terras, *Handbook of Russian Literature*, 517–18.

Glinka, M. I. "Zapiski Mikhaila Ivanovicha Glinki. Chast' chetvertaia." *Russkaia starina*, no. 2 (1870): 419–62.

Glinka-Volzhskii, A. S., and G. A. Leman, eds. *Gleb Uspenskii*. Moscow: Gosudarstvennyi literaturnyi musei, 1939.

Goerner, Tatiana. "The Theme of Art and Aesthetics in Dostoevsky's *The Idiot*." *Ulbandus Review* 2, no. 2 (Fall 1982): 79–95.

Gogol [Gogol'], N. V. *Sobranie sochinenii*. 7 vols. Moscow: Khudozhestvennaia literatura, 1984–86.

Golovin, Ivan. "Puteshestvennik nashego vremeni." *Syn otechestva* 5 (1838), section 1: 67–83.

Goncharov, I. A. *Polnoe sobranie sochinenii i pisem*. 20 vols. St. Petersburg: Nauka, 1997–2017.

Goodlad, Lauren M. E., and Julia M. Wright. "Introduction and Keywords." *Romanticism and Victorianism on the Net* 48 (November 2007): 1–26. https://doi.org/10.7202/017435ar.

Gorsuch, Anne E. *All This Is Your World: Soviet Tourism at Home and Abroad after Stalin*. Oxford: Oxford University Press, 2011.

———. "'There's No Place Like Home': Soviet Tourism in Late Stalinism." *Slavic Review* 62, no. 4 (Winter 2003): 760–85.

———. "Time Travelers: Soviet Tourists to Eastern Europe." In Gorsuch and Koenker, *Turizm*, 205–26.

Gorsuch, Anne E., and Diane P. Koenker, eds. *Turizm: The Russian and East European Tourist under Capitalism and Socialism*. Ithaca and London: Cornell University Press, 2006.

Goscilo, Helena. "Women's Space and Women's Place in Contemporary Russian Fiction." In *Gender and Russian Literature: New Perspectives*, edited by Rosalind J. Marsh, 326–47. Cambridge: Cambridge University Press, 1996.

Goscilo-Kostin, Helena. "Tolstoyan Fare: Credo à la Carte." *Slavonic and East European Review* 62, no. 4 (October 1984): 481–95.

Grant, Bruce. *The Captive and the Gift: Cultural Histories of Sovereignty in Russia and the Caucasus*. Ithaca and London: Cornell University Press, 2009.

Greenfeld, Liah. "The Formation of the Russian National Identity: The Role of Status Insecurity and *Ressentiment*." *Comparative Studies in Society and History* 32, no. 3 (July 1990): 549–59.

Greenleaf, Monika. *Pushkin and Romantic Fashion: Fragment, Elegy, Orient, Irony*. Stanford: Stanford University Press, 1994.

Greenwood, Davydd J. "Culture by the Pound: An Anthropological Perspective on Tourism as Cultural Commoditization." In Smith, *Hosts and Guests*, 171–85.

Greenwood, E. B. *Tolstoy. The Comprehensive Vision*. London: J. M. Dent, 1975.

Grigor'ev, A. A. *Izbrannye proizvedeniia*. Leningrad: Sovetskii pisatel', 1959.

Grigorovich, D. V. *Polnoe sobranie sochinenii*. 12 vols. 3rd ed. St. Petersburg: A. F. Marks, 1896.

Grüner, Frank. "'Russia's Battle against the Foreign': The Anti-Cosmopolitanism Paradigm in Russian and Soviet Ideology." *European Review of History* 17, no. 3 (June 2010): 445–72.

Gustafson, Richard F. *Leo Tolstoy: Resident and Stranger. A Study in Fiction and Theology*. Princeton: Princeton University Press, 1986.

Hamburg, G. M. "Language and Conservative Politics in Alexandrine Russia." In Offord et al., *French and Russian in Imperial Russia*, 118–38.

———. "Tolstoy's Spirituality." In Orwin, *Anniversary Essays on Tolstoy*, 138–58.

Hammarberg, Gitta. *From the Idyll to the Novel: Karamzin's Sentimentalist Prose*. Cambridge: Cambridge University Press, 1991.

Hannerz, Ulf. "Cosmopolitans and Locals in World Culture." In Delanty and Inglis, *Cosmopolitanism*, 3:26–38.

Harkness, Kristen M. "Mariia Iakunchikova and the Roots of Decadence in Late Nineteenth-Century Russian Modernism." In *Decadence, Degeneration, and the End: Studies in the European Fin de Siècle*, edited by Marja Härmänmaa and Christopher Nissen, 145–63. New York: Palgrave Macmillan, 2014.

Haywood, Richard Mowbray. *Russia Enters the Railway Age, 1842–1855*. Boulder, CO and New York: East European Monographs; distributed by Columbia University Press, 1998.

Hazlitt, William. "On Going a Journey" (1822). In William Hazlitt, *Table Talk; or, Original Essays*, 249–61. New York: Chelsea House, 1983.

Hellman, Ben. *Fairy Tales and True Stories: The History of Russian Literature for Children and Young People (1574–2010)*. Leiden, the Netherlands: Loninlkijke Brill NV, 2013.

Helmers, Rutger. *Not Russian Enough? Nationalism and Cosmopolitanism in Nineteenth-Century Russian Opera*. Rochester: University of Rochester Press, 2014.

Herman, David. "Stricken by Infection: Art and Adultery in *Anna Karenina* and *Kreutzer Sonata*." *Slavic Review* 56, no. 1 (Spring 1997): 15–36.

Herzen [Gertsen], A. I. *Sobranie sochinenii*. 30 vols. Moscow: Izdatel'stvo Akademii nauk, 1954–65.

———. *Letters from France and Italy, 1847–1851*, edited and translated by Judith E. Zimmerman. Pittsburgh and London: University of Pittsburgh Press, 1995.

Hokanson, Katya. "Pushkin's Captive Crimea: Imperialism in *The Fountain of Bakhchisarai*." In *Russian Subjects: Empire, Nation, and the Culture of the Golden Age*, edited by Monika Greenleaf and Stephan Moeller-Sally, 123–48. Evanston: Northwestern University Press, 1998.

———. "Russian Woman Travelers in Central Asia and India." *Russian Review* 70, no. 1 (January 2011): 1–19.

———. *Writing at Russia's Border*. Toronto, Buffalo, London: University of Toronto Press, 2008.

Holland, Kate. "Dostoevsky's Journalism in the 1870s." In Martinsen and Maiorova, *Dostoevsky in Context*, 288–94.

Hommaire de Hell, Adèle and Xavier. "Les Steppes de la mer Caspienne, le Caucase, la Crimée et la Russie méridionale" (1843–45). In *Le Voyage en Russie. Anthologie des voyageurs français aux XVIIIe et XIXe siècles*, edited by Claude de Grève, 764–70. Paris: Robert Laffont, 1990.

Horne, Donald, *The Great Museum: The Re-presentation of History*. London: Pluto Press, 1984.

———. *The Intelligent Tourist*. Smithfield, NSW: Margaret Gee Publishing, 1992.

Hosking, Geoffrey. *Russia: People and Empire, 1552–1917*. Cambridge, MA: Harvard University Press, 1997.

Hudspith, Sarah. *Dostoevsky and the Idea of Russianness: A New Perspective on Unity and Brotherhood*. London and New York: Routledge Curzon, 2004.

Hulme, Peter. *Colonial Encounters: Europe and the Native Caribbean, 1492–1797*. London and New York: Routledge, 1992 [1986].

Hulme, Peter, and Tim Youngs, eds. *The Cambridge Companion to Travel Writing*. Cambridge: Cambridge University Press, 2002.

Humphrey, Caroline. "Cosmopolitanism and Kosmopolitizm in the Political Life of Soviet Citizens." *Focaal* 44 (Winter 2004): 138–52.

Iakovlev, Vladimir. *Italiia: Pis'ma iz Venetsii, Rima i Neapolia*. St. Petersburg: Tipografiia Koroleva, 1855.

Iampol'skii, I. G., ed. *Poety 1860-kh godov*. Leningrad: Sovetskii pisatel', 1968.

Ivanits, Linda. *Dostoevsky and the Russian People*. Cambridge: Cambridge University Press. 2008.

Ivanov [sic]. "Poezdka na obnovlennye kavkazskie vody." *Russkii vestnik* (October 1875): 678–721.

Jackson, Robert Louis. *Dostoevsky's Quest for Form: A Study of His Philosophy of Art*. New Haven: Yale University Press, 1966.

Jahn, Gary R. "Brother or Other: Tolstoy's Equivocal Surrender to the Concept of Brotherhood." In Donskov and Woodsworth, *Tolstoy and the Concept of Brotherhood*, 71–87.

Jakobson, Roman. "Le mythe de la France en Russie." In *Russie, folie, poésie*, edited and presented by Tzvetan Todorov and translated by Nancy Huston, Marc B. de Launay, and André Markowicz, 157–68. Paris: Seuil, 1986.

James, Henry [Genri Dzhems]. "Venetsiia." *Istoricheskii vestnik* 14 (December 1883): 576–96.

Jones, Malcolm V. "Dostoyevsky and Europe: Travels in the Mind." *Renaissance and Modern Studies* 24 (1980): 38–57.

Jones, W. Gareth. "Introduction." In *Tolstoi and Britain*, edited by W. Gareth Jones, 1–30. Oxford: Berg, 1995.

K. N. M. [sic]. "Obshchestvennoe vospitanie v Anglii. Iz dorozhnoi zapisnoi knigi." *Russkii vestnik* (January 1874): 244–68.

Kahn, Andrew. "Introduction: Karamzin and the Creation of a Readership." In Nikolai Karamzin, *Letters of a Russian Traveller*, translated by Andrew Kahn, 1–19. Oxford: Voltaire Foundation, 2003.

———. "Nikolai Karamzin's Discourses of Enlightenment." Ibid., 459–552.

Kagan-Kans, Eva. *Hamlet and Don Quixote: Turgenev's Ambivalent Vision*. The Hague: Mouton, 1975.

Kappeler, Andreas. *The Russian Empire: A Multiethnic History*. Translated by Alfred Clayton. Harlow: Pearson Education Limited, 2001.

Karamzin, N. M. *Pis'ma russkogo puteshestvennika*. Edited by Iu. M. Lotman, N. A. Marchenko, and B. A. Uspenskii. Leningrad: Nauka, 1984.

———. *Sochineniia*. 2 vols. Leningrad: Khudozhestvannaia literatura, 1984.

Karlinsky, Simon. "Two Pushkin Studies." *California Slavic Studies* 2 (1963): 96–120.

———. *Anton Chekhov's Life and Thought. Selected Letters and Commentary*. Translated by Michael Henry Heim in collaboration with Karlinsky. Berkeley: University of California Press, 1973.

Kasitsyn, D. "Iz dorozhnykh nabliudenii. Ot Berlina do Rima." *Russkii vestnik* (April 1876): 714–27.

———. "Iz nabliudenii v Rime." *Russkii vestnik* (February 1877): 489–523.

Katz, Michael R. Introduction to *Polinka Saks* and *The Story of Aleksei Dmitrich*, by A. V. Druzhinin. Translated by Michael R. Katz, 1–13. Evanston: Northwestern University Press, 1992.

Kauchtschischwili, Nina. *L'Italia nella vita e nell'opera di P. A. Vjàzemskij*. Milan: Socièta editrice vita e pensiero, 1964.

Kelly, Catriona. *A History of Russian Women's Writing, 1820–1992*. Oxford: Oxford University Press, 1994.

———. "Pushkin's Vicarious Grand Tour: A Neo-Sociological Interpretation of 'K vel'mozhe' (1830)." *Slavonic and East European Review* 77, no. 1 (January 1999): 1–29.

———. *Refining Russia: Advice Literature, Polite Culture, and Gender from Catherine to Yeltsin*. Oxford: Oxford University Press, 2001.

Khodarkovsky, Michael. *Bitter Choices: Loyalty and Betrayal in the Russian Conquest of the North Caucasus*. Ithaca and London: Cornell University Press, 2011.

Khomiakov, A. S. *O starom i novom: Stat'i i ocherki*. Moscow: Sovremennik, 1988.

———. *Sochineniia*. 3 vols. 2nd ed. Moscow: Tipografiia Lebedeva and Universitetskaia tipografiia M. Katkova, 1878–82.

———. *Stikhotvoreniia i dramy*. 2nd ed. Leningrad: Sovetskii pisatel', 1969.

King, Charles. *The Ghost of Freedom: A History of the Caucasus*. Oxford: Oxford University Press, 2008.

Kivelson, Valerie A. and Joan Neuberger, eds. *Picturing Russia: Explorations in Visual Culture*. New Haven and London: Yale University Press, 2008.

Kleespies, Ingrid. "East West Home Is Best: The Grand Tour in D. I. Fonvizin's *Pis'ma iz Francii* and N M. Karamzin's *Pis'ma russkogo putešestvennika*." *Russian Literature* 52 (2002), nos. 1–3 (2002): 251–69.

———. *A Nation Astray: Nomadism and National Identity in Russian Literature*. DeKalb: Northern Illinois University Press, 2012.

Kleingeld, Pauline. "Christoph Martin Wieland." In *The Bloomsbury Dictionary of Eighteenth-Century German Philosophers*, edited by Heiner F. Klemme and Manfred Kuehn, 849–52. London: Continuum Publishers, 2010.

———. "Six Varieties of Cosmopolitanism in late Eighteenth-Century Germany." In Delanty and Inglis, *Cosmopolitanism*, 1:172–90.

Klioutchkine, Konstantine. "Between Sacrifice and Indulgence: Nikolai Nekrasov as a Model for the Intelligentsia." *Slavic Review* 66, no. 1 (Spring 2007): 45–62.

———. "Sentimental'naia kommertsiia: *Pis'ma russkogo puteshestvennika* N. M. Karamzina." *Novoe literaturnoe obozrenie* 25 (1997): 84–98.

Knapp, Liza. "The Development of Style and Theme in Tolstoy." In Orwin, *The Cambridge Companion to Tolstoy*, 161–75.

———. "The Setting." In Knapp and Mandelker, *Approaches to Teaching Tolstoy's* Anna Karenina, 24–34.

Knapp, Liza, and Amy Mandelker, eds. *Approaches to Teaching Tolstoy's* Anna Karenina. New York: Modern Language Association of America, 2003.

Knight, Nathaniel. "Imperiia napokaz: Vserossiiskaia ethnograficheskaia vystavka 1867 goda." *Novoe literaturnoe obozrenie* 51, no. 5 (2001): 111–31.

Koenker, Diane P. *Club Red: Vacation Travel and the Soviet Dream*. Ithaca and London: Cornell University Press, 2013.

———. "The Proletarian Tourist in the 1930s: Between Mass Excursion and Mass Escape." In Gorsuch and Koenker, *Turizm*, 119–40.

———. "Travel to Work, Travel to Play: On Russian Tourism, Travel, and Leisure." *Slavic Review* 62, no. 4 (2003): 657–65.

Kokorev, V. A. "Iz putevykh zametok. Pis'mo k S. A. Khrulevu." In *Utro. Literaturnyi sbornik*, edited by M. P. Pogodin and colleagues at *Moskvitianin*, 197–210. Moscow: Tipografiia Barfknekhta, 1859.

Kolonitskii, Boris I. "Antibourgeois Propaganda and Anti-'*Burzhui*' Consciousness in 1917." *Russian Review* 53, no. 2 (April 1994): 183–96

Komolova, N. P. *Italiia v russkoi kul'ture Serebrianogo veka. Vremena i sud'by*. Moscow: Nauka, 2005.

Kovarskii, N. A. "Poeziia I. P. Miatleva." In Miatlev, *Stikhovoreniia*, 5–47.

Kovarsky, Gina. "The Moral Education of the Reader." In Knapp and Mandelker, *Approaches to Teaching Tolstoy's* Anna Karenina, 166–72.

Kovalevskii, P. M. *Etiudy puteshestvennika: Italiia, Shveitsariia, puteshestvenniki i puteshestvie*. St. Petersburg: Tipografiia V. Golovina, 1864.

———. "Kartiny Italii: Venetsiia." *Otechestvennye zapiski* 116, no. 2 (February 1858), section 1: 441–96.

———. "Putevye vpechatleniia ipokhondrika." *Sovremennik* 77, no. 10 (October 1859), section 1: 359–76.

Kozelsky, Mara. *Christianizing the Crimea: Shaping Sacred Space in the Russian Empire and Beyond*. DeKalb: Northern Illinois University Press, 2010.

———. *Crimea in War and Transformation*. New York: Oxford University Press, 2018.

Kramskoi, I. N. *Pis'ma, stat'i*. 2 vols. Moscow: Iskusstvo, 1965–66.

Kretovich I. "Don, Kavkaz, i Krym. (Iz putevykh vospominanii)." *Vestnik Evropy* (June 1868): 723–768.

Krippendorf, Jost. *The Holiday Makers: Understanding the Impact of Travel and Leisure*. Translated by Vera Andrassy. London: Heinemann, 1987.

Kropotkin, Peter. *Memoirs of a Revolutionist*. Edited by James Allen Rogers. Garden City: Anchor Books, 1962.

Kudriavtsev, P. N. [Nestroev, pseud.]. "Bel'veder (iz putevykh zapisok Russkogo)." *Otechestvennye zapiski* 45, no. 3 (1846), section 8: 14–42.

Kuehn, Julia. "Colonial Cosmopolitanism: Albert Smith and Rudyard Kipling in Victorian Hong Kong." *Studies in Travel Writing* 19, no. 3 (2015): 224–43.

Kupreianova, E. N. "Vyrazhenie esteticheskikh vozzrenii i nravstvennykh iskanii L. Tolstogo v romane *Anna Karenina*." *Russkaia literatura*, no. 3 (1960): 117–36.

Kuprin, A. I. *Sobranie sochinenii*. 6 vols. Moscow: Gosudarstvennoe izdatel'stvo khudozhestvennoi literatury, 1957–58.

Kushakov, A. V. "M. L. Mikhailov." In Mikhailov, *Sobranie stikhotvorenii*, 5–44. Leningrad: Sovetskii pisatel', 1953.

Lachinova, E. [E. Khamar-Dabanov, pseud.]. *Prodelki na Kavkaze*. 2 vols. St. Petersburg: K. Zhernakov, 1844.

Lambert, E. *Sons against Fathers: Studies in Russian Radicalism and Revolution*. Oxford: Oxford University Press, 1965.

Lanz, Kenneth. *The Dostoevsky Encyclopedia*. Westport: Greenwood Publishing Group, 2004.

Lavrov, P. L. [P. U.] "Russkii turist 40-vykh godov." *Delo*, no. 8 (1877): 24–40.

Layton, Susan. "Imagining a Chechen Military Aristocracy: The Story of the Georgian Princesses Held Hostage by Shamil." *Central Asian Survey* 23, no. 2 (June 2004): 183–203.

———. *Russian Literature and Empire: Conquest of the Caucasus from Pushkin to Tolstoy*. Cambridge: Cambridge University Press, 1994.

———. "Russian Military Tourism: The Crisis of the Crimean War Period." In Gorsuch and Koenker, *Turizm*, 43–63.

———. "Tolstoi and Lidia Veselitskaia's *Mimi at the Spa*: The *Fin de Siècle* Tourist Adulteress." In *Tolstoi: Art and Influence*, edited by Joe Andrew and Robert Reid. Leiden: Brill, 2021.

Lazzerini, Edward. "The Crimea under Russian Rule: 1783 to the Great Reforms." In *Russian Colonial Expansion to 1917*, edited by Michael Rywkin, 123–38. London and New York: Mansell Publishing Limited, 1988.

———. "Local Accommodation and Resistance to Colonialism in Nineteenth-Century Crimea." In Brower and Lazzerini, *Russia's Orient*, 169–87.

Leatherbarrow, W. J. "*The Devils* in the Context of Dostoevsky's Life and Works." In *Dostoevsky's "The Devils": A Critical Companion*, edited by W. J. Leatherbarrow, 3–59. Evanston: Northwestern University Press, 1999.

———. "Introduction: Dostoevskii and Britain." In *Dostoevsky and Britain*, edited by W. J. Leatherbarrow, 1–38. Oxford and Providence, RI: Berg, 1995.

Lebedev, Iu. V. "L. N. Tolstoi na puti k 'Voine i miru' (Sevastopol' i 'Sevastopol'skie rasskazy')." *Russkaia literatura* 4 (1976): 61–82.

LeBlanc, Ronald. "Levin Visits Anna: The Iconography of Harlotry." *Tolstoy Studies Journal* 3 (1990): 1–20.

———. "Unpalatable Pleasures: Tolstoy, Food, and Sex." *Tolstoy Studies Journal* 6 (1993): 1–32.

Leikin, N. A. *Gde apel'siny zreiut. Iumoristicheskoe opisanie puteshestviia suprugov Nikolaia Ivanovicha i Glafiry Semenovny Ivanovykh po Riv'ere i Italii*. St. Petersburg: S. P. Iakovlev, 1898.

———. *Nashi zagranitsei. Iumoristicheskoe opisanie poezdki suprugov Nikolaia Ivanovicha i Glafiry Semenovny Ivanovykh v Parizh i obratno*. 16th ed. St. Petersburg: S. P. Iakovlev, 1899.

Leont'ev, K. N. "Ispoved' muzha." In *Polnoe sobranie sochinenii*, vol. 2. St. Petersburg: Izdatel'stvo "Vladimir Dal'," 2000–2014.

———. "O vsemirnoi liubvi. Rech' F. M. Dostoevskogo na Pushkinskom prazdnike." In *Vostok, Rossiia i slavianstvo. Filosofskaia i politicheskaia publitsistika. Dukhovnaia proza, 1872–1891*, 312–329. Moscow: Izdatel'stvo Respublika, 1996.

Lermontov, M. Iu. *Sobranie sochinenii*. 4 vols. Moscow: Khudozhestvennaia literatura, 1983–84.

Leskov, N. S. *Polnoe sobranie sochinenii*. 30 vols. Moscow: Terra, 1996–2016.

Letkova, E. "Otdykh." *Russkaia mysl'*, no. 1 (1896): 180–96.

Levin, Iu. D. "Poeziia M. L. Mikhailova." In M. L. Mikhailov, *Sobranie stikhotvorenii*, 5–43. Leningrad: Sovetskii pisatel', 1969.

Levin, Sh. M. *Ocherki po istorii russkoi obshchestvennoi mysli: Vtoraia polovina XIX–nachalo XX veka*. Leningrad: Nauka, 1974.

Levitskii, G. P. "Pereselenie tatar iz Kryma v Turtsiiu." *Vestnik Evropy* (October 1882): 596–639.

Likhachev, D. S. "Vstupitel'noe slovo na torzhestvennom vechere posviashchennom 225-letnemu iubileiu so dnia rozhdeniia N. M. Karamzina." In *N. M. Karamzin. Pro i Contra. Lichnost' i tvorchestvo N. M. Karamzina v otsenke russkikh pisatelei, kritikov, issledovatelei: antologiia*, compiled, introduced, and commentary by L. A. Sapchenko, 746–48. St. Petersburg: RKhGI, 2006.

Lincoln, W. Bruce. *Nicholas I: Emperor and Autocrat of All the Russias*. Bloomington: Indiana University Press, 1980.

Lipson, Leopol'd. *Pervoe v Rossii predpriiatie dlia obshchestvennykh puteshestvii, 1885–86*. St. Petersburg: V. Kene, 1885.

Livanov, F. V. *Putevoditel' po Krymu s istoricheskim opisaniem dostoprimechatel'nostei*. Moscow: Tipografiia T. Ris, 1875.

Lloyd, David W. *Battlefield Tourism: Pilgrimage and the Commemoration of the Great War in Britain, Australia and Canada, 1919–1939*. London: Berg, 1998.

Lo Gatto, Ettore. *Russi in Italia*. Rome: Editori riuniti, 1971.

Lomachevskii, D. "Novaia Voenno-Gruzinskaia Doroga." *Russkii vestnik* (July 1857), section 1: 164–73.

Lonergan, Jennifer. "M. L. Mikhailov and Russian Radical Ideas About Women, 1847–1865." PhD diss., University of Bristol, 1995.

Lotman, Jurij. *The Structure of the Artistic Text*. Translated by Ronald Vroon. Ann Arbor: University of Michigan Press, 1977.

———. *Universe of the Mind: A Semiotic Theory of Culture*. Translated by Ann Shukman, introduction by Umberto Eco. London and New York: I. B. Tauris, 1990.

Lounsbery, Anne. *Life Is Elsewhere: Symbolic Geography in the Russian Provinces, 1800–1917*. Ithaca and London: Northern Illinois University Press, an imprint of Cornell University Press, 2019.

———. *Thin Culture, High Art: Gogol, Hawthorne and Authorship in Nineteenth-Century Russia and America*. Cambridge, MA: Harvard University Press, 2007.

Lovell, Stephen. *Summerfolk: A History of the Dacha, 1710–2000*. Ithaca and London: Cornell University Press, 2003.

Lywood, William George. "From Russia's Orient to Russia's Riviera: Reimagining the Black Sea Coast/Caucasus from Romantic Literature to Early Tourist Guidebooks." MA thesis, Ohio State University, 2009. Available at rave.ohiolink.edu/etdc/view?acc_num=osu1236622370. Ohio State University, 2009.

MacCannell, Dean. *The Ethics of Sightseeing*. Berkeley, Los Angeles, London: University of California Press, 2011.

———. *The Tourist: A New Theory of the Leisure Class*. 3rd ed. Berkeley, Los Angeles, and London: University of California Press, 2013 [1976].

Maguire, Robert A. *Exploring Gogol*. Stanford: Stanford University Press, 1994.

Maikov, A. N. *Polnoe sobranie sochinenii*. 4 vols. St. Petersburg: A. F. Marks, 1914.

Maikov, V. N. "Stikhotvoreniia Kol'tsova." *Otechestvennye zapiski* 49, nos. 11–12 (1846), section 5: 1–38 (article 1) and 39–70 (article 2).

Maiorova, Olga. "Avseenko, Vasilii Grigor'evich." In *Russkie pisateli. 1800–1917. Biograficheskii slovar'*, 1:20–22. Moscow: Sovetskaia entsiklopediia, 1989.

———. "Empire." In Martinsen and Maiorova, *Dostoevsky in Context*, 86–97.

———. *From the Shadow of Empire: Defining the Russian Nation through Cultural Mythology, 1855–1870*. Madison: University of Wisconsin Press, 2010.

Maisuradze, Giorgi, and Franziska Thun-Hohenstein. *Sonniges Georgien. Figuren des Nationalen im Sowjetimperium*. Berlin: Kulturverlag Kadmos, 2015.

Mal'gin, Andrei. *Russkaia Riv'era: Kurorty, turizm i otdykh v Krymu v epokhu Imperii, konets XVIII–nachalo XX v.* Simferopol': SONAT, 2004.

Malia, Martin. *Alexander Herzen and the Birth of Russian Socialism*. New York: The Universal Library, Grosset and Dunlap, 1961.

Mandelker, Amy. *Framing Anna Karenina: Tolstoy, the Woman Question and the Victorian Novel*. Columbus: Ohio State University Press, 1993.

Mann, Iu. V. *Russkaia filosofskaia estetika, 1820–1830-e gody*. Moscow: Iskusstvo, 1969.

Manning, Paul. *Strangers in a Strange Land: Occidentalist Publics and Orientalist Geographies in Nineteenth-Century Georgian Imaginaries*. Boston: Academic Studies Press, 2012.

Manuilov, V. A., ed. *Lermontovskaia entsiklopediia*. Moscow: Izdatel'stvo Sovetskoi entsiklopedii, 1981.

Markov, E. L. "Burzhuaznye idealy." *Russkaia rech'* (July 1879): 200–238.

———. *Chernozemnye polia*. az.lib.ru/markow_e_l/indexvote.shtml.

———. "Literaturnaia khandra." *Russkaia rech'* (February 1879): 233–60.

———. *Ocherki Kryma. Kartiny krymskoi zhizni, prirody i istorii*. 2nd ed. St. Petersburg and Moscow: Tovarishchestvo M. O. Vol'f, 1884 [St. Petersburg: Tipografiia K. N. Plotnika, 1872]. 3rd edition. St. Petersburg and Moscow: Tovarishchestvo M. O. Vol'f, 1902. Repr. Simferopol': Terra-AiTi, 2015.

———. "Romanist-psikhiatr." *Russkaia rech'* (May 1879): 243–75 and (June 1879): 151–206.

———. "S chuzhoi storony (Pis'ma v Shchigrovskii uezd)." *Otechestvennye zapiski* 184 (June 1869), section 1: 225–50; ibid. 185 (July 1869), section 1: 1–32.

———. "Tsaritsa Adriatiki. Iz puteshestviia po evropeiskomu iugu." *Vestnik Evropy* (January 1903): 216–75.

———. "Vena, Miunkhen, Venetsiia. Khudozhestvennye zametki." *Vestnik Evropy* (August 1873): 433–68.

Markovitch, Milan I. *Jean-Jacques Rousseau et Tolstoï*. Geneva: Slatkine Reprints, 1975 [1928].

Martinsen, Deborah A., and Olga Maiorova, eds. *Dostoevsky in Context*. Cambridge: Cambridge University Press, 2015.

Matich, Olga. *Erotic Utopia: The Decadent Imagination in Russia's Fin de Siècle*. Madison: University of Wisconsin Press, 2005.

Maurer, Eva. "*Al'pinizm* as Mass Sport and Elite Recreation: Soviet Mountaineering Camps under Stalin." In Gorsuch and Koenker, *Turizm*, 141–62.

McReynolds, Louise. "The Prerevolutionary Russian Tourist: Commercialization in the Nineteenth Century." In Gorsuch and Koenker, *Turizm*, 17–42.

———. *Russia at Play: Leisure Activities at the End of the Tsarist Era*. Ithaca and London: Cornell University Press, 2003.

Mech, A. "Voenno-ossetinskaia doroga na Kavkaze." *Russkaia mysl'*, no. 7 (1900), section 2: 24–46.

Medzhibovskaya, Inessa. *Tolstoy and the Religious Culture of His Time: A Biography of a Long Conversion, 1845–1887*. Lanham, MD: Lexington Books, 2008.

Mel'gunov, N. A. [N. Livenskii, pseud.]. "Turisty voobshche, i osobenno russkie. (Pis'mo v redaktsiiu iz-za granitsy)." *Otechestvennye zapiski* 123 (March 1859), section 6: 1–11.

Mel'nitskii, V. P. "Pereezdy po Rossii v 1852 godu." *Sovremennik* 39, no. 6 (June 1853), section 6: 127–47; ibid. 40, no. 7 (July 1853), section 6: 1–37; and ibid. no. 8 (August 1853), section 6: 135–69.

Merezhkovskii, D. S. *Stikhotvoreniia i poemy.* St. Petersburg: Akademicheskii proekt, 2000.

Miatlev, I. P. *Sensatsii i zamechaniia gospozhi Kurdiukovoi za granitseiu, dan l'etranzhe.* 2nd ed. St. Petersburg: Izdanie A. S. Suvorina, 1907.

———. *Stikhotvoreniia. Sensatsii i zamechaniia gospozhi Kurdiukovoi.* 2nd ed. Leningrad: Sovetskii pisatel', 1969.

Mignolo, Walter D. "The Many Faces of Cosmo-polis: Border Thinking and Critical Cosmopolitanism." In Delanty and Inglis, *Cosmopolitanism*, 2:58–83.

Mikhailov, M. M. "Iz Berlina." *Sovremennik* 87, no. 5 (May 1861), section 2: 199–212.

———. *Londonskie zametki. Sovremennik* 76–77, nos. 6–9 (June, July, August, September 1859), section 3: 219–38, 1–16, 221–36, 145–56.

———. *Parizhskie pis'ma. Sovremennik* 72–73, nos. 9–12 (September, October, November, December 1858), section 1: 269–90, 447–69, 181–224, 615–29; and ibid. 73, nos. 1–2 (January, February 1859), section 3: 163–82; 215–36.

———. *Sobranie stikhotvorenii.* Leningrad: Sovetskii pisatel', 1969.

———. *Sochinenia.* 3 vols. Moscow: Gosudarstvennoe izdatel'stvo khudozhestvennoi literatury, 1958.

Mikhailovskii, N. K. *Literaturnye vospominaniia i sovremennaia smuta.* 2 vols. St. Petersburg: Tipografiia V. M. Vol'fa, 1900.

———. *Sochineniia.* 6 vols. St. Petersburg: Tipografiia V. M. Vol'fa, 1896–97.

Milkova, Stiliana. "From Rome to Paris to Rome: Reversing the Grand Tour in Gogol's 'Rome.'" *Slavic and East European Journal* 59, no. 4 (2015): 493–516.

Miller, Robin Feuer. "Dostoevsky and the Tale of Terror." In Leatherbarrow, *Dostoevsky and Britian*, 139–58.

Miltchina, Vera, and Alexander Ospovat, eds. *Les Russes découvrent la France au XVIIIe et au XIXe siècle.* Translated by Camille Lambert, H. Guterman, H. Mongault, and G. Struve. Paris: Éditions du Globe and Moscow: Progress, 1989.

Mirsky, D. S. *A History of Russian Literature from Its Beginnings to 1900.* New York: Vintage Books, 1958.

Moers, Ellen. *The Dandy: Brummell to Beerbohm.* London: Secker and Warburg, 1960.

Møller, Peter Ulf. *Postlude to* The Kreutzer Sonata. *Tolstoj and the Debate on Sexual Morality in Russian Literature in the 1890s.* Translated by John Kendal. Leiden: E. J. Brill, 1988.

Monas, Sidney. "The Twilit Middle Class of Nineteenth-Century Russia." In Clowes, Kassow, and West, *Between Tsar and People*, 28-37.

Mondry, Henrietta. *Pure, Strong and Sexless: The Peasant Woman's Body and Gleb Uspensky.* Amsterdam and New York: Rodopi, 2006.

Mordovtsev, D. L. "Po Ispanii." *Vestnik Evropy* (January 1884): 73–118.

Morson, Gary Saul. *The Boundaries of Genre: Dostoevsky's* Diary of a Writer *and the Traditions of Literary Utopia*. Evanston: Northwestern University Press, 1981.

———. "The Reader as Voyeur: Tolstoi and the Poetics of Didactic Fiction." *Canadian-American Slavic Studies* 12, no. 4 (Winter 1978): 465–80.

———. "Work and the Authentic Life in Tolstoy." *Tolstoy Studies Journal* 9 (1997): 36–48.

Morson, Gary Saul, and Caryl Emerson. *Mikhail Bakhtin: Creation of a Prosaics*. Stanford: Stanford University Press, 1990.

Moser, Charles A. *Esthetics as Nightmare: Russian Literary Theory, 1855–1870*. Princeton: Princeton University Press, 1989.

———. "Saltykov-Shchedrin, Mikhail Evgrafovch." In Terras, *Handbook of Russian Literature*, 381–83.

Moskvich, Anna. *Prakticheskii putevoditel' po Krymu s prilozheniem*. 2nd ed. Yalta: Tipografiia N. P. Petrova, 1889.

Moskvich, Grigorii. *Illiustrirovannyi prakticheskii putevoditel' po Kavkazu*. 8th ed. Odessa: Tipografiia L. Nitche, 1903.

Murav'ev, A. N. *Gruziia i Armeniia*. 2 vols. St. Petersburg: Tipografiia III Otdeleniia, 1848.

Murray, John, III. *A Handbook for Travellers in Russia, Poland, and Finland, including the Crimea, Caucasus, Siberia, and Central Asia*. 4th and 5th eds. London: John Murray, 1888, 1893.

———. *A Handbook for Travellers in Switzerland, and the Alps of Savoy and Piedmont*. 6th ed. London: John Murray, 1854.

Nabokov, Vladimir. Commentary on *Eugene Onegin: A Novel in Verse*, by Alexander Pushkin, translated by Vladimir Nabokov, 1:89–334, 2.1:5–547, 2.2:3–384. Princeton: Princeton University Press, Bollinger Paperback, 1981.

Nash, Dennison. "Tourism as a Form of Imperialism." In Smith, *Hosts and Guests*, 37–52.

"Nashi v Parizhe." *Iskra*, no. 22 (June 14, 1863): 301–11.

Nava, Mica. "Cosmopolitan Modernity: Everyday Imaginaries and the Register of Difference." *Theory, Culture & Society* 19, nos. 1–2 (2000): 81–99.

Nechaev, S. D. "Otryvki iz putevykh zapisok o iugo-vostochnoi Rossii." *Moskovskii telegraf* 7, no. 1 (1826), section 1: 26–41.

Nekrasov, N. A. *Polnoe sobranie sochinenii i pisem*. 15 vols. Leningrad: Nauka, 1981–2000.

Nikitenko, A. V. "Rafaeleva sistinskaia madonna." *Russkii vestnik* (October 1857), section 1: 586–97.

Nivat, Georges. "L'Italie de Blok et celle de Gumilev." *Revue des études slaves* 54, nos. 3–4 (1982): 697–709.

Norman, John O. "Pavel Tretiakov and Merchant Art Patronage, 1850–1900." In Clowes, Kassow, and West, *Between Tsar and People*, 93–107.

Novopolin, G. S. *Pornograficheskii element v russkoi literature*. St. Petersburg: Tipografiia M. M. Stasiulevicha, 1909.

Nugent, Thomas. *The Grand Tour, or A Journey through the Netherlands, Germany, Italy, and France*. 4 vols. 3rd ed. London: J. Rivington and Sons, 1778.

Offord, Derek. "Dostoevsky and Chernyshevsky." *Slavonic and East European Review* 57, no. 4 (October 1979): 509–30.

———. *Journeys to a Graveyard: Perceptions of Europe in Classical Russian Travel Writing*. Dordrecht: Springer, 2005.

———. "Linguistic Gallophobia in Russian Comedy." In Offord, Ryazanova-Clark, Rjéoutski, and Argent, *French and Russian in Imperial Russia*, 79–99.

"Literature and Ideas in Russia after the Crimean War: the 'Plebeian' Writers." In *Ideology in Russian Literature*, edited by Richard Freeborn and Jane Grayson, 47-78. London: Palgrave, 1990.

———. *Portraits of Early Russian Liberals. A Study of the Thought of T. N. Granovsky, V. P. Botkin, P. V. Annenkov, A. V. Druzhinin and K. D. Kavelin*. Cambridge: Cambridge University Press, 1985.

Offord, Derek, Lara Ryazanova-Clark, Vladislav Rjéoutski, and Gesine Argent, eds. *French and Russian in Imperial Russia*, vol. 2: *Language Attitudes and Identity*. Edinburgh: Edinburgh University Press, 2015.

Ogarev, N. P. *Izbrannye stikhotvoreniia i poemy*. Moscow: Gosudarstvennoe izdatel'stvo khudozhestvennoi literatury, 1938.

Okenfuss, Max J. "The Cultural Transformation of Petr Tolstoi." In *Russia and the West in the Eighteenth Century*, edited by A. G. Cross, 228–37. Newtonville, MA: Oriental Research Partners, 1983.

Opul'skii, A. *L. N. Tolstoi na Kavkaze (Literaturno-kraevedcheskii ocherk)*. Ordzhonikidze: Severo-Ossetinskoe kniznoe izdatel'stvo, 1960.

Orwin, Donna Tussing, ed. *Anniversary Essays on Tolstoy*. Cambridge: Cambridge University Press, 2010.

———, ed. *The Cambridge Companion to Tolstoy*. Cambridge: Cambridge University Press, 2002.

———. "The Return to Nature: Tolstoyan Echoes in *The Idiot*." *Russian Review* 58, no. 1 (January 1999): 87–102.

———. *Tolstoy's Art and Thought, 1847–1880*. Princeton: Princeton University Press, 1993.

Osborne, Suzanne. "When a House is Not a Home: The Alien Residences of *Effi Briest* and *Anna Karenina*." *Tolstoy Studies Journal* 5 (1992): 67–77.

Ousby, Ian. *The Englishman's England: Taste, Travel and the Rise of Tourism*. Cambridge: Cambridge University Press, 1990.

P_ii, N. "Sovremennye izvestiia zagranichnye." *Moskvitianin* 5, no. 17 (1853), section 7: 67–86.

Page, Tanya. "Sentimentalism." In Terras, *Handbook of Russian Literature*, 395–97.

Panaeva, A. Ia. [N. Stanitskii, pseud.]. *Russkie v Italii*. *Sovremennik* 67, no. 2 (February 1858), section 1: 291–365.

Papernyi, Z. "Perechityvaia Belinskogo." *Literaturnaia gazeta*, June 5, 1948, 1.

Patyk, Lynn Ellen. "The Byronic Terrorist: Boris Savinkov's Literary Self-Mythologization." In *Just Assassins? The Culture of Terrorism in Russia*, edited by Anthony Anemone, 163–90. Evanston: Northwestern University Press, 2010.

Pavlova, Karolina. *Polnoe sobranie stikhotvorenii*. Leningrad: Sovetskii pisatel', 1939.

Peace, Richard. "Dostoevsky and the 'Golden Age.'" *Dostoevsky Studies* 3 (1982): 61–78.

Pearson, Irene. "Raphael as Seen by Russian Writers from Zhukovsky to Turgenev." *Slavic and East European Review* 59, no. 3 (July 1981): 346–69.

Pemble, John. *The Mediterranean Passion: Victorians and Edwardians in the South*. New York: Clarendon Press, Oxford University Press, 1987.

Perlina, Nina. "Dostoevsky and His Fellow Prisoners from the House of the Dead." In Ransel and Shallcross, *Polish Encounters, Russian Identity*, 100–109.

Peterson, Dale E. *The Clement Vision: Poetic Realism in Turgenev and James*. Port Washington, New York and London: Kennikat Press, 1975.

Petrushevskii, A. "Russkaia Riv'era." *Vestnik Evropy* (March 1892): 620–49.

Phillips, Timothy S. J. "Health and Sociability: The Cultural Significance of the Resort in Nineteenth-Century Russia." DPhil diss., Oxford University, 2005.

Pipes, Richard. "The Background and Growth of Karamzin's Political Ideas down to 1810." In *Karamzin's Memoir on Ancient and Modern Russia*, translation and analysis by Richard Pipes, 3–92. Cambridge, MA: Harvard University Press, 1959.

Pisarev, D. I. *Polnoe sobranie sochinenii i pisem*. 12 vols. Moscow: Nauka, 2000–2013.

Plekhanov, G. V. "Rech' na mogile A. I. Gertsena v Nitstse 7 aprelia 1912 g." In *A. I. Gertsena v russkoi kritike. Sbornik statei*, introduction and commentary by V. Putnitsev, 243–47. 2nd ed. Moscow: Gosudarstvennoe izdatel'stvo khudozhestvennoi literatury, 1953.

Pogodin, M. P., ed. *Aleksei Petrovich Ermolov. Materialy dlia ego biografii*. 4 vols. Moscow: Katkov and Co., 1864.

———. *God v chuzhikh kraiakh, 1839. Dorozhnyi dnevnik*. 4 vols. Moscow: Universitetskaia tipografiia, 1844.

———. [M. P.]. "Otryvki iz zagranichnykh pisem." *Moskvitianin*, no. 1 (January 1854), section 7: 7–24.

———. *Povesti*. 3 vols. Moscow: Tipografiia S. Selivanovskogo, 1832.

Pollock, Sheldon. "Cosmopolitanism and Vernacular in History." In Breckenridge, Pollock, Bhabha, and Chakrabarty, *Cosmopolitanism*, 15–53.

Pollock, Sheldon, Homi K. Bhabha, Carol A. Breckenridge, and Dipesh Chakrabarty. "Cosmopolitanisms." In Breckenridge, Pollock, Bhabha, and Chakrabarty, *Cosmopolitanism*, 1–14.

Polonskii, Ia. P. *Polnoe sobranie sochinenii*. 4 vols. St. Petersburg: Tipografiia R. Golike, 1885–86.

———. *Stikhotvoreniia*. Leningrad: Sovetskii pisatel', 1954.

———. "Tiflisskie sakli." *Sovremennik* 40, no. 7 (July 1853), section 1: 5–52. Repr. in Polonskii, *Polnoe sobranie sochinenii*, 3:410–86.

Pomerantsev, Il'ia. "Dostoevskii i literatura puteshestvii." *Russian Literature* 47 (2000): 93–109.

Pope, Alexander. *The Dunciad*. Vol. 5 of *The Poems of Alexander Pope*. 11 vols. Edited by James Sutherland. 3rd ed. London and New York, Routledge, 1993.

Popov, P. S., ed. *Pis'ma k A. V. Druzhininu, 1850–1863*. Moscow: Gosudarstvennyi literaturnyi muzei, 1948.

Pratt, Mary Louise. *Imperial Eyes: Travel Writing and Transculturation*. London and New York: Routledge, 1992.

Prutkov, Koz'ma. *Sochineniia*. Moscow: Khudozhestvennaia literatura, 1960.

Pushkin, A. S. *Eugene Onegin: A Novel in Verse*. 2 vols. Translated with commentary by Vladimir Nabokov. Princeton: Princeton University Press, Bollinger Paperback, 1981.

———. *Sobranie sochinenii*. 10 vols. Gosudarstvennoe izdatel'stvo khudozhestvennoi literatury, 1959–62.

Pypin, A. N. "P. V. Annenkov." *Vestnik Evropy* (March 1892): 301–43.

———. Review of *Ocherki Kryma*, by Evgeny Markov. *Vestnik Evropy* (January 1884): 424–25.

Radozhitskii, I. T. "Doroga ot reki Dona do Georgievska na prostranstve 500 verst." *Otechestvennye zapiski* 15, no. 41 (1823): 343–75.

———. "Progulka k kavkazskim mineral'nym vodam." *Otechestvennye zapiski* 16, no. 44 (1823): 387–411.

Raeff, Marc. *Origins of the Russian Intelligentsia. The Eighteenth-Century Nobility*. New York and London: Harcourt Brace Jovanovich, 1966.

———, ed. *Russian Intellectual History: An Anthology*. New York, Chicago, and Burlingame: Harcourt, Brace, and World, 1966.

Ram, Harsha. *The Imperial Sublime: A Russian Poetics of Empire*. Madison: University of Wisconsin Press, 2006.

———. "Modernism on the Periphery: Literary Life in Postrevolutionary Tbilisi." *Kritika: Explorations in Russian and Eurasian History* 5, no. 2 (Spring 2004): 367–82.

Randolph, John. *The House in the Garden: The Bakunin Family and the Romance of Russian Idealism*. Ithaca and London: Cornell University Press, 2007.

Ransel, David L., and Bozena Shallcross, eds. *Polish Encounters, Russian Identity*. Bloomington: Indiana University Press, 2005.

Rayfield, Donald. *Understanding Chekhov: A Critical Study of Chekhov's Prose and Drama*. Madison: University of Wisconsin Press, 1999.

Reid, Robert. "Ethnotope and Imperium in Nineteenth-Century Russian Literature." In *Why Europe? Problems of Culture and Identity*, vol. 1, *Political and Historical Dimensions*, edited by Joe Andrew, Malcolm Crook, and Michael Waller, 182–201. London: Palgrave Macmillan, 2000.

Reiser, S. A., and V. V. Zhdanov, eds. *N. A. Dobroliubov v vospominaniiakh ego sovremennikov*. Leningrad: Khudozhestvennaia literatura, 1961

Renner, Andreas. "Defining a Russian Nation: Mikhail Katkov and the 'Invention' of National Politics." *Slavonic and East European Review* 81, no. 4 (October 2003): 659–82.

Reyfman, Irina. "Service Ranks." In Martinsen and Maiorova, *Dostoevsky in Context*, 98–105.

Riasanovsky, Nicholas V. *Nicholas I and Official Nationality in Russia, 1825–1855*. Berkeley: University of California Press, 1959.

———. *Russia and the West in the Teaching of the Slavophiles: A Study of Romantic Ideology*. Cambridge, MA: Harvard University Press, 1952.

———. *The Image of Peter the Great in Russian History and Thought*. Oxford: Oxford University Press, 1992.

Rieber, Alfred J. "The Sedimentary Society." In Clowes, Kassow, and West, *Between Tsar and People*, 343–66.

Ripp, Victor. *Turgenev's Russia: From* Notes of a Hunter *to* Fathers and Sons. Ithaca and London: Cornell University Press, 1980.

Risaliti, Renato. *Russi in Italia tra Settecento e Novecento*. Turin: Centro interuniversitario di ricerche sul viaggio in Italia, 2010.

Roberts, Mary Louise. "Gender, Consumption, and Commodity Culture." *American Historical Review* 103, no. 3 (1998): 817–44.

Robbins, Bruce. *Feeling Global: Internationalism in Distress*. New York: New York University Press, 1999.

Roboli, T. "Literatura puteshestviia." In *Russian Prose*, edited by B. Eikhenbaum and Iu. Tynianov, 42–73. The Hague: Mouton, 1963 [1926].

Rogger, Hans. *National Consciousness in Eighteenth-Century Russia*. Cambridge, MA: Harvard University Press, 1960.

Romanov, Grand Duke Konstantin [K. R., pseud.]. "V Venetsii." *Vestnik Evropy* (September 1882): 287–90.

Rosenfeld, Sophia. "Citizens of Nowhere in Particular." In Delanty and Inglis, *Cosmopolitanism*, 1:191–215.

Rostopchina, E. P. *Stikhotvoreniia, Proza, Pis'ma*. Moscow: Sovetskaia Rossiia, 1986.

Rozen, A. E. *Zapiski dekabrista*. Leipzig: Duncker and Humblot, 1870.

Ruane, Christine. *The Empire's New Clothes: A History of the Russian Fashion Industry, 1700–1917*. New Haven: Yale University Press, 2009.

———. "European Fashion in Russia." In Kivelson and Neuberger, *Picturing Russia*, 119–23.

Saburov, Ia. "Poezdka v Saratov, Astrakhan' i na Kavkaz." *Moskovskii nabliudatel'*, no. 2 (May 1835): 176–229.

Said, Edward W. *Culture and Imperialism*. London: Chatto and Widus, 1993.

———. *Orientalism*. New York: Vintage Books, 1979.

Saltykov-Shchedrin, M. E. *Sobranie sochinenii*. 20 vols. Moscow: Khodozhestvennaia literatura, 1965–77.

Sandler, Stephanie. *Commemorating Pushkin: Russia's Myth of a National Poet*. Stanford: Stanford University Press, 2004.

Saylor, Ian. "*Anna Karenina* and *Don Giovanni*: The Vengeance Motif in Oblonsky's Dream." *Tolstoy Studies Journal* 8 (1995–96): 112–16.

Scanlan, James P. *Dostoevsky the Thinker*. Ithaca and London: Cornell University Press, 2002.

Schivelbusch, Wolfgang. *The Railway Journey: The Industrialization of Time and Space in the Nineteenth Century*. Oakland: University of California Press, 1986 [1979].

Schlereth, Thomas J. *The Cosmopolitan Ideal in Enlightenment Thought*. South Bend, IN: University of Notre Dame Press, 1977.

Schönle, Andreas. *Authenticity and Fiction in the Russian Literary Journey, 1790–1840*. Cambridge, MA: Harvard University Press, 2000.

———. "Broken History and Crumbling Stones: The Romantic Conception of Architectural Preservation." *Slavic Review* 71, no. 4 (Winter 2012): 745–65.

———. "Garden of the Empire: Catherine's Appropriation of the Crimea." *Slavic Review* 60, no. 1 (Spring 2001): 1–23.

Sears, John F. *Sacred Places: American Tourist Attractions in the Nineteenth Century*. New York: Oxford University Press, 1989.

Seeley, Frank Friedeberg. *Turgenev: A Reading of his Fiction*. Cambridge: Cambridge University Press, 1991.

Shasko, Philip. "Nikolai Alexandrovich Mel'gunov on the Reformation and the Work Ethic." *Comparative Studies in Society and History* 9, no. 3 (April 1967): 256–65.

Shatz, Marshall. "Bakunin, Turgenev and *Rudin*." In *The Golden Age of Russian Literature and Thought: Selected Papers from the Fourth World Congress for Soviet and East European Studies, Harrogate, 1990*, edited by Derek Offord, 103–14. Basingstoke: Macmillan Press, 1992.

Shelgunov, N. V., L. P. Shelgunova, and M. L. Mikhailov. *Vospominaniia*. 2 vols. Edited by E. Vilenskaia, E. Ol'khovskii, and L. Roitberg. Moscow: Khudozhestvennaia literatura, 1967.

Shevelenko, Irina. "Imperiia i natsiia v voobrazhenii russkogo modernizma." *Ab Imperio* 3 (2009): 171–206.

Shevyrev, S. P. *Poezdka v Kirillo-Belozerskii monastyr': Vakatsionnye dni professora S. Shevyreva v 1847 godu*. 2 vols. Moscow: Universitetskaia tipografiia, 1850.

Shishkin, Mikhail. *Russkaia Shveitsariia. Literaturno-istoricheskii putevoditel'*. Moscow: Vagrius, 2006.

Shishkina, O. P. *Zametki i vospominaniia russkoi puteshestvennitsy po Rossii v 1845 godu*. 2 vols. St. Petersburg: Tipografiia Vtorogo Otdeleniia Sobstvennoi E. I. V. Kantseliarii, 1848.

Shishkov, A. A. ["Otryvki iz nezakonchennogo romana o 'kakhetinskom bunte' 1812 goda."] In *Russkie pisateli o Gruzii*, edited by Vano Shaduri, 429–64. Tbilisi: Zaria Vostoka, 1948.

Shneyder, Vadim. "Myshkin's Million: Merchants, Capitalists, and the Economic Imaginary in *The Idiot*." *Russian Review* 77, no. 2 (March 2018): 241–58.

Shtukaturova, N. M. *Kostium narodov Rossii v grafike 18–20 vekov*. Moscow: VRIB Soiuzreklamkul'tura, 1990.

Siegel, Kristi, and Toni B. Wulff. "Travel as Spectacle: The Illusion of Knowledge and Sight." In *Issues in Travel Writing: Empire, Spectacle, and Displacement*, edited by Kristi Siegel, 109–22. New York: Peter Lang, 2002.

Šilbajoris, Rimvydas. *Tolstoy's Aesthetics and His Art*. Columbus: Slavica Publishers, 1990.

Skabichevskii, A. M. *N. A. Dobroliubov. Ego zhizn' i literaturnaia deiatel'nost'*. St. Petersburg: Tipografiia tovarishchestva "Obshchestvennaia pol'za," 1894.

Skatov, N. N. *Nekrasov*. Moscow: Molodaia gvardiia, 1994.

Slezkine, Yuri. *Arctic Mirrors: Russia and the Small Peoples of the North*. Ithaca and London: Cornell University Press, 1994.

———. "Naturalists versus Nations: Eighteenth-Century Russian Scholars Confront Ethnic Diversity." In Brower and Lazzerini, *Russia's Orient*, 27–57.

Sloane, David A. "Anna Reading and Women Reading in Russian Literature." In Knapp and Mandelker, *Approaches to Teaching Tolstoy's* Anna Karenina, 124–30.

Slobin, Greta. "Turgenev Finds a Home in Russia Abroad." In *Turgenev: Art, Ideology, and Legacy*, edited by Robert Reid and Joe Andrew, 189–216. Amsterdam and New York: Rodopi, 2010.

Smith, Valene L. ed. *Hosts and Guests: The Anthropology of Tourism*. 2nd ed. Philadelphia: University of Pennsylvania Press, 1989.

Smith, Virginia Llewellyn. *Anton Chekhov and the Lady with the Dog*. Oxford: Oxford University Press, 1973.

Sokolov, D. N. *Progulka po Krymu s tsel'iu oznakomit'sia s nim*. Odessa: Tipografiia L. Nitche, 1869.

Solov'ev, N. I. "Trud i naslazhdenie. Stat'a tret'ia i poslednaia." *Otechestvennye zapiski* 167 (July 1866), section 1: 125–49.

Solov'ev, V. S. *Pis'ma*. 4 vols. St. Petersburg: Tipografiia tovarishchestva "Obshchestvennaia pol'za," 1908–1923.

Spivak, Gayatri Chakravorty. "Cultural Talks in the Hot Peace: Revisiting the 'Global Village.'" In Cheah and Robbins, *Cosmopolitics*, 329–48.

Stasov, V. V. *Izbrannye sochineniia*. 3 vols. Moscow: Iskusstvo, 1952.

Stavrou, Theofanis George. *Art and Culture in Nineteenth-Century Russia*. Bloomington: Indiana University Press, 1983.

Stebbins, Robert A. "Cultural Tourism as Serious Leisure." *Annals of Tourism Research* 23, no. 4 (October 1996): 948–50.

Sterne, Lawrence. *A Sentimental Journey and Other Writings*. Edited by Ian Jack and Tim Parnell. Oxford: Oxford University Press, 2003 [1768].

Stites, Richard. "M. L. Mikhailov and the Emergence of the Woman Question." *Canadian-American Slavic Studies* 3, no. 2 (January 1969): 178–99.

———. *The Women's Liberation Movement in Russia: Feminism, Nihilism, and Bolshevism, 1860–1930*. Princeton: Princeton University Press, 1978.

Strakhov, N. N. *Biografiia, pis'ma i zametki iz zapisnoi knizhki F. M. Dostoevskogo*. St. Petersburg: Izdanie A. S. Suvorina, 1883.

———. *Bor'ba s zapadom v nashei literature: Istoricheskie i kriticheskie ocherki*. 3 vols. St. Petersburg: Tipografiia S. Dobrodeeva, 1882, 1883, 1896.

———. "Iz poezdki v Italiiu." In *Bratskaia pomoshch' postradavshim semeistvam Bosnii i Gertsegoviny*, 75–92. St. Petersburg: Izdanie Peterburgskogo otdela slavianskogo komiteta, tipografiia A. A. Kraevskogo, 1876.

———. *Kriticheskie stat'i ob I. S. Turgeneve i L. N. Tolstom (1862–1885)*. St. Petersburg: Tipografiia brat. Panteleevykh, 1885.

———. "Krymskie vpechatleniia." *Niva*, no. 2 (1870): 23–25.

Struve, G. "Iskusstvo i positivism." *Russkii vestnik* (May 1875): 357–79.

Sumarokov, Petr [P. S.]. "Pis'ma s Kavkaza." *Moskovskii telegraf* 33, no. 11 (June 1830): 313–39.

Sunderland, Willard. "Imperial Space: Territorial Thought and Practice in the Eighteenth Century." In Burbank, von Hagen, and Remnev, *Russian Empire*, 33–66.

———. "Russians into Yakuts? 'Going Native' and Problems of Russian National Identity in the Siberian North, 1870s–1914." *Slavic Review* 55, no. 4 (Winter 1996): 806–25.

———. "Shop Signs, Monuments, Souvenirs: Views of the Empire in Everyday Life." In Kivelson and Neuberger, *Picturing Russia*, 104–8.

Suny, Ronald Grigor. "Images of the Armenians in the Russian Empire." In *The Armenian Image in History and Literature*, edited by Richard G. Hovannisian, 105–37. Malibu: Undena, 1981.

———. "The Empire Strikes Out: Imperial Russia, National Identity, and Theories of Empire." In *A State of Nations: Empire and Nation-Making in the Age of Lenin and Stalin*, edited by Ronald Suny and Terry Martin, 23–66. Oxford: Oxford University Press, 2001.

Suslova, Polina. *Diary*. In Dostoevskii, *The Gambler*, translated by Victor Terras, edited by Edward Wasiolek, 201–302. Chicago: University of Chicago Press, 1972.

Swift, Mark Stanley. "Chekhov's 'Ariadna': A Portrait of Psychopathy and Sin." *Slavonic and East European Review* 86, no 1 (January 2008): 26–57.

Swinglehurst, Edmund. *Cook's Tours: The Story of Popular Travel*. Poole, UK: Sterling Publishing Co., 1982.

Tagirova, Nailya. "Mapping the Empire's Economic Regions from the Nineteenth to the Early Twentieth Century." In Burbank, von Hagen, and Remnev, *Russian Empire*, 125–38.

Taruskin, Richard. *Defining Russia Musically: Historical and Hermeneutical Essays*. Princeton: Princeton University Press, 1997.

Tchaikovskii, P. I. See Chaikovskii.

Terras, Victor. *Belinskij and Russian Literary Criticism: The Heritage of Organic Aesthetics* Madison: University of Wisconsin Press, 1974.

———, ed. *Handbook of Russian Literature*. New Haven: Yale University Press, 1985.

Thackeray, William Makepeace. *The Book of Snobs*. Teddington: The Echo Library, 2006 [1848].

——— [Michael Angelo Titmarsh, pseud.]. *The Kickleburys on the Rhine*. 3rd ed. London: Smith, Elder, 1851.

———. *Angliiskie turisty* [Russian translation by A. Butakov of Thackeray's *The Kickleburys on the Rhine*]. *Otechestvennye zapiski* 76 (June1851), section 8: 106–44.

Tissot, Victor. "Putevye vpechatleniia. Poezdka po Lombardii i Avstrii." *Russkii vestnik* (August 1877): 767–814.

Tiutcheva, A. F. *Pri dvore dvukh imperatorov. Vospominaniia, dnevnik, 1853–1882*. 2 vols. Cambridge, MA: Oriental Research Partners, 1975 [1928–29].

Tolstoi, A. K. *Sobranie sochinenii*. 4 vols. Moscow: Khudozhestvennaia literatura, 1963–64.

Tolstoi, L. N. *Anna Karenina*. Translated by Richard Pevear and Larissa Volokhonsky. London: Penguin Books, 2001.

———. *Polnoe sobranie sochinenii*. 90 vols. Moscow: Gosudarstvennoe izdatel'stvo khudozhestvennoi literatury, 1928–58.

———. *Tolstoy's Letters*. Edited and translated by R. F. Christian. 2 vols. New York: Charles Scriber's Sons, 1978.

Toporov, V. N. "Italiia v Peterburge." In *Italiia i slavianskii mir. Sovestsko-ital'ianskii simpozium in honorem Professore Ettore Lo Gatto*, edited by N. M. Kurennaia, L. A. Sofronova, and V. A. Khorev, 49–81. Moscow: Akademiia nauk, 1990.

Totleben, E. I. "O vyselenii tatar iz Kryma v 1860 godu." *Russkaia starina* 78 (June 1893): 531–50.

"Tourism Development to Continue in Russia's North Caucasus: Medvedev." Defunct website (en.rian.ru/business). Downloaded copy available from author.

Towner, John. "Approaches to Tourism History." *Annals of Tourism Research* 15 (1988): 47–62.

Trachtenberg, Alan. "Forward." In Schivelbusch, *The Railway Journey*, xiii–xvi.

Trollope, Frances. *The Robertses on Their Travels*. 3 vols. London: Henry Colburn, 1846.

Tur, Evgeniia [E. V. Salhias de Tournemir, pseud.]. "Krymskie pis'ma." https://lavanda.life/o-kryme/444-krymskie-pisma-evgenii-tur.html.

———. *Na rubezhe*. *Russkii vestnik* (October 1857): 803–84; ibid. (November 1857): 37–104; 331–68.

Turgenev, I. S. *Polnoe sobranie sochinenii i pisem*. 28 vols. Moscow and Leningrad: Akademiia nauk, 1960–68.

———. *Sobranie sochinenii*. 12 vols. Moscow: Khudozhestvennaia literatura, 1975–79.

Turner, Louis, and John Ash. *The Golden Hordes. International Tourism and the Pleasure Periphery*. New York: St. Martin's Press, 1976.

Turoma, Sanna. *Brodsky Abroad: Empire, Tourism, Nostalgia*. Madison: University of Wisconsin Press, 2010.

Turovskii, N. F. "Dnevnik poezdki po Rossii v 1841 godu." Edited by A. Golombievskii. *Russkaia starina* 155 (September 1913): 492-528.

Ungurianu, Dan. *Plotting History: The Russian Historical Novel in the Imperial Age*. Madison: University of Wisconsin Press, 2007.

Urbain, Jean-Didier. *L'idiot du voyage. Histoires de touristes*. 2nd ed. Paris: Plon, 2002 [1991].

U. [N. G. Ustrialov?]. "Zametki o khoziaistvennom polozhenii Rossii." *Russkii vestnik* 41 (September 1862): 238–86.

Urry, John. *Consuming Places*. London and New York: Routledge, 1995.

———. *The Tourist Gaze: Leisure and Travel in Contemporary Society*. London: Sage, 1996 [1990].

Uspenskii, G. I. *Polnoe sobranie sochinenii*. 14 vols. Moscow: Akademiia nauk, 1940–54.

Vainshtein, Olga. "Russian Dandyism: Constructing a Man of Fashion." In *Russian Masculinities in History and Culture*, edited by Barbara Evans Clements, Rebecca Friedman, and Dan Healey, 51–75. Basingstoke: Palgrave, 2002.

Vaiskopf, Mikhail and Rita Giuliani, eds. *Gogol' i Italiia / Gogol' e Italia*. Moscow: Rossiskii gosudarstvennyi gumanitarnyi universitet, 2004.

Valkenier, Elizabeth Kridl. "The Intelligentsia and Art." In Stavrou, *Art and Culture in Nineteenth-Century Russia*, 153–71.

———. *Russian Realist Art. The State and Society: The Peredvizhniki and Their Tradition*. Ann Arbor: University of Michigan Press, 1977.

Van der Veer, Peter. "Colonial Cosmopolitanism." In Vertovec and Cohen, *Conceiving Cosmopolitanism*, 165–79.

Veblen, Thorstein. *The Theory of the Leisure Class*. Edited with introduction by Martha Banta. Oxford: Oxford University Press, 2007 [1899].

Venturi, Franco. *Roots of Revolution: A History of the Populist and Socialist Movements in Nineteenth-Century Russia*. Translated by Francis Haskell. New York: Grosset and Dunlap, 1966.

Verderevskii, E. A. *Ot Zaural'ia do Zakavkaz'ia. Iumoristicheskie, sentimental'nye i prakticheskie pis'ma s dorogi*. Moscow: V. Gautier, 1857.

———. *Plen u Shamilia*. St. Petersburg: Tipografiia Koroleva, 1856.

Vertovec, Steven, and Robin Cohen, eds. *Conceiving Cosmopolitanism: Theory, Context, and Practice*. Oxford and New York: Oxford University Press, 2002.

Veselitskaia, L. I. [V. Mikulich, pseud.]. *Dusia. Rasskazy*. Petrograd: Novoe vremia, 1915.

———. *Mimochka*. Moscow: Eksmo, 2003.

———. *Mimochka na vodakh*. *Vestnik Evropy* (February 1891): 526–68; ibid. (March 1891): 30–60.

———. "Mimochka—nevesta". *Vestnik Evropy* (September 1883): 5–30

———. *Mimochka otravilas'*. *Vestnik Evropy* (September 1893): 112–40; ibid. (October 1893): 417–54.

———. *Rasskazy*. St. Petersburg: Izdanie A. S. Suvorina, 1911.

Viazemskii, P. A. *Polnoe sobranie sochinenii*. 11 vols. St. Petersburg: Tipografiia M. M. Stasiulevicha, 1878–87.

———. *Stikhotvoreniia*. Moscow: Sovetskaia Rossia, 1978.

———. *V doroge i doma. Sobranie stikhotvorenii*. Moscow: Tipografiia Bakhmeteva, 1862.

Visson, Lynn. "Kasha vs. Cachet Blanc: The Gastronomic Dialectics of Russian Literature." In *Russianness: Studies on a Nation's Identity: In Honor of Rufus Mathewson, 1918–1978*, edited by Robert L. Belknap, 60–73. Ann Arbor: Ardis, 1990.

Voronov, N. "Chernomorskie pis'ma." *Russkii vestnik* (March 1857), section 3: 87–93.

Walicki, Andrzej. *The Slavophile Controversy: History of a Conservative Utopia in Nineteenth-Century Russian Thought*. Translated by Hilda Andrews-Rusiecka. Oxford: Clarendon Press, 1975.

———. "The Slavophile Thinkers and the Polish Question in 1863." In Ransel and Shallcross, *Polish Encounters, Russian Identity*, 89–99.

Walkowitz, Rebecca L. *Cosmopolitan Style: Modernism beyond the Nation*. New York: Columbia University Press, 2007.

Walton, John K. "British Tourism between Industrialization and Globalization: An Overview." In Berghoff, Korte, Schneider, and Harvey, *The Making of Modern Tourism*, 109–31.

Watson, Nicola J. *The Literary Tourist: Readers and Places in Romantic and Victorian Britain*. Basingstoke and New York: Palgrave Macmillan, 2006.

———. "Shakespeare on the Tourist Trail." In *The Cambridge Companion to Shakespeare and Popular Culture*, edited by Robert Shaughnessy, 199–226. Cambridge: Cambridge University Press, 2007.

Weir, Justin. *Leo Tolstoy and the Alibi of Narrative*. New Haven: Yale University Press, 2011.

———. "Tolstoy Sees the Truth but Waits: The Consequences of Aesthetic Vision in *Anna Karenina*." In Knapp and Mandelker, *Approaches to Teaching Tolstoy's* Anna Karenina, 173–79.

West, Sally. *I Shop in Moscow: Advertising and the Creation of Consumer Culture in Late Tsarist Russia*. DeKalb: Northern Illinois University Press, 2011.

Williams, Brian Glyn. *The Crimean Tatars: The Diaspora Experience and the Forging of a Nation*. Leiden: Brill, 2001.

———. *The Crimean Tatars. From Soviet Genocide to Putin's Conquest*. London: Hurst and Co., 2015.

———. "'*Hijra*' and Forced Migration from Nineteenth-Century Russia to the Ottoman Empire. A Critical Analysis of the Great Tatar Emigration of 1860–1861." *Cahiers du monde russe* 41, no. 1 (2000): 79–108.

Winks, Robin W., and Joan Neuberger. *Europe and the Making of Modernity, 1815–1914*. New York and Oxford: Oxford University Press, 2005.

Winner, Thomas. *Chekhov and His Prose*. New York, Chicago, and San Francisco: Holt, Rinehart and Winston, 1966.

Wirtschafter, Elise Kimerling. "The Groups Between: *Raznochintsy*, Intelligentsia, Professionals." In Dominic Lieven, *The Cambridge History of Russia*, 2:245–63. Cambridge: Cambridge University Press, 2006.

———. *Social Identity in Imperial Russia*. Ithaca and London: Cornell University Press, 1997.

Witcomb, Andrea. *Re-Imagining the Museum: Beyond the Mausoleum*. London and New York: Routledge, 2003.

Woehrlin, William F. *Chernyshevskii: The Man and the Journalist*. Cambridge, MA: Harvard University Press, 1971.

Wohlgemut, Esther. *Romantic Cosmopolitanism*. Basingstoke: Palgrave Macmillan, 2009.

Wortman, Richard S. *The Crisis of Russian Populism*. Cambridge: Cambridge University Press, 1967.

———. *Scenarios of Power: Myth and Ceremony in Russian Monarchy*. 2 vols. Princeton: Princeton University Press, 1995, 2000.

———. *Visual Texts, Ceremonial Texts, Texts of Exploration: Collected Articles on the Representation of Russian Monarchy*. Boston: Academic Studies Press, 2013.

Veblen, Thorstein. *The Theory of the Leisure Class*. Edited with introduction and notes by Martha Banta. Oxford: Oxford University Press, 2009 [1899].

Young, Sarah J. "The Crystal Palace." In Martinsen and Maiorova, *Dostoevsky in Context*, 176–84.

Zagoskin, M. N. *Sochineniia*. 2 vols. Moscow: Khudozhestvennaia literatura, 1987.

———. *Moskva i moskvichi. Zapiski Bogdana Il'icha Bel'skogo*. Moscow: Moskovskii rabochii, 1988.

Zaionchkovskii, P. A., ed. *Istoriia dorevoliutsionnoi Rossii v dnevnikakh i vospominaniiakh*. 5 vols. in 13 parts. Moscow: Kniga, 1976–89.

Zalesov, Pavel. *Kartinki Kavkaza. Putevye vospominaniia i zametki o kavkazskoi prirode*. Moscow: Tipografiia E. Lissner and Iu. Roman, 1883.

Zarin, S. "Gruzinskaia Voennaia Doroga. Putevye vpechatleniia." *Zvezda*, no. 15 (1893): 285–90.

Zelenetskii, K. P. "Kavkazskie mineral'nye vody v 1852." *Moskvitianin* 2, no. 6 (1853), section 7: 41–70.

Zhukova, M. S. [M. Zh-k-va]. *Ocherki iuzhnoi Frantsii i Nitstsy. Iz dorozhnykh zapisok 1840 i 1842 godov*. 2 vols. St. Petersburg: Izdatel'stvo A. Ivanova, 1844.

Zisserman, A. L. "Desiat' let na Kavkaze. Stat'ia pervaia." *Sovremennik* 47, no. 9 (September 1854), section 5: 1–42.

Zorin, Andrei. "Krym v istorii russkogo samosoznaniia." *Novoe literaturnoe obozrenie* 31 (1998): 125–43.

Index

Page numbers in bold refer to illustrations.

CPSIA information can be obtained
at www.ICGtesting.com
Printed in the USA
LVHW020706080523
746255LV00001B/11

9 781644 694206